Bearing Design in Machinery

MECHANICAL ENGINEERING
A Series of Textbooks and Reference Books

Founding Editor

L. L. Faulkner

Columbus Division, Battelle Memorial Institute
and Department of Mechanical Engineering
The Ohio State University
Columbus, Ohio

1. *Spring Designer's Handbook*, Harold Carlson
2. *Computer-Aided Graphics and Design*, Daniel L. Ryan
3. *Lubrication Fundamentals*, J. George Wills
4. *Solar Engineering for Domestic Buildings*, William A. Himmelman
5. *Applied Engineering Mechanics: Statics and Dynamics*, G. Boothroyd and C. Poli
6. *Centrifugal Pump Clinic*, Igor J. Karassik
7. *Computer-Aided Kinetics for Machine Design*, Daniel L. Ryan
8. *Plastics Products Design Handbook, Part A: Materials and Components; Part B: Processes and Design for Processes*, edited by Edward Miller
9. *Turbomachinery: Basic Theory and Applications*, Earl Logan, Jr.
10. *Vibrations of Shells and Plates*, Werner Soedel
11. *Flat and Corrugated Diaphragm Design Handbook*, Mario Di Giovanni
12. *Practical Stress Analysis in Engineering Design*, Alexander Blake
13. *An Introduction to the Design and Behavior of Bolted Joints*, John H. Bickford
14. *Optimal Engineering Design: Principles and Applications*, James N. Siddall
15. *Spring Manufacturing Handbook*, Harold Carlson
16. *Industrial Noise Control: Fundamentals and Applications*, edited by Lewis H. Bell
17. *Gears and Their Vibration: A Basic Approach to Understanding Gear Noise*, J. Derek Smith
18. *Chains for Power Transmission and Material Handling: Design and Applications Handbook*, American Chain Association
19. *Corrosion and Corrosion Protection Handbook*, edited by Philip A. Schweitzer
20. *Gear Drive Systems: Design and Application*, Peter Lynwander
21. *Controlling In-Plant Airborne Contaminants: Systems Design and Calculations*, John D. Constance
22. *CAD/CAM Systems Planning and Implementation*, Charles S. Knox
23. *Probabilistic Engineering Design: Principles and Applications*, James N. Siddall
24. *Traction Drives: Selection and Application*, Frederick W. Heilich III and Eugene E. Shube
25. *Finite Element Methods: An Introduction*, Ronald L. Huston and Chris E. Passerello

26. *Mechanical Fastening of Plastics: An Engineering Handbook*, Brayton Lincoln, Kenneth J. Gomes, and James F. Braden
27. *Lubrication in Practice: Second Edition*, edited by W. S. Robertson
28. *Principles of Automated Drafting*, Daniel L. Ryan
29. *Practical Seal Design*, edited by Leonard J. Martini
30. *Engineering Documentation for CAD/CAM Applications*, Charles S. Knox
31. *Design Dimensioning with Computer Graphics Applications*, Jerome C. Lange
32. *Mechanism Analysis: Simplified Graphical and Analytical Techniques*, Lyndon O. Barton
33. *CAD/CAM Systems: Justification, Implementation, Productivity Measurement*, Edward J. Preston, George W. Crawford, and Mark E. Coticchia
34. *Steam Plant Calculations Manual*, V. Ganapathy
35. *Design Assurance for Engineers and Managers*, John A. Burgess
36. *Heat Transfer Fluids and Systems for Process and Energy Applications*, Jasbir Singh
37. *Potential Flows: Computer Graphic Solutions*, Robert H. Kirchhoff
38. *Computer-Aided Graphics and Design: Second Edition*, Daniel L. Ryan
39. *Electronically Controlled Proportional Valves: Selection and Application*, Michael J. Tonyan, edited by Tobi Goldoftas
40. *Pressure Gauge Handbook*, AMETEK, U.S. Gauge Division, edited by Philip W. Harland
41. *Fabric Filtration for Combustion Sources: Fundamentals and Basic Technology*, R. P. Donovan
42. *Design of Mechanical Joints*, Alexander Blake
43. *CAD/CAM Dictionary*, Edward J. Preston, George W. Crawford, and Mark E. Coticchia
44. *Machinery Adhesives for Locking, Retaining, and Sealing*, Girard S. Haviland
45. *Couplings and Joints: Design, Selection, and Application*, Jon R. Mancuso
46. *Shaft Alignment Handbook*, John Piotrowski
47. *BASIC Programs for Steam Plant Engineers: Boilers, Combustion, Fluid Flow, and Heat Transfer*, V. Ganapathy
48. *Solving Mechanical Design Problems with Computer Graphics*, Jerome C. Lange
49. *Plastics Gearing: Selection and Application*, Clifford E. Adams
50. *Clutches and Brakes: Design and Selection*, William C. Orthwein
51. *Transducers in Mechanical and Electronic Design*, Harry L. Trietley
52. *Metallurgical Applications of Shock-Wave and High-Strain-Rate Phenomena*, edited by Lawrence E. Murr, Karl P. Staudhammer, and Marc A. Meyers
53. *Magnesium Products Design*, Robert S. Busk
54. *How to Integrate CAD/CAM Systems: Management and Technology*, William D. Engelke
55. *Cam Design and Manufacture: Second Edition*; with cam design software for the IBM PC and compatibles, disk included, Preben W. Jensen
56. *Solid-State AC Motor Controls: Selection and Application*, Sylvester Campbell
57. *Fundamentals of Robotics*, David D. Ardayfio
58. *Belt Selection and Application for Engineers*, edited by Wallace D. Erickson
59. *Developing Three-Dimensional CAD Software with the IBM PC*, C. Stan Wei
60. *Organizing Data for CIM Applications*, Charles S. Knox, with contributions by Thomas C. Boos, Ross S. Culverhouse, and Paul F. Muchnicki

61. *Computer-Aided Simulation in Railway Dynamics*, by Rao V. Dukkipati and Joseph R. Amyot
62. *Fiber-Reinforced Composites: Materials, Manufacturing, and Design*, P. K. Mallick
63. *Photoelectric Sensors and Controls: Selection and Application*, Scott M. Juds
64. *Finite Element Analysis with Personal Computers*, Edward R. Champion, Jr., and J. Michael Ensminger
65. *Ultrasonics: Fundamentals, Technology, Applications: Second Edition, Revised and Expanded*, Dale Ensminger
66. *Applied Finite Element Modeling: Practical Problem Solving for Engineers*, Jeffrey M. Steele
67. *Measurement and Instrumentation in Engineering: Principles and Basic Laboratory Experiments*, Francis S. Tse and Ivan E. Morse
68. *Centrifugal Pump Clinic: Second Edition, Revised and Expanded*, Igor J. Karassik
69. *Practical Stress Analysis in Engineering Design: Second Edition, Revised and Expanded*, Alexander Blake
70. *An Introduction to the Design and Behavior of Bolted Joints: Second Edition, Revised and Expanded*, John H. Bickford
71. *High Vacuum Technology: A Practical Guide*, Marsbed H. Hablanian
72. *Pressure Sensors: Selection and Application*, Duane Tandeske
73. *Zinc Handbook: Properties, Processing, and Use in Design*, Frank Porter
74. *Thermal Fatigue of Metals*, Andrzej Weronski and Tadeusz Hejwowski
75. *Classical and Modern Mechanisms for Engineers and Inventors*, Preben W. Jensen
76. *Handbook of Electronic Package Design*, edited by Michael Pecht
77. *Shock-Wave and High-Strain-Rate Phenomena in Materials*, edited by Marc A. Meyers, Lawrence E. Murr, and Karl P. Staudhammer
78. *Industrial Refrigeration: Principles, Design and Applications*, P. C. Koelet
79. *Applied Combustion*, Eugene L. Keating
80. *Engine Oils and Automotive Lubrication*, edited by Wilfried J. Bartz
81. *Mechanism Analysis: Simplified and Graphical Techniques, Second Edition, Revised and Expanded*, Lyndon O. Barton
82. *Fundamental Fluid Mechanics for the Practicing Engineer*, James W. Murdock
83. *Fiber-Reinforced Composites: Materials, Manufacturing, and Design, Second Edition, Revised and Expanded*, P. K. Mallick
84. *Numerical Methods for Engineering Applications*, Edward R. Champion, Jr.
85. *Turbomachinery: Basic Theory and Applications, Second Edition, Revised and Expanded*, Earl Logan, Jr.
86. *Vibrations of Shells and Plates: Second Edition, Revised and Expanded*, Werner Soedel
87. *Steam Plant Calculations Manual: Second Edition, Revised and Ex panded*, V. Ganapathy
88. *Industrial Noise Control: Fundamentals and Applications, Second Edition, Revised and Expanded*, Lewis H. Bell and Douglas H. Bell
89. *Finite Elements: Their Design and Performance*, Richard H. MacNeal
90. *Mechanical Properties of Polymers and Composites: Second Edition, Revised and Expanded*, Lawrence E. Nielsen and Robert F. Landel
91. *Mechanical Wear Prediction and Prevention*, Raymond G. Bayer

92. *Mechanical Power Transmission Components*, edited by David W. South and Jon R. Mancuso
93. *Handbook of Turbomachinery*, edited by Earl Logan, Jr.
94. *Engineering Documentation Control Practices and Procedures*, Ray E. Monahan
95. *Refractory Linings Thermomechanical Design and Applications*, Charles A. Schacht
96. *Geometric Dimensioning and Tolerancing: Applications and Techniques for Use in Design, Manufacturing, and Inspection*, James D. Meadows
97. *An Introduction to the Design and Behavior of Bolted Joints: Third Edition, Revised and Expanded*, John H. Bickford
98. *Shaft Alignment Handbook: Second Edition, Revised and Expanded*, John Piotrowski
99. *Computer-Aided Design of Polymer-Matrix Composite Structures*, edited by Suong Van Hoa
100. *Friction Science and Technology*, Peter J. Blau
101. *Introduction to Plastics and Composites: Mechanical Properties and Engineering Applications*, Edward Miller
102. *Practical Fracture Mechanics in Design*, Alexander Blake
103. *Pump Characteristics and Applications*, Michael W. Volk
104. *Optical Principles and Technology for Engineers*, James E. Stewart
105. *Optimizing the Shape of Mechanical Elements and Structures*, A. A. Seireg and Jorge Rodriguez
106. *Kinematics and Dynamics of Machinery*, Vladimír Stejskal and Michael Valášek
107. *Shaft Seals for Dynamic Applications*, Les Horve
108. *Reliability-Based Mechanical Design*, edited by Thomas A. Cruse
109. *Mechanical Fastening, Joining, and Assembly*, James A. Speck
110. *Turbomachinery Fluid Dynamics and Heat Transfer*, edited by Chunill Hah
111. *High-Vacuum Technology: A Practical Guide, Second Edition, Revised and Expanded*, Marsbed H. Hablanian
112. *Geometric Dimensioning and Tolerancing: Workbook and Answerbook*, James D. Meadows
113. *Handbook of Materials Selection for Engineering Applications*, edited by G. T. Murray
114. *Handbook of Thermoplastic Piping System Design*, Thomas Sixsmith and Reinhard Hanselka
115. *Practical Guide to Finite Elements: A Solid Mechanics Approach*, Steven M. Lepi
116. *Applied Computational Fluid Dynamics*, edited by Vijay K. Garg
117. *Fluid Sealing Technology*, Heinz K. Muller and Bernard S. Nau
118. *Friction and Lubrication in Mechanical Design*, A. A. Seireg
119. *Influence Functions and Matrices*, Yuri A. Melnikov
120. *Mechanical Analysis of Electronic Packaging Systems*, Stephen A. McKeown
121. *Couplings and Joints: Design, Selection, and Application, Second Edition, Revised and Expanded,* Jon R. Mancuso
122. *Thermodynamics: Processes and Applications*, Earl Logan, Jr.
123. *Gear Noise and Vibration*, J. Derek Smith
124. *Practical Fluid Mechanics for Engineering Applications,* John J. Bloomer
125. *Handbook of Hydraulic Fluid Technology*, edited by George E. Totten
126. *Heat Exchanger Design Handbook*, T. Kuppan

127. *Designing for Product Sound Quality*, Richard H. Lyon
128. *Probability Applications in Mechanical Design*, Franklin E. Fisher and Joy R. Fisher
129. *Nickel Alloys,* edited by Ulrich Heubner
130. *Rotating Machinery Vibration: Problem Analysis and Troubleshooting,* Maurice L. Adams, Jr.
131. *Formulas for Dynamic Analysis,* Ronald L. Huston and C. Q. Liu
132. *Handbook of Machinery Dynamics*, Lynn L. Faulkner and Earl Logan, Jr.
133. *Rapid Prototyping Technology: Selection and Application,* Kenneth G. Cooper
134. *Reciprocating Machinery Dynamics: Design and Analysis*, Abdulla S. Rangwala
135. *Maintenance Excellence: Optimizing Equipment Life-Cycle Decisions,* edited by John D. Campbell and Andrew K. S. Jardine
136. *Practical Guide to Industrial Boiler Systems*, Ralph L. Vandagriff
137. *Lubrication Fundamentals: Second Edition, Revised and Expanded*, D. M. Pirro and A. A. Wessol
138. *Mechanical Life Cycle Handbook: Good Environmental Design and Manufacturing,* edited by Mahendra S. Hundal
139. *Micromachining of Engineering Materials,* edited by Joseph McGeough
140. *Control Strategies for Dynamic Systems: Design and Implementation,* John H. Lumkes, Jr.
141. *Practical Guide to Pressure Vessel Manufacturing,* Sunil Pullarcot
142. *Nondestructive Evaluation: Theory, Techniques, and Applications,* edited by Peter J. Shull
143. *Diesel Engine Engineering: Thermodynamics, Dynamics, Design, and Control,* Andrei Makartchouk
144. *Handbook of Machine Tool Analysis,* Ioan D. Marinescu, Constantin Ispas, and Dan Boboc
145. *Implementing Concurrent Engineering in Small Companies,* Susan Carlson Skalak
146. *Practical Guide to the Packaging of Electronics: Thermal and Mechanical Design and Analysis*, Ali Jamnia
147. *Bearing Design in Machinery: Engineering Tribology and Lubrication,* Avraham Harnoy
148. *Mechanical Reliability Improvement: Probability and Statistics for Experimental Testing,* R. E. Little
149. *Industrial Boilers and Heat Recovery Steam Generators: Design, Applications, and Calculations*, V. Ganapathy

Additional Volumes in Preparation

Reliability Verification, Testing, and Analysis in Engineering Design, Gary S. Wasserman

Mechanical Properties of Engineering Materials, Wole Soboyejo

Industrial Noise Control and Acoustics, Randall F. Barron

The CAD Guidebook: A Basic Manual for Understanding and Improving Computer-Aided Design, Stephen J. Schoonmaker

Fundamental Mechanics of Fluids: Third Edition, I. G. Currie

HVAC Water Chillers and Cooling Towers: Fundamentals, Application, and Operations, Herbert W. Stanford III

Handbook of Turbomachinery: Second Edition, Revised and Expanded, Earl Logan, Jr., and Ramendra Roy

Progressing Cavity Pumps, Downhole Pumps, and Mudmotors, Lev Nelik

Gear Noise and Vibration: Second Edition, Revised and Expanded, J. Derek Smith

Intermediate Heat Transfer, Kau-Fui Vincent Wong

Mechanical Engineering Software

Spring Design with an IBM PC, Al Dietrich

Mechanical Design Failure Analysis: With Failure Analysis System Software for the IBM PC, David G. Ullman

Bearing Design in Machinery

Engineering Tribology and Lubrication

Avraham Harnoy
New Jersey Institute of Technology
Newark, New Jersey

MARCEL DEKKER, INC. NEW YORK • BASEL

ISBN: 0-8247-0703-6

This book is printed on acid-free paper.

Headquarters
Marcel Dekker, Inc.
270 Madison Avenue, New York, NY 10016
tel: 212-696-9000; fax: 212-685-4540

Eastern Hemisphere Distribution
Hutgasse 4, Postfach 812, CH-4001 Basel, Switzerland
tel: 41-61-260-6300; fax: 41-61-260-6333

World Wide Web
http://www.dekker com

The publisher offers discounts on this book when ordered in bulk quantities. For more information, write to Special Sales/Professional Marketing at the headquarters address above.

Current printing (last digit):
10 9 8 7 6 5 4 3 2 1

PRINTED IN THE UNITED STATES OF AMERICA

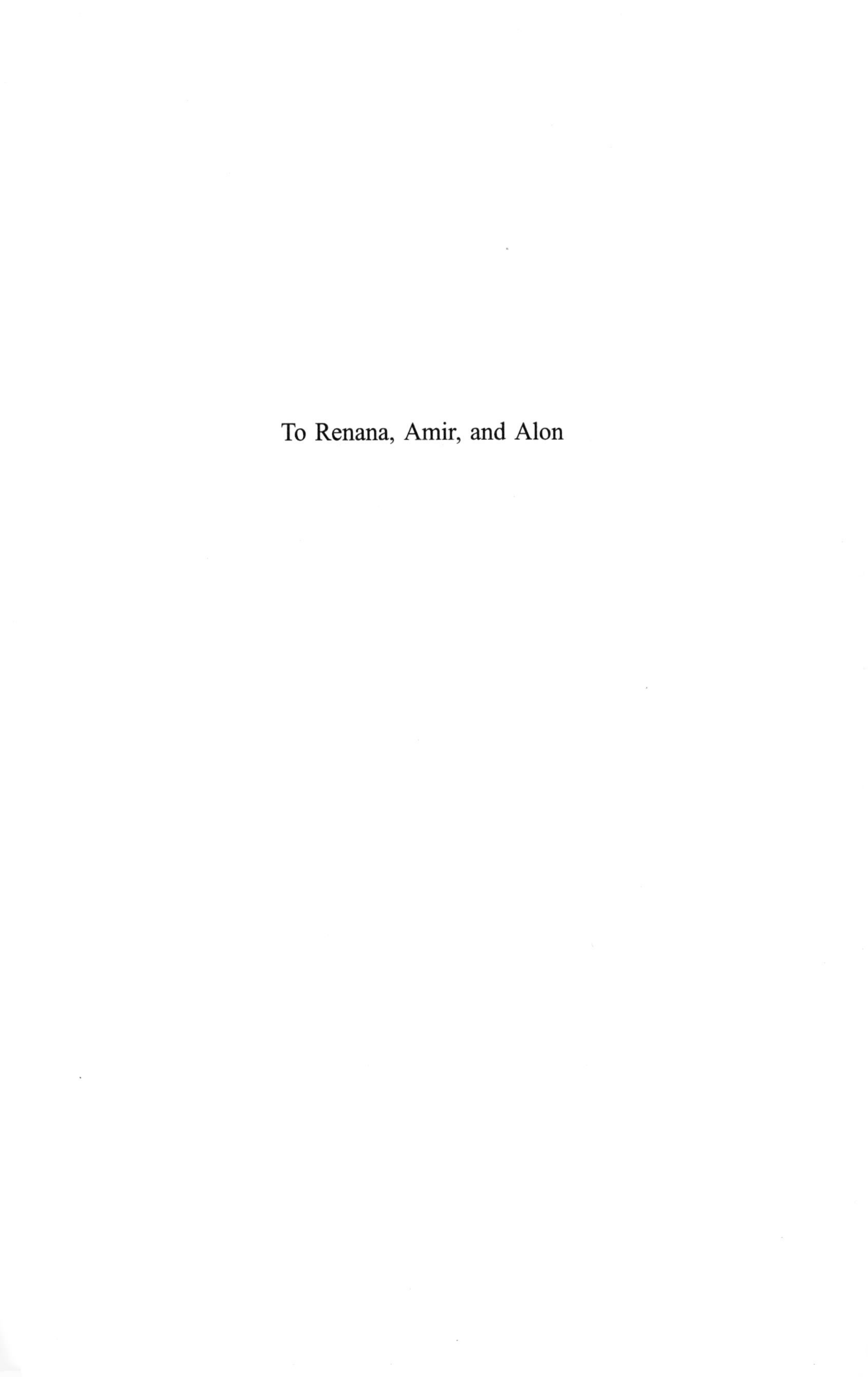

To Renana, Amir, and Alon

Preface

Most engineering schools offer senior courses in bearing design in machinery. These courses are offered under various titles, such as Tribology, Bearings and Bearing Lubrication, and Advanced Machine Design. This book is intended for use as a textbook for these and similar courses for undergraduate students and for self-study by engineers involved in design, maintenance, and development of machinery. The text includes many examples of problems directly related to important design cases, which are often encountered by engineers. In addition, students will find this book useful as a reference for design projects and machine design courses.

Engineers have already realized that there is a need for a basic course and a textbook for undergraduate students that does not focus on only one bearing type, such as a hydrodynamic bearing or a rolling-element bearing, but presents the big picture—an overview of all bearing types. This course should cover the fundamental aspects of bearing selection, design, and tribology. Design engineers require much more knowledge for bearing design than is usually taught in machine design courses.

This book was developed to fill this need. The unique approach of this text is that it is not intended only for scientists and graduate students, but it is specifically tailored as a basic practical course for engineers. For this purpose, the traditional complex material of bearing design was simplified and presented in a methodical way that is easily understood, and illustrated by many examples.

However, this text also includes chapters for advanced studies, to upgrade the text for graduate-level courses.

Engineering schools continually strive to strengthen the design component of engineering education, in order to meet the need of the industry, and this text is intended to satisfy this requirement. Whenever an engineer faces the task of designing a machine, his first questions are often which bearings to select and how to arrange them, and how to house, lubricate and seal the bearings. Appropriate bearing design is essential for a reliable machine operation, because bearings wear out and fail by fatigue, causing a breakdown in machine operation.

I have used the material in this book for many years to teach a tribology course for senior undergraduate students and for an advanced course, Bearings and Bearing Lubrication, for graduate students. The book has benefited from the teaching experience and constructive comments of the students over the years.

The first objective of this text is to present the high-level theory in bearing design in a simplified form, with an emphasis on the basic physical concepts. For example, the hydrodynamic fluid film theory is presented in basic terms, without resorting to complex fluid dynamic derivations. The complex mathematical integration required for solving the pressure wave in fluid-film bearings is replaced in many cases by a simple numerical integration, which the students and engineers may prefer to perform with the aid of a personal computer. The complex calculations of contact stresses in rolling-element bearings are also presented in a simplified practical form for design engineers.

The second objective is that the text be self-contained, and the explanation of the material be based on first principles. This means that engineers of various backgrounds can study this text without prerequisite advanced courses.

The third objective is not to dwell only on theory and calculations, but rather to emphasize the practical aspects of bearing design, such as bearings arrangement, high-temperature considerations, tolerances, and material selection. In the past, engineers gained this expert knowledge only after many years of experience. This knowledge is demonstrated in this text by a large number of drawings of design examples and case studies from various industries. In addition, important economical considerations are included. For bearing selection and design, engineers must consider the initial cost of each component as well as the long-term maintenance expenses.

The fourth objective is to encourage students to innovate design ideas and unique solutions to bearing design problems. For this purpose, several case studies of interesting and unique solutions are included in this text.

In the last few decades, there has been remarkable progress in machinery and there is an ever-increasing requirement for better bearings that can operate at higher speeds, under higher loads, and at higher temperatures. In response to this need, a large volume of experimental and analytical research has been conducted that is directly related to bearing design. Another purpose of this text is to make the vast amount of accumulated knowledge readily available to engineers.

In many cases, bearings are selected by using manufacturers' catalogs of rolling-element bearings. However, as is shown in this text, rolling bearings are only one choice, and depending on the application, other bearing types can be more suitable or more economical for a specific application. This book reviews the merits of other bearing types to guide engineers.

Bearing design requires an interdisciplinary background. It involves calculations that are based on the principles of fluid mechanics, solid mechanics, and material science. The examples in the book are important to show how all these engineering principles are used in practice. In particular, the examples are necessary for self-study by engineers, to answer the questions that remain after reading the theoretical part of the text.

Extensive use is made of the recent development in computers and software for solving basic bearing design problems. In the past, engineers involved in bearing design spent a lot of time and effort on analytical derivations, particularly on complicated mathematical integration for calculating the load capacity of hydrodynamic bearings. Recently, all this work was made easier by computer-aided numerical integration. The examples in this text emphasize the use of computers for bearing design.

Chapter 1 is a survey of the various bearing types; the advantages and limitations of each bearing type are discussed. The second chapter deals with lubricant viscosity, its measurement, and variable viscosity as a function of temperature and pressure. Chapter 3 deals with the characteristics of lubricants, including mineral and synthetic oils and greases, as well as the many additives used to enhance the desired properties.

Chapters 4–7 deal with the operation of fluid-film bearings. The hydrodynamic lubrication theory is presented from first principles, and examples of calculations of the pressure wave and load capacity are included. Chapter 8 deals with the use of charts for practical bearing design procedures, and estimation of the operation temperature of the oil. Chapter 9 presents practical examples of widely used hydrodynamic bearings that overcome the limitations of the common hydrodynamic journal bearings. Chapter 10 covers the design of hydrostatic pad bearings in which an external pump generates the pressure. The complete hydraulic system is discussed.

Chapter 11 deals with bearing materials. The basic principles of practical tribology (friction and wear) for various materials are introduced. Metals and nonmetals such as plastics and ceramics as well as composite materials are included.

Chapters 12 and 13 deal with rolling element bearings. In Chapter 12, the calculations of the contact stresses in rolling bearings and elastohydrodynamic lubrication are presented with practical examples. In Chapter 13, the practical aspects of rolling bearing lubrication are presented. In addition, the selection of rolling bearings is outlined, with examples. Most important, the design considerations of bearing arrangement are discussed, and examples provided. Chapter 14 covers the subject of bearing testing under static and dynamic conditions.

Chapter 15 deals with hydrodynamic journal bearings under dynamic load. It describes the use of computers for solving the trajectory of the journal center under dynamic conditions. Chapters 16 and 17 deal with friction characteristics and modeling of dynamic friction, which has found important applications in control of machines with friction. Chapter 18 presents a unique case study of composite bearing—hydrodynamic and rolling-element bearing in series. Chapter 19 deals with viscoelastic (non-Newtonian) lubricants, such as the VI improved oils, and Chapter 20 describes the operation of natural human joints as well as the challenges in the development of artificial joint implants.

I acknowledge the constructive comments of many colleagues and engineers involved in bearing design, and the industrial publications and advice provided by the members of the Society of Tribology and Lubrication Engineers. Many graduates who had taken this course have already used the preliminary notes for actual design and provided valuable feedback and important comments.

I am grateful to my graduate and undergraduate students, whose valuable comments were instrumental in making the text easily understood. Many solved problems were added because the students felt that they were necessary for unambiguous understanding of the many details of bearing design. Also, I wish to express my appreciation to Ted Allen and Marcel Dekker, Inc., for the great help and support with this project.

I acknowledge all the companies that provided materials and drawings, in particular, FAG and SKF. I am also pleased to thank the graduate students Simon Cohn and Max Roman for conducting experiments that are included in the text, helping with drawings, and reviewing examples, and Gaurav Dave, for help with the artwork.

Special thanks to my son, Amir Harnoy, who followed the progress of the writing of this text, and continually provided important suggestions. Amir is a mechanical project engineer who tested the text in actual designs for the aerospace industry. Last but not least, particular gratitude to my wife, Renana, for help and encouragement during the long creation of this project.

Avraham Harnoy

Contents

Symbols

NOMENCLATURE FOR HYDRODYNAMIC BEARINGS

$\vec{a}$ = acceleration vector
a = tan α, slope of inclined plane slider
B = length of plane slider (x direction) (Fig. 4-5)
C = radial clearance
c = specific heat
e = eccentricity
F = external load
F_f = friction force
F(t) = time dependent load; having components $F_x(t)$, $F_y(t)$
h = variable film thickness
h_n = h_{min}, minimum film thickness
h_0 = film thickness at a point of peak pressure
L = length of the sleeve (z direction) (Fig. 7-1); width of a plane slider (z direction) (Fig. 4-5)
m = mass of journal
N = bearing speed [RPM]
n = bearing speed [rps]
O; O_1 = sleeve and journal centers, respectively (Fig. 6-1)

p = pressure wave in the fluid film
P = average pressure
PV = bearing limit (product of average pressure times sliding velocity)
q = constant flow rate in the clearance (per unit of bearing length)
R = journal radius
R_1 = bearing bore radius
t = time
$\bar{t}$ = ωt, dimensionless time
U = journal surface velocity
V = sliding velocity
VI = viscosity index (Eq. 2-5)
W = bearing load carrying capacity, W_x, W_y, components
α = slope of inclined plane slider, or variable slope of converging clearance
α = viscosity-pressure exponent, Eq. 2-6
β = h_2/h_1, ratio of maximum and minimum film thickness in plane slider
ε = eccentricity ratio, e/C
ϕ = Attitude angle, Fig. 1-3
λ = relaxation time of the fluid
ρ = density
θ = angular coordinates (Figs. 1-3 and 9-1)
$\tau_{xy}, \tau_{yz}, \tau_{xz}$ = shear stresses
$\sigma_x . \sigma_y, \sigma_z$ = tensile stresses
ω = angular velocity of the journal
μ = absolute viscosity
μ_o = absolute viscosity at atmospheric pressure
ν = kinematic viscosity, μ/ρ

NOMENCLATURE FOR HYDROSTATIC BEARINGS

A_e = effective bearing area (Eq. 10-25)
B = width of plate in unidirectional flow
d_i = inside diameter of capillary tube
$\dot{E}_h$ = hydraulic power required to pump the fluid through the bearing and piping system
$\dot{E}_f$ = mechanical power provided by the drive (electrical motor) to overcome the friction torque (Eq. 10.15)
$\dot{E}_t$ = total power of hydraulic power and mechanical power required to maintain the operation of hydrostatic bearing (Eq. 10-18)
h_0 = clearance between two parallel, concentric disks
H_p = head of pump = $H_d - H_s$

H_d = discharge head (Eq. 10-51)
H_s = suction head (Eq. 10-52)
k = bearing stiffness (Eq. 10-23)
K = parameter used to calculate stiffness of bearing = $3\kappa A_e Q$
L = length of rectangular pad
l_c = length of capillary tube
p_d = pump discharge pressure
p_r = recess pressure
p_s = supply pressure (also pump suction pressure)
Δp = pressure loss along the resistance
Q = flow rate
R = disk radius
R_0 = radius of a round recess
R_f = flow resistance = $\Delta p/Q$
R_{in} = resistance of inlet flow restrictor
T_m = mechanical torque of motor
V = fluid velocity
W = load capacity
Z = height
η_1 = efficiency of motor
η_2 = efficiency of pump
κ = constant that depends on bearing geometry (Eq. 10-27)
β = ratio of recess pressure to the supply pressure, p_r/p_s
μ = fluid viscosity
γ = specific weight of fluid

NOMENCLATURE FOR ROLLING ELEMENT BEARINGS

a = half width of rectangular contact area (Fig. 12-8)
a, b = small and large radius, respectively, of an ellipsoidal contact area
d = rolling element diameter
d_i, d_o = inside and outside diameters of a ring
E_{eq} = equivalent modulus of elasticity [N/m^2]
$\hat{E}$ = elliptical integral, defined by Eq. 12-28 and estimated by Eq. 12.19
F_c = centrifugal force of a rolling element
h_c = central film thickness
h_{min}, h_n = minimum film thickness
k = ellipticity-parameter, b/a , estimated by Eq. 12.17
L = An effective length of a line contact between two cylinders
m_r = mass of a rolling element (ball or cylinder)

n_r = number of rolling elements around the bearing
p = pressure distribution
p_{max} = maximum Hertz pressure at the center of contact area (Eq. 12-15)
q_a = parameter to estimate, $\hat{E}$, defined in Eq. 12-18
r = deep groove radius
R_1, R_2 = radius of curvatures of two bodies in contact
R_{1x}, R_{2x} = radius of curvatures, in plane y, z, of two bodies in contact
R_{1y}, R_{2y} = radius of curvatures, in plane x, z, of two bodies in contact
$R_{eq,}$ = equivalent radius of curvature
R_r = race-conformity ratio, r/d
R_s = equivalent surface roughness at the contact (Eq. 12-38)
R_{s1} and R_{s2} = surface roughness of two individual surfaces in contact
R_x = equivalent contact radius (Eqs. 12-5, 12-6)
R_d = curvature difference defined by Eq. 12-27
t^* = parameter estimated by Eq. 12.25 for calculating τ_{yz} in Eq. 12-24
$\hat{T}$ = elliptical integral, defined by Eq. 12.28 and estimated by Eq. 12-22
U_C = velocity of a rolling element center (Eq. 12-31)
U_r = rolling velocity (Eq. 12-35)
$\overline{W}$ = dimensionless bearing load carrying capacity
W = load carrying capacity
W_i, W_o = resultant normal contact forces of the inner and outer ring races in angular contact bearing
W_{max} = maximum load on a single rolling element
N = bearing speed [RPM]
α = viscosity-pressure exponent
α = linear thermal-expansion coefficient
α_r = radius ratio = R_y/R_x
Λ = a ratio of a film thickness and size of surface asperities, R_s (Eq. 12-39)
δ_m = maximum deformation of the roller in a normal direction to the contact area (Eq. 12-7, 12-21)
ξ = ratio of rolling to sliding velocity
$\tau_{xy}, \tau_{yz}, \tau_{xz}$ = shear stresses
$\sigma_x, \sigma_y, \sigma_z$ = tensile stresses
μ_0 = absolute viscosity of the lubricant at atmospheric pressure
ν = Poisson's ratio
ω = angular speed
ω_C = angular speed of the center of a rolling element (or cage) [rad/s]
ρ = density

1

Classification and Selection of Bearings

1.1 INTRODUCTION

Moving parts in machinery involve relative sliding or rolling motion. Examples of relative motion are linear sliding motion, such as in machine tools, and rotation motion, such as in motor vehicle wheels. Most bearings are used to support rotating shafts in machines. Rubbing of two bodies that are loaded by a normal force (in the direction normal to the contact area) generates energy losses by friction and wear. Appropriate bearing design can minimize friction and wear as well as early failure of machinery. The most important objectives of bearing design are to extend bearing life in machines, reduce friction energy losses and wear, and minimize maintenance expenses and downtime of machinery due to frequent bearing failure. In manufacturing plants, unexpected bearing failure often causes expensive loss of production. Moreover, in certain cases, such as in aircraft, there are very important safety considerations, and unexpected bearing failures must be prevented at any cost.

During the past century, there has been an ever-increasing interest in the friction and wear characteristics of various bearing designs, lubricants, and materials for bearings. This scientific discipline, named *Tribology*, is concerned with the friction, lubrication, and wear of interacting surfaces in relative motion. Several journals are dedicated to the publication of original research results on this subject, and several books have been published that survey the vast volume of research in tribology. The objectives of the basic research in tribology are similar to those of bearing design, focusing on the reduction of friction and wear. These efforts resulted in significant advances in bearing technology during the past century. This improvement is particularly in lubrication, bearing materials, and the introduction of rolling-element bearings and bearings supported by lubrication films. The improvement in bearing technology resulted in the reduction of friction, wear, and maintenance expenses, as well as in the longer life of machinery.

The selection of a proper bearing type for each application is essential to the reliable operation of machinery, and it is an important component of machine design. Most of the maintenance work in machines is in bearing lubrication as well as in the replacement of damaged or worn bearings. The appropriate selection of a bearing type for each application is very important to minimize the risk of early failure by wear or fatigue, thereby to secure adequate bearing life. There are other considerations involved in selection, such as safety, particularly in aircraft. Also, cost is always an important consideration in bearing selection—the designer should consider not only the initial cost of the bearing but also the cost of maintenance and of the possible loss of production over the complete life cycle of the machine.

Therefore, the first step in the process of bearing design is the selection of the bearing type for each application. In most industries there is a tradition concerning the type of bearings applied in each machine. However, a designer should follow current developments in bearing technology; in many cases, selection of a new bearing type can be beneficial. Proper selection can be made from a variety of available bearing types, which include rolling-element bearings, dry and boundary lubrication bearings, as well as hydrodynamic and hydrostatic lubrication bearings. An additional type introduced lately is the electromagnetic bearing. Each bearing type can be designed in many different ways and can be made of various materials, as will be discussed in the following chapters.

It is possible to reduce the size and weight of machines by increasing their speed, such as in motor vehicle engines. Therefore, there is an increasing requirement for higher speeds in machinery, and the selection of an appropriate bearing type for this purpose is always a challenge. In many cases, it is the limitation of the bearing that limits the speed of a machine. It is important to select a bearing that has low friction in order to minimize friction-energy losses, equal to the product of friction torque and angular speed. Moreover, friction-energy losses are dissipated in the bearing as heat, and it is essential to prevent bearing overheating. If the temperature of the sliding surfaces is too close to the melting point of the bearing material, it can cause bearing failure. In the following chapters, it will be shown that an important task in the design process is the prevention of bearing overheating.

1.1.1 Radial and Thrust Bearings

Bearings can also be classified according to their geometry related to the relative motion of elements in machinery. Examples are journal, plane-slider, and spherical bearings. A journal bearing, also referred to as a sleeve bearing, is widely used in machinery for rotating shafts. It consists of a bushing (sleeve) supported by a housing, which can be part of the frame of a machine. The shaft (journal) rotates inside the bore of the sleeve. There is a small clearance between the inner diameter of the sleeve and the journal, to allow for free rotation. In contrast, a plane-slider bearing is used mostly for linear motion, such as the slides in machine tools.

A bearing can also be classified as a radial bearing or a thrust bearing, depending on whether the bearing load is in the radial or axial direction, respectively, of the shaft. The shafts in machines are loaded by such forces as reactions between gears and tension in belts, gravity, and centrifugal forces. All the forces on the shaft must be supported by the bearings, and the force on the bearing is referred to as a *bearing load*. The load on the shaft can be divided into radial and axial components. The axial component (also referred to as *thrust load*) is in the direction of the shaft axis (see Fig. 1-1), while the *radial load* component is in the direction normal to the shaft axis. In Fig. 1-1, an example of a loaded shaft is presented. The reaction forces in helical gears have radial and axial components. The component F_a is in the axial direction, while all the other components are radial loads. Examples of solved problems are included at the end of this chapter. Certain bearings, such as conical roller bearings, shown in Fig. 1-1, or angular ball bearings, can support radial as well as thrust forces. Certain other bearings, however, such as hydrodynamic journal bearings, are applied only for radial loads, while the hydrodynamic thrust bearing supports

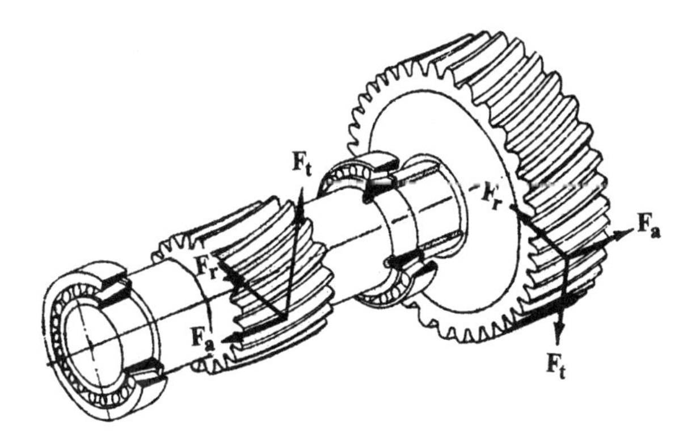

FIG. 1-1 Load components on a shaft with helical gears.

only axial loads. A combination of radial and thrust bearings is often applied to support the shaft in machinery.

1.1.2 Bearing Classification

Machines could not operate at high speed in their familiar way without some means of reducing friction and the wear of moving parts. Several important engineering inventions made it possible to successfully operate heavily loaded shafts at high speed, including the rolling-element bearing and hydrodynamic, hydrostatic, and magnetic bearings.

1. *Rolling-element bearings* are characterized by rolling motion, such as in ball bearings or cylindrical rolling-element bearings. The advantage of rolling motion is that it involves much less friction and wear, in comparison to the sliding motion of regular sleeve bearings.
2. The term *hydrodynamic bearing* refers to a sleeve bearing or an inclined plane-slider where the sliding plane floats on a thin film of lubrication. The fluid film is maintained at a high pressure that supports the bearing load and completely separates the sliding surfaces. The lubricant can be fed into the bearing at atmospheric or higher pressure. The pressure wave in the lubrication film is generated by hydrodynamic action due to the rapid rotation of the journal. The fluid film acts like a viscous wedge and generates high pressure and load-carrying capacity. The sliding surface floats on the fluid film, and wear is prevented.
3. In contrast to hydrodynamic bearing, *hydrostatic bearing* refers to a configuration where the pressure in the fluid film is generated by an external high-pressure pump. The lubricant at high pressure is fed into the bearing recesses from an external pump through high-pressure tubing. The fluid, under high pressure in the bearing recesses, carries the load and separates the sliding surfaces, thus preventing high friction and wear.
4. A recent introduction is the *electromagnetic bearing*. It is still in development but has already been used in some unique applications. The concept of operation is that a magnetic force is used to support the bearing load. Several electromagnets are mounted on the bearing side (stator poles). The bearing load capacity is generated by the magnetic field between rotating laminators, mounted on the journal, and stator poles, on the stationary bearing side. Active feedback control keeps the journal floating without any contact with the bearing surface. The advantage is that there is no contact between the sliding surfaces, so wear is completely prevented as long as there is magnetic levitation.

Further description of the characteristics and applications of these bearings is included in this and the following chapters.

1.2 DRY AND BOUNDARY LUBRICATION BEARINGS

Whenever the load on the bearing is light and the shaft speed is low, wear is not a critical problem and a sleeve bearing or plane-slider lubricated by a very thin layer of oil (boundary lubrication) can be adequate. Sintered bronzes with additives of other elements are widely used as bearing materials. Liquid or solid lubricants are often inserted into the porosity of the material and make it self-lubricated. However, in heavy-duty machinery—namely, bearings operating for long periods of time under heavy load relative to the contact area and at high speeds—better bearing types should be selected to prevent excessive wear rates and to achieve acceptable bearing life. Bearings from the aforementioned list can be selected, namely, rolling-element bearings or fluid film bearings.

In most applications, the sliding surfaces of the bearing are lubricated. However, bearings with dry surfaces are used in unique situations where lubrication is not desirable. Examples are in the food and pharmaceutical industries, where the risk of contamination by the lubricant forbids its application. The sliding speed, V, and the average pressure in the bearing, P, limit the use of dry or boundary lubrication. For plastic and sintered bearing materials, a widely accepted limit criterion is the product PV for each bearing material and lubrication condition. This product is proportional to the amount of friction-energy loss that is dissipated in the bearing as heat. This is in addition to limits on the maximum sliding velocity and average pressure. For example, a self-lubricated sintered bronze bearing has the following limits:

Surface velocity limit, V, is 6 m/s, or 1180 ft/min
Average surface-pressure limit, P, is 14 MPa, or 2000 psi
PV limit is 110,000 psi-ft/min, or 3.85×10^6 Pa-m/s

In comparison, bearings made of plastics have much lower PV limit. This is because the plastics have a low melting point; in addition, the plastics are not good conductors of heat, in comparison to metals. For these reasons, the PV limit is kept at relatively low values, in order to prevent bearing failure by overheating. For example, Nylon 6, which is widely used as a bearing material, has the following limits as a bearing material:

Surface velocity limit, V, is 5 m/s
Average surface-pressure limit, P, is 6.9 MPa
PV limit is 105×10^3 Pa-m/s

Remark. In hydrodynamic lubrication, the symbol for surface velocity of a rotating shaft is U, but for the PV product, sliding velocity V is traditionally used.

Conversion to SI Units.

$$1 \text{ lbf/in.}^2 \text{ (psi)} = 6895 \text{ N/m}^2 \text{ (Pa)}$$
$$1 \text{ ft/min} = 0.0051 \text{ m/s}$$
$$1 \text{ psi-ft/min} = 6895 \times 0.0051 = 35 \text{ Pa-m/s} = 35 \times 10^{-6} \text{ MPa-m/s}$$

An example for calculation of the PV value in various cases is included at the end of this chapter. The PV limit is much lower than that obtained by multiplying the maximum speed and maximum average pressure due to the load capacity. The reason is that the maximum PV is determined from considerations of heat dissipation in the bearing, while the average pressure and maximum speed can be individually of higher value, as long as the product is not too high. If the maximum PV is exceeded, it would usually result in a faster-than-acceptable wear rate.

1.3 HYDRODYNAMIC BEARING

An inclined plane-slider is shown in Fig. 1-2. It carries a load F and has horizontal velocity, U, relative to a stationary horizontal plane surface. The plane-slider is inclined at an angle α relative to the horizontal plane. If the surfaces were dry, there would be direct contact between the two surfaces, resulting in significant friction and wear. It is well known that friction and wear can be reduced by lubrication. If a sufficient quantity of lubricant is provided and the

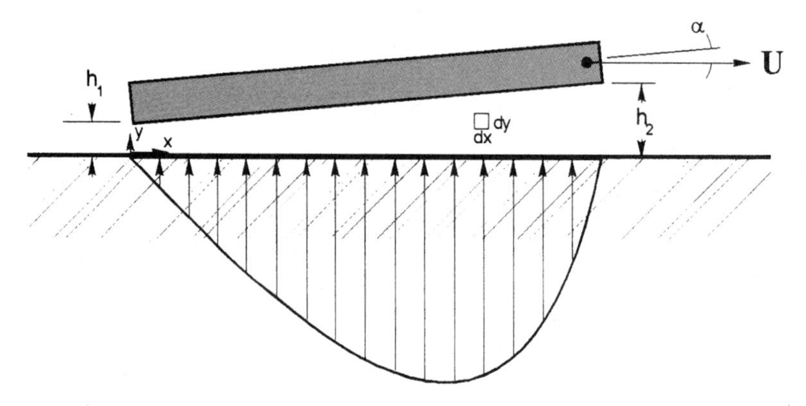

FIG. 1-2 Hydrodynamic lubrication of plane-slider.

sliding velocity is high, the surfaces would be completely separated by a very thin lubrication film having the shape of a fluid wedge. In the case of complete separation, full hydrodynamic lubrication is obtained. The plane-slider is inclined, to form a converging viscous wedge of lubricant as shown in Fig. 1-2. The magnitudes of h_1 and h_2 are very small, of the order of only a few micrometers. The clearance shown in Fig. 1-2 is much enlarged.

The lower part of Fig. 1-2 shows the pressure distribution, p (pressure wave), inside the thin fluid film. This pressure wave carries the slider and its load. The inclined slider, floating on the lubricant, is in a way similar to water-skiing, although the physical phenomena are not identical. The pressure wave inside the lubrication film is due to the fluid viscosity, while in water-skiing it is due to the fluid inertia. The generation of a pressure wave in hydrodynamic bearings can be explained in simple terms, as follows: The fluid adheres to the solid surfaces and is dragged into the thin converging wedge by the high shear forces due to the motion of the plane-slider. In turn, high pressure must build up in the fluid film in order to allow the fluid to escape through the thin clearances.

A commonly used bearing in machinery is the hydrodynamic journal bearing, as shown in Fig. 1-3. Similar to the inclined plane-slider, it can support a radial load without any direct contact between the rotating shaft (journal) and the bearing sleeve. The viscous fluid film is shaped like a wedge due to the eccentricity, e, of the centers of the journal relative to that of bearing bore. As with the plane-slider, a pressure wave is generated in the lubricant, and a thin fluid

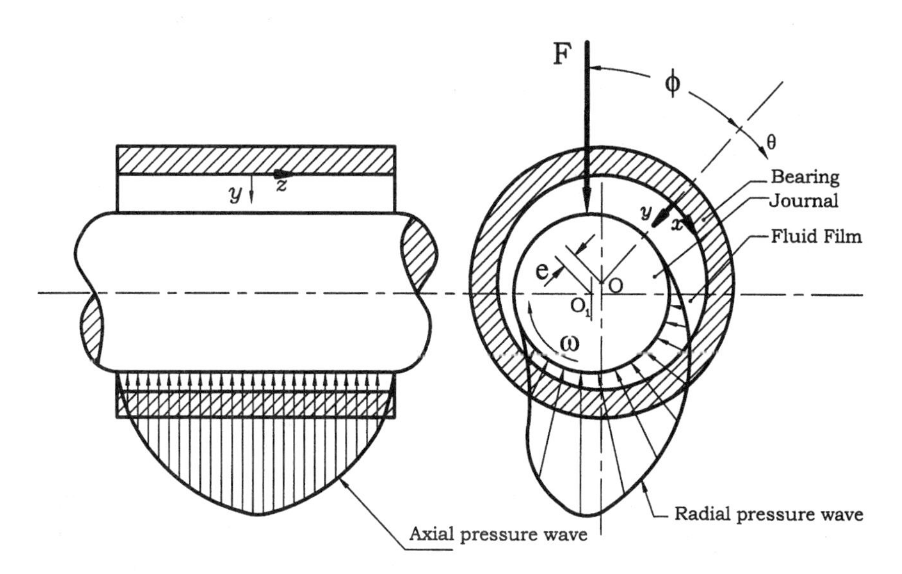

FIG. 1-3 Hydrodynamic journal bearing.

film completely separates the journal and bearing surfaces. Due to the hydrodynamic effect, there is low friction and there is no significant wear as long as a complete separation is maintained between the sliding surfaces.

The pressure wave inside the hydrodynamic film carries the journal weight together with the external load on the journal. The principle of operation is the uneven clearance around the bearing formed by a small eccentricity, e, between the journal and bearing centers, as shown in Fig. 1-3. The clearance is full of lubricant and forms a thin fluid film of variable thickness. A pressure wave is generated in the converging part of the clearance. The resultant force of the fluid film pressure wave is the load-carrying capacity, W, of the bearing. For bearings operating at steady conditions (constant journal speed and bearing load), the load-carrying capacity is equal to the external load, F, on the bearing. But the two forces of action and reaction act in opposite directions.

In a hydrodynamic journal bearing, the load capacity (equal in magnitude to the bearing force) increases with the eccentricity, e, of the journal. Under steady conditions, the center of the journal always finds its equilibrium point, where the load capacity is equal to the external load on the journal. Figure 1-3 indicates that the eccentricity displacement, e, of the journal center, away from the bearing center, is not in the vertical direction but at a certain attitude angle, ϕ, from the vertical direction. In this configuration, the resultant load capacity, due to the pressure wave, is in the vertical direction, opposing the vertical external force. The fluid film pressure is generated mostly in the converging part of the clearance, and the attitude angle is required to allow the converging region to be below the journal to provide the required lift force in the vertical direction and, in this way, to support the external load.

In real machinery, there are always vibrations and disturbances that can cause occasional contact between the surface asperities (surface roughness), resulting in severe wear. In order to minimize this risk, the task of the engineer is to design the hydrodynamic journal bearing so that it will operate with a minimum lubrication-film thickness, h_n, much thicker than the size of the surface asperities. Bearing designers must keep in mind that if the size of the surface asperities is of the order of magnitude of 1 micron, the minimum film thickness, h_n, should be 10–100 microns, depending on the bearing size and the level of vibrations expected in the machine.

1.3.1 Disadvantages of Hydrodynamic Bearings

One major disadvantage of hydrodynamic bearings is that a certain minimum speed is required to generate a full fluid film that completely separates the sliding surfaces. Below that speed, there is mixed or boundary lubrication, with direct contact between the asperities of the rubbing surfaces. For this reason, even if the bearing is well designed and successfully operating at the high rated speed of the machine, it can be subjected to excessive friction and wear at low speed, such as during starting and stopping of journal rotation. In particular, hydrodynamic bearings undergo severe wear during start-up, when they accelerate from zero speed, because static friction is higher than dynamic friction.

A second important disadvantage is that hydrodynamic bearings are completely dependent on a continuous supply of lubricant. If the oil supply is interrupted, even for a short time for some unexpected reason, it can cause overheating and sudden bearing failure. It is well known that motor vehicle engines do not last a long time if run without oil. In that case, the hydrodynamic bearings fail first due to the melting of the white metal lining on the bearing. This risk of failure is the reason why hydrodynamic bearings are never used in critical applications where there are safety concerns, such as in aircraft engines. Failure of a motor vehicle engine, although it is highly undesirable, does not involve risk of loss of life; therefore, hydrodynamic bearings are commonly used in motor vehicle engines for their superior performance and particularly for their relatively long operation life.

A third important disadvantage is that the hydrodynamic journal bearing has a low stiffness to radial displacement of the journal (low resistance to radial run-out), particularly when the eccentricity is low. This characteristic rules out the application of hydrodynamic bearings in precision machines, e.g., machine tools. Under dynamic loads, the low stiffness of the bearings can result in dynamic instability, particularly with lightly loaded high-speed journals. The low stiffness causes an additional serious problem of bearing whirl at high journal speeds. The bearing whirl phenomenon results from instability in the oil film, which often results in bearing failure.

Further discussions of the disadvantages of journal bearing and methods to overcome these drawbacks are included in the following chapters.

1.4 HYDROSTATIC BEARING

The introduction of externally pressurized hydrostatic bearings can solve the problem of wear at low speed that exists in hydrodynamic bearings. In hydrostatic bearings, a fluid film completely separates the sliding surfaces at all speeds, including zero speed. However, hydrostatic bearings involve higher cost in comparison to hydrodynamic bearings. Unlike hydrodynamic bearings, where the pressure wave in the oil film is generated inside the bearing by the rotation of the journal, an external oil pump pressurizes the hydrostatic bearing. In this way, the hydrostatic bearing is not subjected to excessive friction and wear rate at low speed.

The hydrostatic operation has the advantage that it can maintain complete separation of the sliding surfaces by means of high fluid pressure during the starting and stopping of journal rotation. Hydrostatic bearings are more expensive than hydrodynamic bearings, since they require a hydraulic system to pump and circulate the lubricant and there are higher energy losses involved in the circulation of the fluid. The complexity and higher cost are reasons that hydrostatic bearings are used only in special circumstances where these extra expenses can be financially justified.

Girard introduced the principle of the hydrostatic bearing in 1851. Only much later, in 1923, did Hodgekinson patent a hydrostatic bearing having wide recesses and fluid pumped into the recesses at constant pressure through flow restrictors. The purpose of the flow restrictors is to allow bearing operation and adequate bearing stiffness when all the recesses are fed at constant pressure from one pump. The advantage of this system is that it requires only one pump without flow dividers for distributing oil at a constant flow rate into each recess.

Whenever there are many recesses, the fluid is usually supplied at constant pressure from one central pump. The fluid flows into the recesses through flow restrictors to improve the radial stiffness of the bearing. A diagram of such system is presented in Fig. 1-4. From a pump, the oil flows into several recesses around the bore of the bearing through capillary flow restrictors. From the recesses, the fluid flows out in the axial direction through a thin radial clearance, h_o, between the journal and lands (outside the recesses) around the circumference of the two ends of the bearing. This thin clearance creates a resistance to the outlet flow from each bearing recess. This outlet resistance, at the lands, is essential to maintain high pressure in each recess around the bearing. This resistance at the outlet varies by any small radial displacement of the journal due to the bearing load. The purpose of supplying the fluid to the recesses through flow restrictors is to make the bearing stiffer under radial force; namely, it reduces radial displacement (radial run-out) of the journal when a radial load is applied. The following is an explanation for the improved stiffness provided by flow restrictors.

When a journal is displaced in the radial direction from the bearing center, the clearances at the lands of the opposing recesses are no longer equal. The resistance to the flow from the opposing recesses decreases and increases, respectively (the resistance is inversely proportional to h_o^3). This results in unequal flow rates in the opposing recesses. The flow increases and decreases, respectively. An important characteristic of a flow restrictor, such as a capillary tube, is that its pressure drop increases with flow rate. In turn, this causes the pressures in the opposing recesses to decrease and increase, respectively. The bearing load capacity resulting from these pressure differences acts in the opposite direction to the radial load on the journal. In this way, the bearing supports the journal with minimal radial displacement. In conclusion, the introduction of inlet flow restrictors increases the bearing stiffness because only a very small radial displacement of the journal is sufficient to generate a large pressure difference between opposing recesses.

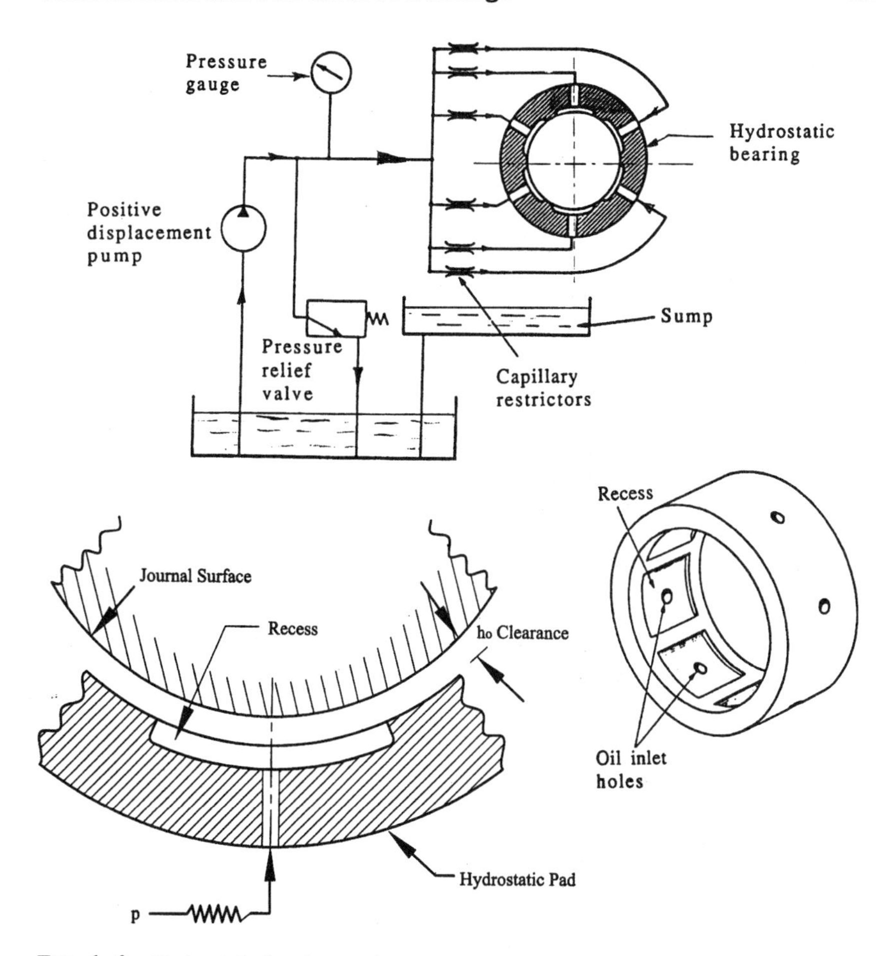

FIG. 1-4 Hydrostatic bearing system.

In summary, the primary advantage of the hydrostatic bearing, relative to all other bearings, is that the surfaces of the journal and bearing are separated by a fluid film at all loads and speeds. As a result, there is no wear and the sliding friction is low at low speeds. A second important advantage of hydrostatic bearings is their good stiffness to radial loads. Unlike hydrodynamic bearings, high stiffness is maintained under any load, from zero loads to the working loads, and at all speeds, including zero speed.

The high stiffness to radial displacement makes this bearing suitable for precision machines, for example, precise machine tools. The high bearing stiffness is important to minimize any radial displacement (run-out) of the journal. In addition, hydrostatic journal bearings operate with relatively large clearances (compared to other bearings); and therefore, there is not any significant run-out that results from uneven surface finish or small dimensional errors in the internal bore of the bearing or journal.

1.5 MAGNETIC BEARING

A magnetic bearing is shown in Fig. 1-5. The concept of operation is that a magnetic field is applied to support the bearing load. Several electromagnets are

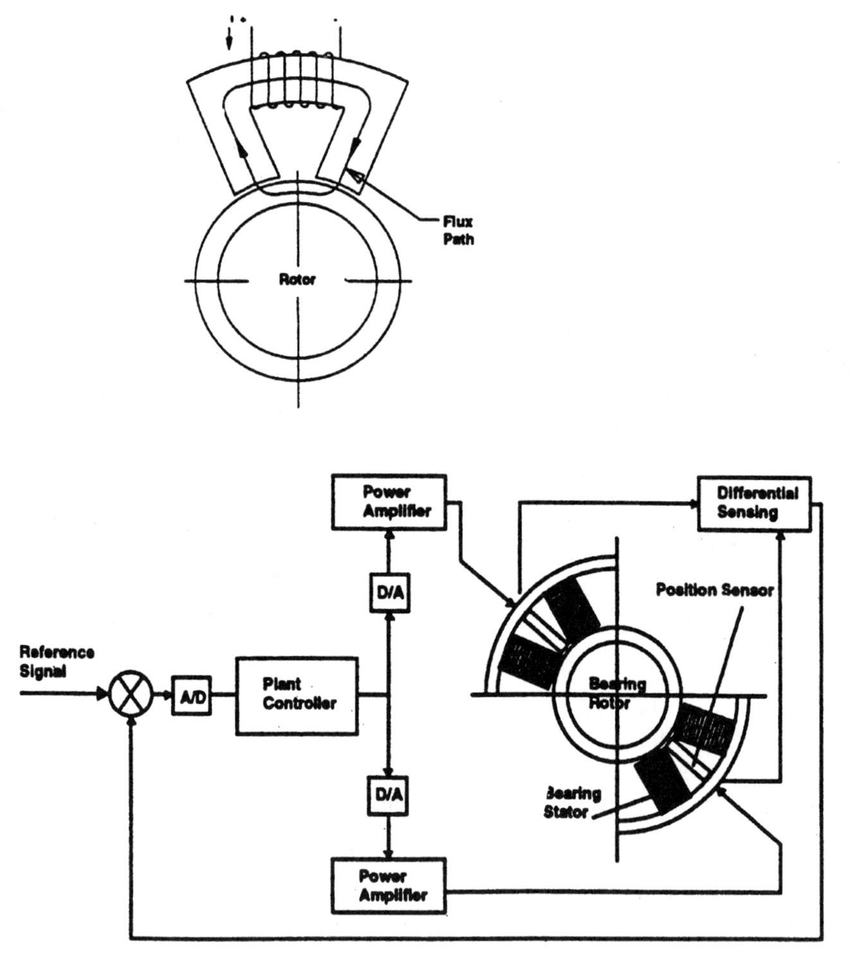

FIG. 1-5 Concept of magnetic bearing. Used by permission of Resolve Magnetic Bearings.

mounted on the bearing side (stator poles). Electrical current in the stator poles generates a magnetic field. The load-carrying capacity of the bearing is due to the magnetic field between the rotating laminators mounted on the journal and the coils of the stator poles on the stationary bearing side.

Active feedback control is required to keep the journal floating without its making any contact with the bearing. The control entails on-line measurement of the shaft displacement from the bearing center, namely, the magnitude of the eccentricity and its direction. The measurement is fed into the controller for active feedback control of the bearing support forces in each pole in order to keep the journal close to the bearing center. This is achieved by varying the magnetic field of each pole around the bearing. In this way, it is possible to control the magnitude and direction of the resultant magnetic force on the shaft. This closed-loop control results in stable bearing operation.

During the last decade, a lot of research work on magnetic bearings has been conducted in order to optimize the performance of the magnetic bearing. The research work included optimization of the direction of magnetic flux, comparison between electromagnetic and permanent magnets, and optimization of the number of magnetic poles. This research work has resulted in improved load capacity and lower energy losses. In addition, research has been conducted to improve the design of the control system, which resulted in a better control of rotor vibrations, particularly at the critical speeds of the shaft.

1.5.1 Disadvantages of Magnetic Bearings

Although significant improvement has been achieved, there are still several disadvantages in comparison with other, conventional bearings. The most important limitations follow.

a. Electromagnetic bearings are relatively much more expensive than other noncontact bearings, such as the hydrostatic bearing. In most cases, this fact makes the electromagnetic bearing an uneconomical alternative.
b. Electromagnetic bearings have less damping of journal vibrations in comparison to hydrostatic oil bearings.
c. In machine tools and other manufacturing environments, the magnetic force attracts steel or iron chips.
d. Magnetic bearings must be quite large in comparison to conventional noncontact bearings in order to generate equivalent load capacity. An acceptable-size magnetic bearing has a limited static and dynamic load capacity. The magnetic force that supports static loads is limited by the saturation properties of the electromagnet core material. The maximum magnetic field is reduced with temperature. In addition, the dynamic

load capacity of the bearing is limited by the available electrical power supply from the amplifier.

e. Finally, electromagnetic bearings involve complex design problems to ensure that the heavy spindle, with its high inertia, does not fall and damage the magnetic bearing when power is shut off or momentarily discontinued. Therefore, a noninterrupted power supply is required to operate the magnetic bearing, even at no load or at shutdown conditions of the system. In order to secure safe operation in case of accidental power failure or support of the rotor during shutdown of the machine, an auxiliary bearing is required. Rolling-element bearings with large clearance are commonly used. During the use of such auxiliary bearings, severe impact can result in premature rolling-element failure.

1.6 ROLLING-ELEMENT BEARINGS

Rolling-element bearings, such as ball, cylindrical, or conical rolling bearings, are the bearings most widely used in machinery. Rolling bearings are often referred to as *antifriction bearings*. The most important advantage of rolling-element bearings is the low friction and wear of rolling relative to that of sliding.

Rolling bearings are used in a wide range of applications. When selected and applied properly, they can operate successfully over a long period of time. Rolling friction is lower than sliding friction; therefore, rolling bearings have lower friction energy losses as well as reduced wear in comparison to sliding bearings. Nevertheless, the designer must keep in mind that the life of a rolling-element bearing can be limited due to fatigue. Ball bearings involve a point contact between the balls and the races, resulting in high stresses at the contact, often named *hertz stresses*, after Hertz (1881), who analyzed for the first time the stress distribution in a point contact.

When a rolling-element bearing is in operation, the rolling contacts are subjected to alternating stresses at high frequency that result in metal fatigue. At high speed, the centrifugal forces of the rolling elements, high temperature (due to friction-energy losses) and alternating stresses all combine to reduce the fatigue life of the bearing. For bearings operating at low and medium speeds, relatively long fatigue life can be achieved in most cases. But at very high speeds, the fatigue life of rolling element bearings can be too short, so other bearing types should be selected. Bearing speed is an important consideration in the selection of a proper type of bearing. High-quality rolling-element bearings, which involve much higher cost, are available for critical high-speed applications, such as in aircraft turbines.

Over the last few decades, a continuous improvement in materials and the methods of manufacturing of rolling-element bearings have resulted in a significant improvement in fatigue life, specifically for aircraft applications. But the trend in modern machinery is to increase the speed of shafts more and more in order to reduce the size of machinery. Therefore, the limitations of rolling-element bearings at very high speeds are expected to be more significant in the future.

The fatigue life of a rolling bearing is a function of the magnitude of the oscillating stresses at the contact. If the stresses are low, the fatigue life can be practically unlimited. The stresses in dry contact can be calculated by the theory of elasticity. However, the surfaces are usually lubricated, and there is a very thin lubrication film at very high pressure separating the rolling surfaces. This thin film prevents direct contact and plays an important role in wear reduction. The analysis of this film is based on the elastohydrodynamic (EHD) theory, which considers the fluid dynamics of the film in a way similar to that of hydrodynamic bearings.

Unlike conventional hydrodynamic theory, EHD theory considers the elastic deformation in the contact area resulting from the high-pressure distribution in the fluid film. In addition, in EHD theory, the lubricant viscosity is considered as a function of the pressure, because the pressures are much higher than in regular hydrodynamic bearings. Recent research work has considered the thermal effects in the elastohydrodynamic film. Although there has been much progress in the understanding of rolling contact, in practice the life of the rolling-element bearing is still estimated by means of empirical equations. One must keep in mind the statistical nature of bearing life. Rolling bearings are selected to have a very low probability of premature failure. The bearings are designed to have a certain predetermined life span, such as 10 years. The desired life span should be determined before the design of a machine is initiated.

Experience over many years indicates that failure due to fatigue in rolling bearings is only one possible failure mode among many other, more frequent failure modes, due to various reasons. Common failure causes include bearing overheating, misalignment errors, improper mounting, corrosion, trapped hard particles, and not providing the bearing with proper lubrication (oil starvation or not using the optimum type of lubricant). Most failures can be prevented by proper maintenance, such as lubrication and proper mounting of the bearing. Fatigue failure is evident in the form of spalling or flaking at the contact surfaces of the races and rolling elements. It is interesting to note that although most rolling bearings are selected by considering their fatigue life, only 5% to 10% of the bearings actually fail by fatigue.

At high-speed operation, a frequent cause for rolling bearing failure is overheating. The heat generated by friction losses is dissipated in the bearing, resulting in uneven temperature distribution in the bearing. During operation, the temperature of the rolling bearing outer ring is lower than that of the inner ring. In turn, there is uneven thermal expansion of the inner and outer rings, resulting in thermal stresses in the form of a tight fit and higher contact stresses between the rolling elements and the races. The extra contact stresses further increase the level of friction and the temperature. This sequence of events can lead to an unstable closed-loop process, which can result in bearing failure by seizure. Common rolling-element bearings are manufactured with an internal clearance to reduce this risk of thermal seizure.

At high temperature the fatigue resistance of the metal is deteriorating. Also, at high speed the centrifugal forces increase the contact stresses between the rolling elements and the outer race. All these effects combine to reduce the fatigue life at very high speeds. Higher risk of bearing failure exists whenever the product of bearing load, F, and speed, n, is very high. The friction energy is dissipated in the bearing as heat. The power loss due to friction is proportional to the product Fn, similar to the product PV in a sleeve bearing. Therefore, the temperature rise of the bearing relative to the ambient temperature is also proportional to this product. In conclusion, load and speed are two important parameters that should be considered for selection and design purposes. In addition to friction-energy losses, bearing overheating can be caused by heat sources outside the bearing, such as in the case of engines or steam turbines.

In aircraft engines, only rolling bearings are used. Hydrodynamic or hydrostatic bearings are not used because of the high risk of a catastrophic (sudden) failure in case of interruption in the oil supply. In contrast, rolling bearings do not tend to catastrophic failure. Usually, in case of initiation of damage, there is a warning noise and sufficient time to replace the rolling bearing before it completely fails. For aircraft turbine engines there is a requirement for ever increasing power output and speed. At the very high speed required for gas turbines, the centrifugal forces of the rolling elements become a major problem. These centrifugal forces increase the hertz stresses at the outer-race contacts and shorten the bearing fatigue life.

The centrifugal force is proportional to the second power of the angular speed. Similarly, the bearing size increases the centrifugal force because of its larger rolling-element mass as well as its larger orbit radius. The DN value (rolling bearing bore, in millimeters, times shaft speed, in revolutions per minute, RPM) is used as a measure for limiting the undesired effect of the centrifugal forces in rolling bearings. Currently, the centrifugal force of the rolling elements is one important consideration for limiting aircraft turbine engines to 2 million DN.

Hybrid bearings, which have rolling elements made of silicon nitride and rings made of steel, have been developed and are already in use. One important advantage of the hybrid bearing is that the density of silicon nitride is much lower than that of steel, resulting in lower centrifugal force. In addition, hybrid bearings have better fatigue resistance at high temperature and are already in use for many industrial applications. Currently, intensive tests are being conducted in hybrid bearings for possible future application in aircraft turbines. However, due to the high risk in this application, hybrid bearings must pass much more rigorous tests before actually being used in aircraft engines.

Thermal stresses in rolling bearings can also be caused by thermal elongation of the shaft. In machinery such as motors and gearboxes, the shaft is supported by two bearings at the opposite ends of the shaft. The friction energy in the bearings increases the temperature of the shaft much more than that of the housing of the machine. It is important to design the mounting of the bearings with a free fit in the housing on one side of the shaft. This bearing arrangement is referred to as a *locating/floating* arrangement; it will be explained in Chapter 13. This arrangement allows for a free thermal expansion of the shaft in the axial direction and elimination of the high thermal stresses that could otherwise develop.

Rolling-element bearings generate certain levels of noise and vibration, in particular during high-speed operation. The noise and vibrations are due to irregular dimensions of the rolling elements and are also affected by the internal clearance in the bearing.

1.7 SELECTION CRITERIA

In comparison to rolling-element bearings, limited fatigue life is not a major problem for hydrodynamic bearings. As long as a full fluid film completely separates the sliding surfaces, the life of hydrodynamic bearings is significantly longer than that of rolling bearings, particularly at very high speeds.

However, hydrodynamic bearings have other disadvantages that make other bearing types the first choice for many applications. Hydrodynamic bearings can be susceptible to excessive friction and wear whenever the journal surface has occasional contact with the bearing surface and the superior fluid film lubrication is downgraded to boundary or mixed lubrication. This occurs at low operating speeds or during starting and stopping, since hydrodynamic bearings require a certain minimum speed to generate an adequate film thickness capable of completely separating the sliding surfaces.

According to the theory that is discussed in the following chapters, a very thin fluid film is generated inside a hydrodynamic bearing even at low journal speed. But in practice, due to surface roughness or vibrations and disturbances, a certain minimum speed is required to generate a fluid film of sufficient thickness that occasional contacts and wear between the sliding surfaces are prevented. Even at high journal speed, surface-to-surface contact may occur because of unexpected vibrations or severe disturbances in the system.

An additional disadvantage of hydrodynamic bearings is a risk of failure if the lubricant supply is interrupted, even for a short time. A combination of high speed and direct contact is critical, because heat is generated in the bearing at a very fast rate. In the case of unexpected oil starvation, the bearing can undergo a catastrophic (sudden) failure. Such catastrophic failures are often in the form of bearing seizure (welding of journal and bearing) or failure due to the melting of the bearing lining material, which is often a white metal of low melting temperature. Without a continuous supply of lubricant, the temperature rises because of the high friction from direct contact.

Oil starvation can result from several causes, such as failure of the oil pump or the motor. In addition, the lubricant can be lost due to a leak in the oil system. The risk of a catastrophic failure in hydrodynamic journal bearings is preventing their utilization in important applications where safety is involved, such as in aircraft engines, where rolling-element bearings with limited fatigue life are predominantly used.

For low-speed applications and moderate loads, plain sleeve bearings with boundary lubrication can provide reliable long-term service and can be an adequate alternative to rolling-element bearings. In most industrial applications, these bearings are made of bronze and lubricated by grease or are self-lubricated sintered bronze. For light-duty applications, plastic bearings are widely used. As long as the product of the average pressure and speed, PV, is within the specified design values, the two parameters do not generate excessive temperature.

If plain sleeve bearings are designed properly, they wear gradually and do not pose the problem of unexpected failure, such as fatigue failure in rolling-element bearings. When they wear out, it is possible to keep the machine running for a longer period before the bearing must be replaced. This is an important advantage in manufacturing machinery, because it prevents the financial losses involved in a sudden shutdown. Replacement of a plain sleeve bearing can, at least, be postponed to a more convenient time (in comparison to a rolling bearing). In manufacturing, unexpected shutdown can result in expensive loss of production. For sleeve bearings with grease lubrication or oil-impregnated porous metal bearings, the manufacturers provide tables of maximum speed and load as well as maximum PV value, which indicate the limits for each bearing material. If these limits are not exceeded, the temperature will not be excessive, resulting in a reliable operation of the bearing. A solved problem is included at the end of this chapter.

Sleeve bearings have several additional advantages. They can be designed so that it is easier to mount and replace them, in comparison to rolling bearings. Sleeve bearings can be of split design so that they can be replaced without removing the shaft. Also, sleeve bearings can be designed to carry much higher loads, in comparison to rolling bearings, where the load is limited due to the high "hertz" stresses. In addition, sleeve bearings are usually less sensitive than rolling bearings to dust, slurry, or corrosion caused by water infiltration.

However, rolling bearings have many other advantages. One major advantage is their relatively low-cost maintenance. Rolling bearings can operate with a minimal quantity of lubrication. Grease-packed and sealed rolling bearings are very convenient for use in many applications, since they do not require further lubrication. This significantly reduces the maintenance cost.

In many cases, machine designers select a rolling bearing only because it is easier to select from a manufacturer's catalogue. However, the advantages and disadvantages of each bearing type must be considered carefully for each application. Bearing selection has long-term effects on the life of the machine as well as on maintenance expenses and the economics of running the machine over its full life cycle. In manufacturing plants, loss of production is a dominant consideration. In certain industries, unplanned shutdown of a machine for even 1 hour may be more expensive than the entire maintenance cost or the cost of the best bearing. For these reasons, in manufacturing, bearing failure must be prevented without consideration of bearing cost. In aviation, bearing failure can result in the loss of lives; therefore, careful bearing selection and design are essential.

1.8 BEARINGS FOR PRECISION APPLICATIONS

High-precision bearings are required for precision applications, mostly in machine tools and measuring machines, where the shaft (referred to as the *spindle* in machine tools) is required to run with extremely low radial or axial run-out. Therefore, precision bearings are often referred to as *precision spindle bearings*. Rolling bearings are widely used in precision applications because in most cases they provide adequate precision at reasonable cost.

High-precision rolling-element bearings are manufactured and supplied in several classes of precision. The precision is classified by the maximum allowed tolerance of spindle run-out. In machine tools, spindle run-out is undesirable because it results in machining errors. Radial spindle run-out in machine tools causes machining errors in the form of deviation from roundness, while axial run-out causes manufacturing errors in the form of deviation from flat surfaces. Rolling-element bearing manufacturers use several tolerance classifications, but the most common are the following three tolerance classes of precision spindle bearings (FAG 1986):

Precision class	Maximum run-out (μm)
1. High-precision rolling-element bearings	2.0
2. Special-precision bearings	1.0
3. Ultraprecision bearings	0.5

Detailed discussion of rolling-element bearing precision is included in Chapter 13. Although rolling-element bearings are widely used in high-precision machine tools, there is an increasing requirement for higher levels of precision. Rolling-element bearings always involve a certain level of noise and vibrations, and there is a limit to their precision. The following is a survey of other bearing types, which can be alternatives for high precision applications

1.9 NONCONTACT BEARINGS FOR PRECISION APPLICATIONS

Three types of noncontact bearings are of special interest for precision machining, because they can run without any contact between the sliding surfaces in the bearing. These noncontact bearings are hydrostatic, hydrodynamic, and electromagnetic bearings. The bearings are noncontact in the sense that there is a thin clearance of lubricant or air between the journal (spindle in machine tools) and the sleeve. In addition to the obvious advantages of low friction and the absence of wear, other characteristics of noncontact bearings are important for ultra-high-precision applications. One important characteristic is the isolation of the spindle from vibrations. Noncontact bearings isolate the spindle from sources of vibrations in the machine or even outside the machine. Moreover, direct contact friction can induce noise and vibrations, such as in stick-slip friction; therefore, noncontact bearings offer the significant advantage of smooth operation for high-precision applications. The following discussion makes the case that hydrostatic bearings are the most suitable noncontact bearing for high-precision applications such as ultra-high-precision machine tools.

The difference between hydrodynamic and hydrostatic bearings is that, for the first, the pressure is generated inside the bearing clearance by the rotation action of the journal. In contrast, in a hydrostatic bearing, the pressure is supplied by an external pump. Hydrodynamic bearings have two major disadvantages that rule them out for use in machine tools: (a) low stiffness at low loads, and (b) at low speeds, not completely noncontact, since the fluid film thickness is less than the size of surface asperities.

In order to illustrate the relative advantage of hydrostatic bearings, it is interesting to compare the nominal orders of magnitude of machining errors in the form of deviation from roundness. The machining errors result from spindle run-out. Higher precision can be achieved by additional means to isolate the spindle from external vibrations, such as from the driving motor. In comparison to rolling-element bearings, experiments in hydrostatic-bearings indicated the following machining errors in the form of deviation from roundness by machine tools with a spindle supported by hydrostatic bearings (see Donaldson and Patterson, 1983 and Rowe, 1967):

Precison class	Machining error (μm)
1. Regular hydrostatic bearing	0.20
2. When vibrations are isolated from the drive	0.05

Experiments indicate that it is important to isolate the spindle from vibrations from the drive. Although a hydrostatic bearing is supported by a fluid film, the film has relatively high stiffness and a certain amount of vibrations can pass through, so additional means for isolation of vibrations is desirable. The preceding figures illustrate that hydrostatic bearings can increase machining precision, in comparison to precision rolling bearings, by one order of magnitude. The limits of hydrostatic bearing technology probably have not been reached yet.

1.10 BEARING SUBJECTED TO FREQUENT STARTS AND STOPS

In addition to wear, high start-up friction in hydrodynamic journal bearings increases the temperature of the journal much more than that of the sleeve, and there is a risk of bearing seizure. There is uneven thermal expansion of the journal and bearing, and under certain circumstances the clearance can be completely eliminated, resulting in bearing seizure. Bearing seizure poses a higher risk than wear, since the failure is catastrophic. This is the motivation for much research aimed at reducing start-up friction.

According to hydrodynamic theory, a very thin fluid film is generated even at low journal speed. But in practice, due to surface roughness, vibrations, and disturbances, a certain high minimum speed is required to generate an adequate film thickness so that occasional contacts and wear between the sliding surfaces are prevented. The most severe wear occurs during starting because the journal is accelerated from zero velocity, where there is relatively high static friction. The lubricant film thickness increases with speed and must be designed to separate the journal and sleeve completely at the rated speed of the machine. During starting, the speed increases, the fluid film builds up its thickness, and friction is reduced gradually.

In applications involving frequent starts, rolling element bearings are usually selected because they are less sensitive to wear during start-up and stopping. But this is not always the best solution, because rolling bearings have a relatively short fatigue-life when the operating speed is very high. In Chapter 18, it is shown that it is possible to solve these problems by using a "*composite bearing*," which is a unique design of hydrodynamic and rolling bearings in series (Harnoy and Rachoor, 1993).

Manufacturers continually attempt to increase the speed of machinery in order to reduce its size. The most difficult problem is a combination of high operating speed with frequent starting and stopping. At very high speed, the life of the rolling-element bearing is short, because fatigue failure is partly determined by the number of cycles, and high speed results in reduced life (measured in hours). In addition to this, at high speed the centrifugal forces of the rolling elements (balls or rollers) increase the fatigue stresses. Furthermore, the temperature of the bearing rises at high speeds; therefore, the fatigue resistance of the material deteriorates. The centrifugal forces and temperature exacerbate the problem and limit the operating speed at which the fatigue life is acceptable. Thus the two objectives, longer bearing life and high operating speed, are in conflict when rolling-element bearings are used. Hydrodynamic bearings operate well at high speeds but are not suitable for frequent-starting applications.

Replacing the hydrodynamic bearing with an externally pressurized hydrostatic bearing can eliminate the wear and friction during starting and stopping. But a hydrostatic bearing is uneconomical for many applications, since it requires an oil pump system, an electric motor, and flow restrictors in addition to the regular bearing system. An example of a unique design of a composite bearing—hydrodynamic and rolling bearings in series—is described in chapter 18. This example is a low-cost solution to the problem involved when high-speed machinery are subjected to frequent starting and stopping.

In conclusion, the designer should keep in mind that the optimum operation of the rolling bearing is at low and moderate speeds, while the best performance of the hydrodynamic bearing is at relatively high speeds. Nevertheless, in aviation, high-speed rolling-element bearings are used successfully. These are expensive high-quality rolling bearings made of special steels and manufactured by unique processes for minimizing impurity and internal microscopic cracks. Materials and manufacturing processes for rolling bearings are discussed in Chapter 13.

1.11 EXAMPLE PROBLEMS

Example Problem 1-1

PV Limits

Consider a shaft supported by two bearings, as shown in Fig. 1-6. The two bearings are made of self-lubricated sintered bronze. The bearing on the left side is under radial load, $F_r = 1200$ lbf, and axial load, $F_a = 0.5F_r$. (The bearing on the right supports only radial load). The journal diameter is $D = 1$ inch, and the

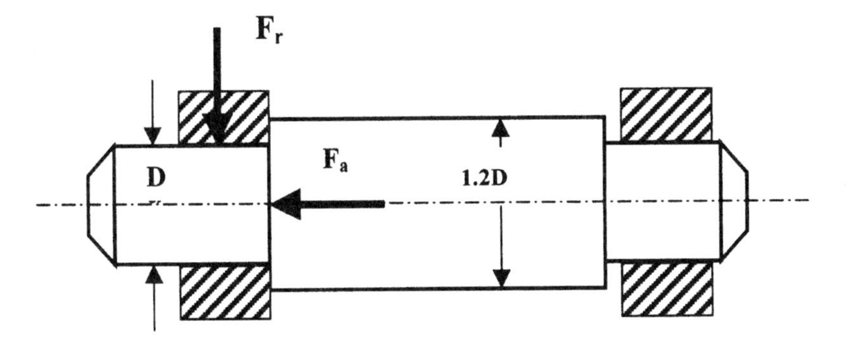

FIG. 1-6 Journal bearing under radial and thrust load.

bearing length $L = D$. The thrust load is supported against a shaft shoulder of diameter $D_1 = 1.2D$. The shaft speed is $N = 1000$ RPM.

Sintered bronze has the following limits:

Surface velocity limit, V, is 6 m/s, or 1180 ft/min.
Surface pressure limit, P, is 14 MPa, or 2000 psi.
PV limit is 110,000 psi-ft/min, or 3.85×10^6 Pa-m/s

a. For the left-side bearing, find the P, V, and PV values for the thrust bearing (in imperial units) and determine if this thrust bearing can operate with a sintered bronze bearing material.
b. For the left-side bearing, also find the P, V, and PV values for the radial bearing (in imperial units) and determine if the radial bearing can operate with sintered bronze bearing material.

Summary of data for left bearing:

$$F_r = 1200 \text{ lbf}$$
$$F_a = 0.5F_r = 600 \text{ lbf}$$
$$D = 1 \text{ in.} = 0.083 \text{ ft (journal diameter)}$$
$$D_1 = 1.2 \text{ in.} = 0.1 \text{ ft (shoulder diameter)}$$
$$N = 1000 \text{ RPM}$$

Solution

a. Thrust Bearing

Calculation of Average Pressure, P. The average pressure, P, in the axial direction is,

$$P = \frac{F_a}{A}$$

where

$$A = \frac{\pi}{4}(D_1^2 - D^2)$$

This is the shoulder area that supports the thrust load. Substituting yields

$$P = \frac{4F_a}{\pi(D_1^2 - D^2)} = \frac{4 \times 600}{\pi(1.2^2 - 1^2)} = 1736 \text{ psi}$$

This is within the allowed limit of $P_{\text{allowed}} = 2000$ psi.

b. Calculation of Average Surface Velocity of Thrust Bearing, V_{th}. The average velocity of a thrust bearing is at the average diameter, $(D_1 + D)/2$:

$$V_{\text{th}} = \omega R_{\text{av}} = \omega \frac{D_{\text{av}}}{2}$$

where $\omega = 2\pi N$ rad/min. Substituting yields

$$V_{\text{th}} = 2\pi N R_{\text{av}} = \frac{2\pi N(D_1 + D)}{4} = 0.5\pi N(D_1 + D)$$

Substitution in the foregoing equation yields

$$V_{\text{th}} = 0.5\pi \times 1000 \text{ rev/min } (0.1 + 0.083) \text{ ft} = 287.5 \text{ ft/min}$$

This is well within the allowed limit of $V_{\text{allowed}} = 1180$ ft/min.

c. Calculation of Actual Average PV Value for the Thrust Bearing:

$$PV = 1736 \text{ psi} \times 287.5 \text{ ft/min} = 500 \times 10^3 \text{ psi-ft/min}$$

Remark. The imperial units for PV are of pressure, in psi, multiplied by velocity, in ft/min.

Conclusion. Although the limits of the velocity and pressure are met, the PV value exceeds the allowed limit for self-lubricated sintered bronze bearing material, where the PV limit is 110,000 psi-ft/min.

b. Radial Bearing

Calculation of Average Pressure

$$P = \frac{F_r}{A}$$

where $A = LD$ is the projected area of the bearing. Substitution yields

$$P = \frac{F_r}{LD} = \frac{1200 \text{ lbf}}{1 \text{ in.} \times 1 \text{ in.}} = 1200 \text{ psi}$$

Calculation of Journal Surface Velocity. The velocity is calculated as previously; however, this time the velocity required is the velocity at the surface of the 1-inch shaft, $D/2$.

$$V_r = \omega \frac{D}{2} = 2\pi n \frac{D}{2} = \pi \times 1000 \text{ rev/min} \times 0.083 \text{ ft} = 261 \text{ ft/min}$$

Calculation of Average PV Value:

$$PV = 1200 \text{ psi} \times 261 \text{ ft/min} = 313 \times 10^3 \text{ psi-ft/min}$$

In a similar way to the thrust bearing, the limits of the velocity and pressure are met; however, the PV value exceeds the allowed limit for sintered bronze bearing material, where the PV limit is 110,000 psi-ft/min.

Example Problem 1-2

Calculation of Bearing Forces

In a gearbox, a spur gear is mounted on a shaft at equal distances from two supporting bearings. The shaft and gear turn together at a speed of 600 RPM. The gearbox is designed to transmit a maximum power of 5 kW. The gear pressure angle is $\phi = 20^\circ$. The diameter of the gear pitch circle is $d_p = 5$ in.

Remark. The gear pressure angle ϕ (PA) is the angle between the line of force action (normal to the contact area) and the direction of the velocity at the pitch point (see Fig. 1-7). Two standard pressure angles ϕ for common involute gears are $\phi = 20^\circ$ and $\phi = 14.5^\circ$. Detailed explanation of the geometry of gears is included in many machine design textbooks, such as *Machine Design*, by Deutschman et al. (1975), or *Machine Design*, by Norton (1996).

a. Find the reaction force on each of the two bearings supporting the shaft.
b. The ratio of the two bearings' length and bore diameter is $L/D = 0.5$. The bearings are made of sintered bronze material ($PV = 110{,}000$ psi-ft/min). Find the diameter and length of each bearing that is required in order not to exceed the PV limit.

Solution

a. Reaction Forces

Given:

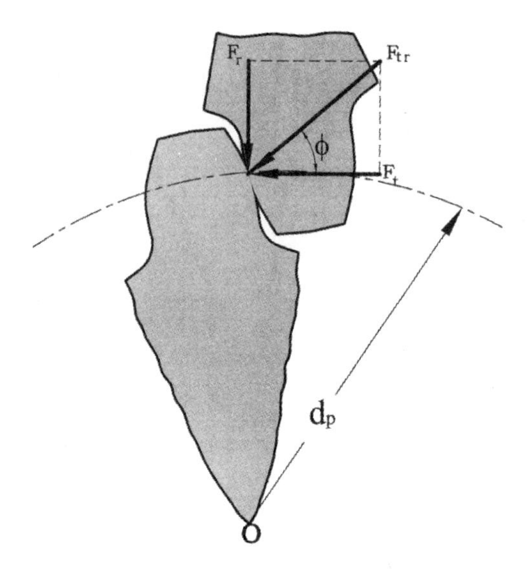

FIG. 1-7 Gear pressure angle.

Rotational speed $N = 600\,\text{RPM}$
Power $\dot{E} = 5000$ W
Diameter of pitch circle $d_p = 5$ in.
Pressure angle $\phi = 20^\circ$
$PV_{\text{allowed}} = 110{,}000$ psi-ft/min
$L/D = 0.5$ (the bore diameter of the bearing, D, is very close to that of the journal, d)

Conversion Factors:

$$1\text{psi} = 6895\ \text{N/m}^2$$
$$1\ \text{ft/min} = 5.08 \times 10^{-3}\ \text{m/s}$$
$$1\ \text{psi-ft/min} = 35\ \text{N/m}^2\text{-m/s}$$

The angular velocity, ω, of the journal is:

$$\omega = \frac{2\pi N}{60} = \frac{2\pi 600}{60} = 52.83\ \text{rad/s}$$

Converting the diameter of the pitch circle to SI units,

$$d_p = 5\ \text{in.} \times 0.0254\ \text{m/in.} = 0.127\ \text{m}$$

The tangential force, F_t, acting on the gear can now be derived from the power, $\dot{E}$:

$$\dot{E} = T\omega$$

where the torque is

$$T = \frac{F_t d_p}{2}$$

Substituting into the power equation:

$$\dot{E} = \frac{F_t d_p \omega}{2}$$

and solving for F_t and substituting yields

$$F_t = \frac{2\dot{E}}{d_p \omega} = \frac{2 \times 5000 \text{ Nm/s}}{0.127 \text{ m} \times 62.83 \text{ rad/s}} = 1253.2 \text{ N}$$

In spur gears, the resultant force acting on the gear is $F = F_{tr}$ (Fig. 1-7)

$$\cos \phi = \frac{F_t}{F}$$

so

$$F = \frac{F_t}{\cos \phi} = \frac{1253.2}{\cos 20^\circ} = 1333.6 \text{ N}$$

The resultant force, F, acting on the gear is equal to the radial component of the force acting on the bearing. Since the gear is equally spaced between the two bearings supporting the shaft, each bearing will support half the load, F. Therefore, the radial reaction, W, of each bearing is

$$W = \frac{F}{2} = \frac{1333.6}{2} = 666.8 \text{ N}$$

b. Bearing Dimensions

The average bearing pressure, P, is

$$P = \frac{F}{A}$$

Here, $A = LD$, where D is the journal diameter and A is the projected area of the contact surface of journal and bearing surface,

$$P = \frac{F}{LD}$$

The velocity of shaft surface, V, is

$$V = \frac{D}{2}\omega$$

Therefore,

$$PV = \frac{W}{LD}\frac{D}{2}\omega$$

Since $L/D = 0.5$, $L = 0.5D$, substituting and simplifying yields

$$PV = \frac{W}{0.5D^2}\frac{\omega D}{2} = \frac{W}{D}\omega$$

The PV limit for self-lubricated sintered bronze is given in English units, converted to SI units, the limit is

$$110{,}000 \text{ psi-ft/min} \times 35 \text{ N/m}^2\text{-m/s} = 3{,}850{,}000 \text{ Pa-m/s}$$

Solving for the journal diameter, D, and substituting yields the diameter of the bearing:

$$D = \frac{W\omega}{PV} = \frac{666.8 \text{ N} \times 62.83 \text{ rad/s}}{3.85 \times 10^6 \text{ Pa} \cdot \text{m/s}} = 0.011 \text{ m}, \qquad \text{or} \qquad \text{D} = 11 \text{ mm}$$

The length of the bearing, L, is

$$L = 0.5D = 0.5 \times 11 \text{ mm} = 5.5 \text{ mm}$$

The resulting diameter, based on a PV calculation, is very small. In actual design, the journal is usually of larger diameter, based on strength-of-material considerations, because the shaft must have sufficient diameter for transmitting the torque from the drive.

Example Problem 1-3

Calculation of Reaction Forces

In a gearbox, one helical gear is mounted on a shaft at equal distances from two supporting bearings. The *helix angle* of the gear is $\psi = 30°$, and the pressure angle (PA) is $\phi = 20°$. The shaft speed is 3600 RPM. The gearbox is designed to transmit maximum power of 20 kW. The diameter of the pitch circle of the gear is equal to 5 in. The right-hand-side bearing is supporting the total thrust load. Find the axial and radial loads on the right-hand-side bearing and the radial load on the left-side bearing.

Solution

The angular velocity of the shaft, ω, is:

$$\omega = \frac{2\pi N}{60} = \frac{2\pi 3600}{60} = 377 \text{ rad/s}$$

Torque produced by the gear is $T = F_t d_p/2$. Substituting this into the power equation, $\dot{E} = T\omega$, yields:

$$\dot{E} = \frac{F_t d_p}{2}\omega$$

Solving for the tangential force, F_t, results in

$$F_t = \frac{2\dot{E}}{d_p\omega} = \frac{2 \times 20{,}000 \text{ N-m/s}}{0.127 \text{ m} \times 377 \text{ rad/s}} = 836 \text{ N}$$

Once the tangential component of the force is solved, the resultant force, F, and the thrust load (axial force), F_a, can be calculated as follows:

$$F_a = F_t \tan\psi$$
$$F_a = 836 \text{ N} \times \tan 30^\circ = 482 \text{ N}$$

and the radial force component is:

$$\begin{aligned} F_r &= F_t \tan\phi \\ &= 836 \text{ N} \times \tan 20^\circ \\ &= 304 \text{ N} \end{aligned}$$

The force components, F_t and F_r, are both in the direction normal to the shaft centerline. The resultant of these two gear force components, F_{tr}, is cause for the radial force component in the bearings. The resultant, F_{tr}, is calculated by the equation (Fig. 1-7)

$$F_{tr} = \sqrt{F_t^2 + F_r^2} = \sqrt{836^2 + 304^2} = 890 \text{ N}$$

The resultant force, F_{tr}, on the gear is supported by the two bearings. It is a radial bearing load because it is acting in the direction normal to the shaft centerline. Since the helical gear is mounted on the shaft at equal distances from both bearings, each bearing will support half of the radial load,

$$W_r = \frac{F_{tr}}{2} = \frac{890 \text{ N}}{2} = 445 \text{ N}$$

However, the thrust load will act only on the right-hand bearing:

$$F_a = 482 \text{ N}$$

Example Problem 1-4

Calculation of Reaction Forces

In a gearbox, two helical gears are mounted on a shaft as shown in Fig. 1-1. The helix angle of the two gears is $\psi = 30^\circ$, and the pressure angle (PA) is $\phi = 20^\circ$. The shaft speed is 3600 RPM. The gearbox is designed to transmit a maximum power of 10 kW. The pitch circle diameter of the small gear is equal to 5 in. and that of the large gear is of 15 in.

a. Find the axial reaction force on each of the two gears and the resultant axial force on each of the two bearings supporting the shaft.
b. Find the three load components on each gear, F_t, F_r, and F_a.

Solution

Given:

Helix angle	$\psi = 30^\circ$
Pressure angle	$\phi = 20^\circ$
Rotational speed	$N = 3600$ RPM
Power	10 kW
Diameter of pitch circle (small)	$d_{P1} = 5$ in.
Diameter of pitch circle (large)	$d_{P2} = 15$ in.

Small Gear

a. *Axial Reaction Forces.*

The first step is to solve for the tangential force acting on the small gear, F_t. It can be derived from the power, E, and shaft speed:

$$\dot{E} = T\omega$$

where the torque is $T = F_t d_p/2$. The angular speed ω in rad/s is

$$\omega = \frac{2\pi N}{60} = \frac{2\pi \times 3600}{60} = 377 \text{ rad/s}$$

Substituting into the power equation yields

$$\dot{E} = \frac{F_t d_p \omega}{2}$$

The solution for F_t acting on the small gear is given by

$$F_t = \frac{2\dot{E}}{d_p \omega}$$

The pitch diameter is 5 in., or $d_p = 0.127$ m. After substitution, the tangential force is

$$F_t = \frac{2 \times 10{,}000 \text{ W}}{(0.127 \text{ m}) \times (377 \text{ rad/s})} = 418 \text{ N}$$

The radial force on the gear (F_r in Fig 1-1) is:

$$F_r = F_t \tan\phi$$
$$F_r = 418\text{ N} \times \tan 20^\circ$$
$$F_r = 152\text{ N}$$

b. Calculation of the Thrust Load, F_a:

The axial force on the gear is calculated by the equation,

$$F_a = F_t \tan\psi$$
$$= 418\text{ N} \times \tan 30^\circ$$
$$= 241\text{ N}$$

$$F_t = 418\text{ N}$$
$$F_r = 152\text{ N}$$
$$F_a = 241\text{ N}$$

Large gear

The same procedure is used for the large gear, and the results are:

$$F_t = 140\text{ N}$$
$$F_r = 51\text{ N}$$
$$F_a = 81\text{ N}$$

Thrust Force on a Bearing

One bearing supports the total thrust force on the shaft. The resultant thrust load on one bearing is the difference of the two axial loads on the two gears, because the thrust reaction forces in the two gears are in opposite directions (see Fig. 1-1):

$$F_a\text{ (bearing)} = 241 - 81 = 180\text{ N}$$

Problems

1-1 Figure 1-4 shows a drawing of a hydrostatic journal bearing system that can support only a radial load. Extend this design and sketch a hydrostatic bearing system that can support combined radial and thrust loads.

1-2 In a gearbox, a spur gear is mounted on a shaft at equal distances from two supporting bearings. The shaft and mounted gear turn together at a speed of 3600 RPM. The gearbox is designed to transmit a maximum power of 3 kW. The gear contact angle is $\phi = 20^\circ$. The pitch diameter of the gear is $d_p = 30$ in. Find the radial force on each of the two bearings supporting the shaft.

The ratio of the two bearings' length and diameter is $L/D = 0.5$. The bearings are made of acetal resin material with the following limits:

Surface velocity limit, V, is 5 m/s.
Average surface-pressure limit, P, is 7 MPa.
PV limit is 3000 psi-ft/min.

Find the diameter of the shaft in order not to exceed the stated limits.

1-3 A bearing is made of Nylon sleeve. Nylon has the following limits as a bearing material:

Surface velocity limit, V, is 5 m/s.
Average surface-pressure limit, P, is 6.9 MPa.
PV limit is 3000 psi-ft/min.

The shaft is supported by two bearings, as shown in Fig. 1-6. The bearing on the left side is under a radial load $F_r = 400\ N$ and an axial load $F_a = 200\ N$. (The bearing on the left supports the axial force.) The journal diameter is d, and the bearing length $L = d$. The thrust load is supported against a shaft shoulder of diameter $D = 1.2d$. The shaft speed is $N = 800$ RPM. For the left-side bearing, find the minimum journal diameter d that would result in P, V, and PV below the allowed limits, in the radial and thrust bearings.

1-4 In a gearbox, one helical gear is mounted on a shaft at equal distances from two supporting bearings. The helix angle of the gear is $\psi = 30°$, and the pressure angle (PA) is $\phi = 20°$. The shaft speed is 1800 RPM. The gearbox is designed to transmit a maximum power of 12 kW. The diameter of the pitch circle of the gear is 5 in. The right-hand-side bearing is supporting the total thrust load. Find the axial and radial load on the right-hand-side bearing and the radial load on the left-side bearing.

1-5 In a gearbox, two helical gears are mounted on a shaft as shown in Fig. 1-1. The helix angle of the two gears is $\phi = 30°$, and the pressure angle (PA) is $\phi = 20°$. The shaft speed is 3800 RPM. The gearbox is designed to transmit a maximum power of 15 kW. The pitch circle diameters of the two gears are 5 in. and 15 in. respectively.

a. Find the axial reaction force on each of the two gears and the resultant axial force on each of the two bearings supporting the shaft.
b. Find the three load components on each gear, F_t , F_r, and F_a.

2

Lubricant Viscosity

2.1 INTRODUCTION

For hydrodynamic lubrication, the viscosity, μ, is the most important characteristic of a fluid lubricant because it has a major role in the formation of a fluid film. However, for boundary lubrication the lubricity characteristic is important. The viscosity is a measure of the fluid's resistance to flow. For example, a low-viscosity fluid flows faster through a capillary tube than a fluid of higher viscosity. High-viscosity fluids are thicker, in the sense that they have higher internal friction to the movement of fluid particles relative to one another.

Viscosity is sensitive to small changes in temperature. The viscosity of mineral and synthetic oils significantly decreases (the oils become thinner) when their temperature is raised. The higher viscosity is restored after the oils cool down to their original temperature. The viscosity of synthetic oils is relatively less sensitive to temperature variations (in comparison to mineral oils). But the viscosity of synthetic oils also decreases with increasing temperature.

During bearing operation, the temperature of the lubricant increases due to the friction, in turn, the oil viscosity decreases. For hydrodynamic bearings, the most important property of the lubricant is its viscosity at the operating bearing temperature. One of the problems in bearing design is the difficulty of precisely predicting the final temperature distribution and lubricant viscosity in the fluid film of the bearing. For a highly loaded bearing combined with slow speed, oils of relatively high viscosity are applied; however, for high-speed bearings, oils of relatively low viscosity are usually applied.

The bearing temperature always rises during operation, due to friction-energy loss that is dissipated in the bearing as heat. However, in certain applications, such as automobile engines, the temperature rise is much greater, due to the heat of combustion. In these cases, the lubricant is subject to very large variations of viscosity due to changes in temperature. A large volume of research and development work has been conducted by engine and lubricant manufacturers to overcome this problem in engines and other machinery, such as steam turbines, that involve a high-temperature rise during operation.

Minimum viscosity is required to secure proper hydrodynamic lubrication when the engine is at an elevated temperature. For this purpose, a lubricant of high viscosity at ambient temperature must be selected. This must result in high viscosity during starting of a car engine, particularly on cold winter mornings, causing heavy demand on the engine starter and battery. For this reason, lubricants with less sensitive viscosity to temperature variations would have a distinct advantage. This has been the motivation for developing the multigrade oils, which are commonly used in engines today. An example of a multigrade oil that is widely used in motor vehicle engines is SAE 5W-30. The viscosity of SAE 5W-30 in a cold engine is about that of the low-viscosity oil SAE 5W, while its viscosity in the hot engine during operation is about that of the higher-viscosity oil SAE 30. The viscosity of synthetic oils is also less sensitive to temperature variations, in comparison to regular mineral oils, and the development of synthetic oils during recent years has been to a great extent motivated by this advantage.

It is important to mention that the viscosity of gases, such as air, reveals an opposite trend, increasing with a rise in temperature. This fact is important in the design of air bearings. However, one must bear in mind that the viscosity of air is at least three orders of magnitude lower than that of mineral oils.

2.2 SIMPLE SHEAR FLOW

In Fig. 2-1, simple shear flow between two parallel plates is shown. One plate is stationary and the other has velocity U in the direction parallel to the plate. The fluid is continuously sheared between the two parallel plates. There is a sliding motion of each layer of molecules relative to the adjacent layers in the x direction. In simple shear, the viscosity is the resistance to the motion of one layer of molecules relative to another layer. The shear stress, τ, between the layers increases with the shear rate, U/h (a measure of the relative sliding rate of adjacent layers). In addition, the shear stress, τ, increases with the internal friction between the layers; that is, the shear stress is proportional to the viscosity, μ, of the fluid.

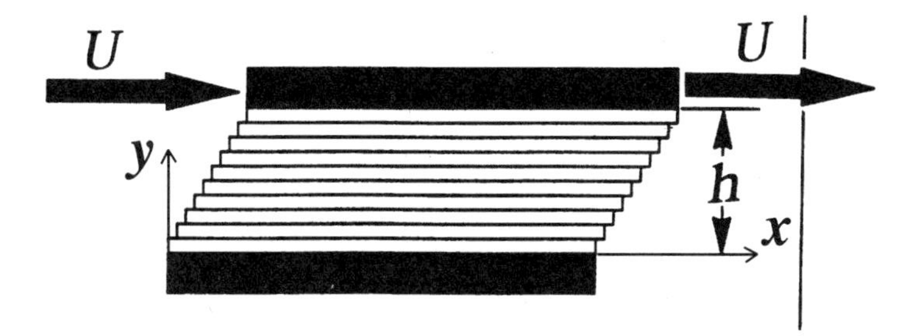

FIG. 2-1 Simple shear flow.

Most lubricants, including mineral and synthetic oils, demonstrate a linear relationship between the shear stress and the shear rate. A similar linear relationship holds in the air that is used in air bearings. Fluids that demonstrate such linear relationships are referred to as *Newtonian fluids*. In simple shear flow, $u = u(y)$, the shear stress, τ, is proportional to the shear rate. The shear rate between two parallel plates without a pressure gradient, as shown in Fig. 2-1, is U/h. But in the general case of simple shear flow, $u = u(y)$, the local shear rate is determined by the velocity gradient du/dy, where u is the fluid velocity component in the x direction and the gradient du/dy is in respect to y in the normal direction to the sliding layers.

One difference between solids and viscous fluids is in the relation between the stress and strain components. In the case of simple shear, the shear stress, τ, in a solid material is proportional to the deformation, shear strain. Under stress, the material ceases to deform when a certain elastic deformation is reached. In contrast, in viscous fluids the deformation rate continues (the fluid flows) as long as stresses are applied to the fluid. An example is a simple shear flow of viscous fluids where the shear rate, du/dy, continues as long as the shear stress, τ, is applied. For liquids, a shear stress is required to overcome the cohesive forces between the molecules in order to maintain the flow, which involves continuous relative motion of the molecules. The proportionality coefficient for a solid is the shear modulus, while the proportionality coefficient for the viscous fluid is the viscosity, μ. In the simple shear flow of a Newtonian fluid, a linear relationship between the shear stress and the shear rate is given by

$$\tau = \mu \frac{du}{dy} \tag{2-1}$$

In comparison, a similar linear equation for elastic simple-shear deformation of a solid is:

$$\tau = G \frac{de_x}{dy} \tag{2-2}$$

where e_x is the displacement (one-time displacement in the x direction) and G is the elastic shear modulus of the solid.

Whenever there is viscous flow, shear stresses must be present to overcome the cohesive forces between the molecules. In fact, the cohesive forces and shear stress decrease with temperature, which indicates a decrease in viscosity with an increase in temperature. For the analysis of hydrodynamic bearings, it is approximately assumed that the viscosity, μ, is a function of the temperature only. However, the viscosity is a function of the pressure as well, although this becomes significant only at very high pressure. Under extreme conditions of very high pressures, e.g., at point or line contacts in rolling-element bearings or gears, the viscosity is considered a function of the fluid pressure.

2.3 BOUNDARY CONDITIONS OF FLOW

The velocity gradient at the solid boundary is important for determining the interaction forces between the fluid and the solid boundary or between the fluid and a submerged body. The velocity gradient, du/dy, at the boundary is proportional to the shear stress at the wall. An important characteristic of fluids is that the fluid adheres to the solid boundary (there are some exceptions; oil does not adhere to a Teflon wall). For most surfaces, at the boundary wall, the fluid has identical velocity to that of the boundary, referred to as the *non-slip condition*. The intermolecular attraction forces between the fluid and solid are relatively high, resulting in slip only of one fluid layer over the other, but not between the first fluid layer and the solid wall. Often, we use the term *friction between the fluid and the solid*, but in fact, the viscous friction is only between the fluid and itself, that is, one fluid layer slides relative to another layer, resulting in viscous friction losses.

Near the solid boundary, the first fluid layer adheres to the solid surface while each of the other fluid layers is sliding over the next one, resulting in a velocity gradient. Equation (2-1) indicates that the slope of the velocity profile (velocity gradient) is proportional to the shear stress. Therefore, the velocity gradient at the wall is proportional to the shear stress on the solid boundary. Integration of the viscous shear forces on the boundary results in the total force caused by shear stresses. This portion of the drag force is referred to as the *skin friction force* or *viscous drag force* between the fluid and a submerged body. The other portion of the drag force is the *form drag*, which is due to the pressure distribution on the surface of a submerged body.

Oils are practically incompressible. This property simplifies the equations, because the fluid density, ρ, can be assumed to be constant, although this assumption cannot be applied to air bearings. Many of the equations of fluid mechanics, such as the Reynolds number, include the ratio of viscosity to density,

ρ, of the fluid. Since this ratio is frequently used, this combination has been given the name *kinematic viscosity*, ν:

$$\nu = \frac{\mu}{\rho} \tag{2-3}$$

The viscosity μ is often referred to as absolute viscosity as a clear distinction from kinematic viscosity.

2.4 VISCOSITY UNITS

The SI unit of pressure, p, as well as shear of stress, τ, is the Pascal (Pa) = Newtons per square meter [N/m^2]. This is a small unit; a larger unit is the kilopascal (kPa) = 10^3 Pa.

In the imperial (English) unit system, the common unit of pressure, p, as well as of shear stress, τ, is lbf per square inch (psi).

The conversion factors for pressure and stress are:

$$1 \text{ Pa} = 1.4504 \times 10^{-4} \text{ psi}$$
$$1 \text{ kPa} = 1.4504 \times 10^{-1} \text{ psi}$$

2.4.1 SI Units

From Eq. (2-1) it can be seen that the SI unit of μ is [N-s/m^2] or [Pa-s]:

$$\mu = \frac{\tau}{du/dy} = \frac{\text{N/m}^2}{(\text{m/s})/\text{m}} = \text{N-s/m}^2$$

The SI units of kinematic viscosity, ν, are [m^2/s]:

$$\nu = \frac{\mu}{\rho} = \frac{\text{N-s/m}^2}{\text{N-s}^2/\text{m}^4} = \text{m}^2/\text{s}$$

2.4.2 cgs Units

An additional cgs unit for absolute viscosity, μ, is the poise [dyne-s/cm^2]. The unit of dyne-seconds per square centimeter is the poise, while the centipoise (one hundredth of a poise) has been widely used in bearing calculations, but now has been gradually replaced by SI units.

The cgs unit for kinematic viscosity, ν, is the stokes (St) [cm^2/s]; a smaller unit is the centistokes (cSt), cSt = 10^{-2} stokes. The unit cSt is equivalent to [mm^2/s].

2.4.3 Imperial Units

In the imperial (English) unit system, the unit of absolute viscosity, μ, is the reyn [lbf-s/in.^{2}] (named after Osborne Reynolds) in pounds (force)-seconds per square inch. The imperial unit for kinematic viscosity, ν, is [in.^{2}/s], square inches per second.

Conversion list of absolute viscosity units, μ:

1 centipoise $= 1.45 \times 10^{-7}$ reyn
1 centipoise $= 0.001$ N-s/m^2
1 centipoise $= 0.01$ poise
1 reyn $= 6.895 \times 10^3$ N-s/m^2
1 reyn $= 6.895 \times 10^6$ centipoise
1 N-s/m^2 $= 10^3$ centipoise
1 N-s/m^2 $= 1.45 \times 10^{-4}$ reyn

2.4.4 Saybolt Universal Seconds (SUS)

In addition to the preceding units, a number of empirical viscosity units have been developed. Empirical viscosity units are a measure of the flow time of oil in a laboratory test instrument of standard geometry. The most common empirical viscosity unit in the United States is the Saybolt universal second (SUS). This Saybolt viscosity is defined as the time, in seconds, required to empty out a volume of 60 cm^3 of fluid through a capillary opening in a Saybolt viscometer. There are equations to convert this Saybolt viscosity to other kinematic viscosities, and the fluid density is required for further conversion to absolute viscosity, μ.

The SUS is related to a standard viscometer (ASTM specification D 88). This unit system is widely used in the United States by commercial oil companies. The following equation converts t (in SUS) into kinematic viscosity, ν, in centistoke (cSt) units:

$$\nu(\text{cSt}) = 0.22t - \frac{180}{t} \tag{2-4}$$

Lubrication engineers often use the conversion chart in Fig. 2-2 to convert from kinematic viscosity to absolute viscosity, and vice versa. Also, the chart is convenient for conversion between the unit systems.

2.5 VISCOSITY–TEMPERATURE CURVES

A common means to determine the viscosity at various temperatures is the ASTM viscosity–temperature chart (ASTM D341). An example of such a chart is given

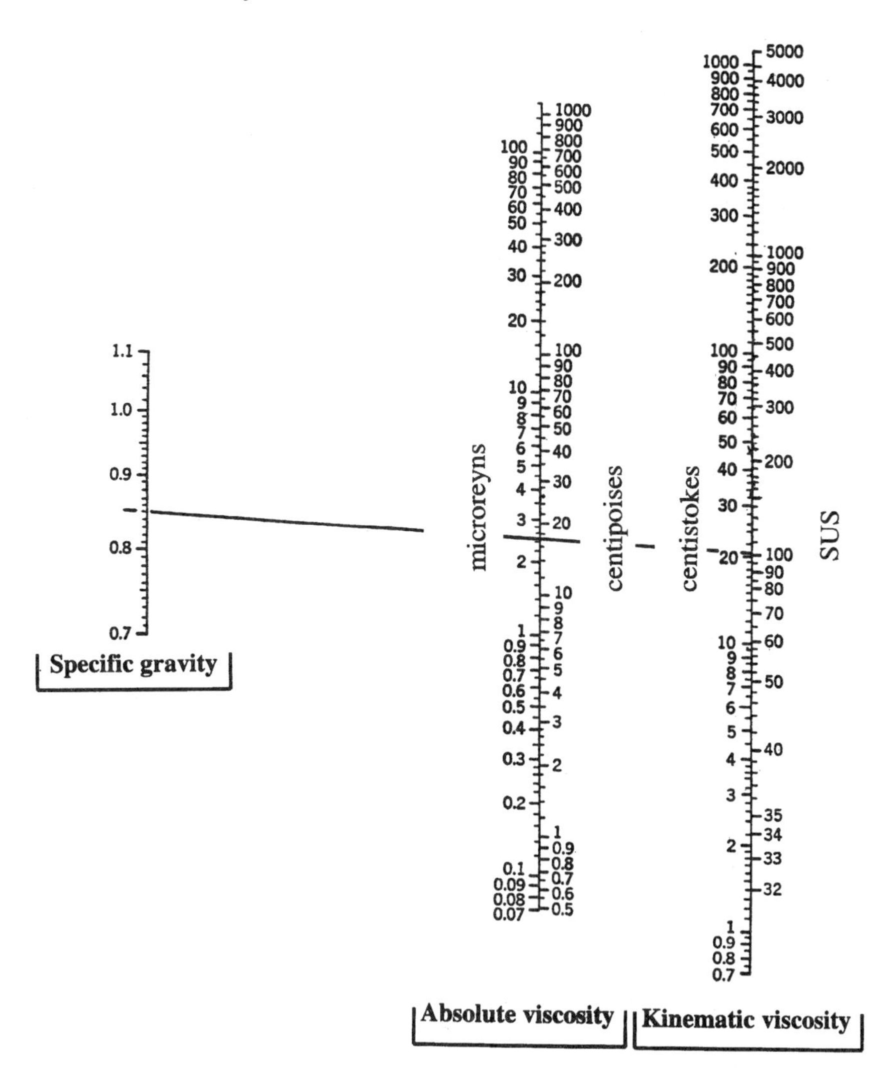

FIG. 2-2 Viscosity conversion chart from Saybolt universal seconds.

in Fig. 2-3. The viscosity of various types of mineral oils is plotted as a function of temperature. These curves are used in the design of hydrodynamic journal bearings.

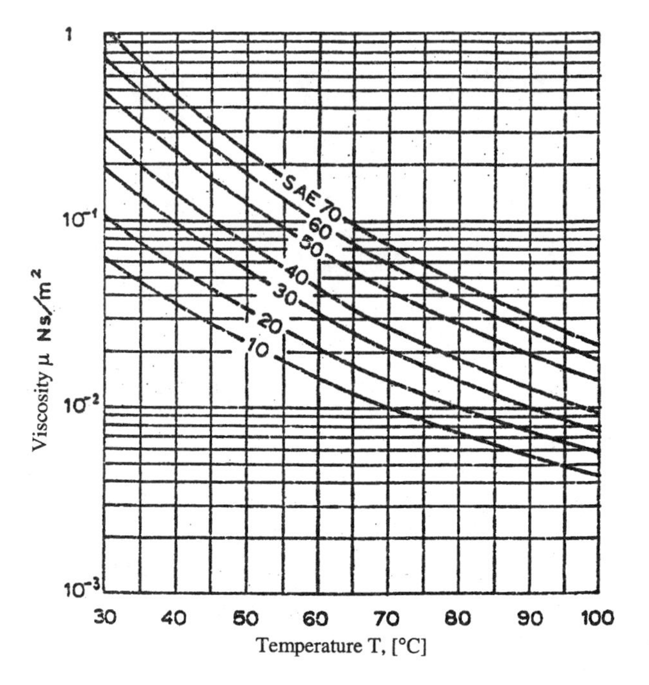

FIG. 2-3 Viscosity–temperature chart.

2.6 VISCOSITY INDEX

Lubricants having a relatively low rate of change of viscosity versus temperature are desirable, particularly in automotive engines. The viscosity index (VI) is a common empirical measure of the level of decreasing viscosity when the temperature of oils increases. The VI was introduced as a basis for comparing Pennsylvania and Gulf Coast crude oils. The Pennsylvania oil exhibited a relatively low change of viscosity with temperature and has been assigned a VI of 100, while a certain Gulf Coast oil exhibited a relatively high change in viscosity with temperature and has been assigned a VI of 0. The viscosity-versus-temperature curve of all other oils has been compared with the Pennsylvania and Gulf Coast oils. The viscosity index of all other oils can be determined from the slope of their viscosity–temperature curve, in comparison to $VI = 0$ and $VI = 100$ oils, as illustrated in Fig. 2-4. Demonstration of the method for determining the viscosity index from various viscosity–temperature curves is presented schematically in Fig. 2-4. The viscosity index of any type of oil is determined by the following equation:

$$VI = \frac{L - U}{L - H} \times 100 \tag{2-5}$$

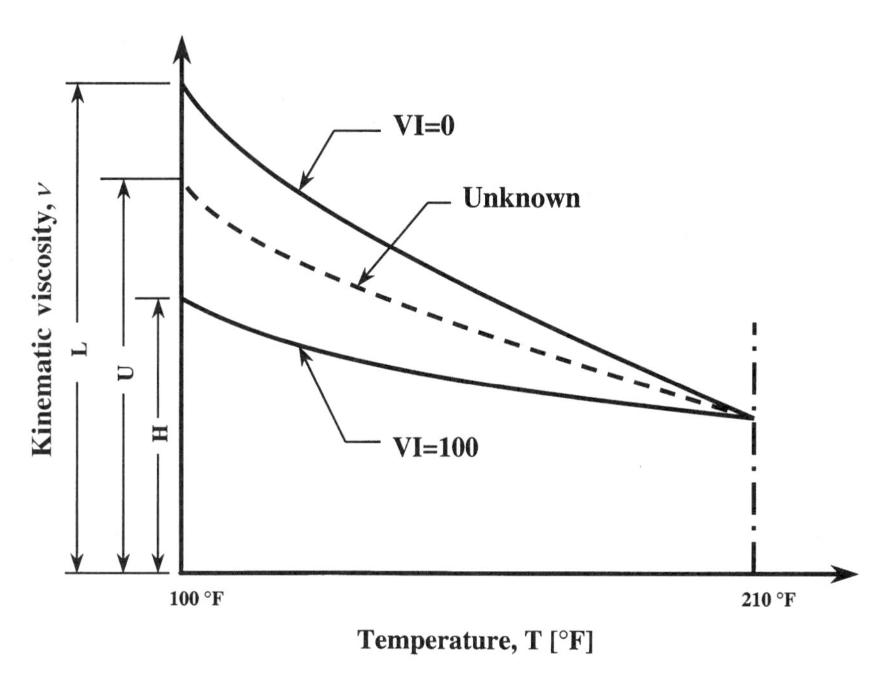

FIG. 2-4 Illustration of the viscosity index.

Here, L is the kinematic viscosity at 100°F of VI = 0 oil, H is the kinematic viscosity at 100°F of the VI = 100, and U is the kinematic viscosity at 100°F of the newly tested oil. High-viscosity-index oils of 100 or above are usually desirable in hydrodynamic bearings, because the viscosity is less sensitive to temperature variations and does not change so much during bearing operation.

2.7 VISCOSITY AS A FUNCTION OF PRESSURE

The viscosity of mineral oils as well as synthetic oils increases with pressure. The effect of the pressure on the viscosity of mineral oils is significant only at relatively high pressure, such as in elastohydrodynamic lubrication of point or line contacts in gears or rolling-element bearings. The effect of pressure is considered for the analysis of lubrication only if the maximum pressure exceeds 7000 kPa (about 1000 psi). In the analysis of hydrodynamic journal bearings, the viscosity–pressure effects are usually neglected. However, in heavily loaded hydrodynamic bearings, the eccentricity ratio can be relatively high. In such cases, the maximum pressure (near the region of minimum film thickness) can be above 7000 kPa. But in such cases, the temperature is relatively high at this region, and this effect tends to compensate for any increase in viscosity by pressure. However, in elastohydrodynamic lubrication of ball bearings, gears, and rollers, the maximum pressure is much higher and the increasing viscosity must be considered in the analysis. Under very high pressure, above 140,000 kPa (20,000 psi), certain oils become plastic solids.

Barus (1893) introduced the following approximate exponential relation of viscosity, μ, versus pressure, p:

$$\mu = \mu_0 e^{\alpha p} \tag{2-6}$$

Here, μ_0 is the absolute viscosity under ambient atmospheric pressure and α is the pressure-viscosity coefficient, which is strongly dependent on the operating temperature.

Values of the pressure-viscosity coefficient, α [m^2/N], for various lubricants have been measured. These values are listed in Table 2-1.

A more accurate equation over a wider range of pressures has been proposed by Rhoelands (1966) and recently has been used for elastohydrodynamic analysis. However, since the Barus equation has a simple exponential form, it has been the basis of most analytical investigations.

TABLE 2-1 Pressure-Viscosity Coefficient, α [m^2/N], for Various Lubricants

	Temperature, t_m		
Fluid	38°C	99°C	149°C
Ester	1.28×10^{-8}	0.987×10^{-8}	0.851×10^{-8}
Formulated ester	1.37	1.00	0.874
Polyalkyl aromatic	1.58	1.25	1.01
Synthetic paraffinic oil	1.77	1.51	1.09
Synthetic paraffinic oil	1.99	1.51	1.29
Synthetic paraffinic oil plus antiwear additive	1.81	1.37	1.13
Synthetic paraffinic oil plus antiwear additive	1.96	1.55	1.25
C-ether	1.80	0.980	0.795
Superrefined naphthenic mineral oil	2.51	1.54	1.27
Synthetic hydrocarbon (traction fluid)	3.12	1.71	0.937
Fluorinated polyether	4.17	3.24	3.02

Source: Jones et al., 1975.

2.8 VISCOSITY AS A FUNCTION OF SHEAR RATE

It has been already mentioned that Newtonian fluids exhibit a linear relationship between the shear stress and the shear rate, and that the viscosity of Newtonian fluids is constant and independent of the shear rate. For regular mineral and synthetic oils this is an adequate assumption, but this assumption is not correct for greases. Mineral oils containing additives of long-chain polymers, such as multigrade oils, are *non-Newtonian* fluids, in the sense that the viscosity is a function of the shear rate. These fluids demonstrate shear-thinning characteristics; namely, the viscosity decreases with the shear rate. The discipline of *rheology* focuses on the investigation of the flow characteristics of non-Newtonian fluids, and much research work has been done investigating the rheology of lubricants.

The following approximate power-law equation is widely used to describe the viscosity of non-Newtonian fluids:

$$\mu = \mu_0 \left| \frac{\partial u}{\partial y} \right|^{n-1} \qquad (0 < n < 1) \tag{2-7}$$

The equation for the shear stress is

$$\tau = \mu_0 \left| \frac{\partial u}{\partial y} \right|^{n-1} \frac{\partial u}{\partial y} \tag{2-8}$$

An absolute value of the shear-rate is used because the shear stress can be positive or negative, while the viscosity remains positive. The shear stress, τ, has the same sign as the shear rate according to Eq. (2-8).

2.9 VISCOELASTIC LUBRICANTS

Polymer melts as well as liquids with additives of long-chain polymers in solutions of mineral oils demonstrate viscous as well as elastic properties and are referred to as *viscoelastic fluids*. Experiments with viscoelastic fluids show that the shear stress is not only a function of the instantaneous shear-rate but also a memory function of the shear-rate history. If the shear stress is suddenly eliminated, the shear rate will decrease slowly over a period of time. This effect is referred to as *stress relaxation*. The relaxation of shear stress takes place over a certain average time period, referred to as the *relaxation time*. The characteristics of such liquids are quite complex, but in principle, the Maxwell model of a spring and a dashpot (viscous damper) in series can approximate viscoelastic behavior. Under extension, the spring has only elastic force while the dashpot has only viscous resistance force. According to the Maxwell model, in a simple shear flow,

$u = u(y)$, the relation between the shear stress and the shear rate is described by the following equation:

$$\tau + \lambda \frac{d\tau}{dt} = \mu \frac{du}{dy} \tag{2-9}$$

Here, λ is the relaxation time (having units of time). The second term with the relaxation time describes the fluid stress-relaxation characteristic in addition to the viscous characteristics of Newtonian fluids.

As an example: In Newtonian fluid flow, if the shear stress, τ, is sinusoidal, it will result in a sinusoidal shear rate in phase with the shear stress oscillations. However, according to the Maxwell model, there will be a phase lag between the shear stress, τ, and the sinusoidal shear rate. Analysis of hydrodynamic lubrication with viscoelastic fluids is presented in Chapter 19.

Problems

2-1a A hydrostatic circular pad comprises two parallel concentric disks, as shown in Fig. 2-5. There is a thin clearance, h_0 between the disks. The upper disk is driven by an electric motor (through a mechanical drive) and has a rotation angular speed ω. For the rotation, power is required to overcome the viscous shear of fluid in the clearance. Derive the expressions for the torque, T, and the power, $\dot{E}_f$, provided by the drive (electric motor) to overcome the friction due to viscous shear in the clearance. Consider only the viscous friction in the thin clearance, h_0, and neglect the friction in the circular recess of radius R_0.

For deriving the expression of the torque, find the shear stresses and torque, dT, of a thin ring, dr, and integrate in the boundaries from R_0 to R. For the power, use the equation, $\dot{E}_f = T\omega$. Show that the results of the derivations are:

$$T_f = \frac{\pi}{2} \mu \frac{R^4}{h_0} \left(1 - \frac{R_0^4}{R^4}\right) \omega \tag{P2-1a}$$

$$\dot{E}_f = \frac{\pi}{2} \mu \frac{R^4}{h_0} \left(1 - \frac{R_0^4}{R^4}\right) \omega^2 \tag{P2-1b}$$

2-1b A hydrostatic circular pad as shown in Fig. 2-5 operates as a viscometer with a constant clearance of $h_0 = 200$ mm between the disks. The disk radius is $R = 200$ mm, and the circular recess radius is $R_0 = 100$ mm. The rotation speed of the upper disk is 600 RPM. The lower disk is mounted on a torque-measuring device, which reads a torque of 250 N-m. Find the fluid viscosity in SI units.

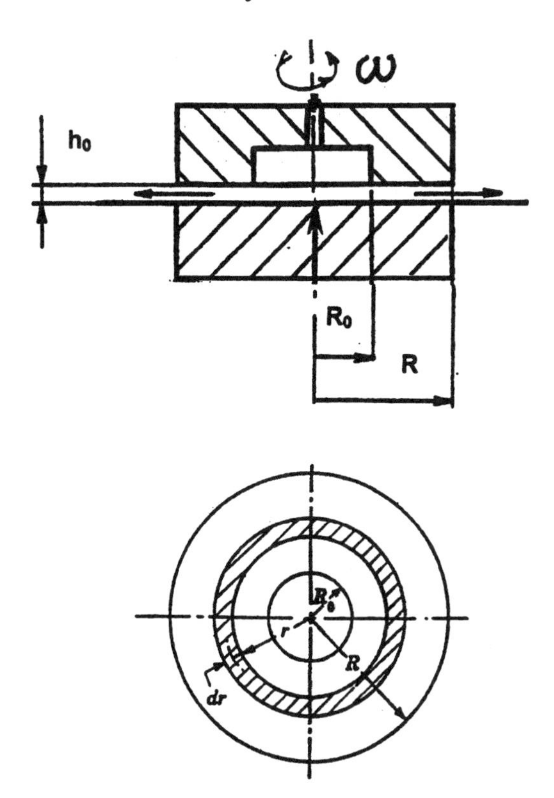

FIG. 2-5 Parallel concentric disks.

2-2 Find the viscosity in Reyns and the kinematic viscosity in centistoke (cSt) units and Saybolt universal second (SUS) units for the following fluids:

a. The fluid is mineral oil, SAE 10, and its operating temperature is 70°C. The lubricant density is $\rho = 860\ \text{kg/m}^3$.
b. The fluid is air, its viscosity is $\mu = 2.08 \times 10^{-5}\ \text{N-s/m}^2$, and its density is $\rho = 0.995\ \text{kg/m}^3$.
c. The fluid is water, its viscosity is $\mu = 4.04 \times 10^{-4}\ \text{N-s/m}^2$, and its density is $\rho = 978\ \text{kg/m}^3$.

2-3 Derive the equations for the torque and power loss of a journal bearing that operates without external load. The journal and bearing are concentric with a small radial clearance, C, between them. The diameter of the shaft is D and the bearing length is L. The shaft turns at a speed of 3600 RPM inside the bushing. The diameter of the shaft is $D = 50$ mm, while the radial clearance $C = 0.025$ mm. (In journal bearings, the ratio of radial clearance, C, to the shaft radius is of the order of 0.001.) The bearing length is $L = 0.5D$. The viscosity of the oil in the clearance is 120 Saybolt seconds, and its density is $\rho = 890\ \mathrm{kg/m^3}$.

a. Find the torque required for rotating the shaft, i.e., to overcome the viscous-friction resistance in the thin clearance.
b. Find the power losses for viscous shear inside the clearance (in watts).

2-4 A journal is concentric in a bearing with a very small radial clearance, C, between them. The diameter of the shaft is D and the bearing length is L. The fluid viscosity is μ and the relaxation time of the fluid (for a Maxwell fluid) is λ. The shaft has sinusoidal oscillations with sinusoidal hydraulic friction torque on the fluid film:

$$M_f = M_0 \sin \omega t$$

This torque will result in a sinusoidal shear stress in the fluid.

a. Neglect fluid inertia, and find the equation for the variable shear stress in the fluid.
b. Find the maximum shear rate (amplitude of the sinusoidal shear rate) in the fluid for the two cases of a Newtonian and a Maxwell fluid.
c. In the case of a Maxwell fluid, find the phase lag between shear rate and the shear stress.

3

Fundamental Properties of Lubricants

3.1 INTRODUCTION

Lubricants are various substances placed between two rubbing surfaces in order to reduce friction and wear. Lubricants can be liquids or solids, and even gas films have important applications. Solid lubricants are often used to reduce dry or boundary friction, but we have to keep in mind that they do not contribute to the heat transfer of the dissipated friction energy. Greases and waxes are widely used for light-duty bearings, as are solid lubricants such as graphite and molybdenum disulphide (MoS_2). In addition, coatings of polymers such as PTFE (Teflon) and polyethylene can reduce friction and are used successfully in light-duty applications.

However, liquid lubricants are used in much larger quantities in industry and transportation because they have several advantages over solid lubricants. The most important advantages of liquid lubricants are the formation of hydrodynamic films, the cooling of the bearing by effective convection heat transfer, and finally their relative convenience for use in bearings.

Currently, the most common liquid lubricants are *mineral oils*, which are made from petroleum. Mineral oils are blends of *base oils* with many different additives to improve the lubrication characteristics. Base oils (also referred to as *mineral oil base stocks*) are extracted from crude oil by a vacuum distillation process. Later, the oil passes through cleaning processes to remove undesired components. Crude oils contain a mixture of a large number of organic compounds, mostly hydrocarbons (compounds of hydrogen and carbon). Various other compounds are present in crude oils. Certain hydrocarbons are suitable for lubrication; these are extracted from the crude oil as base oils.

Mineral oils are widely used because they are available at relatively low cost (in comparison to synthetic lubricants). The commercial mineral oils are various base oils (comprising various hydrocarbons) blended to obtain the desired properties. In addition, they contain many additives to improve performance, such as oxidation inhibitors, rust-prevention additives, antifoaming agents, and high-pressure agents. A long list of additives is used, based on each particular application. The most common oil additives are discussed in this chapter.

During recent years, synthetic oils have been getting a larger share of the lubricant market. The synthetic oils are more expensive, and they are applied only whenever the higher cost can be financially justified. Blends of mineral and synthetic base oils are used for specific applications where unique lubrication characteristics are required. Also, greases are widely used, particularly for the lubrication of rolling-element bearings and gears.

3.2 CRUDE OILS

Most lubricants use mineral oil base stocks, made from crude oil. Each source of crude oil has its own unique composition or combination of compounds, resulting in a wide range of characteristics as well as appearance. Various crude oils have different colors and odors, and have a variety of viscosities as well as other properties. Crude oils are a mixture of hydrocarbons and other organic compounds. But they also contain many other compounds with various elements, including sulfur, nitrogen, and oxygen. Certain crude oils are preferred for the manufacture of lubricant base stocks because they have a desirable composition. Certain types of hydrocarbons are desired and extracted from crude oil to prepare lubricant base stocks. Desired components in the crude oil are saturated hydrocarbons, such as paraffin and naphthene compounds. Base oil is manufactured by means of distillation and extraction processes to remove undesirable components.

In the modern refining of base oils, the crude oil is first passed through an atmospheric-pressure distillation. In this unit, lighter fractions, such as gases, gasoline, and kerosene, are separated and removed. The remaining crude oil passes through a second vacuum distillation, where the lubrication oil components are separated. The various base oils are cleaned from the undesired components by means of solvent extraction. The base oil is dissolved in a volatile solvent in order to remove the wax as well as many other undesired components. Finally, the base oil is recovered from the solvent and passed through a process of hydrogenation to improve its oxidation stability.

3.3 BASE OIL COMPONENTS

Base oil components are compounds of hydrogen and carbon referred to as *hydrocarbon compounds*. The most common types are paraffin and naphthene compounds. Chemists refer to these two types as *saturated mineral oils*, while the third type, the *aromatic compounds* are *unsaturated*. Saturated mineral oils have proved to have better oxidation resistance, resulting in lubricants with long life and minimum sludge. A general property required of all mineral oils (as well as other lubricants) is that they be able to operate and flow at low temperature (low pour point). For example, if motor oils became too thick in cold weather, it would be impossible to start our cars.

In the past, Pennsylvania crude oil was preferred, because it contains a higher fraction of paraffin hydrocarbons, which have the desired lubrication characteristics. Today, however, it is feasible to extract small desired fractions of base oils from other crude oils, because modern refining processes separate all crude oils into their many components, which are ultimately used for various applications. But even today, certain crude oils are preferred for the production of base oils. The following properties are the most important in base-oil components.

3.3.1 Viscosity Index

The viscosity index (VI), already discussed in Chapter 2, is a common measure to describe the relationship of viscosity, μ, versus temperature, T. The curve of log μ versus log T is approximately linear, and the slope of the curve indicates the sensitivity of the viscosity to temperature variations. The viscosity index number is inversely proportional to the slope of the viscosity–temperature ($\mu-T$) curve in logarithmic coordinates. A high VI number is desirable, and the higher the VI number the flatter the $\mu-T$ curve, that is, the lubricant's viscosity is less sensitive to changes in temperature. Most commercial lubricants contain additives that serve as *VI improvers* (they increase the VI number by flattening the $\mu-T$ curve). In the old days, only the base oil determined the VI number. Pennsylvania oil was considered to have the best thermal characteristic and was assigned the highest VI, 100. But today's lubricants contain VI improvers, such as long-chain polymer additives or blends of synthetic lubricants with mineral oils, that can have high-VI numbers approaching 200. In addition, it is important to use high-VI base oils in order to achieve high-quality thermal properties of this order. Paraffins are base oil components with a relatively high VI number (Pennsylvania oil has a higher fraction of paraffins.) The naphthenes have a medium-to-high VI, while the aromatics have a low VI.

3.3.2 Pour Point

This is a measure of the lowest temperature at which the oil can operate and flow. This property is related to viscosity at low temperature. The pour point is determined by a standard test: The pour point is the lowest temperature at which a certain flow is observed under a prescribed, standard laboratory test. A low pour point is desirable because the lubricant can be useful in cold weather conditions. Paraffin is a base-oil component that has medium-to-high pour point, while naphthenes and aromatics have a desirably low pour point.

3.3.3 Oxidation Resistance

Oxidation inhibitors are meant to improve the oxidation resistance of lubricants for high-temperature applications. A detailed discussion of this characteristic is included in this chapter. However, some base oils have a better oxidation resistance for a limited time, depending on the operation conditions. Base oils having a higher oxidation resistance are desirable and are preferred for most applications. The base-oil components of paraffin and naphthene types have a relatively good oxidation resistance, while the aromatics exhibit poorer oxidation resistance.

The paraffins have most of the desired properties. They have a relatively high VI and relatively good oxidation stability. But paraffins have the disadvantage of a relatively higher pour point. For this reason, naphthenes are also widely used in blended mineral oils. Naphthenes also have good oxidation resistance, but their only drawback is a low-to-medium VI.

The aromatic base-oil components have the most undesirable characteristics, a low VI and low oxidation resistance, although they have desirably low pour points. In conclusion, each component has different characteristics, and lubricant manufacturers attempt to optimize the properties for each application via the proper blending of the various base-oil components.

3.4 SYNTHETIC OILS

A variety of synthetic base oils are currently available for engineering applications, including lubrication and heat transfer fluids. The most widely used are poly-alpha olefins (PAOs), esters, and polyalkylene glycols (PAGs). The PAOs and esters have different types of molecules, but both exhibit good lubrication properties. There is a long list of synthetic lubricants in use, but these three types currently have the largest market penetration.

The acceptance of synthetic lubricants in industry and transportation has been slow, for several reasons. The cost of synthetic lubricants is higher (it can be 2–100 times higher than mineral base oils). Although the initial cost of synthetic lubricants is higher, in many cases the improvement in performance and the longer life of the oil makes them an attractive long-term economic proposition. Initially, various additives (such as antiwear and oxidation-resistance additives) for mineral oils were adapted for synthetic lubricants. But experience indicated that such additives are not always compatible with the new lubricants. A lot of research has been conducted to develop more compatible additives, resulting in a continuous improvement in synthetic lubricant characteristics. There are other reasons for the slow penetration of synthetic lubricants into the market, the major one being insufficient experience with them. Industry has been reluctant to take the high risk of the breakdown of manufacturing machinery and the loss of production. Synthetic lubricants are continually penetrating the market for motor vehicles; their higher cost is the only limitation for much wider application.

The following is a list of the most widely used types of synthetic lubricants in order of their current market penetration:

1. Poly-alpha olefins (PAOs)
2. Esters
3. Polyalkylene glycols (PAGs)
4. Alkylated aromatics
5. Polybutenes
6. Silicones
7. Phosphate esters
8. PFPEs
9. Other synthetic lubricants for special applications.

3.4.1 Poly-alpha Olefins (PAOs)

The PAO lubricants can replace, or even be applied in combination with, mineral oils. The PAOs are produced via polymerization of olefins. Their chemical composition is similar to that of paraffins in mineral oils. In fact, they are synthetically made pure paraffins, with a narrower molecular weight distribution in comparison with paraffins extracted from crude oil. The processing causes a chemical linkage of olefins in a paraffin-type oil. The PAO lubricants have a reduced volatility, because they have a narrow molecular weight range, making them superior in this respect to parrafinic mineral oils derived from crude oil, which have much wider molecular weight range. A fraction of low-molecular-weight paraffin (light fraction) is often present in mineral oils derived from crude oil. This light fraction in mineral oils causes an undesired volatility, whereas this fraction is not present in synthetic oils. Most important, PAOs have a high viscosity index (the viscosity is less sensitive to temperature variations) and much better low-temperature characteristics (low pour point) in comparison to mineral oils.

3.4.2 Esters

This type of lubricant, particularly polyol esters (for example, pentaerithritol and trimethyrolpropane) is widely used in aviation fluids and automotive lubricants. Also, it is continually penetrating the market for industrial lubricants. Esters comprise two types of synthetic lubricants. The first type is dibasic acid esters, which are commonly substituted for mineral oils and can be used in combination with mineral oils. The second type is hindered polyol esters, which are widely used in high-temperature applications, where mineral oils are not suitable.

3.4.3 Polyalkylene Glycols (PAGs)

This type of base lubricant is made of linear polymers of ethylene and propylene oxides. The PAGs have a wide range of viscosity, including relatively high viscosity (in comparison to mineral oils) at elevated temperatures. The polymers can be of a variety of molecular weights. The viscosity depends on the range of the molecular weight of the polymer. Polymers of higher molecular weight exhibit higher viscosity. Depending on the chemical composition, these base fluids can be soluble in water or not. These synthetic lubricants are available in a very wide range of viscosities—from 55 to 300,000 SUS at 100°F (12–65,000 centistoke at 38°C). The viscosity of these synthetic base oils is less sensitive to temperature change in comparison to petroleum oils. The manufacturers provide viscosity vs. temperature charts that are essential for any lubricant application. In addition, polyalkylene-glycols base polymers have desirably low pour points in comparison to petroleum oils. Similar to mineral oils, they usually contain a wide range of additives to improve oxidation resistance, lubricity, as well as other lubrication characteristics. The additives must be compatible with the various synthetic oils.

Figure 3-1 presents an example of viscosity vs. temperature charts, for several polyalkylene-glycol base oils. The dotted line is a reference curve for petroleum base oil (mineral oil). It is clear that the negative slope of the synthetic oils is less steep in comparison to that of the mineral oil. It means that the viscosity of synthetic oils is less sensitive to a temperature rise. In fact, polyalkylene-glycol base oils can reach the highest viscosity index. The viscosity index of polyalkylene-glycols is between 150 and 290, while the viscosity index of commercial mineral oils ranges from 90 to 140. In comparison, the viscosity index of commercial polyol esters ranges from 120 to 180.

Another important property is the change of viscosity with pressure, which is more moderate in certain synthetic oils in comparison to mineral oils. This characteristic is important in the lubrication of rolling bearings and gears (EHD lubrication). The change of viscosity under pressure is significant only at very high pressures, such as the point or line contact of rolling elements and races. Figure 3-2 presents an example of viscosity vs. pressure charts, for several

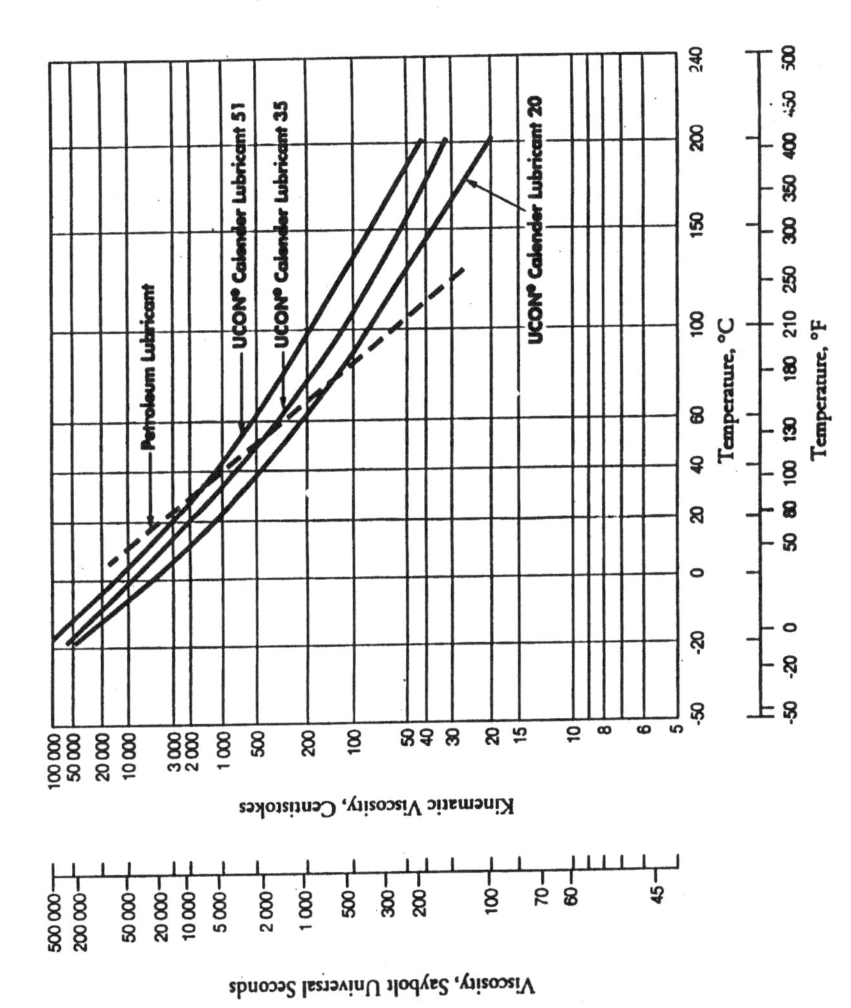

FIG. 3-1 Viscosity vs. temperature charts of commercial polyalkylene-glycol lubricants. (Used by permission of Union Carbide Corp.)

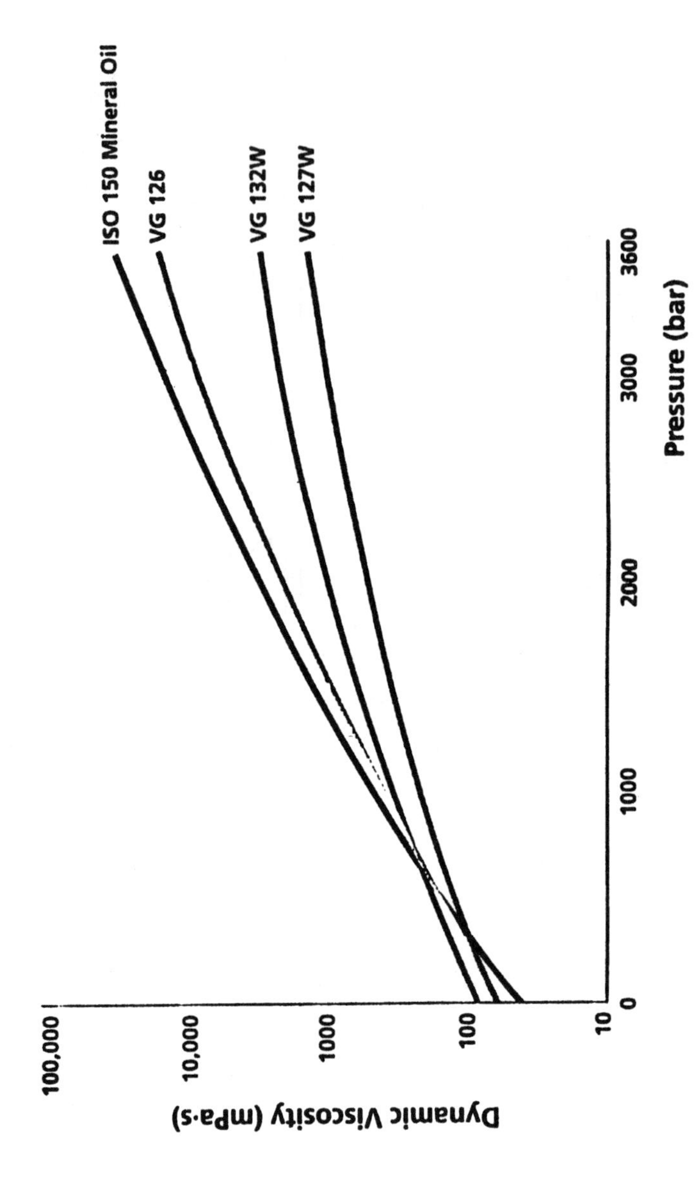

FIG. 3-2 Viscosity vs. pressure charts of commercial polyalkylene-glycol lubricants. (Used by permission of ICI Performance Chemicals.)

commercial polyalkylene-glycols as compared with a mineral oil. This chart is produced by tests that are conducted using a high-pressure viscometer.

3.4.4 Synthetic Lubricants for Special Applications

There are several interesting lubricants produced to solve unique problems in certain applications. An example is the need for a nonflammable lubricant for safety in critical applications. Halocarbon oils (such as polychlorotrifluoroethylene) can prove a solution to this problem because they are inert and nonflammable and at the same time they provide good lubricity. However, these lubricants are not for general use because of their extremely high cost. These lubricants were initially used to separate uranium isotopes during World War II.

In general, synthetic oils have many advantages, but they have some limitations as well: low corrosion resistance and incompatibility with certain seal materials (they cause swelling of certain elastomers). However, the primary disadvantage of synthetic base oils is their cost. They are generally several times as expensive in comparison to regular mineral base oils. As a result, they are substituted for mineral oils only when there is financial justification in the form of significant improvement in the lubrication performance or where a specific requirement must be satisfied. In certain applications, the life of the synthetic oil is longer than that of mineral oil, due to better oxidation resistance, which may result in a favorable cost advantage over the complete life cycle of the lubricant.

3.4.5 Summary of Advantages of Synthetic Oils

The advantages of synthetic oils can be summarized as follows: Synthetic oils are suitable for applications where there is a wide range of temperature. The most important favorable characteristics of these synthetic lubricants are: (a) their viscosity is less sensitive to temperature variations (high VI), (b) they have a relatively low pour point, (c) they have relatively good oxidation resistance; and (d) they have the desired low volatility. On the other hand, these synthetic lubricants are more expensive and should be used only where the higher cost can be financially justified. Concerning cost, we should consider not only the initial cost of the lubricant but also the overall cost. If a synthetic lubricant has a longer life because of its better oxidation resistance, it will require less frequent replacement. Whenever the oil serves for a longer period, there are additional savings on labor and downtime of machinery. All this should be considered when estimating the cost involved in a certain lubricant. Better resistance to oxidation is an important consideration, particularly where the oil is exposed to relatively high temperature.

3.5 GREASES

Greases are made of mineral or synthetic oils. The grease is a suspension of oil in soaps, such as sodium, calcium, aluminum, lithium, and barium soaps. Other thickeners, such as silica and treated clays, are used in greases as well. Greases are widely used for the lubrication of rolling-element bearings, where very small quantities of lubricant are required. Soap and thickeners function as a sponge to contain the oil. Inside the operating bearing, the sponge structure is gradually broken down, and the grease is released at a very slow rate. The oil slowly bleeds out, continually providing a very thin lubrication layer on the bearing surfaces. The released oil is not identical to the original oil used to make the grease. The lubrication layer is very thin and will not generate a lubrication film adequate enough to separate the sliding surfaces, but it is effective only as a boundary lubricant, to reduce friction and wear.

In addition to rolling bearings, greases are used for light-duty journal bearings or plane-sliders. Inside the bearing, the grease gradually releases small quantities of oil. This type of lubrication is easy to apply and reduces the maintenance cost. For journal or plane-slider bearings, greases can be applied only for low PV values, where boundary lubrication is adequate. The oil layer is too thin to play a significant role in cooling the bearing or in removing wear debris.

For greases, the design of the lubrication system is quite simple. Grease systems and their maintenance are relatively inexpensive. Unlike liquid oil, grease does not easily leak out. Therefore, in all cases where grease is applied there is no need for tight seals. A complex oil bath method with tight seals must be used only for oil lubrication. But for grease, a relatively simple labyrinth sealing (without tight seals) with a small clearance can be used, and this is particularly important where the shaft is not horizontal (such as in a vertical shaft). The drawback of tight seals on a rotating shaft is that the seals wear out, resulting in frequent seal replacement. Moreover, tight seals yield friction-energy losses that add heat to the bearing. Also, in grease lubrication, there is no need to maintain oil levels, and relubrication is less frequent in comparison to oil.

When rolling elements in a bearing come in contact with the grease, the thickener structure is broken down gradually, and a small quantity of oil slowly bleeds out to form a very thin lubrication layer on the rolling surfaces.

A continuous supply of a small amount of oil is essential because the thin oil layer on the bearing surface is gradually evaporated or deteriorated by oxidation. Therefore, bleeding from the grease must be continual and sufficient; that is, the oil supply should meet the demand. After the oil in the grease is depleted, new grease must be provided via repeated lubrication of the bearing. Similar to liquid oils, greases include many protective additives, such as rust and oxidation inhibitors.

The temperature of the operating bearing is the most important factor for selecting a grease type. The general-purpose grease covers a wide temperature range for most practical purposes. This range is from −400°C to 1210°C (−400°F to 2500°F). But care must be exercised at very high or very low operating temperatures, where low-temperature greases or extreme high-temperature greases should be applied. It would be incorrect to assume that grease suitable for a high temperature would also be successful at low temperatures, because high-temperature grease will be too hard for low-temperature applications. Greases made of sodium and mixed sodium–calcium soaps greases are suitable as general-purpose greases, although calcium soap is limited to rather low temperatures. For applications requiring water resistance, such as centrifugal pumps, calcium, lithium, and barium soap greases and the nonsoap greases are suitable. Synthetic oils are used to make greases for extremely low or extremely high temperatures. It is important to emphasize that different types of grease should not be mixed, particularly greases based on mineral oil with those based on synthetic oils. Bearings must be thoroughly cleaned before changing to a different grease type.

3.5.1 Grease Groups

a. *General-purpose greases*: These greases can operate at temperatures from −40°C to 121°C (−40°F to 250°F).
b. *High-temperature greases*: These greases can operate at temperatures from −18°C to 149°C (0°F to 300°F).
c. *Medium-temperature greases*: These greases can operate at temperatures from 0°C to 93°C (32°F to 200°F).
d. *Low-temperature greases*: These greases operate at temperatures as low as −55°C (−67°F) and as high as 107°C (225°F).
e. *Extremely high-temperature greases*: These greases can operate at temperatures up to 230°C to (450°F).

These five groups are based only on operating temperature. Other major characteristics that should be considered for the selection of grease for each application include consistency, oxidation resistance, water resistance, and melting point. There are grease types formulated for unique operating conditions, such as heavy loads, high speeds, and highly corrosive or humid environments. Grease manufacturers should be consulted, particularly for heavy-duty applications or severe environments. In the case of dust environments, the grease should be replaced more frequently to remove contaminants from the bearing. Greases for miniature bearings for instruments require a lower contamination level than standard greases.

Grease characteristics are specified according to standard tests. For example, the consistency (hardness) of grease, an important characteristic, is determined according to the ASTM D-217 standard penetration test. This test is conducted at 25°C by allowing a cone to penetrate into the grease for 5 seconds, higher penetration means softer grease. Standard worked penetration is determined by repeating the test after working the grease in a standard grease worker for 60 strokes. Prolonged working is testing after 100,000 strokes. The normal worked penetration for general-purpose grease is approximately between 250 and 350. Roller bearings require softer grease (ASTM worked penetration above 300) to reduce the rolling resistance. Other characteristics, such as oxidation stability, dropping point, and dirt count, apply in the same way to grease for roller bearings or ball bearings.

The selection of grease depends on the operating conditions, particularly the bearing temperature. The oxidation stability is an important selection criterion at high temperature. Oxidation stability is determined according to the ASTM D-942 standard oxidation test. The sensitivity of grease oxidation to temperature is demonstrated by the fact that a rise of 8°C (14°F) nearly doubles the oxidation rate. Commercial high-temperature greases are usually formulated with oxidation inhibitors to provide adequate oxidation resistance at high temperature.

3.6 ADDITIVES TO LUBRICANTS

Lubricants include a long list of additives to improve their characteristics. Lubricating oils are formulated with additives to protect equipment surfaces, enhance oil properties, and to protect the lubricant from degradation. Manufacturers start with blends of base oils with the best characteristics and further improve the desired properties by means of various additives. The following is a general discussion of the desired properties of commercial lubricants and the most common additives.

3.6.1 Additives to Improve the Viscosity Index

Multigrade oils, such as SAE 10W-40, contain significant amounts of additives that improve the viscosity index. Chapter 2 discusses the advantage of flattening the viscosity–temperature curve by using viscosity index improvers (VI improvers). These additives are usually long-chain polymeric molecules. They have a relatively high molecular weight, on the order of 25,000–500,000 molecular weight units, which is three orders of magnitude larger than that of the base-oil molecules. Examples of VI improvers are ethylene-propylene copolymers, polymethacrylates, and polyisobutylenes.

It is already recognized in the discipline of multiphase flow that small solid particles (such as spheres) in suspension increase the apparent viscosity of the base fluid (the suspension has more resistance to flow). Moreover, the viscosity increases with the diameter of the suspended particles.

In a similar way, additives of long-chain polymer molecules in a solution of mineral oils increase the apparent viscosity of the base oil. The long-chain molecules coil up into a spherical shape and play a similar role to that of a suspension of solid spheres. However, the diameter of the coils increases with the temperature and tends to raise the apparent viscosity more at higher temperatures. At higher temperatures, polymeric molecules are more soluble in the base because they interact better with it. In turn, the large molecules will uncoil at higher temperature, resulting in a larger coil diameter, and the viscosity of the lubricant increases. On the other hand, at lower temperatures, the polymeric molecules tend to coil up, their diameter decreases, and, in turn, the viscosity of the oil is reduced. This effect tends to diminish the stronger effect of viscosity reduction with increasing temperature of base oils.

In summary: When polymer additives are dissolved in base oils, the viscosity of the solution is increased, but the rise in viscosity is much greater at high temperatures than at low temperatures. In conclusion, blending oils with long chain polymers results in a desirable flattening of the viscosity–temperature curve.

The long-chain molecules in multigrade oils gradually tear off during operation due to high shear rates in the fluid. This reduces the viscosity of the lubricant as well as the effectiveness of polymers as VI improvers. This phenomenon, often referred to as *degradation*, limits the useful life of the lubricant. Permanent viscosity loss in thickened oils occurs when some of the polymer molecules break down under high shear rates. The shorter polymer molecules contribute less as VI improvers. The resistance to this type of lubricant degradation varies among various types of polymer molecules. Polymers having more resistance to degradation are usually selected. The advantage of synthetic oils is that they have relatively high VI index, without the drawback of degradation. In certain lubricants, synthetic oils are added to improve the VI index, along with polymer additives. Long-chain polymer additives together with blends of synthetic lubricants can improve significantly the VI numbers of base oils. High VI numbers of about 200 are usually obtained for multigrade oils, and maximum value of about 400 for synthetic oils.

3.6.1.1 Viscosity–Shear Effects

The long-chain molecule polymer solution of mineral oils is a non-Newtonian fluid. There is no more linearity between the shear stress and the shear-rate. Fluids that maintain the same viscosity at various shear rates are called *Newtonian fluids*. This is true of most single-viscosity-grade oils. However, multigrade oils are non-Newtonian fluids, and they lose viscosity under high rates of shear. This loss can be either temporary or permanent. In addition to the long-term effect of degradation, there is an immediate reduction of viscosity at high shear rates. This temporary viscosity loss is due to the elongation and orientation of the polymer molecules in the direction of flow. In turn, there is less internal friction and flow-induced reduction of the viscosity of the lubricant. When the oil is no longer subjected to high shear rates, the molecules return to their preferred spherical geometry, and their viscosity recovers. Equations (2-7) and (2-8) describe such non-Newtonian characteristics via a power-law relation between the shear rate and stress.

3.6.1.2 Viscoelastic Fluids

In addition to the foregoing nonlinearity, long-chain polymer solutions exhibit viscoelastic properties. Viscoelastic flow properties can be described by the Maxwell equation [Eq. (2-9)].

3.6.2 Oxidation Inhibitors

Oxidation can take place in any oil, mineral or synthetic, at elevated temperature whenever the oil is in contact with oxygen in the air. Oil oxidation is undesirable because the products of oxidation are harmful chemical compounds, such as organic acids, that cause corrosion. In addition, the oxidation products contribute to a general deterioration of the properties of the lubricant. Lubricant degradation stems primarily from thermal and mechanical energy. Lubricant degradation is catalyzed by the presence of metals and oxygen.

The organic acids, products of oil oxidation, cause severe corrosion of the steel journal and the alloys used as bearing materials. The oil circulates, and the corrosive lubricant can damage other parts of the machine. In addition, the oxidation products increase the viscosity of the oils as well as forming sludge and varnish on the bearing and journal surfaces. Excessive oil oxidation can be observed by a change of oil color and also can be recognized by the unique odors of the oxidation products.

At high temperature, oxygen reacts with mineral oils to form hydroperoxides and, later, organic acids. The oxidation process is considerably faster at elevated temperature; in fact, the oxidation rate doubles for a nearly 10°C rise in oil temperature. It is very important to prevent or at least to slow down this undesirable process. Most lubricants include additives of oxidation inhibitors, particularly in machines where the oil serves for relatively long periods of time and is exposed to high temperatures, such as steam turbines and motor vehicle engines. The oxidation inhibitors improve the lubricant's desirable characteristic of oxidation resistance, in the sense that the chemical process of oxidation becomes very slow.

Radical scavengers, peroxide decomposers, and metal deactivators are used as inhibitors of the oil degradation process. Two principle types of antioxidants that act as radical scavengers are aromatic amines and hindered phenolics. The mechanistic behavior of these antioxidants explains the excellent performance of the *aromatic amine* type under high-temperature oxidation conditions and the excellent performance of the *hindered phenolic* type under low-temperature oxidation conditions. Appropriate combinations of both types allow for optimum protection across the widest temperature range. Other widely used additives combine the two properties of oxidation and corrosion resistance, e.g., zinc dithiophosphates and sulfurized olefins. There are several companies that have specialized in the research in and development of oxidation inhibitors. Lubricants in service for long periods of time at elevated temperature, such as engine oils, must include oxidation inhibitors to improve their oxidation resistance. As mentioned earlier, synthetic oils without oxidation inhibitors have better oxidation resistance, but they also must include oxidation inhibitors when used in high-temperature applications, such as steam turbines and engines.

For large machines and in manufacturing it is important to monitor the lubricant for depletion of the oxidation inhibitors and possible initiation of corrosion, via periodic laboratory tests. For monitoring the level of acidity during operation, the *neutralization number* is widely used. The rate of increasing acidity of a lubricating oil is an indication of possible problems in the operation conditions. If the acid content of the oil increases too fast, it can be an indication of contamination by outside sources, such as penetration of acids in chemical plants. Oils containing acids can also be easily diagnosed by their unique odor in comparison to regular oil. In the laboratory, standard tests ASTM D 664 and ASTM D 974 are used to measure the amount of acid in the oil.

3.6.3 Pour-Point Depressants

The pour point is an important characteristic whenever a lubricant is applied at low temperatures, such as when starting a car engine on winter mornings when the temperature is at the freezing point. The oil can solidify at low temperature; that is, it will loose its fluidity. Saturated hydrocarbon compounds of the paraffin and naphthene types are commonly used, since they have a relatively low pour-point temperature. Pour-point depressants are oil additives, which were developed to lower the pour-point temperature. Also, certain synthetic oils were developed that can be applied in a wide range of temperatures and have a relatively very low pour point.

3.6.4 Antifriction Additives

A bearing operating with a full hydrodynamic film has low friction and a low wear rate. The lubricant viscosity is the most important characteristic for maintaining effective hydrodynamic lubrication operation. However, certain bearings are designed to operate under boundary lubrication conditions, where there is direct contact between the asperities of the rubbing surfaces. The asperities deform under the high contact pressure; due to adhesion between the two surfaces, there is a relatively high friction coefficient. Measurements of the friction coefficient, f, versus the sliding velocity U (*Stribeck curve*) indicate a relatively high friction coefficient at low sliding velocity, in the boundary lubrication region (see Chapter 16). In this region, the friction is reducing at a steep negative slope with velocity. Under such conditions of low velocity, there is a direct contact of the surface asperities. The *antifriction* characteristic of the lubricant, often referred to as *oil lubricity*, can be very helpful in reducing high levels of friction. A wide range of oil additives has been developed to improve the antifriction characteristics and to reduce the friction coefficient under boundary lubrication conditions.

Much more research work is required to fully understand the role of antifriction additives in reducing boundary lubrication friction. The current explanation is that the additives are absorbed and react with the metal surface and its oxides to form thin layers of low-shear-strength material. The layers are compounds of long-chain molecules such as alcohol, amines, and fatty acids. A common antifriction additive is oleic acid, which reacts with iron oxide to form a thin layer of iron-oleate soap. Antiwear additives such as zinc dialkyldithiophosphate (ZDDP) are also effective in friction reduction.

Theory postulates that the low shear strength of the various long-chain molecular layers, as well as the soap film, results in a lower friction coefficient. The thin layer can be compared to a deck of cards that slide easily, relative to each other, in a parallel direction to that of the two rubbing surfaces. But at the same time, the long-chain molecular layers can hold very high pressure in the direction normal to the rubbing surfaces. The thin layers on the surface can reduce the shear force required for relative sliding of the asperities of the two surfaces; thus it reduces the friction coefficient.

The friction coefficient, f, in boundary lubrication is usually measured in friction-testing machines, such as four-ball or pin-on-disk testing machines. But these friction measurements for liquid lubricants are controversial because of the steep slope of the $f-U$ curve. Moreover, it is not a "clean" measurement of the effect of an antifriction additive. The friction reduction is a combination of two effects, the fluid viscosity combined with the surface treatment by the antifriction additive. A much better measurement is to record the complete Stribeck curve, which clearly indicates the friction in the various lubrication regions. The antifriction performance of various oil additives is tested under conditions of boundary lubrication. A reduction in the maximum friction coefficient is an indication of the effectiveness in improving the antifriction characteristics of the base mineral oil. Experiments with steel sliding on steel indicate a friction coefficient in the range of 0.10–0.15 when lubricated only with a regular mineral oil. However, the addition of 2% oleic acid to the oil reduces the friction coefficient to the range of 0.05–0.08. Lubricants having good antifriction characteristics have considerable advantages, even for hydrodynamic bearings, such as the reduction of friction during the start-up of machinery.

3.6.5 Solid Colloidal Dispersions

Recent attempts to reduce boundary lubrication friction include the introduction of very small microscopic solid particles (powders) in the form of colloidal dispersions in the lubricant. More tests are required to verify the effectiveness of colloidal dispersions. These antifriction additives are suspensions of very fine solid particles of graphite, PTFE (Teflon), or MoS_2, and the particle sizes are much less than 1 μm. More research is required, on the one hand, for testing the magnitude of the reduction in friction and, on the other hand, for accurately explaining the antifriction mechanism of solid colloidal dispersions in the lubricant. Theory postulates that these solid additives form a layer of solid particles on the substrate surface. The particles are physically attracted to the surface by adhesion and form a thin protective film that can shear easily but at the same time can carry the high pressure at the contact between the surface asperities (in a similar way to surface layers formed by antifriction liquid additives).

3.6.6 Antiwear Additives

The main objective of antiwear oil additives is to reduce the wear rate in sliding or rolling motion under boundary lubrication conditions. An additional important advantage of antiwear additives is that they can reduce the risk of a catastrophic bearing failure, such as seizure, of sliding or rolling-element bearings. The explanation for the protection mechanism is similar to that for antifriction layers. Antiwear additives form thin layers of organic, metal-organic, or metal salt film on the surface. This thin layer formed on the surface is sacrificed to protect the metal. The antiwear additives form a thin layer that separates the rubbing surfaces and reduces the adhesion force at the contact between the peaks of the asperities of the two surfaces. Oil tests have indicated that wear debris in the oil contain most of the antiwear-layer material.

Zinc dialkyldithio-phosphate (ZDDP) is an effective, widely used antiwear additive. It is applied particularly in automotive engines, as well as in most other applications including hydraulic fluids. Zinc is considered a hazardous waste material, and there is an effort to replace this additive by more environmentally friendly additives. After ZDDP decomposes, several compounds are generated of metal-organic, zinc sulfide, or zinc phosphate. The compounds react with the surface of steel shafts and form iron sulfide or iron phosphate, which forms an antiwear film on the surface. These antiwear films are effective in boundary lubrication conditions. Additional types of antiwear additives are various phosphate compounds, organic phosphates, and various chlorine compounds. Various antiwear additives are commonly used to reduce the wear rate of sliding as well as rolling-element bearings.

The effectiveness of antiwear additives can be measured on various commercial wear-testing machines, such as four-ball or pin-on-disk testing machines (similar to those for friction testing). The operating conditions must be close to those in the actual operating machinery. The rate of material weight loss is an indication of the wear rate. Standard tests, for comparison between various lubricants, should operate under conditions described in ASTM G 99-90.

We have to keep in mind that laboratory friction-testing machines do not always accurately correlate with the conditions in actual industrial machinery. However, it is possible to design experiments that simulate the operating conditions and measure wear rate under situations similar to those in industrial machines. The results are useful in selecting the best lubricant as well as the antifriction additives for minimizing friction and wear for any specific application. Long-term lubricant tests are often conducted on site on operating industrial machines. However, such tests are over a long period, and the results are not always conclusive, because the conditions in practice always vary with time. By means of on-site tests, in most cases it is impossible to compare the performance of several lubricants, or additives, under identical operation conditions.

3.6.7 Corrosion Inhibitors

Chemical contaminants can be generated in the oil or enter into the lubricant from contaminated environments. Corrosive fluids often penetrate through the seals into the bearing and cause corrosion inside the bearing. This problem is particularly serious in chemical plants where there is a corrosive environment, and small amounts of organic or inorganic acids usually contaminate the lubricant and cause considerable corrosion. Also, organic acids from the oil oxidation process can cause severe corrosion in bearings. Organic acids from oil oxidation must be neutralized; otherwise, the acids degrade the oil and cause corrosion. Oxygen reacts with mineral oils at high temperature. The oil oxidation initially forms hydroperoxides and, later, organic acids. White metal (babbitt) bearings as well as the steel in rolling-element bearings are susceptible to corrosion by acids. It is important to prevent oil oxidation and contain the corrosion damage by means of corrosion inhibitors in the form of additives in the lubricant.

In addition to acids, water can penetrate through seals into the oil (particularly in water pumps) and cause severe corrosion. Water can get into the oil from the outside or by condensation. Penetration of water into the oil can cause premature bearing failure in hydrodynamic bearings and particularly in standard rolling-element bearings. Water in the oil is a common cause for corrosion. Only a very small quantity of water is soluble in the oils, about 80 PPM (parts per million); above this level, even a small quantity of water that is not in solution is harmful. The presence of water can be diagnosed by the unique hazy color of the oil. Water acts as a catalyst and accelerates the oil oxidation process. Water is the cause for corrosion of many common bearing metals and particularly steel shafts; for example, water reacts with steel to form rust (hydrated iron oxide). Therefore in certain applications that involve water penetration, stainless steel shafts and rolling bearings are used. Rust inhibitors can also help in reducing corrosion caused by water penetration.

In rolling-element bearings, the corrosion accelerates the fatigue process, referred to as *corrosion fatigue*. The corrosion introduces small cracks in the metal surface that propagate into the metal via oscillating fatigue stresses. In this way, water promotes contact fatigue in rolling-element bearings. It is well known that water penetration into the bearings is often a major problem in centrifugal pumps; wherever it occurs, it causes an early bearing failure, particularly for rolling-element bearings, which involve high fatigue stresses.

Rust inhibitors are oil additives that are absorbed on the surfaces of ferrous alloys in preference to water, thus preventing corrosion. Also, metal deactivators are additives that reduce nonferrous metal corrosion. Similar to rust inhibitors, they are preferentially absorbed on the surface and are effective in protecting it from corrosion. Examples of rust inhibitors are oil-soluble petroleum sulfonates and calcium sulfonate, which can increase corrosion protection.

3.6.8 Antifoaming Additives

Foaming of liquid lubricants is undesirable because the bubbles deteriorate the performance of hydrodynamic oil films in the bearing. In addition, foaming adversely affects the oil supply of lubrication systems (it reduces the flow rate of oil pumps). Also, the lubricant can overflow from its container (similar to the use of liquid detergent without antifoaming additives in a washing machine). The function of antifoaming additives is to increase the interfacial tension between the gas and the lubricant. In this way, the bubbles collapse, allowing the gas to escape.

Problems

3-1 Find the viscosity of the following three lubricants at 20°C and 100°C:

a. SAE 30
b. SAE 10W-30
c. Polyalkylene glycol synthetic oil

List the three oils according to the sensitivity of viscosity to temperature, based on the ratio of viscosity at 20°C to viscosity at 100°C.

3-2 Explain the advantages of synthetic oils in comparison to mineral oil. Suggest an example application where there is a justification for using synthetic oil of higher cost.

3-3 List five of the most widely used synthetic oils. What are the most important characteristics of each of them?

3-4 Compare the advantages of using greases versus liquid lubricants. Suggest two example applications where you would prefer to use grease for lubrication and two examples where you would prefer to use liquid lubricant. Justify your selection in each case.

3-5 a. Explain the process of oil degradation by oxidation.
b. List the factors that determine the oxidation rate.
c. List the various types of oxidation inhibitors.

4

Principles of Hydrodynamic Lubrication

4.1 INTRODUCTION

A hydrodynamic plane-slider is shown in Fig. 1-2 and the widely used hydrodynamic journal bearing is shown in Fig. 1-3. Hydrodynamic lubrication is the fluid dynamic effect that generates a lubrication fluid film that completely separates the sliding surfaces. The fluid film is in a thin clearance between two surfaces in relative motion. The hydrodynamic effect generates a hydrodynamic *pressure wave* in the fluid film that results in *load-carrying capacity*, in the sense that the fluid film has sufficient pressure to carry the external load on the bearing. The pressure wave is generated by a wedge of viscous lubricant drawn into the clearance between the two converging surfaces or by a squeeze-film action.

The thin clearance of a plane-slider and a journal bearing has the shape of a thin converging wedge. The fluid adheres to the solid surfaces and is dragged into the converging clearance. High shear stresses drag the fluid into the wedge due to the motion of the solid surfaces. In turn, high pressure must build up in the fluid film before the viscous fluid escapes through the thin clearance. The pressure wave in the fluid film results in a load-carrying capacity that supports the external load on the bearing. In this way, the hydrodynamic film can completely separate the sliding surfaces, and, thus, wear of the sliding surfaces is prevented. Under steady conditions, the hydrodynamic load capacity, W, of a bearing is equal to the external load, F, on the bearing, but it is acting in the opposite direction. The hydrodynamic theory of lubrication solves for the fluid velocity, pressure wave, and resultant load capacity.

Experiments and hydrodynamic analysis indicated that the hydrodynamic load capacity is proportional to the sliding speed and fluid viscosity. At the same time, the load capacity dramatically increases for a thinner fluid film. However, there is a practical limit to how much the bearing designer can reduce the film thickness. A very thin fluid film is undesirable, particularly in machines with vibrations. Whenever the hydrodynamic film becomes too thin, it results in occasional contact of the surfaces, which results in severe wear. Picking the optimum film-thickness is an important decision in the design process; it will be discussed in the following chapters.

Tower (1880) conducted experiments and demonstrated for the first time the existence of a pressure wave in a hydrodynamic journal bearing. Later, Reynolds (1886) derived the classical theory of hydrodynamic lubrication. A large volume of analytical and experimental research work in hydrodynamic lubrication has subsequently followed the work of Reynolds. The classical theory of Reynolds and his followers is based on several assumptions that were adopted to simplify the mathematical derivations, most of which are still applied today. Most of these assumptions are justified because they do not result in a significant deviation from the actual conditions in the bearing. However, some other classical assumptions are not realistic but were necessary to simplify the analysis. As in other disciplines, the introduction of computers permitted complex hydrodynamic lubrication problems to be solved by numerical analysis and have resulted in the numerical solution of such problems under realistic conditions without having to rely on certain inaccurate assumptions.

At the beginning of the twentieth century, only long hydrodynamic journal bearings had been designed. The length was long in comparison to the diameter, $L > D$; long-bearing theory of Reynolds is applicable to such bearings. Later, however, the advantages of a short bearing were recognized. In modern machinery, the bearings are usually short, $L < D$; short-bearing theory is applicable. The advantage of a long bearing is its higher load capacity in comparison to a short bearing. Moreover, the load capacity of a long bearing is even much higher per unit of bearing area. In comparison, the most important advantages of a short bearing that make it widely used are: (a) better cooling due to faster circulation of lubricant; (b) less sensitivity to misalignment; and (c) a compact design.

Simplified models are commonly used in engineering to provide insight and simple design tools. Hydrodynamic lubrication analysis is much simplified if the bearing is assumed to be *infinitely long* or *infinitely short*. But for a finite-length bearing, there is a three-dimensional flow that requires numerical solution by computer. In order to simplify the analysis, long journal bearings, $L > D$, are often solved as infinitely long bearings, while short bearings, $L < D$, are often solved as infinitely short bearings.

4.2 ASSUMPTIONS OF HYDRODYNAMIC LUBRICATION THEORY

The first assumption of hydrodynamic lubrication theory is that the fluid film flow is laminar. The flow is laminar at low Reynolds number (Re). In fluid dynamics, the Reynolds number is useful for estimating the ratio of the inertial and viscous forces. For a fluid film flow, the expression for the Reynolds number is

$$\mathrm{Re} = \frac{U\rho h}{\mu} = \frac{Uh}{\nu} \tag{4-1}$$

Here, h is the average magnitude of the variable film thickness, ρ is the fluid density, μ is the fluid viscosity, and ν is the kinematic viscosity. The transition from laminar to turbulent flow in hydrodynamic lubrication initiates at about $\mathrm{Re} = 1000$, and the flow becomes completely turbulent at about $\mathrm{Re} = 1600$. The Reynolds number at the transition can be lower if the bearing surfaces are rough or in the presence of vibrations. In practice, there are always some vibrations in rotating machinery.

In most practical bearings, the Reynolds number is sufficiently low, resulting in laminar fluid film flow. An example problem is included in Chapter 5, where Re is calculated for various practical cases. That example shows that in certain unique applications, such as where water is used as a lubricant (in certain centrifugal pumps or in boats), the Reynolds number is quite high, resulting in turbulent fluid film flow.

Classical hydrodynamic theory is based on the assumption of a linear relation between the fluid stress and the strain-rate. Fluids that demonstrate such a linear relationship are referred to as *Newtonian fluids* (see Chapter 2). For most lubricants, including mineral oils, synthetic lubricants, air, and water, a linear relationship between the stress and the strain-rate components is a very close approximation. In addition, liquid lubricants are considered to be *incompressible*. That is, they have a negligible change of volume under the usual pressures in hydrodynamic lubrication.

Differential equations are used for theoretical modeling in various disciplines. These equations are usually simplified under certain conditions by disregarding terms of a relatively lower order of magnitude. Order analysis of the various terms of an equation, under specific conditions, is required for determining the most significant terms, which capture the most important effects. A term in an equation can be disregarded and omitted if it is lower by one or several orders of magnitude in comparison to other terms in the same equation. Dimensionless analysis is a useful tool for determining the relative orders of magnitude of the terms in an equation. For example, in fluid dynamics, the dimensionless Reynolds number is a useful tool for estimating the ratio of inertial and viscous forces.

In hydrodynamic lubrication, the fluid film is very thin, and in most practical cases the Reynolds number is low. Therefore, the effect of the inertial forces of the fluid (ma) as well as gravity forces (mg) are very small and can be neglected in comparison to the dominant effect of the viscous stresses. This assumption is applicable for most practical hydrodynamic bearings, except in unique circumstances.

The fluid is assumed to be continuous, in the sense that there is continuity (no sudden change in the form of a step function) in the fluid flow variables, such as shear stresses and pressure distribution. In fact, there are always very small air bubbles in the lubricant that cause discontinuity. However, this effect is usually negligible, unless there is a massive fluid foaming or *fluid cavitation* (formation of bubbles when the vapor pressure is higher than the fluid pressure). In general, classical fluid dynamics is based on the continuity assumption. It is important for mathematical derivations that all functions be continuous and differentiable, such as stress, strain-rate, and pressure functions.

The following are the basic ten assumptions of classical hydrodynamic lubrication theory. The first nine were investigated and found to be justified, in the sense that they result in a negligible deviation from reality for most practical oil bearings (except in some unique circumstances). The tenth assumption however, has been introduced only for the purpose of simplifying the analysis.

Assumptions of classical hydrodynamic lubrication theory

1. The flow is laminar because the Reynolds number, Re, is low.
2. The fluid lubricant is continuous, Newtonian, and incompressible.
3. The fluid adheres to the solid surface at the boundary and there is no fluid slip at the boundary; that is, the velocity of fluid at the solid boundary is equal to that of the solid.
4. The velocity component, v, across the thin film (in the y direction) is negligible in comparison to the other two velocity components, u and w, in the x and z directions, as shown in Fig. 1-2.
5. Velocity gradients along the fluid film, in the x and z directions, are small and negligible relative to the velocity gradients across the film because the fluid film is thin, i.e., $du/dy \gg du/dx$ and $dw/dy \gg dw/dz$.
6. The effect of the curvature in a journal bearing can be ignored. The film thickness, h, is very small in comparison to the radius of curvature, R, so the effect of the curvature on the flow and pressure distribution is relatively small and can be disregarded.

7. The pressure, p, across the film (in the y direction) is constant. In fact, pressure variations in the y direction are very small and their effect is negligible in the equations of motion.
8. The force of gravity on the fluid is negligible in comparison to the viscous forces.
9. Effects of fluid inertia are negligible in comparison to the viscous forces. In fluid dynamics, this assumption is usually justified for low-Reynolds-number flow.

These nine assumptions are justified for most practical hydrodynamic bearings. In contrast, the following additional tenth assumption has been introduced only for simplification of the analysis.

10. The fluid viscosity, μ, is constant.

It is well known that temperature varies along the hydrodynamic film, resulting in a variable viscosity. However, in view of the significant simplification of the analysis, most of the practical calculations are still based on the assumption of a constant equivalent viscosity that is determined by the average fluid film temperature. The last assumption can be applied in practice because it has already been verified that reasonably accurate results can be obtained for regular hydrodynamic bearings by considering an equivalent viscosity. The average temperature is usually determined by averaging the temperature of the bearing inlet and outlet lubricant. Various other methods have been suggested to calculate the equivalent viscosity.

A further simplification of the analysis can be obtained for very long and very short bearings. If a bearing is very long, the flow in the axial direction (z direction) can be neglected, and the three-dimensional flow reduces to a much simpler two-dimensional flow problem that can yield a closed form of analytical solution.

A long journal bearing is where the bearing length is much larger than its diameter, $L \gg D$, and a short journal bearing is where $L \ll D$. If $L \gg D$, the bearing is assumed to be infinitely long; if $L \ll D$, the bearing is assumed to be infinitely short.

For a journal bearing whose length L and diameter D are of a similar order of magnitude, the analysis is more complex. This three-dimensional flow analysis is referred to as a *finite-length bearing analysis*. Computer-aided numerical analysis is commonly applied to solve for the finite bearing. The results are summarized in tables that are widely used for design purposes (see Chapter 8).

4.3 HYDRODYNAMIC LONG BEARING

The coordinates of a long hydrodynamic journal bearing are shown in Fig. 4-1. The velocity components of the fluid flow, u, v, and w are in the x, y, and z directions, respectively. A journal bearing is long if the bearing length, L, is much larger than its diameter, D. A plane-slider (see Fig. 1-2) is long if the bearing width, L, in the z direction is much larger than the length, B, in the x direction (the direction of the sliding motion), or $L \gg B$.

In addition to the ten classical assumptions, there is an additional assumption for a long bearing—it can be analyzed as an infinitely long bearing. The pressure gradient in the z direction (axial direction) can be neglected in comparison to the pressure gradient in the x direction (around the bearing). The pressure is assumed to be constant along the z direction, resulting in two-dimensional flow, $w = 0$.

In fact, in actual long bearings there is a side flow from the bearing edge, in the z direction, because the pressure inside the bearing is higher than the ambient pressure. This side flow is referred to as an *end effect*. In addition to flow, there are other end effects, such as capillary forces. But for a long bearing, these effects are negligible in comparison to the constant pressure along the entire length.

4.4 DIFFERENTIAL EQUATION OF FLUID MOTION

The following analysis is based on first principles. It does not use the Navier–Stokes equations or the Reynolds equation and does not require in-depth knowledge of fluid dynamics. The following self-contained derivation can help in understanding the physical concepts of hydrodynamic lubrication.

An additional merit of a derivation that does not rely on the Navier–Stokes equations is that it allows extending the theory to applications where the Navier–Stokes equations do not apply. An example is lubrication with non-Newtonian fluids, which cannot rely on the classical Navier–Stokes equations because they

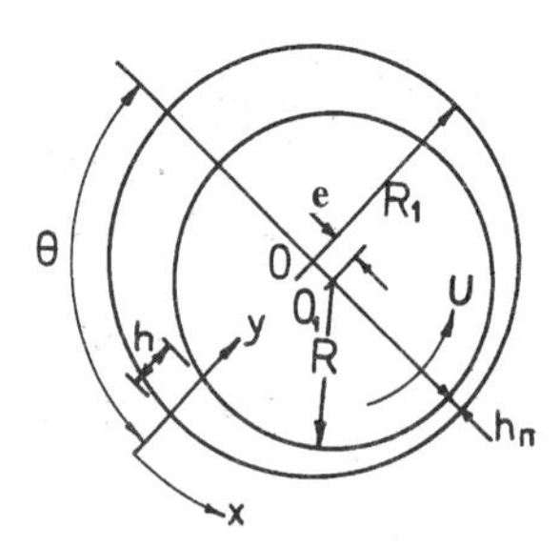

FIG. 4-1 Coordinates of a long journal bearing.

assume the fluid is Newtonian. Since the following analysis is based on first principles, a similar derivation can be applied to non-Newtonian fluids (see Chapter 19).

The following hydrodynamic lubrication analysis includes a derivation of the differential equation of fluid motion and a solution for the flow and pressure distribution inside a fluid film. The boundary conditions of the velocity and the conservation of mass (or the equivalent conservation of volume for an incompressible flow) are considered for this derivation.

The equation of the fluid motion is derived by considering the balance of forces acting on a small, infinitesimal fluid element having the shape of a rectangular parallelogram of dimensions dx and dy, as shown in Fig. 4-2. This elementary fluid element inside the fluid film is shown in Fig. 1-2. The derivation is for a two-dimensional flow in the x and y directions. In an infinitely long bearing, there is no flow or pressure gradient in the z direction. Therefore, the third dimension of the parallelogram (in the z direction) is of unit length (1).

The pressure in the x direction and the shear stress, τ, in the y direction are shown in Fig. 4-2. The stresses are subject to continuous variations. A relation between the pressure and shear-stress gradients is derived from the balance of forces on the fluid element. The forces are the product of stresses, or pressures, and the corresponding areas. The fluid inertial force (ma) is very small and is therefore neglected in the classical hydrodynamic theory (see assumptions listed earlier), allowing the derivation of the following force equilibrium equation in a similar way to a static problem:

$$(\tau + d\tau)dx \cdot 1 - \tau\, dx \cdot 1 = (p + dp) \cdot dy \cdot 1 = p\, dy \cdot 1 \tag{4-2}$$

Equation (4-2) reduces to

$$d\tau\, dx = dp\, dy \tag{4-3}$$

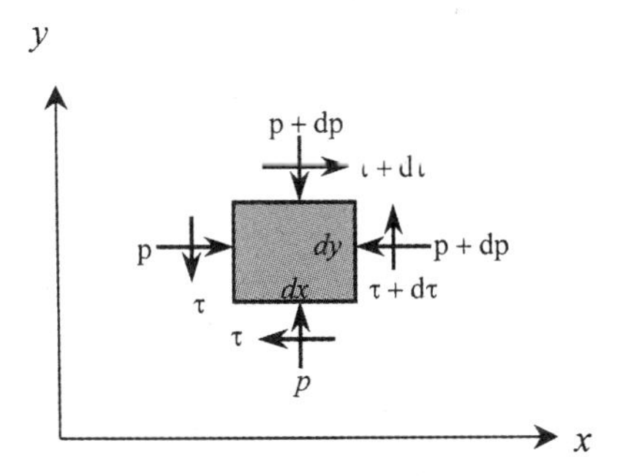

FIG. 4-2 Balance of forces on an infinitesimal fluid element.

After substituting the full differential expression $d\tau = (\partial\tau/\partial y)dy$ in Eq. (4-3) and substituting the equation $\tau = \mu\,(\partial u/\partial y)$ for the shear stress, Eq. (4-3) takes the form of the following differential equation:

$$\frac{dp}{dx} = \mu \frac{\partial^2 u}{\partial y^2} \tag{4-4}$$

A partial derivative is used because the velocity, u, is a function of x and y. Equation (4-4), is referred to as the *equation of fluid motion*, because it can be solved for the velocity distribution, u, in a thin fluid film of a hydrodynamic bearing.

Comment. In fact, it is shown in Chapter 5 that the complete equation for the shear stress is $\tau = \mu(du/dy + dv/dx)$. However, according to our assumptions, the second term is very small and is neglected in this derivation.

4.5 FLOW IN A LONG BEARING

The following simple solution is limited to a fluid film of steady geometry. It means that the geometry of the fluid film does not vary with time relative to the coordinate system, and it does not apply to time-dependent fluid film geometry such as a bearing under dynamic load. A more universal approach is possible by using the Reynolds equation (see Chapter 6). The Reynolds equation applies to all fluid films, including time-dependent fluid film geometry.

Example Problems 4-1

Journal Bearing

In Fig. 4-1, a journal bearing is shown in which the bearing is stationary and the journal turns around a stationary center. Derive the equations for the fluid velocity and pressure gradient.

The variable film thickness is due to the journal eccentricity. In hydrodynamic bearings, $h = h(x)$ is the variable film thickness around the bearing. The coordinate system is attached to the stationary bearing, and the journal surface has a constant velocity, $U = \omega R$, in the x direction.

Solution

The coordinate x is along the bearing surface curvature. According to the assumptions, the curvature is disregarded and the flow is solved as if the boundaries were a straight line.

Equation (4-4) can be solved for the velocity distribution, $u = u(x, y)$. Following the assumptions, variations of the pressure in the y direction are negligible (Assumption 6), and the pressure is taken as a constant across the film thickness because the fluid film is thin. Therefore, in two-dimensional flow of a long bearing, the pressure is a function of x only. In order to simplify the solution for the velocity, u, the following substitution is made in Eq. (4-4):

$$2m(x) = \frac{1}{\mu}\frac{dp}{dx} \tag{4-5}$$

where $m(x)$ is an unknown function of x that must be solved in order to find the pressure distribution. Equation 4-4 becomes

$$\frac{\partial u^2}{\partial y^2} = 2m(x) \tag{4-6}$$

Integrating Eq. (4-6) twice yields the following expression for the velocity distribution, u, across the fluid film (n and k are integration constants):

$$u = my^2 + ny + k \tag{4-7}$$

Here, m, n, and k are three unknowns that are functions of x only. Three equations are required to solve for these three unknowns. Two equations are obtained from the two boundary conditions of the flow at the solid surfaces, and the third equation is derived from the continuity condition, which is equivalent to the conservation of mass of the fluid (or conservation of volume for incompressible flow).

The fluid adheres to the solid wall (no slip condition), and the fluid velocity at the boundaries is equal to that of the solid surface. In a journal bearing having a stationary bearing and a rotating journal at surface speed $U = \omega R$ (see Fig. 4-1), the boundary conditions are

$$\begin{aligned} &\text{at } y = 0: && u = 0 \\ &\text{at } y = h(x): && u = U\cos\alpha \approx U \end{aligned} \tag{4-8}$$

The slope between the tangential velocity U and the x direction is very small; therefore, $\cos\alpha \approx 1$, and we can assume that at $y = h(x)$, $u \approx U$.

The third equation, which is required for the three unknowns, m, n, and k, is obtained from considerations of conservation of mass. For an infinitely long bearing, there is no flow in the axial direction, z; therefore, the amount of mass flow through each cross section of the fluid film is constant (the cross-sectional plane is normal to the x direction). Since the fluid is incompressible, the volume flow rate is also constant at any cross section. The constant-volume flow rate, q, per unit of bearing length is obtained by integration of the velocity component, u, along the film thickness, as follows:

$$q = \int_0^h u\, dy = \text{constant} \tag{4-9}$$

Equation (4-9) is applicable only for a steady fluid film geometry that does not vary with time.

The pressure wave around the journal bearing is shown in Fig. 1-3. At the peak of the pressure wave, $dp/dx = 0$, and the velocity distribution, $u = u(y)$, at that point is linear according to Eq. (4-4). The linear velocity distribution in a simple shear flow (in the absence of pressure gradient) is shown in Fig. 4-3. If the film thickness at the peak pressure point is $h = h_0$, the flow rate, q, per unit length is equal to the area of the velocity distribution triangle:

$$q = \frac{Uh_0}{2} \tag{4-10}$$

The two boundary conditions of the velocity as well as the conservation of mass condition form the following three equations, which can be solved for m, n and k:

$$\begin{aligned} 0 &= m0^2 + n0 + k \Rightarrow k = 0 \\ U &= mh^2 + nh \\ \frac{Uh_0}{2} &= \int_0^h (my^2 + ny)\, dy \end{aligned} \tag{4-11}$$

After solving for m, n, and k and substituting these values into Eq. (4-7), the following equation for the velocity distribution is obtained:

$$u = 3U\left(\frac{1}{h^2} - \frac{h_0}{h^3}\right)y^2 + U\left(\frac{3h_0}{h^2} - \frac{2}{h}\right)y \tag{4-12}$$

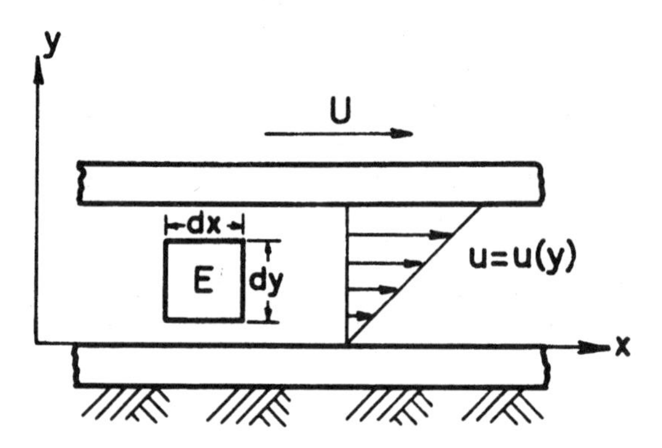

FIG. 4-3 Linear velocity distribution for a simple shear flow (no pressure gradient).

From the value of m, the expression for the pressure gradient, dp/dx, is solved [see Eq. (4-5)]:

$$\frac{dp}{dx} = 6U\mu\frac{h - h_0}{h^3} \tag{4-13}$$

Equation (4-13) still contains an unknown constant, h_0, which is the film thickness at the peak pressure point. This will be solved from additional information about the pressure wave. Equation (4-13) can be integrated for the pressure wave.

Example Problem 4-2

Inclined Plane Slider

As discussed earlier, a steady-fluid-film geometry (relative to the coordinates) must be selected for a simple derivation of the pressure gradient. The second example is of an inclined plane-slider having a configuration as shown in Fig. 4-4. This example is of a converging viscous wedge similar to that of a journal bearing; however, the lower part is moving in the x direction and the upper plane is stationary while the coordinates are stationary. This bearing configuration is selected because the geometry of the clearance (and fluid film) does not vary with time relative to the coordinate system.

Find the velocity distribution and the equation for the pressure gradient in the inclined plane-slider shown in Fig 4-4.

Solution

In this case, the boundary conditions are:

$$\begin{aligned} &\text{at } y = 0: && u = U \\ &\text{at } y = h(x): && u = 0 \end{aligned}$$

In this example, the lower boundary is moving and the upper part is stationary. The coordinates are stationary, and the geometry of the fluid film does not vary

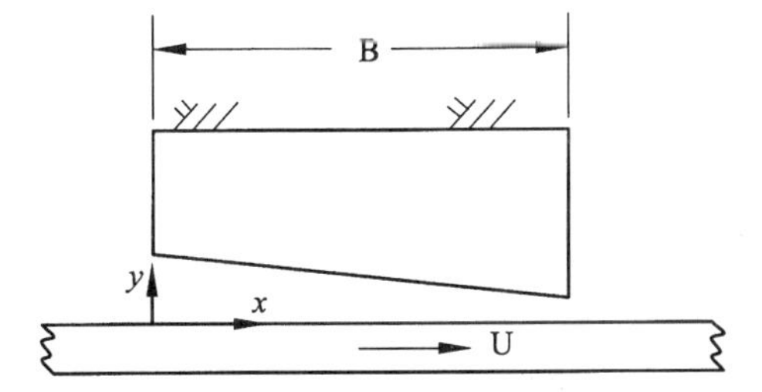

FIG. 4-4 Inclined plane-slider (converging flow in the x direction).

relative to the coordinates. In this case, the flow rate is constant, in a similar way to that of a journal bearing. This flow rate is equal to the area of the velocity distribution triangle at the point of peak pressure, where the clearance thickness is h_0. The equation for the constant flow rate is

$$q = \int_0^h u \, dy = \frac{h_0 U}{2}$$

The two boundary conditions of the velocity and the constant flow-rate condition form the three equations for solving for m, n, and k:

$$U = m0^2 + n0 + k \quad \Rightarrow \quad k = U$$

$$0 = mh^2 + nh + U$$

$$\frac{Uh_0}{2} = \int_0^h (my^2 + ny + U)dy$$

After solving for m, n, and k and substituting these values into Eq. (4-7), the following equation for the velocity distribution is obtained:

$$u = 3U\left(\frac{1}{h^2} - \frac{h_0}{h^3}\right)y^2 + U\left(\frac{3h_0}{h^2} - \frac{4}{h}\right)y + U$$

From the value of m, an identical expression to Eq. (4-13) for the pressure gradient, dp/dx, is obtained for $\partial h/\partial x < 0$ (a converging slope in the x direction):

$$\frac{dp}{dx} = 6U\mu\frac{h - h_0}{h^3} \qquad \text{for } \frac{\partial h}{\partial x} < 0 \text{ (negative slope)}$$

This equation applies to a converging wedge where the coordinate x is in the direction of a converging clearance. It means that the clearance reduces along x as shown in Fig. 4-4.

In a converging clearance near $x = 0$, the clearance slope is negative, $\partial h/\partial x < 0$. This means that the pressure increases near $x = 0$. At that point, $h > h_0$, resulting in $dp/dx > 0$.

If we reverse the direction of the coordinate x, the expression for the pressure gradient would have an opposite sign:

$$\frac{dp}{dx} = 6U\mu\frac{h_0 - h}{h^3} \qquad \text{for } \frac{\partial h}{\partial x} > 0 \text{ (positive slope)}$$

This equation applies to a plane-slider, as shown in Fig 4-5, where the coordinate x is in the direction of increasing clearance. The unknown constant, h_0, will be determined from the boundary conditions of the pressure wave.

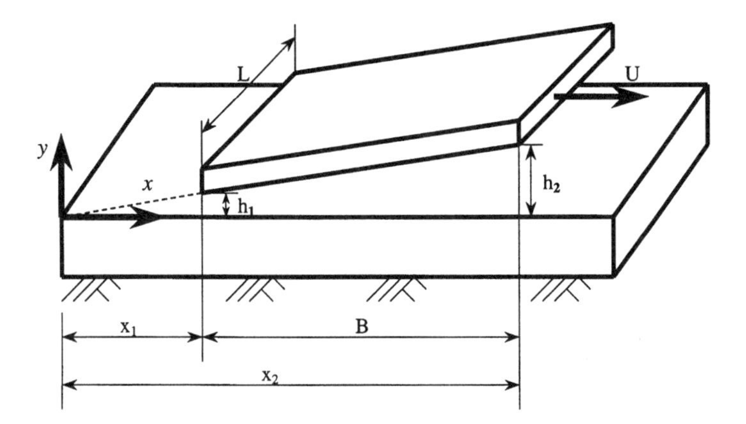

FIG. 4-5 Inclined plane-slider (x coordinate in the direction of a diverging clearance).

4.6 PRESSURE WAVE

4.6.1 Journal Bearing

The pressure wave along the x direction is solved by integration of Eq. (4-13). The two unknowns, h_0, and the integration constant are solved from the two boundary conditions of the pressure wave. In a plane-slider, the locations at the start and end of the pressure wave are used as pressure boundary conditions. These locations are not obvious when the clearance is converging and diverging, such as in journal bearings, and other boundary conditions of the pressure wave are used for solving h_0. Integrating the pressure gradient, Eq. (4-13), results in the following equation for a journal bearing:

$$p = 6\mu U \int_0^x \frac{h - h_0}{h^3}\, dx + p_0 \tag{4-14a}$$

Here, the pressure p_0 represents the pressure at $x = 0$. In a journal bearing, the lubricant is often fed into the clearance through a hole in the bearing at $x = 0$. In that case, p_0 is the supply pressure.

4.6.2 Plane-Slider

In the case of an inclined slider, p_0 is the atmospheric pressure. Pressure is commonly measured with reference to atmospheric pressure (gauge pressure), resulting in $p_0 = 0$ for an inclined slider.

The pressure wave, $p(x)$, can be solved for any bearing geometry, as long as the film thickness, $h = h(x)$, is known. The pressure wave can be solved by analytical or numerical integration. The analytical integration of complex functions has been a challenge in the past. However, the use of computers makes numerical integration a relatively easy task.

An inclined plane slider is shown in Fig. 4-5, where the inclination angle is α. The fluid film is equivalent to that in Fig. 4-4, although the x is in the opposite direction, $\partial h/\partial x > 0$, and the pressure wave is

$$p = 6\mu U \int_{x_1}^{x} \frac{h_0 - h}{h^3}\, dx \tag{4-14b}$$

In order to have concise equations, the slope of the plane-slider is substituted by $a = \tan\alpha$, and the variable film thickness is given by the function

$$h(x) = ax \tag{4-15}$$

Here, x is measured from the point of intersection of the plane-slider and the bearing surface. The minimum and maximum film thicknesses are h_1 and h_2, respectively, as shown in Fig. 4-5.

In order to solve the pressure distribution in any converging fluid film, Eq. (4-14) is integrated after substituting the value of h according to Eq. (4-15). After integration, there are two unknowns: the constant h_0 in Eq. (4-10) and the constant of integration, p_o. The two unknown constants are solved for the two boundary conditions of the pressure wave. At each end of the inclined plane, the pressure is equal to the ambient (atmospheric) pressure, $p = 0$. The boundary conditions are:

$$\begin{aligned} &\text{at } h = h_1: \quad p = 0 \\ &\text{at } h = h_2: \quad p = 0 \end{aligned} \tag{4-16}$$

The solution can be analytically performed in closed form or by numerical integration (see Appendix B). The numerical integration involves iterations to find h_0. Hydrodynamic lubrication equations require frequent use of computer programming to perform the trial-and-error iterations. An example of a numerical integration is shown in Example Problem 4-4.

Analytical Solution. For an infinitely long plane-slider, $L \gg B$, analytical integration results in the following pressure wave along the x direction (between $x = h_1/a$ and $x = h_2/a$):

$$p = \frac{6\mu U}{a^3} \frac{(h_1 - ax)(ax - h_2)}{(h_1 + h_2)x^2} \tag{4-17}$$

At the boundaries $h = h_1$ and $h = h_2$, the pressure is zero (atmospheric pressure).

4.7 PLANE-SLIDER LOAD CAPACITY

Once the pressure wave is solved, it is possible to integrate it again to solve for the bearing load capacity, W. For a plane-slider, the integration for the load capacity is according to the following equation:

$$W = L\int_{x_1}^{x_2} p\,dx \tag{4-18}$$

The foregoing integration of the pressure wave can be derived analytically, in closed form. However, in many cases, the derivation of an analytical solution is too complex, and a computer program can perform a numerical integration. It is beneficial for the reader to solve this problem numerically, and writing a small computer program for this purpose is recommended.

An analytical solution for the load capacity is obtained by substituting the pressure in Eq. (4-17) into Eq. (4-18) and integrating in the boundaries between $x_1 = h_1/a$ and $x_2 = h_2/a$. The final analytical expression for the load capacity in a plane-slider is as follows:

$$W = \frac{6\mu ULB^2}{h_2^2}\left(\frac{1}{\beta - 1}\right)^2\left[\ln\beta - \frac{2(\beta - 1)}{\beta + 1}\right] \tag{4-19}$$

where β is the ratio of the maximum and minimum film thickness, h_2/h_1. A similar derivation can be followed for nonflat sliders, such as in the case of a slider having a parabolic surface in Problem 4-2, at the end of this chapter.

4.8 VISCOUS FRICTION FORCE IN A PLANE-SLIDER

The friction force, F_f, is obtained by integrating the shear stress, τ over any cross-sectional area along the fluid film. For convenience, a cross section is selected along the bearing stationary wall, $y = 0$. The shear stress at the wall, τ_w, at $y = 0$ can be obtained via the following equation:

$$\tau_w = \mu\frac{du}{dy}\Big|_{(y=0)} \tag{4-20}$$

The velocity distribution can be substituted from Eq. (4-12), and after differentiation of the velocity function according to Eq. (4-20), the shear stress at the wall, $y = 0$, is given by

$$\tau_w = \mu U\left(\frac{3h_0}{h^2} - \frac{2}{h}\right) \tag{4-21}$$

The friction force, F_f, for a long plane-slider is obtained by integration of the shear stress, as follows:

$$F_f = L \int_{x_1}^{x_2} \tau \, dx \tag{4-22}$$

4.8.1 Friction Coefficient

The bearing friction coefficient, f, is defined as the ratio of the friction force to the bearing load capacity:

$$f = \frac{F_f}{W} \tag{4-23}$$

An important objective of a bearing design is to minimize the friction coefficient. The friction coefficient is usually lower with a thinner minimum film thickness, h_n (in a plane-slider, $h_n = h_1$). However, if the minimum film thickness is too low, it involves the risk of severe wear between the surfaces. Therefore, the design involves a compromise between a low-friction requirement and a risk of severe wear. Determination of the desired minimum film thickness, h_n, requires careful consideration. It depends on the surface finish of the sliding surfaces and the level of vibrations and disturbances in the machine. This part of the design process is discussed in the following chapters.

4.9 FLOW BETWEEN TWO PARALLEL PLATES

Example Problem 4-3

Derive the equation of the pressure gradient in a unidirectional flow inside a thin clearance between two stationary parallel plates as shown in Fig. 4-6. The flow is parallel, in the x direction only. The constant clearance between the plates is h_0, and the rate of flow is Q, and the x axis is along the center of the clearance.

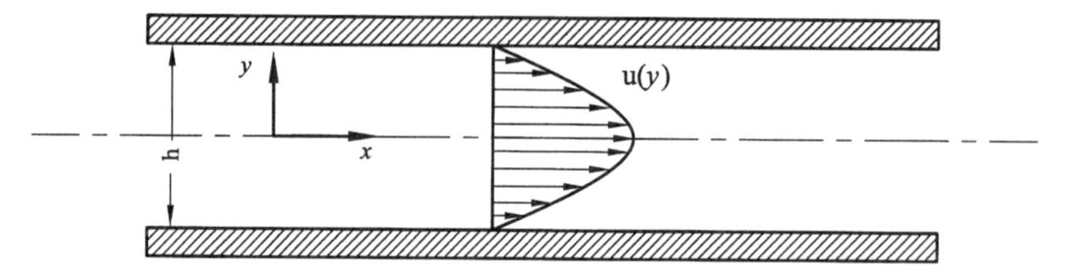

FIG. 4-6 Flow between two parallel plates.

Solution

In a similar way to the solution for hydrodynamic bearing, the parallel flow in the x direction is derived from Eq. (4-4), repeated here as Eq. (4-24):

$$\frac{dp}{dx} = \mu \frac{\partial^2 u}{\partial y^2} \tag{4-24}$$

This equation can be rewritten as

$$\frac{\partial u^2}{\partial y^2} = \frac{1}{\mu}\frac{dp}{dx} \tag{4-25}$$

The velocity profile is solved by a double integration. Integrating Eq. (4-25) twice yields the expression for the velocity u:

$$u = \frac{1}{2\mu}\frac{dp}{dx} y^2 + ny + k \tag{4-26}$$

Here, n and k are integration constants obtained from the two boundary conditions of the flow at the solid surfaces (no-slip condition).

The boundary conditions at the wall of the two plates are:

$$\text{at } y = \pm \frac{h_0}{2}: \quad u = 0 \tag{4-27}$$

The flow is symmetrical, and the solution for n and k is

$$n = 0, \qquad k = -m\left(\frac{h_0}{2}\right)^2 \tag{4-28}$$

The flow equation becomes

$$u = \frac{1}{2\mu}\frac{dp}{dx}\left(y^2 - \frac{h^2}{4}\right) \tag{4-29}$$

The parabolic velocity distribution is shown in Fig. 4-6. The pressure gradient is obtained from the conservation of mass. For a parallel flow, there is no flow in the z direction. For convenience, the y coordinate is measured from the center of the clearance. The constant-volume flow rate, Q, is obtained by integrating the velocity component, u, along the film thickness, as follows:

$$Q = 2L \int_0^{h/2} u \, dy \tag{4-30}$$

Here, L is the width of the parallel plates, in the direction normal to the flow (in the z direction). Substituting the flow in Eq. (4-29) into Eq. (4-30) and integrating yields the expression for the pressure gradient as a function of flow rate, Q:

$$\frac{dp}{dx} = -\frac{12\mu}{bh_0^3}Q \tag{4-31}$$

This equation is useful for the hydrostatic bearing calculations in Chapter 10. The negative sign means that a negative pressure slope in the x direction is required for a flow in the same direction.

4.10 FLUID FILM BETWEEN A CYLINDER AND A FLAT PLATE

There are important applications of a full fluid film at the rolling contact of a cylinder and a flat plate and at the contact of two parallel cylinders. Examples are cylindrical rolling bearings, cams, and gears. A very thin fluid film that separates the surfaces is shown in Fig. 4-7. In this example, the cylinder is stationary and the flat plate has a velocity U in the x direction. In Chapter 6, this problem is extended to include rolling motion of the cylinder over the plate.

The problem of a cylinder and a flat plate is a special case of the general problem of contact between two parallel cylinders. By using the concept of equivalent radius (see Chapter 12), the equations for a cylinder and a flat plate can be extended to that for two parallel cylinders.

Fluid films at the contacts of rolling-element bearings and gear teeth are referred to as *elastohydrodynamic* (EHD) films. The complete analysis of a fluid film in actual rolling-element bearings and gear teeth is quite complex. Under load, the high contact pressure results in a significant elastic deformation of the contact surfaces as well as a rise of viscosity with pressure (see Chapter 12).

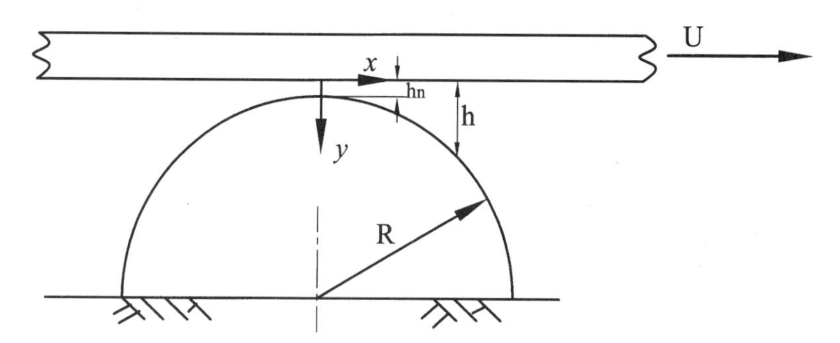

FIG. 4.7 Fluid film between a cylinder and a flat-plate.

However, the following problem is for a light load where the solid surfaces are assumed to be rigid and the viscosity is constant.

The following problem considers a plate and a cylinder with a minimum clearance, h_{min}. In it we consider the case of a light load, where the elastic deformation is very small and can be disregarded (cylinder and plate are assumed to be rigid). In addition, the values of maximum and minimum pressures are sufficiently low, and there is no fluid cavitation. The viscosity is assumed to be constant. The cylinder is stationary, and the flat plate has a velocity U in the x direction as shown in Fig. 4-7.

The cylinder is long in comparison to the film length, and the long-bearing analysis can be applied.

4.10.1 Film Thickness

The film thickness in the clearance between a flat plate and a cylinder is given by

$$h(\theta) = h_{min} + R(1 - \cos\theta) \tag{4-32}$$

where θ is a cylinder angle measured from the minimum film thickness at $x = 0$.

Since the minimum clearance, h_{min}, is very small (relative to the cylinder radius), the pressure is generated only at a very small region close to the minimum film thickness, where $x \ll R$, or $x/R \ll 1$.

For a small ratio of x/R, the equation of the clearance, h, can be approximated by a parabolic equation. The following expression is obtained by expanding Eq. (4-32) for h into a Taylor series and truncating terms that include powers higher than $(x/R)^2$. In this way, the expression for the film thickness h can be approximated by

$$h(x) = h_{min} + \frac{x^2}{2R} \tag{4-33a}$$

4.10.2 Pressure Wave

The pressure wave can be derived from the expression for the pressure gradient, dp/dx, in Eq. (4-13). The equation is

$$\frac{dp}{dx} = 6\mu U \frac{h_0 - h}{h^3}$$

The unknown h_0 can be replaced by the unknown x_0 according to the equation

$$h_0(x) = h_{min} + \frac{x_0^2}{2R} \tag{4-33b}$$

After substituting the value of h according to Eqs. (4-33), the solution for the pressure wave can be obtained by the following integration:

$$p(x) = 24\mu UR^2 \int_{-\infty}^{x} \frac{x_0^2 - x^2}{(2Rh_{\min} + x^2)^3}\, dx \tag{4-34}$$

The unknown x_0 is solved by the following boundary conditions of the pressure wave:

$$\begin{aligned} &\text{at } x = -\infty: \quad p = 0 \\ &\text{at } x = \infty: \quad p = 0 \end{aligned} \tag{4-35}$$

For numerical integration, the infinity can be replaced by a relatively large finite value, where pressure is very small and can be disregarded.

Remark: The result is an antisymmetrical pressure wave (on the two sides of the minimum film thickness), and there will be no resultant load capacity. In actual cases, the pressures are high, and there is a cavitation at the diverging side. A solution that considers the cavitation with realistic boundary conditions is presented in Chapter 6.

4.11 SOLUTION IN DIMENSIONLESS TERMS

If we perform a numerical integration of Eq. (4-34) for solving the pressure wave, the solution would be limited to a specific bearing geometry of cylinder radius R and minimum clearance $h_{\min}$. The numerical integration must be repeated for a different bearing geometry.

For a universal solution, there is obvious merit to performing a solution in dimensionless terms. For conversion to dimensionless terms, we normalize the x coordinate by dividing it by $\sqrt{2Rh_{\min}}$ and define a dimensionless coordinate as

$$\bar{x} = \frac{x}{\sqrt{2Rh_{\min}}} \tag{4-36}$$

In addition, a dimensionless clearance ratio is defined:

$$\bar{h} = \frac{h}{h_{\min}} \tag{4-37}$$

The equation for the variable clearance ratio as a function of the dimensionless coordinate becomes

$$\bar{h} = 1 + \bar{x}^2 \tag{4-38}$$

Let us recall that the unknown h_0 is the fluid film thickness at the point of peak pressure. It is often convenient to substitute it by the location of the peak pressure, x_0, and the dimensionless relation then is

$$\bar{h}_0 = 1 + \bar{x}_0^2 \tag{4-39}$$

In addition, if the dimensionless pressure is defined as

$$\bar{p} = \frac{h_{\min}^2}{\sqrt{2Rh_{\min}}} \frac{1}{6\mu U} p \tag{4-40}$$

then the following integration gives the dimensionless pressure wave:

$$\bar{p} = \int d\bar{p} = \int_{-\infty}^{\bar{x}} \frac{\bar{x}^2 - \bar{x}_0^2}{(1 + \bar{x}^2)^3} \, d\bar{x} \tag{4-41}$$

For a numerical integration of the pressure wave according to Eq. (4-41), the boundary $\bar{x} = -\infty$ is replaced by a relatively large finite dimensionless value, where pressure is small and can be disregarded, such as $\bar{x} = -4$. An example of numerical integration is given in Example Problem 4.

In a similar way, for a numerical solution of the unknown x_0 and the load capacity, it is possible to replace infinity with a finite number, for example, the following mathematical boundary conditions of the pressure wave of a full fluid film between a cylinder and a plane:

$$\begin{aligned} &\text{at } \bar{x} = -\infty: \quad \bar{p} = 0 \\ &\text{at } \bar{x} = \infty: \quad \bar{p} = 0 \end{aligned} \tag{4-42a}$$

These conditions are replaced by practical numerical boundary conditions:

$$\begin{aligned} &\text{at } \bar{x} = -4: \quad \bar{p} = 0 \\ &\text{at } \bar{x} = +4: \quad \bar{p} = 0 \end{aligned} \tag{4-42b}$$

These practical numerical boundary conditions do not introduce a significant error because a significant hydrodynamic pressure is developed only near the minimum fluid film thickness, at $\bar{x} = 0$.

Example Problem 4-4

Ice Sled

An ice sled is shown in Fig. 4-8. On the left-hand side, there is a converging clearance that is formed by the geometry of a quarter of a cylinder. A flat plate continues the curved cylindrical shape. The flat part of the sled is parallel to the flat ice. It is running parallel over the flat ice on a thin layer of water film of a constant thickness h_0.

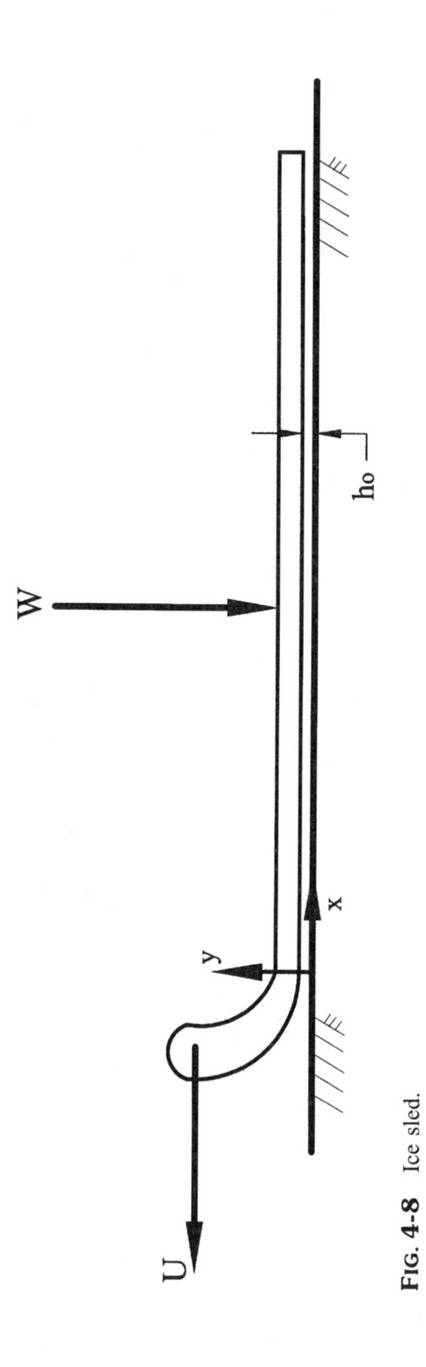

Fig. 4-8 Ice sled.

Derive and plot the pressure wave at the entrance region of the fluid film and under the flat ice sled as it runs over the ice at velocity U. Derive the equation of the sled load capacity.

Solution

The long-bearing approximation is assumed for the sled similar to the converging slope in Fig. 4-4. The fluid film equation for an infinitely long bearing is

$$\frac{dp}{dx} = 6\mu U \frac{h - h_0}{h^3}$$

In the parallel region, $h = h_0$, and it follows from the foregoing equation that

$$\frac{dp}{dx} = 0 \qquad \text{(along the parallel region of constant clearance)}$$

In this case, $h_0 = h_{min}$, and the equation for the variable clearance at the converging region is

$$h(x) = h_0 + \frac{x^2}{2R}$$

This means that for a wide sled, the pressure is constant within the parallel region. This is correct only if $L \gg B$, where L is the bearing width (in the z direction) and B is along the sled in the x direction.

Fluid flow in the converging region generates the pressure, which is ultimately responsible for supporting the load of the sled. Through the adhesive force of viscous shear, the fluid is dragged into the converging clearance, creating the pressure in the parallel region.

Substituting $h(x)$ into the pressure gradient equation yields

$$\frac{dp}{dx} = 6\mu U \frac{\left(h_0 + \dfrac{x^2}{2R}\right) - h_0}{\left(h_0 + \dfrac{x^2}{2R}\right)^3}$$

or

$$\frac{dp}{dx} = 24\mu U R^2 \frac{x^2}{(2Rh_0 + x^2)^3}$$

Applying the limits of integration and the boundary condition, we get the following for the pressure distribution:

$$p(x) = 24\mu U R^2 \int_{-\infty}^{x} \frac{x^2}{(2Rh_0 + x^2)^3}\, dx$$

This equation can be integrated analytically or numerically. For numerical integration, since a significant pressure is generated only at a low x value, the infinity boundary of integration is replaced by a finite magnitude.

Conversion to a Dimensionless Equation. As discussed earlier, there is an advantage in solving the pressure distribution in dimensionless terms. A regular pressure distribution curve is limited to the specific bearing data of given radius R and clearance h_0. The advantage of a dimensionless curve is that it is universal and applies to any bearing data. For conversion of the pressure gradient to dimensionless terms, we normalize x by dividing by $\sqrt{2Rh_0}$ and define dimensionless terms as follows:

$$\bar{x} = \frac{x}{\sqrt{2Rh_0}}, \qquad \bar{h} = \frac{h}{h_0}, \qquad \bar{h} = 1 + \bar{x}^2$$

Converting to dimensionless terms, the pressure gradient equation gets the form

$$\frac{dp}{dx} = \frac{6\mu U}{h_0^2} \frac{\bar{h} - 1}{(1 + \bar{x}^2)^3}$$

Here, $h_0 = h_{min}$ is the minimum film thickness at $x = 0$. Substituting for the dimensionless clearance and rearranging yields

$$\frac{h_0^2}{\sqrt{2Rh_0} 6\mu U} dp = \frac{\bar{x}^2}{(1 + \bar{x}^2)^3} dx$$

The left-hand side of this equation is defined as the dimensionless pressure. Dimensionless pressure is equal to the following integral:

$$\bar{p} = \frac{h_0^2}{\sqrt{2Rh_0}} \frac{1}{6\mu U} \int_0^p dp = \int_{-\infty}^{x} \frac{\bar{x}^2}{(1 + \bar{x}^2)^3} d\bar{x}$$

Numerical Integration. The dimensionless pressure is solved by an analytical or numerical integration of the preceding function within the specified boundaries (see Appendix B). The pressure p_0 under the flat plate is obtained by integration to the limit $x = 0$:

$$\bar{p} = \frac{h_0^2}{\sqrt{2Rh_0}} \frac{1}{6\mu U} p_0 = \sum_{-3}^{x} \frac{\bar{x}_i^2}{(1 + \bar{x}_i^2)^3} \Delta x_i$$

The pressure is significant only near $x = 0$. Therefore, for the numerical integration, a finite number replaces infinity. The resulting pressure distribution is shown in Fig. 4-9.

Comparison with Analytical Integration. The maximum pressure at $x = 0$ as well as along the constant clearance, $x > 0$, can also be solved by analytical

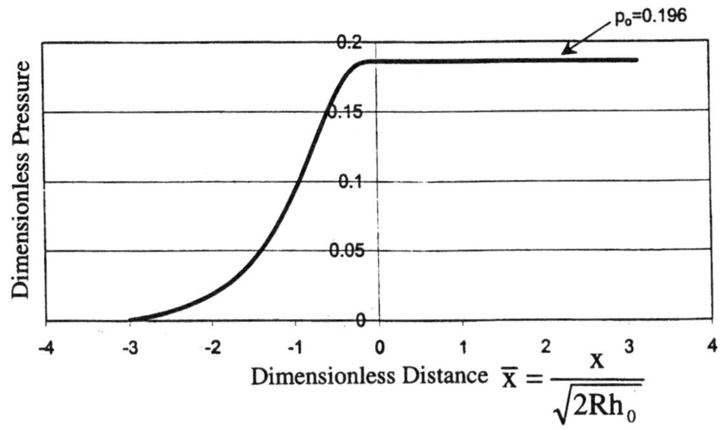

FIG. 4-9 Dimensionless pressure wave.

integration of the following equation:

$$\bar{p}_0 = \int_{-\infty}^{0} \frac{\bar{x}^2}{(1+\bar{x}^2)^3}\, d\bar{x}$$

Using integration tables, the following integral solution is obtained:

$$\bar{p} = -\frac{x}{4(1+x^2)^2} + \frac{x}{8(1+x^2)} + \frac{\arctan(x)}{8}\bigg|_{\infty}^{0} = \frac{\pi}{2 \times 8} = 0.196$$

This result is equal to that obtained by a numerical integration.

Load Capacity. The first step is to find the pressure p_0 from the dimensionless pressure wave, which is equal to 0.196. The converging entrance area is small in comparison to the area under the flat plate. Neglecting the pressure in the entrance region, the equation for the load capacity, W, becomes

$$W = p_0 BL$$

Here, B and L are the dimensions of the flat-plate area.

Calculation of Film Thickness. When the load capacity of the sled W is known, it is possible to solve for the film thickness, $h_0 = h_{\min}$. Substituting in the equation, $W = p_0 BL$, the equation of the constant pressure p_0 under the flat plate as a function of the clearance h_0 can allow us to solve for the constant film thickness. The constant pressure p_0 is derived from its dimensionless counterpart:

$$\bar{p}_0 = \frac{h_0^2}{\sqrt{2Rh_0}} \frac{1}{6\mu U} p_0 = 0.196$$

If the load, cylinder radius, water viscosity, and sled velocity are known, it is possible to solve for the film thickness, h_0.

Example Problem 4-5

Derive the equation for the pressure gradient of a journal bearing if the journal and bearing are both rotating around their stationary centers. The surface velocity of the bearing bore is $U_j, = \omega_j R$, and the surface velocity of the journal is $U_b = \omega_b R_1$.

Solution

Starting from Eq. (4-4):

$$\frac{dp}{dx} = \mu \frac{\partial^2 u}{\partial y^2}$$

and integrating twice (in a similar way to a stationary bearing) yields

$$u = my^2 + ny + k$$

However, the boundary conditions and continuity conditions are as follows:

$$\text{at } y = 0: \quad u = \omega_b R_1$$
$$\text{at } y = h(x): \quad u = \omega_j R$$

In this case, the constant-volume flow rate, q, per unit of bearing length at the point of peak pressure is

$$q = \int_0^h u \, dy = \frac{(\omega_b R_1 + \omega_j R)h_0}{2}$$

Solving for m, n and k and substituting in a similar way to the previous problem while also assuming $R_1 \approx R$, the following equation for the pressure gradient is obtained:

$$\frac{dp}{dx} = 6R(\omega_b + \omega_j)\mu \frac{h - h_0}{h^3}$$

Problems

4-1 A long plane-slider, $L = 200$ mm and $B = 100$ mm, is sliding at velocity of 0.3 m/s. The minimum film thickness is $h_1 = 0.005$ mm and the maximum film thickness is $h_2 = 0.010$ mm. The fluid is SAE 30, and the operating temperature of the lubricant is assumed a constant 30°C.

a. Assume the equation for an infinitely long bearing, and use numerical integration to solve for the pressure wave (use trial and error to solve for x_0). Plot a curve of the pressure distribution $p = p(x)$.

b. Use numerical integration to find the load capacity. Compare this to the load capacity obtained from Eq. (4-19).

c. Find the friction force and the friction coefficient.

4-2 A slider is machined to have a parabolic surface. The slider has a horizontal velocity of 0.3 m/s. The minimum film $h_{min} = 0.020$ mm, and the clearance varies with x according to the following equation:

$$h = 0.020 + 0.01x^2$$

The slider velocity is $U = 0.5$ m/s. The length $L = 300$ mm and the width in the sliding direction $B = 100$ mm. The lubricant is SAE 40 and the temperature is assumed constant, $T = 40°$C. Assume the equation for an infinitely long bearing.

a. Use numerical integration and plot the dimensionless pressure distribution, $p = p(x)$.

b. Use numerical integration to find the load capacity.

c. Find the friction force and the friction coefficient.

4-3 A blade of a sled has the geometry shown in the Figure 4-8. The sled is running over ice on a thin layer of water film. The total load (weight of the sled and person) is 1500 N. The sled velocity is 20 km/h, the radius of the inlet curvature is 30 cm, and the sled length $B = 30$ cm, and width is $L = 100$ cm. The viscosity of water is $\mu = 1.792 \times 10^{-3}$ N-s/m^2.

Find the film thickness ($h_0 = h_{min}$) of the thin water layer shown in Fig. 4-8.

Direction: The clearance between the plate and disk is $h = h_{min} + x^2/2R$, and assume that $p = 0$ at $x = R$.

5

Basic Hydrodynamic Equations

5.1 NAVIER–STOKES EQUATIONS

The pressure distribution and load capacity of a hydrodynamic bearing are analyzed and solved by using classical fluid dynamics equations. In a thin fluid film, the viscosity is the most important fluid property determining the magnitude of the pressure wave, while the effect of the fluid inertia (ma) is relatively small and negligible. Reynolds (1894) introduced classical hydrodynamic lubrication theory. Although a lot of subsequent research has been devoted to this discipline, Reynolds' equation still forms the basis of most analytical research in hydrodynamic lubrication. The Reynolds equation can be derived from the Navier–Stokes equations, which are the fundamental equations of fluid motion.

The derivation of the Navier–Stokes equations is based on several assumptions, which are included in the list of assumptions (Sec. 4.2) that forms the basis of the theory of hydrodynamic lubrication. An important assumption for the derivation of the Navier–Stokes equations is that there is a linear relationship between the respective components of stress and strain rate in the fluid.

In the general case of three-dimensional flow, there are nine stress components referred to as components of the *stress tensor*. The directions of the stress components are shown in Fig 5-1.

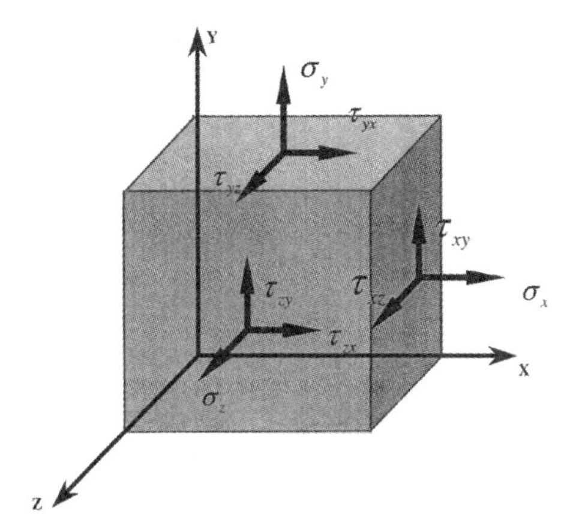

FIG. 5-1 Stress components acting on a rectangular fluid element.

The stress components σ_x, σ_y, σ_z are of tension or compression (if the sign is negative), as shown in Fig. 5-1. However, the mixed components τ_{xy}, τ_{zy}, τ_{xz} are shear stresses parallel to the surfaces.

It is possible to show by equilibrium considerations that the shear components are symmetrical:

$$\tau_{xy} = \tau_{yx}, \qquad \tau_{yz} = \tau_{zy}, \qquad \tau_{xz} = \tau_{zx} \tag{5-1}$$

Due to symmetry, the number of stress components is reduced from nine to six.

In rectangular coordinates the six stress components are

$$\begin{aligned}
\sigma_x &= -p + 2\mu \frac{\partial u}{\partial x} \\
\sigma_y &= -p + 2\mu \frac{\partial v}{\partial y} \\
\sigma_z &= -p + 2\mu \frac{\partial w}{\partial z} \\
\tau_{xy} &= \tau_{yx} = \mu\left(\frac{\partial v}{\partial x} + \frac{\partial u}{\partial y}\right) \\
\tau_{yz} &= \tau_{zy} = \mu\left(\frac{\partial w}{\partial y} + \frac{\partial v}{\partial z}\right) \\
\tau_{zx} &= \tau_{xz} = \mu\left(\frac{\partial u}{\partial z} + \frac{\partial w}{\partial x}\right)
\end{aligned} \tag{5-2}$$

A fluid that can be described by Eq. (5-2) is referred to as *Newtonian fluid*. This equation is based on the assumption of a linear relationship between the stress and strain-rate components. For most lubricants, such a linear relationship is an adequate approximation. However, under extreme conditions, e.g., very high pressure of point or line contacts, this assumption is no longer valid. An assumption that is made for convenience is that the viscosity, μ, of the lubricant is constant. Also, lubrication oils are practically *incompressible*, and this property simplifies the Navier–Stokes equations because the density, ρ, can be assumed to be constant. However, this assumption cannot be applied to air bearings.

Comment. As mentioned earlier, in thin films the velocity component v is small in comparison to u and w, and two shear components can be approximated as follows:

$$
\begin{aligned}
\tau_{xy} &= \tau_{yx} = \mu\left(\frac{\partial v}{\partial x} + \frac{\partial u}{\partial y}\right) \approx \mu\frac{\partial u}{\partial y} \\
\tau_{yz} &= \tau_{zy} = \mu\left(\frac{\partial w}{\partial y} + \frac{\partial v}{\partial z}\right) \approx \mu\frac{\partial w}{\partial y}
\end{aligned}
\tag{5-3}
$$

The Navier–Stokes equations are based on the balance of forces acting on a small, infinitesimal fluid element having the shape of a rectangular parallelogram with dimensions dx, dy, and dz, as shown in Fig. 5-1. The force balance is similar to that in Fig. 4-1; however, the general balance of forces is of three dimensions, in the x, y and z directions. The surface forces are the product of stresses, or pressures, and the corresponding areas.

When the fluid is at rest there is a uniform hydrostatic pressure. However, when there is fluid motion, there are deviatoric normal stresses σ'_x, σ'_y, σ'_z that are above the hydrostatic (average) pressure, p. Each of the three normal stresses is the sum of the average pressure, and the deviatoric normal stress (above the average pressure), as follows:

$$
\sigma_x = -p + \sigma'_x, \qquad \sigma_y = -p + \sigma'_y, \qquad \sigma_z = -p + \sigma'_z \tag{5-4a}
$$

According to Newton's second law, the sum of all forces acting on a fluid element, including surface forces in the form of stresses and body forces such as the gravitational force, is equal to the product of mass and acceleration (ma) of the fluid element. After dividing by the volume of the fluid element, the equations of the force balance become

$$
\begin{aligned}
\rho\frac{du}{dt} &= X - \frac{\partial p}{\partial x} + \frac{\partial \sigma'_x}{\partial x} + \frac{\partial \tau_{xy}}{\partial y} + \frac{\partial_{xz}}{\partial z} \\
\rho\frac{dv}{dt} &= Y - \frac{\partial p}{\partial y} + \frac{\partial \tau_{yx}}{\partial y} + \frac{\partial \sigma'_y}{\partial y} + \frac{\partial \tau_{yz}}{\partial z} \\
\rho\frac{dw}{dt} &= Z - \frac{\partial p}{\partial z} + \frac{\partial \tau_{zx}}{\partial x} + \frac{\partial \tau_{zy}}{\partial y} + \frac{\partial \sigma'_z}{\partial z}
\end{aligned}
\tag{5-4b}
$$

Here, p is the pressure, u, v, and w are the velocity components in the x, y, and z directions, respectively. The three forces X, Y, Z are the components of a body force, per unit volume, such as the gravity force that is acting on the fluid. According to the assumptions, the fluid density, ρ, and the viscosity, μ, are considered constant. The derivation of the Navier–Stokes equations is included in most fluid dynamics textbooks (e.g., White, 1985).

For an incompressible flow, the continuity equation, which is derived from the conservation of mass, is

$$\frac{\partial u}{\partial x} + \frac{\partial v}{\partial y} + \frac{\partial w}{\partial z} = 0 \tag{5-5}$$

After substituting the stress components of Eq. (5-2) into Eq. (5-4b), using the continuity equation (5-5) and writing in full the convective time derivative of the acceleration components, the following Navier–Stokes equations in Cartesian coordinates for a Newtonian incompressible and constant-viscosity fluid are obtained

$$\rho\left(\frac{\partial u}{\partial t} + u\frac{\partial u}{\partial x} + v\frac{\partial u}{\partial y} + w\frac{\partial u}{\partial z}\right) = X - \frac{\partial p}{\partial x} + \mu\left(\frac{\partial^2 u}{\partial x^2} + \frac{\partial^2 u}{\partial y^2} + \frac{\partial^2 u}{\partial z^2}\right) \tag{5-6a}$$

$$\rho\left(\frac{\partial v}{\partial t} + u\frac{\partial v}{\partial x} + v\frac{\partial v}{\partial y} + w\frac{\partial v}{\partial z}\right) = Y = -\frac{\partial p}{\partial y} + \mu\left(\frac{\partial^2 v}{\partial x^2} + \frac{\partial^2 v}{\partial y^2} + \frac{\partial^2 v}{\partial z^2}\right) \tag{5-6b}$$

$$\rho\left(\frac{\partial w}{\partial t} + u\frac{\partial w}{\partial x} + v\frac{\partial w}{\partial y} + w\frac{\partial w}{\partial z}\right) = Z - \frac{\partial p}{\partial z} + \mu\left(\frac{\partial^2 w}{\partial x^2} + \frac{\partial^2 w}{\partial y^2} + \frac{\partial^2 w}{\partial z^2}\right) \tag{5-6c}$$

The Navier–Stokes equations can be solved for the velocity distribution. The velocity is described by its three components, u, v, and w, which are functions of the location (x, y, z) and time. In general, fluid flow problems have four unknowns: u, v, and w and the pressure distribution, p. Four equations are required to solve for the four unknown functions. The equations are the three Navier–Stokes equations, the fourth equation is the continuity equation (5-5).

5.2 REYNOLDS HYDRODYNAMIC LUBRICATION EQUATION

Hydrodynamic lubrication involves a thin-film flow, and in most cases the fluid inertia and body forces are very small and negligible in comparison to the viscous forces. Therefore, in a thin-film flow, the inertial terms [all terms on the left side of Eqs. (5.6)] can be disregarded as well as the body forces X, Y, Z. It is well known in fluid dynamics that the ratio of the magnitude of the inertial terms relative to the viscosity terms in Eqs. (5-6) is of the order of magnitude of the

Reynolds number, Re. For a lubrication flow (thin-film flow), $Re \ll 1$, the Navier–Stokes equations reduce to the following simple form:

$$\frac{\partial p}{\partial x} = \mu\left(\frac{\partial^2 u}{\partial x^2} + \frac{\partial^2 u}{\partial y^2} + \frac{\partial^2 u}{\partial z^2}\right) \tag{5-7a}$$

$$\frac{\partial p}{\partial y} = \mu\left(\frac{\partial^2 v}{\partial x^2} + \frac{\partial^2 v}{\partial y^2} + \frac{\partial^2 v}{\partial z^2}\right) \tag{5-7b}$$

$$\frac{\partial p}{\partial z} = \mu\left(\frac{\partial^2 w}{\partial x^2} + \frac{\partial^2 w}{\partial y^2} + \frac{\partial^2 w}{\partial z^2}\right) \tag{5-7c}$$

These equations indicate that viscosity is the dominant effect in determining the pressure distribution in a fluid film bearing.

The assumptions of classical hydrodynamic lubrication theory are summarized in Chapter 4. The velocity components of the flow in a thin film are primarily u and w in the x and z directions, respectively. These directions are along the fluid film layer (see Fig. 1-2). At the same time, there is a relatively very slow velocity component, v, in the y direction across the fluid film layer. Therefore, the pressure gradient in the y direction in Eq. (5-7b) is very small and can be disregarded.

In addition, Eqs. (5-7a and c) can be further simplified because the order of magnitude of the dimensions of the thin fluid film in the x and z directions is much higher than that in the y direction across the film thickness. The orders of magnitude are

$$\begin{aligned} x &= O(B) \\ y &= O(h) \\ z &= O(L) \end{aligned} \tag{5-8a}$$

Here, the symbol O represents *order of magnitude*. The dimension B is the bearing length along the direction of motion (x direction), and h is an average fluid film thickness. The width L is in the z direction of an inclined slider. In a journal bearing, L is in the axial z direction and is referred to as the *bearing length*.

In hydrodynamic bearings, the fluid film thickness is very small in comparison to the bearing dimensions, $h \ll B$ and $h \ll L$. By use of Eqs. (5-8b), a comparison can be made between the orders of magnitude of the second derivatives of the various terms on the right-hand side of Eq. (5-7a), which are as follows:

$$\frac{\partial^2 u}{\partial y^2} = O\left(\frac{U}{h^2}\right)$$
$$\frac{\partial^2 u}{\partial x^2} = O\left(\frac{U}{B^2}\right) \tag{5-8b}$$
$$\frac{\partial^2 u}{\partial z^2} = O\left(\frac{U}{L^2}\right)$$

In conventional finite-length bearings, the ratios of dimensions are of the following orders:

$$\frac{L}{B} = O(1)$$
$$\frac{h}{B} = L(10^{-3}) \tag{5-9}$$

Equations (5-8) and (5-9) indicate that the order of the term $\partial^2 u/\partial y^2$ is larger by 10^6, in comparison to the order of the other two terms, $\partial^2 u/\partial x^2$ and $\partial^2 u/\partial z^2$. Therefore, the last two terms can be neglected in comparison to the first one in Eq. (5-7a). In the same way, only the term $\partial^2 w/\partial y^2$ is retained in Eq. (5-7c). According to the assumptions, the pressure gradient across the film thickness, $\partial p/\partial y$, is negligible, and the Navier–Stokes equations reduce to the following two simplified equations:

$$\frac{\partial p}{\partial x} = \mu \frac{\partial^2 u}{\partial y^2} \quad \text{and} \quad \frac{\partial p}{\partial z} = \mu \frac{\partial^2 w}{\partial y^2} \tag{5-10}$$

The first equation is identical to Eq. (4-4), which was derived from first principles in Chapter 4 for an infinitely long bearing. In a long bearing, there is a significant pressure gradient only in the x direction; however, for a finite-length bearing, there is a pressure gradient in the x and z directions, and the two Eqs. (5-10) are required for solving the flow and pressure distributions.

The two Eqs. (5-10) together with the continuity Eq. (5-5) and the boundary condition of the flow are used to derive the Reynolds equation. The derivation of the Reynolds equation is included in several books devoted to the analysis of hydrodynamic lubrication see Pinkus (1966), and Szeri (1980). The Reynolds equation is widely used for solving the pressure distribution of hydrodynamic bearings of finite length. The Reynolds equation for Newtonian incompressible and constant-viscosity fluid in a thin clearance between two rigid surfaces of relative motion is given by

$$\frac{\partial}{\partial x}\left(\frac{h^3}{\mu}\frac{\partial p}{\partial x}\right)+\frac{\partial}{\partial z}\left(\frac{h^3}{\mu}\frac{\partial p}{\partial z}\right)=6(U_1-U_2)\frac{\partial h}{\partial x}+6\frac{\partial}{\partial x}(U_1+U_2)+12(V_2-V_1) \tag{5-11}$$

The velocity components of the two surfaces that form the film boundaries are shown in Fig. 5-2. The tangential velocity components, U_1 and U_2, in the x direction are of the lower and upper sliding surfaces, respectively (two fluid film boundaries). The normal velocity components, in the y direction, V_1 and V_2, are of the lower and upper boundaries, respectively. In a journal bearing, these components are functions of x (or angle θ) around the journal bearing.

The right side of Eq. (5-11) must be negative in order to result in a positive pressure wave and load capacity. Each of the three terms on the right-hand side of Eq. (5-11) has a physical meaning concerning the generation of the pressure wave. Each term is an action that represents a specific type of relative motion of the surfaces. Each action results in a positive pressure in the fluid film. The various actions are shown in Fig. 5-3. These three actions can be present in a bearing simultaneously, one at a time or in any other combination. The following are the various actions.

Viscous wedge action: This action generates positive pressure wave by dragging the viscous fluid into a converging wedge.

Elastic stretching or compression of the boundary surface: This action generates a positive pressure by compression of the boundary. The compression of the surface reduces the clearance volume and the viscous fluid is squeezed out, resulting in a pressure rise. This action is

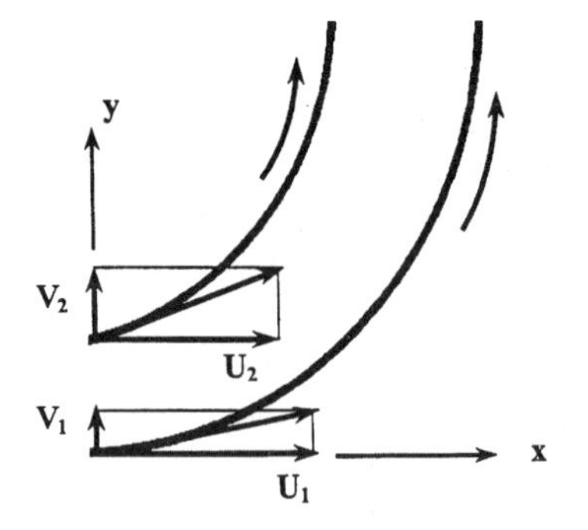

FIG. 5-2 Directions of the velocity components of fluid-film boundaries in the Reynolds equation.

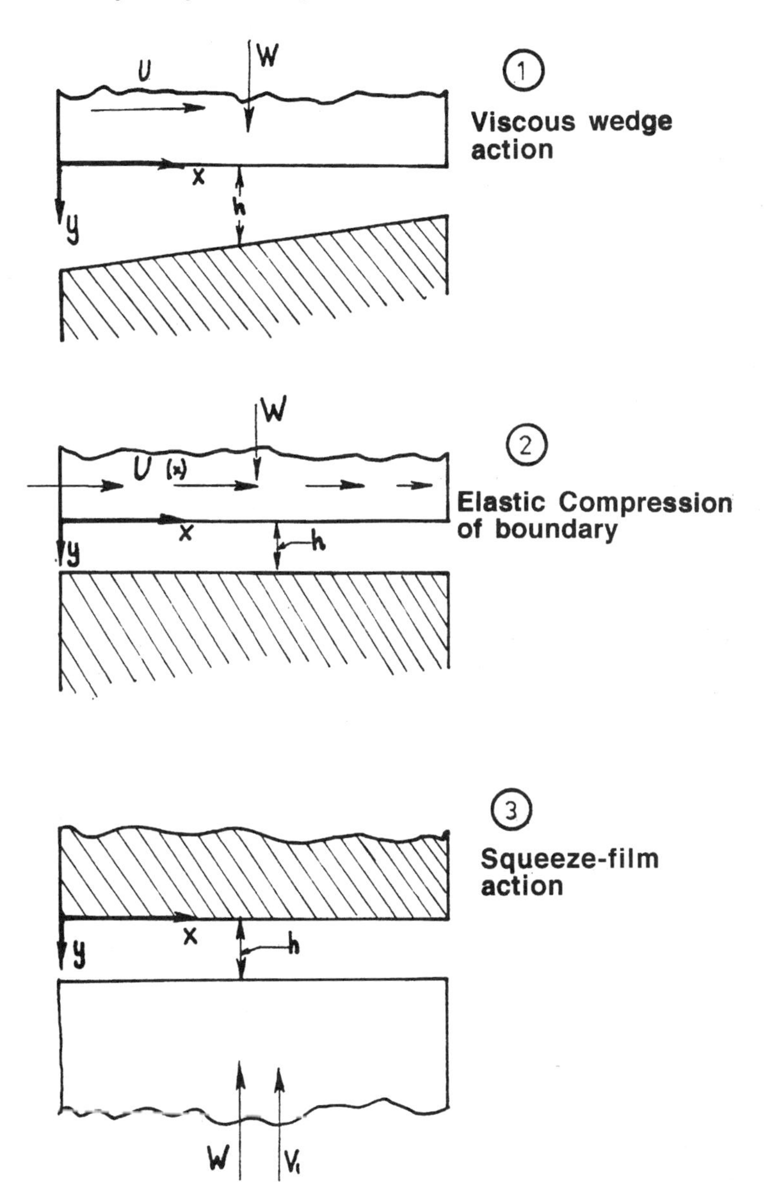

FIG. 5-3 Viscous film actions that result in a positive pressure wave.

negligible in practical rigid bearings. Continuous stretching or compression of the boundaries does not exist in steady-state operation. It can act only as a transient effect, under dynamic condition, for an elastomer bearing material. This action is usually not considered for rigid bearing materials.

Squeeze-film action: The squeezing action generates a positive pressure by reduction of the fluid film volume. The incompressible viscous fluid is squeezed out through the thin clearance. The thin clearance has resistance to the squeeze-film flow, resulting in a pressure buildup to overcome the flow resistance (see Problem 5-3).

In most practical bearings, the surfaces are rigid and there is no stretching or compression action. In that case, the Reynolds equation for an incompressible fluid and constant viscosity reduces to

$$\frac{\partial}{\partial x}\left(\frac{h^3}{\mu}\frac{\partial p}{\partial x}\right) + \frac{\partial}{\partial z}\left(\frac{h^3}{\mu}\frac{\partial p}{\partial z}\right) = 6(U_1 - U_2)\frac{\partial h}{\partial x} + 12(V_2 - V_1) \tag{5-12}$$

As indicated earlier, the two right-hand terms must be negative in order to result in a positive pressure wave. On the right side of the Reynolds equation, the first term of relative sliding motion $(U_1 - U_2)$ describes a viscous wedge effect. It requires inclined surfaces, $\partial h/\partial x$, to generate a fluid film wedge action that results in a pressure wave. Positive pressure is generated if the film thickness reduces in the x direction (negative $\partial h/\partial x$).

The second term on the right side of the Reynolds equation describes a squeeze-film action. The difference in the normal velocity $(V_2 - V_1)$ represents the motion of surfaces toward each other, referred to as *squeeze-film action*. A positive pressure builds up if $(V_2 - V_1)$ is negative and the surfaces are approaching each other. The Reynolds equation indicates that a squeeze-film effect is a viscous effect that can generate a pressure wave in the fluid film, even in the case of parallel boundaries.

It is important to mention that the Reynolds equation is objective, in the sense that the pressure distribution must be independent of the selection of the coordinate system. In Fig. 5-2 the coordinates are stationary and the two surfaces are moving relative to the coordinate system. However, the same pressure distribution must result if the coordinates are attached to one surface and are moving and rotating with it. For convenience, in most problems we select a stationary coordinate system where the x coordinate is along the bearing surface and the y coordinate is normal to this surface. In that case, the lower surface has only a tangential velocity, U_1, and there is no normal component, $V_1 = 0$.

The value of each of the velocity components of the fluid boundary, U_1, U_2, V_1, V_2, depends on the selection of the coordinate system. Surface velocities in a stationary coordinate system would not be the same as those in a moving coordinate system. However, velocity differences on the right-hand side of the Reynolds equation, which represent relative motion, are independent of the selection of the coordinate system. In journal bearings under dynamic conditions, the journal center is not stationary. The velocity of the center must be considered for the derivation of the right-hand side terms of the Reynolds equation.

5.3 WIDE PLANE-SLIDER

The equation of a plane-slider has been derived from first principles in Chapter 4. Here, this equation will be derived from the Reynolds equation and compared to that in Chapter 4.

A plane-slider and its coordinate system are shown in Fig. 1-2. The lower plate is stationary, and the velocity components at the lower wall are $U_1 = 0$ and $V_1 = 0$. At the same time, the velocity at the upper wall is equal to that of the slider. The slider has only a horizontal velocity component, $U_2 = U$, where U is the plane-slider velocity. Since the velocity of the slider is in only the x direction and there is no normal component in the y direction, $V_2 = 0$. After substituting the velocity components of the two surfaces into Eq. (5-11), the Reynolds equation will reduce to the form

$$\frac{\partial}{\partial x}\left(\frac{h^3}{\mu}\frac{\partial p}{\partial x}\right) + \frac{\partial}{\partial z}\left(\frac{h^3}{\mu}\frac{\partial p}{\partial z}\right) = 6U\frac{\partial h}{\partial x} \tag{5-13}$$

For a wide bearing, $L \gg B$, we have $\partial p/\partial z \cong 0$, and the second term on the left side of Eq. (5-13) can be omitted. The Reynolds equation reduces to the following simplified form:

$$\frac{\partial}{\partial x}\left(\frac{h^3}{\mu}\frac{\partial p}{\partial x}\right) = 6U\frac{\partial h}{\partial x} \tag{5-14}$$

For a plane-slider, if the x coordinate is in the direction of a converging clearance (the clearance reduces with x), as shown in Fig. 4-4, integration of Eq. (5-14) results in a pressure gradient expression equivalent to that of a hydrodynamic journal bearing or a negative-slope slider in Chapter 4. The following equation is the expression for the pressure gradient for a converging clearance, $\partial h/\partial x < 0$ (negative slope):

$$\frac{dp}{dx} = 6U\mu\frac{h - h_0}{h^3} \qquad \text{for initial } \frac{\partial h}{\partial x} < 0 \text{ (negative slope in Fig. 4-4)} \tag{5-15}$$

The unknown constant h_0 (constant of integration) is determined from additional information concerning the boundary conditions of the pressure wave. The meaning of h_0 is discussed in Chapter 4—it is the film thickness at the point of a peak pressure along the fluid film. In a converging clearance such as a journal bearing near $x = 0$, the clearance slope is negative, $\partial h/\partial x < 0$. The result is that the pressure increases at the start of the pressure wave (near $x = 0$). At that point, the pressure gradient $dp/dx > 0$ because $h > h_0$.

However, if the x coordinate is in the direction of a diverging clearance (the clearance increases with x), as shown in Fig. (1-2), Eq. (5-15) changes its sign and takes the following form:

$$\frac{dp}{dx} = 6U\mu\frac{h_0 - h}{h^3} \qquad \text{for initial } \frac{\partial h}{\partial x} > 0 \text{ (positive slope in Fig. 4-5)} \tag{5-15}$$

5.4 FLUID FILM BETWEEN A FLAT PLATE AND A CYLINDER

A fluid film between a plate and a cylinder is shown in Fig. 5-4. In Chapter 4, the pressure wave for relative sliding is derived, where the cylinder is stationary and a flat plate has a constant velocity in the x direction. In the following example, the previous problem is extended to a combination of rolling and sliding. In this case, the flat plate has a velocity U in the x direction and the cylinder rotates at an angular velocity ω around its stationary center. The coordinate system (x, y) is stationary.

In Sec. 4.8, it is mentioned that there is a significant pressure wave only in the region close to the minimum film thickness. In this region, the slope between the two surfaces (the fluid film boundaries), as well as between the two surface velocities, is of a very small angle α. For a small α, we can approximate that $\cos\alpha \approx 1$.

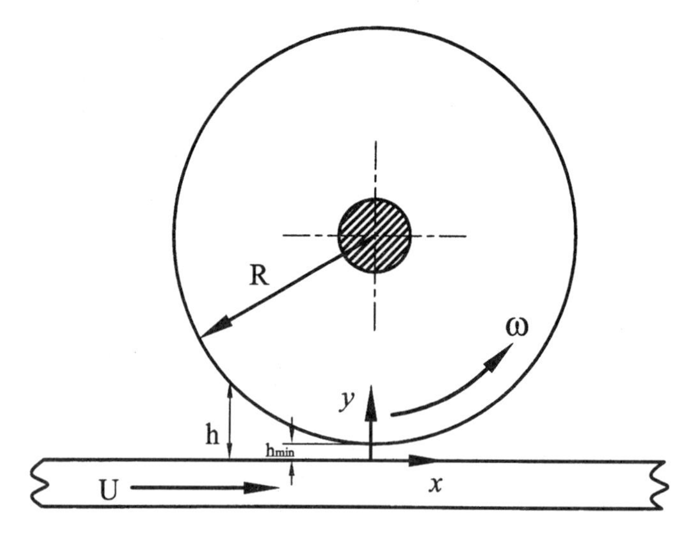

FIG. 5-4 Fluid film between a moving plate and a rotating cylinder.

In Fig. 5-4, the surface velocity of the cylinder is not parallel to the x direction and it has a normal component V_2. The surface velocities on the rotating cylinder surface are

$$U_2 = \omega R \cos\alpha \approx \omega R \qquad V_2 \approx \omega R \frac{\partial h}{\partial x} \tag{5-16}$$

At the same time, on the lower plate there is only velocity U in the x direction and the boundary velocity is

$$U_1 = U \qquad V_1 = 0 \tag{5-17}$$

Substituting Eqs. (5-16) and (5-17) into the right side of Eq. (5-11), yields

$$\begin{aligned} 6(U_1 - U_2)\frac{\partial h}{\partial x} + 12(V_2 - V_1) &= 6(U - \omega R)\frac{\partial h}{\partial x} + 12\omega R \frac{\partial h}{\partial x} \\ &= 6(U + \omega R)\frac{\partial h}{\partial x} \end{aligned} \tag{5-18}$$

The Reynolds equation for a fluid film between a plate and a cylinder becomes

$$\frac{\partial}{\partial x}\left(\frac{h^3}{\mu}\frac{\partial p}{\partial x}\right) + \frac{\partial}{\partial z}\left(\frac{h^3}{\mu}\frac{\partial p}{\partial z}\right) = 6(U + \omega R)\frac{\partial h}{\partial x} \tag{5-19}$$

For a long bearing, the pressure gradient in the axial direction is negligible, $\partial p/\partial z \cong 0$. Integration of Eq. (5-19) yields

$$\frac{dp}{dx} = 6\mu(U + \omega R)\frac{h - h_0}{h^3} \tag{5-20}$$

This result indicates that the pressure gradient, the pressure wave, and the load capacity are proportional to the sum of the two surface velocities in the x direction. The sum of the plate and cylinder velocities is $(U + \omega R)$. In the case of pure rolling, $U = \omega R$, the pressure wave, and the load capacity are twice the magnitude of that generated by pure sliding. Pure sliding is when the cylinder is stationary, $\omega = 0$, and only the plate has a sliding velocity U.

The unknown constant h_0, (constant of integration) is the film thickness at the point of a peak pressure, and it can be solved from the boundary conditions of the pressure wave.

5.5 TRANSITION TO TURBULENCE

For the estimation of the Reynolds number, Re, the average radial clearance, C, is taken as the average film thickness. The Reynolds number for the flow inside the clearance of a hydrodynamic journal bearing is

$$\mathrm{Re} = \frac{U\rho C}{\mu} = \frac{UC}{\nu} \tag{5-21}$$

Here, U is the journal surface velocity, as shown in Fig. 1-2 and ν is the kinematic viscosity $\nu = \mu/\rho$. In most cases, hydrodynamic lubrication flow involves low Reynolds numbers. There are other examples of thin-film flow in fluid mechanics, such as the boundary layer, where Re is low.

The flow in hydrodynamic lubrication is laminar at low Reynolds numbers. Experiments in journal bearings indicate that the transition from laminar to turbulent flow occurs between Re = 1000 and Re = 1600. The value of the Reynolds number at the transition to turbulence is not the same in all cases. It depends on the surface finish of the rotating surfaces as well as the level of vibrations in the bearing. The transition is gradual: Turbulence starts to develop at about Re = 1000; and near Re = 1600, full turbulent behavior is maintained. In hydrodynamic bearings, turbulent flow is undesirable because it increases the friction losses. Viscous friction in turbulent flow is much higher in comparison to laminar flow. The effect of the turbulence is to increase the apparent viscosity; that is, the bearing performance is similar to that of a bearing having laminar flow and much higher lubricant viscosity.

In journal bearings, Taylor vortexes can develop at high Reynolds numbers. The explanation for the initiation of Taylor vortexes involves the centrifugal forces in the rotating fluid inside the bearing clearance. At high Re, the fluid film becomes unstable because the centrifugal forces are high relative to the viscous resistance. Theory indicates that in concentric cylinders, Taylor vortexes would develop only if the inner cylinder is rotating relative to the outer, stationary cylinder.

This instability gives rise to vortexes (Taylor vortexes) in the fluid film. Taylor (1923) published his classical work on the theory of stability between rotating cylinders. According to this theory, a stable laminar flow in a journal bearing is when the Reynolds number is below the following ratio:

$$\mathrm{Re} < 41\left(\frac{R}{C}\right)^{1/2} \tag{5-22}$$

In journal bearings, the order of R/C is 1000; therefore, the limit of the laminar flow is Re = 1300, which is between the two experimental values of Re = 1000 and Re = 1600, mentioned earlier. If the clearance C were reduced, it would extend the Re limit for laminar flow. The purpose of the following example problem is to illustrate the magnitude of the Reynolds number for common journal bearings with various lubricants.

In addition to Taylor vortexes, transition to turbulence can be initiated due to high-Reynolds-number flow, in a similar way to instability in the flow between two parallel plates.

Example Problem 5-1

Calculation of the Reynolds Number

The value of the Reynolds number, Re, is considered for a common hydrodynamic journal bearing with various fluid lubricants. The journal diameter is $d = 50$ mm; the radial clearance ratio is $C/R = 0.001$. The journal speed is 10,000 RPM. Find the Reynolds number for each of the following lubricants, and determine if Taylor vortices can occur.

a. The lubricant is mineral oil, SAE 10, and its operating temperature is 70°C. The lubricant density is $\rho = 860\ \text{kg/m}^3$.
b. The lubricant is air, its viscosity is $\mu = 2.08 \times 10^{-5}\ \text{N-s/m}^2$, and its density is $\rho = 0.995\ \text{kg/m}^3$.
c. The lubricant is water, its viscosity is $\mu = 4.04 \times 10^{-4}\ \text{N-s/m}^2$, and its density is $\rho = 978\ \text{kg/m}^3$.
d. For mineral oil, SAE 10, at 70°C (in part a) find the journal speed at which instability, in the form of Taylor vortices, initiates.

Solution

The journal bearing data is as follows:

Journal speed, $N = 10{,}000$ RPM
Journal diameter $d = 50$ mm, $R = 25 \times 10^{-3}$, and $C/R = 0.001$
$C = 25 \times 10^{-6}$ m

The journal surface velocity is calculated from

$$U = \frac{\pi d N}{60} = \frac{\pi(0.050 \times 10{,}000)}{60} = 26.18\ \text{m/s}$$

a. For estimation of the Reynolds number, the average clearance C is used as the average film thickness, and Re is calculated from (5-22):

$$\text{Re} = \frac{U \rho C}{\mu} < 41 \left(\frac{R}{C}\right)^{1/2}$$

The critical Re for Taylor vortices is

$$\text{Re (critical)} = 41 \left(\frac{R}{C}\right)^{1/2} = 41 \times (1000)^{0.5} = 1300$$

For SAE 10 oil, the lubricant viscosity (from Fig. 2-2) and density are:

Viscosity: $\mu = 0.01\ \text{N-s/m}^2$
Density: $\rho = 860\ \text{kg/m}^3$

The Reynolds number is

$$Re = \frac{U\rho C}{\mu} = \frac{26.18 \times 860 \times 25 \times 10^{-6}}{0.01} = 56.3 \text{ (laminar flow)}$$

This example shows that a typical journal bearing lubricated by mineral oil and operating at relatively high speed is well within the laminar flow region.

b. The Reynolds number for air as lubricant is calculated as follows:

$$Re = \frac{U\rho C}{\mu} = \frac{26.18 \times 0.995 \times 25 \times 10^{-6}}{2.08 \times 10^{-5}}$$
$$= 31.3 \text{ (laminar flow)}$$

c. The Reynolds number for water as lubricant is calculated as follows:

$$Re = \frac{U\rho C}{\mu} = \frac{26.18 \times 978 \times 25 \times 10^{-6}}{4.04 \times 10^{-4}}$$
$$= 1584 \text{ (turbulent flow)}$$

The kinematic viscosity of water is low relative to that of oil or air. This results in relatively high Re and turbulent flow in journal bearings. In centrifugal pumps or bearings submerged in water in ships, there are design advantages in using water as a lubricant. However, this example indicates that water lubrication often involves turbulent flow.

d. The calculation of journal speed where instability in the form of Taylor vortices initiates is obtained from

$$Re = \frac{U\rho C}{\mu} = 41\left(\frac{R}{C}\right)^{1/2} = 41 \times (1000)^{0.5} = 1300$$

Surface velocity U is derived as unknown in the following equation:

$$1300 = \frac{U\rho C}{\mu} = \frac{U \times 860 \times 25 \times 10^{-6}}{0.01} \Rightarrow U = 604.5\,\text{m/s}$$

and the surface velocity at the transition to Taylor instability is

$$U = \frac{\pi d N}{60} = \frac{\pi(0.050)N}{60} = 604.5 \text{ m/s}$$

The journal speed N where instability in the form of Taylor vortices initiates is solved from the preceding equation:

$$N = 231{,}000 \text{ RPM}$$

This speed is above the range currently applied in journal bearings.

Example Problem 5-2

Short Plane-Slider

Derive the equation of the pressure wave in a short plane-slider. The assumption of an infinitely short bearing can be applied where the width L (in the z direction) is very short relative to the length B ($L \ll B$). In practice, an infinitely short bearing can be assumed where $L/B = O(10^{-1})$.

Solution

Order-of-magnitude considerations indicate that in an infinitely short bearing, dp/dx is very small and can be neglected in comparison to dp/dz. In that case, the first term on the left side of Eq. (5.12) is small and can be neglected in comparison to the second term. This omission simplifies the Reynolds equation to the following form:

$$\frac{\partial}{\partial z}\left(\frac{h^3}{\mu}\frac{\partial p}{\partial z}\right) = -6U\frac{\partial h}{\partial x} \tag{5-23}$$

Double integration results in the following parabolic pressure distribution, in the z direction:

$$p = -\frac{3\mu U}{h^3}\frac{dh}{dx}z^2 + C_1 z + C_2 \tag{5-24}$$

The two constants of integration can be obtained from the boundary conditions of the pressure wave. At the two ends of the bearing, the pressure is equal to atmospheric pressure, $p = 0$. These boundary conditions can be written as

$$\text{at } z = \pm\frac{L}{2}: \quad p = 0 \tag{5-25}$$

The following expression for the pressure distribution in a short plane-slider (a function of x and z) is obtained:

$$p(x, z) = -3\mu U\left(\frac{L^2}{4} - z^2\right)\frac{h'}{h^3} \tag{5-26}$$

Here, $h' = \partial h/\partial x$. In the case of a plane-slider, $\partial h/\partial x = -\tan a$, the slope of the plane-slider.

Comment. For a short bearing, the result indicates discontinuity of the pressure wave at the front and back ends of the plane-slider (at $h = h_1$ and $h = h_2$). In fact, the pressure at the front and back ends increases gradually, but this has only a small effect on the load capacity. This deviation from the actual pressure wave is similar to the edge effect in an infinitely long bearing.

5.6 CYLINDRICAL COORDINATES

There are many problems that are conveniently described in cylindrical coordinates, and the Navier–Stokes equations in cylindrical coordinates are useful for that purpose. The three coordinates r, ϕ, z are the radial, tangential, and axial coordinates, respectively, v_r, v_ϕ, v_z are the velocity components in the respective directions. For hydrodynamic lubrication of thin films, the inertial terms are disregarded and the three Navier–Stokes equations for an incompressible, Newtonian fluid in cylindrical coordinates are as follows:

$$\begin{aligned}
\frac{\partial p}{\partial r} &= \mu\left(\frac{\partial^2 v_r}{\partial r^2} + \frac{1}{r}\frac{\partial^2 v_r}{\partial r} - \frac{v_r}{r^2} + \frac{1}{r^2}\frac{\partial^2 v_\phi}{\partial \phi^2} - \frac{2}{r^2}\frac{\partial v_\phi}{\partial \phi} + \frac{\partial^2 v_r}{\partial z^2}\right) \\
\frac{1}{r}\frac{\partial p}{\partial \phi} &= \mu\left(\frac{\partial^2 v_\phi}{\partial r^2} + \frac{1}{r}\frac{\partial v_\phi}{\partial r} - \frac{v_\phi}{r^2} + \frac{1}{r^2}\frac{\partial^2 v_\phi}{\partial \phi^2} + \frac{2}{r^2}\frac{\partial v_r}{\partial \phi} + \frac{\partial^2 v_\phi}{\partial z^2}\right) \\
\frac{\partial p}{\partial z} &= \mu\left(\frac{\partial^2 v_z}{\partial r^2} + \frac{1}{r}\frac{\partial v_z}{\partial r} + \frac{1}{r^2}\frac{\partial^2 v_z}{\partial \phi^2} + \frac{\partial^2 v_z}{\partial z^2}\right)
\end{aligned} \tag{5-27}$$

Here, v_r, v_ϕ, v_z are the velocity components in the radial, tangential, and vertical directions r, ϕ, and z, respectively. The constant density is ρ, the variable pressure is p, and the constant viscosity is μ.

The equation of continuity in cylindrical coordinates is

$$\left(\frac{\partial v_r}{\partial r} + \frac{v_r}{r} + \frac{1}{r}\frac{\partial v_\theta}{\partial \theta} + \frac{\partial v_z}{\partial z}\right) = 0 \tag{5-28}$$

In cylindrical coordinates, the six stress components are

$$\begin{aligned}
\sigma_r &= -p + 2\mu\frac{\partial v_r}{\partial r} \\
\sigma_\phi &= -p + 2\mu\left(\frac{1}{r}\frac{\partial v_\phi}{\partial \phi} + \frac{v_r}{r}\right) \\
\sigma_z &= -\rho + 2\mu\frac{\partial v_z}{\partial z} \\
\tau_{rz} &= \mu\left(\frac{\partial v_r}{\partial z} + \frac{\partial v_z}{\partial r}\right) \\
\tau_{r\phi} &= \mu\left[r\frac{\partial}{\partial r}\left(\frac{v_\phi}{r}\right) + \frac{1}{r}\frac{\partial v_r}{\partial \phi}\right] \\
\tau_{\phi z} &= \mu\left(\frac{\partial v_\phi}{\partial z} + \frac{1}{r}\frac{\partial v_z}{\partial \phi}\right)
\end{aligned} \tag{5-29}$$

5.7 SQUEEZE-FILM FLOW

An example of the application of cylindrical coordinates to the squeeze-film between two parallel, circular, concentric disks is shown in Fig. 5-5. The fluid film between the two discs is very thin. The disks approach each other at a certain speed. This results in a squeezing of the viscous thin film and in a radial pressure distribution in the clearance and a load capacity that resists the motion of the moving disk.

For a thin film, further simplification of the Navier–Stokes equations is similar to that in the derivation of Eq. (5-10). The pressure is assumed to be constant across the film thickness (in the z direction), and the dimension of z is much smaller than that of r or $r\phi$. For a problem of radial symmetry, the Navier–Stokes equations reduce to the following:

$$\frac{dp}{dr} = \mu \frac{\partial^2 v_r}{\partial z^2} \tag{5-30}$$

Here, v_r is the fluid velocity in the radial direction.

Example Problem 5-3

A fluid is squeezed between two parallel, circular, concentric disks, as shown in Fig. 5-5. The fluid film between the two discs is very thin. The upper disk has

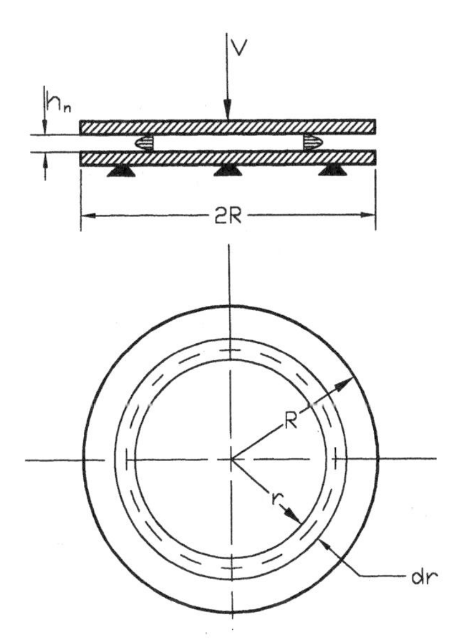

FIG. 5-5 Squeeze-film flow between two concentric, parallel disks.

velocity V, toward the lower disk, and this squeezes the fluid so that it escapes in the radial direction. Derive the equations for the radial pressure distribution in the thin film and the resultant load capacity.

Solution

The first step is to solve for the radial velocity distribution, v_r, in the fluid film. In a similar way to the hydrodynamic lubrication problem in the previous chapter, we can write the differential equation in the form

$$\frac{\partial^2 v_r}{\partial z^2} = \frac{dp}{dr}\frac{1}{\mu} = 2m$$

Here, m is a function of r only; $m = m(r)$ is a substitution that represents the pressure gradient. By integration the preceding equation twice, the following parabolic distribution of the radial velocity is obtained:

$$v_r = mz^2 + nz + k$$

Here, m, n, and k are functions of r only. These functions are solved by the boundary conditions of the radial velocity and the conservation of mass as well as fluid volume for incompressible flow.

The two boundary conditions of the radial velocity are

$$\begin{aligned} &\text{at } z = 0: \quad v_r = 0 \\ &\text{at } z = h: \quad v_r = 0 \end{aligned}$$

In order to solve for the three unknowns m, n and k, a third equation is required; this is obtained from the fluid continuity, which is equivalent to the conservation of mass.

Let us consider a control volume of a disk of radius r around the center of the disk. The downward motion of the upper disk, at velocity V, reduces the volume of the fluid per unit of time. The fluid is incompressible, and the reduction of volume is equal to the radial flow rate Q out of the control volume. The flow rate Q is the product of the area of the control volume πr^2 and the downward velocity V:

$$Q = \pi r^2 V$$

The same flow rate Q must apply in the radial direction through the boundary of the control volume. The flow rate Q is obtained by integration of the radial velocity distribution of the film radial velocity, v_r, along the z direction, multiplied by the circumference of the control volume ($2\pi r$). The flow rate Q becomes

$$Q = 2\pi r \int_0^h v_r \, dz$$

Since the fluid is incompressible, the flow rate of the fluid escaping from the control volume is equal to the flow rate of the volume displaced by the moving disk:

$$V\pi r^2 = 2\pi r \int_0^h v_r \, dz$$

Use of the boundary conditions and the preceding continuity equation allows the solution of m, n, and k. The solution for the radial velocity distribution is

$$v_r = \frac{3rV}{h}\left(\frac{z^2}{h^2} - \frac{z}{h}\right)$$

and the pressure gradient is

$$\Rightarrow \frac{dp}{dr} = -6\mu V \frac{r}{h^3}$$

The negative sign means that the pressure is always decreasing in the r direction. The pressure gradient is a linear function of the radial distance r, and this function can be integrated to solve for the pressure wave:

$$p = \int dp = -\frac{6\mu V}{h^3} \int r \, dr = -\frac{3\mu V}{h^3} r^2 + C$$

Here, C is a constant of integration that is solved by the boundary condition that states that at the outside edge of the disks, the fluid pressure is equal to atmospheric pressure, which can be considered to be zero:

$$\text{at } r = R: \quad p = 0$$

Substituting in the preceding equation yields:

$$0 = -\frac{3\mu V}{h^3} R^2 + C \Rightarrow C = \frac{3\mu V}{h^3} R^2$$

The equation for the radial pressure distribution is therefore

$$\Rightarrow p = -\frac{3\mu V}{h^3} r^2 + \left(\frac{3\mu V}{h^3} R^2\right) = \frac{3\mu V}{h^3}(R^2 - r^2)$$

The pressure has its maximum value at the center of the disk radius:

$$P_{max} = \frac{3\mu V}{h^3}[R^2 - (0)^2] = \frac{3\mu V}{h^3} R^2$$

The parabolic pressure distribution is shown in Fig. 5-6.

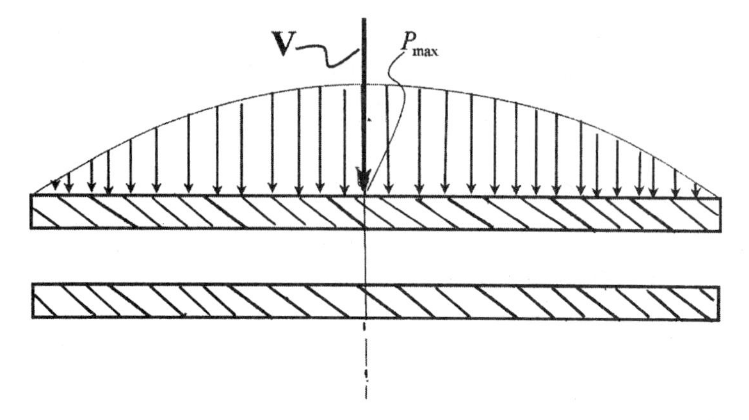

FIG. 5-6 Radial pressure distribution in squeeze-film flow.

Now that the pressure distribution has been solved, the load capacity is obtained by the following integration:

$$W = \int dW = \int_A pdA = \int_0^R \frac{3\mu V}{h^3}(R^2 - r^2)2\pi r dr$$

$$= \frac{6\pi\mu V}{h^3}\left[R^2 \int_0^R r\,dr - \int_0^R r^3\,dr\right]$$

$$\therefore \Rightarrow W = \frac{3\pi\mu V R^4}{2h^3}$$

The load capacity equation indicates that a squeeze-film arrangement can act as a damper that resists the squeezing motion. The load capacity increases dramatically as the film thickness becomes thinner, thus preventing the disks from coming into contact. Theoretically, at $h = 0$ the load capacity is approaching infinity. In practice, there is surface roughness and there will be contact by a finite force.

Example Problem 5-4

Two parallel circular disks of 30-mm diameter, as shown in Fig. 5-5, operate as a damper. The clearance is full of SAE 30 oil at a temperature of 50°C. The damper is subjected to a shock load of 7000 N.

a. Find the film thickness h if the load causes a downward speed of the upper disk of 10 m/s.
b. What is the maximum pressure developed due to the impact of the load?

Solution

a. The viscosity of the lubricant is obtained from the viscosity–temperature chart:

$$\mu_{SAE30@50^\circ C} = 5.5 \times 10^{-2}\ \text{N-s/m}^2$$

The film thickness is derived from the load capacity equation:

$$W = \frac{3\pi\mu V R^4}{2h^3}$$

The instantaneous film thickness, when the disk speed is 10 m/s, is

$$h = \left(\frac{3\pi \times 5.5 \times 10^{-2} \times 10 \times 15 \times 10^{-3}}{2 \times 7000}\right)^{\frac{1}{3}}$$

$$h = 0.266\ \text{mm}$$

b. The maximum pressure at the center is

$$p_{\max} = \frac{3\mu V}{h^3} R^2 = \frac{3(5.5 \times 10^{-2}\ \text{N-s/m}^2)(10\ \text{ms})(15 \times 10^{-3}\ \text{m})^2}{(0.266 \times 10^{-3}\ \text{m})^3}$$

$$= 1.97 \times 10^7\ \text{Pa} = 19.7\ \text{MPa}$$

Problems

5-1 Two long cylinders of radii R_1 and R_2, respectively, have parallel centrelines, as shown in Fig. 5-7. The cylinders are submerged in fluid and are rotating in opposite directions at angular speeds of ω_1 and ω_2, respectively. The minimum clearance between the cylinders is h_n. If the fluid viscosity is μ, derive the Reynolds equation, and write the expression for the pressure gradient around the minimum clearance.

The equation of the clearance is

$$h(x) = h_n + \frac{x^2}{2R_{eq}}$$

For calculation of the variable clearance between two cylinders having a convex contact, the equation for the equivalent radius R_{eq} is

$$\frac{1}{R_{eq}} = \frac{1}{R_1} + \frac{1}{R_2}$$

5-2 Two long cylinders of radii R_1 and R_2, respectively, in concave contact

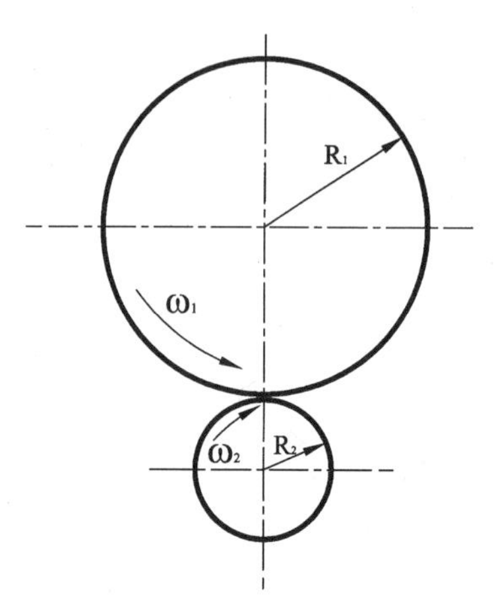

FIG. 5-7 Two parallel cylinders, convex contact.

are shown in Fig. 5-8. The cylinders have parallel centerlines and are in rolling contact. The angular speed of the large (external) cylinder is ω, and the angular speed of the small (internal) cylinder is such that there is rolling without sliding. There is a small minimum clearance between the cylinders, h_n. If the fluid viscosity is μ, derive the Reynolds equation, and write the expression for the pressure gradient around the minimum clearance. For a concave contact (Fig. 5-8) the equivalent radius R_{eq} is

$$\frac{1}{R_{eq}} = \frac{1}{R_1} - \frac{1}{R_2}$$

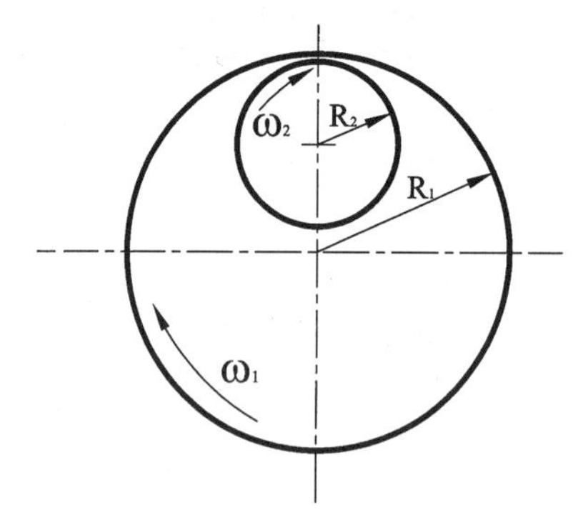

FIG. 5-8 Two parallel cylinders, concave contact.

5-3 The journal diameter of a hydrodynamic bearing is $d = 100$ mm, and the radial clearance ratio is $C/R = 0.001$. The journal speed is $N = 20{,}000$ RPM. The lubricant is mineral oil, SAE 30, at 70°C. The lubricant density at the operating temperature is $\rho = 860$ kg/m^3.

a. Find the Reynolds number.
b. Find the journal speed, N, where instability in the form of Taylor vortices initiates.

5-4 Two parallel circular disks of 100-mm diameter have a clearance of 1 mm between them. Under load, the downward velocity of the upper disk is 2 m/s. At the same time, the lower disk is stationary. The clearance is full of SAE 10 oil at a temperature of 60°C.

a. Find the load on the upper disk that results in the instantaneous velocity of 2 m/s.
b. What is the maximum pressure developed due to that load?

5-5 Two parallel circular disks (see Fig. 5-5) of 200-mm diameter have a clearance of $h = 2$ mm between them. The load on the upper disk is 200 N. The lower disk is stationary, and the upper disk has a downward velocity V. The clearance is full of oil, SAE 10, at a temperature of 60°C.

a. Find the downward velocity of the upper disk at that instant.
b. What is the maximum pressure developed due to that load?

6

Long Hydrodynamic Journal Bearings

6.1 INTRODUCTION

A hydrodynamic journal bearing is shown in Fig. 6-1. The journal is rotating inside the bore of a sleeve with a thin clearance. Fluid lubricant is continuously supplied into the clearance. If the journal speed is sufficiently high, pressure builds up in the fluid film that completely separates the rubbing surfaces.

Hydrodynamic journal bearings are widely used in machinery, particularly in motor vehicle engines and high-speed turbines. The sleeve is mounted in a housing that can be a part of the frame of a machine. For successful operation, the bearing requires high-precision machining. For most applications, the mating surfaces of the journal and the internal bore of the bearing are carefully made with precise dimensions and a good surface finish.

The radial clearance, C, between the bearing bore of radius R_1 and of the journal radius, R, is $C = R_1 - R$ (see Fig. 6-1). In most practical cases, the clearance, C, is very small relative to the journal radius, R. The following order of magnitude is applicable for most design purposes:

$$\frac{C}{R} \approx 10^{-3} \tag{6-1}$$

A long hydrodynamic bearing is where the bearing length, L, is long in comparison to the journal radius ($L \gg R$). Under load, the center of the journal,

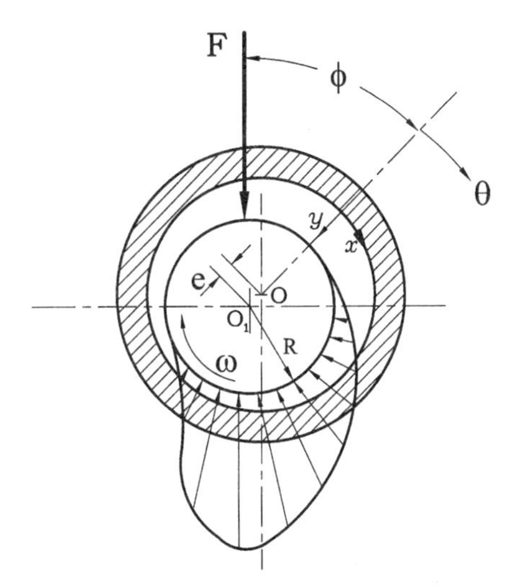

FIG. 6-1 Hydrodynamic journal bearing (clearance exaggerated).

O_1, is displaced in the radial direction relative to the bearing center, O, as indicated in Fig. 6-1. The distance $O - O_1$ is the eccentricity, e, and the dimensionless eccentricity ratio, ε, is defined as

$$\epsilon = \frac{e}{C} \tag{6-2}$$

If the journal is concentric to the bearing bore, there is a uniform clearance around the bearing. But due to the eccentricity, there is a variable-thickness clearance around the bearing. The variable clearance, h, is equal to the lubricant film thickness. The clearance is converging (h decreases) along the region from $\theta = 0$ to $\theta = \pi$, where the clearance has its minimum thickness, h_n. After that, the clearance is diverging (h increases) along the region from $\theta = \pi$ to $\theta = 2\pi$, where the clearance has a maximum thickness, h_m. The clearance, h, is a function of the coordinate θ in the direction of the rotation of the journal. The coordinate θ is measured from the point of maximum clearance on the centerline O–O. An approximate equation for the thin clearance in a journal bearing is

$$h(\theta) = C(1 + \varepsilon \cos \theta) \tag{6-3}$$

The minimum and maximum clearances are h_n and h_m, respectively, along the symmetry line O–O_1. They are derived from Eq. (6-3) as follows:

$$h_n = C(1 - \varepsilon) \qquad \text{and} \qquad h_m = C(1 + \varepsilon) \tag{6-4}$$

Whenever ε approaches 1 (one), h_n approaches zero, and there is an undesirable contact between the journal and bearing bore surfaces, resulting in severe wear, particularly in a high-speed journal bearing. Therefore, for proper design, the bearing must operate at eccentricity ratios well below 1, to allow adequate minimum film thickness to separate the sliding surfaces. Most journal bearings operate under steady conditions in the range from $\varepsilon = 0.6$ to $\varepsilon = 0.8$. However, the most important design consideration is to make sure the minimum film thickness, h_n, will be much higher than the size of surface asperities or the level of journal vibrations during operation.

6.2 REYNOLDS EQUATION FOR A JOURNAL BEARING

In Chapter 4, the hydrodynamic equations of a long journal bearing were derived from first principles. In this chapter, the hydrodynamic equations are derived from the Reynolds equation. The advantage of the present approach is that it can apply to a wider range of problems, such as bearings under dynamic conditions.

Let us recall that the general Reynolds equation for a Newtonian incompressible thin fluid film is

$$\frac{\partial}{\partial x}\left(\frac{h^3}{\mu}\frac{\partial p}{\partial x}\right) + \frac{\partial}{\partial z}\left(\frac{h^3}{\mu}\frac{\partial p}{\partial z}\right) = 6(U_1 - U_2)\frac{\partial h}{\partial x} + 12(V_2 - V_1) \tag{6-5}$$

Here, U_1 and U_2 are velocity components, in the x direction, of the lower and upper sliding surfaces, respectively (fluid film boundaries), while the velocities components V_1 and V_2 are of the lower and upper boundaries, respectively, in the y direction (see Fig. 5-2). The difference in normal velocity $(V_1 - V_2)$ is of relative motion (squeeze-film action) of the surfaces toward each other.

For most journal bearings, only the journal is rotating and the sleeve is stationary, $U_1 = 0$, $V_1 = 0$ (as shown in Fig. 6-1). The second fluid film boundary is at the journal surface that has a velocity $U = \omega R$. However, the velocity U is not parallel to the x direction (the x direction is along the bearing surface).

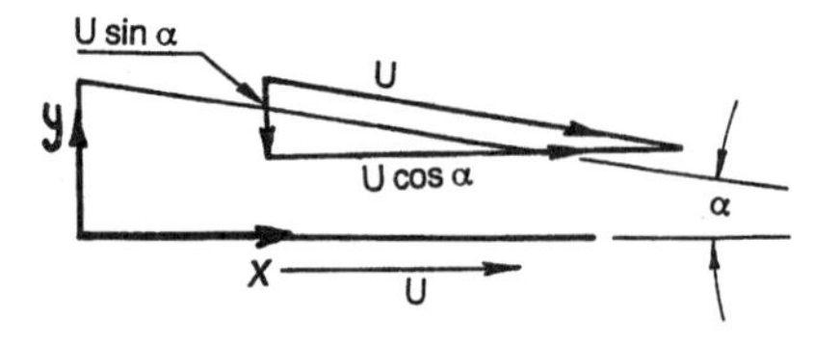

FIG. 6-2 Velocity components of the fluid film boundaries.

Therefore, it has two components (as shown in Fig. 6-2), U_2 and V_2, in the x and y directions, respectively:

$$U_2 = U\cos\alpha$$
$$V_2 = -U\sin\alpha \tag{6-6}$$

Here, the slope α is between the bearing and journal surfaces. In a journal bearing, the slope α is very small; therefore, the following approximations can be applied:

$$U_2 = U\cos\alpha \approx U$$
$$V_2 = -U\sin\alpha \approx -U\tan\alpha \tag{6-7}$$

The slope α can be expressed in terms of the function of the clearance, h:

$$\tan\alpha = -\frac{\partial h}{\partial x} \tag{6-8}$$

The normal component V_2 becomes

$$V_2 \approx U\frac{\partial h}{\partial x} \tag{6-9}$$

After substituting Eqs. (6.7) and (6.9) into the right-hand side of the Reynolds equation, it becomes

$$6(U_1 - U_2)\frac{\partial h}{\partial x} + 12(V_2 - V_1) = 6(0 - U)\frac{\partial h}{\partial x} + 12U\frac{\partial h}{\partial x} = 6U\frac{\partial h}{\partial x} \tag{6-10}$$

The Reynolds equation for a Newtonian incompressible fluid reduces to the following final equation:

$$\frac{\partial}{\partial x}\left(\frac{h^3}{\mu}\frac{\partial p}{\partial x}\right) + \frac{\partial}{\partial z}\left(\frac{h^3}{\mu}\frac{\partial p}{\partial z}\right) = 6U\frac{\partial h}{\partial x} \tag{6-11}$$

For an infinitely long bearing, $\partial p/\partial z \cong 0$; therefore, the second term on the left-hand side of Eq. (6-11) can be omitted, and the Reynolds equation reduces to the following simplified one-dimensional equation:

$$\frac{\partial}{\partial x}\left(\frac{h^3}{\mu}\frac{\partial p}{\partial x}\right) = 6U\frac{\partial h}{\partial x} \tag{6 12}$$

6.3 JOURNAL BEARING WITH ROTATING SLEEVE

There are many practical applications where the sleeve is rotating as well as the journal. In that case, the right-hand side of the Reynolds equation is not the same as for a common stationary bearing. Let us consider an example, as shown in Fig.

6-3, where the sleeve and journal are rolling together like internal friction pulleys. The rolling is similar to that in a cylindrical rolling bearing. The internal sleeve and journal surface are rolling together without slip, and both have the same tangential velocity $\omega_j R = \omega_b R_1 = U$.

The tangential velocities of the film boundaries, in the x direction, of the two surfaces are

$$U_1 = U$$
$$U_2 = U \cos\alpha \approx U \tag{6-13}$$

The normal velocity components, in the y direction, of the film boundaries (the journal and sleeve surfaces) are

$$V_1 = 0$$
$$V_2 \approx U \frac{\partial h}{\partial x} \tag{6-14}$$

The right-hand side of the Reynolds equation becomes

$$6(U_1 - U_2)\frac{\partial h}{\partial x} + 12(V_2 - V_1) = 6(U - U)\frac{\partial h}{\partial x} + 12(U\frac{\partial h}{\partial x} - 0) = 12U\frac{\partial h}{\partial x} \tag{6-15}$$

For the rolling action, the Reynolds equation for a journal bearing with a rolling sleeve is given by

$$\frac{\partial}{\partial x}\left(\frac{h^3}{\mu}\frac{\partial p}{\partial x}\right) + \frac{\partial}{\partial z}\left(\frac{h^3}{\mu}\frac{\partial p}{\partial z}\right) = 12U\frac{\partial h}{\partial x} \tag{6-16}$$

The right-hand side of Eq. (6-16) indicates that the rolling action will result in a doubling of the pressure wave of a common journal bearing of identical geometry as well as a doubling its load capacity. The physical explanation is that the fluid is squeezed faster by the rolling action.

6.4 COMBINED ROLLING AND SLIDING

In many important applications, such as gears, there is a combination of rolling and sliding of two cylindrical surfaces. Also, there are several unique applications where a hydrodynamic journal bearing operates in a combined rolling-and-sliding mode. A combined bearing is shown in Fig. 6-3, where the journal surface velocity is $R\omega_j$ while the sleeve inner surface velocity is $R_1\omega_b$.

The coefficient ξ is the ratio of the rolling and sliding velocity. In terms of the velocities of the two surfaces, the ratio is,

$$\xi = \frac{R_1\omega_b}{R\omega_j} \tag{6-17}$$

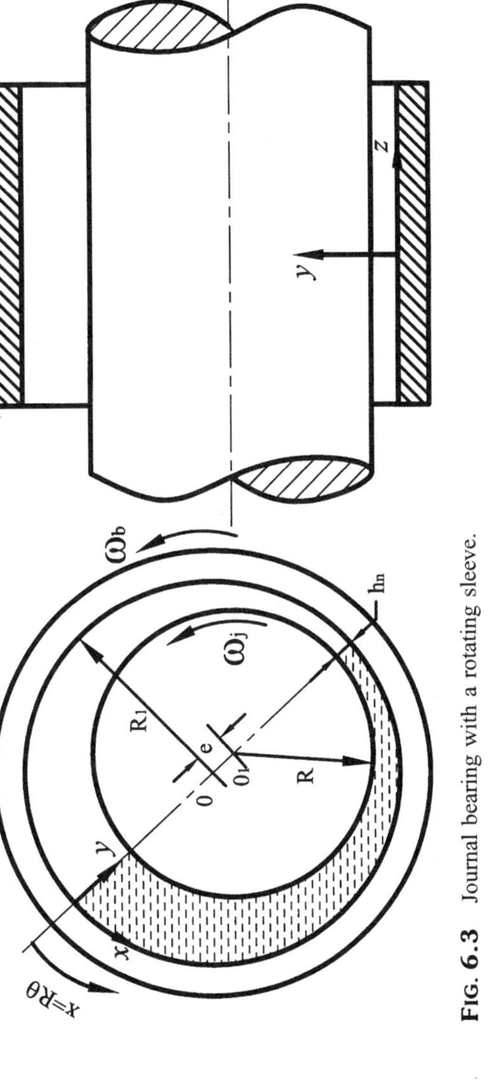

FIG. 6.3 Journal bearing with a rotating sleeve.

The journal surface velocity is $U = R\omega_j$, and the sleeve surface velocity is the product ξU, where ξ is the rolling-to-sliding ratio. The common journal bearing has a pure sliding and $\xi = 0$, while in pure rolling, $\xi = 1$. For all other combinations, $0 < \xi < 1$.

The tangential velocities (in the x direction) of the fluid–film boundaries of the two surfaces are

$$U_1 = R_1\omega_b = \xi R\omega_j$$
$$U_2 = R\omega_j \cos\alpha \approx R\omega_j \tag{6-18}$$

The normal components, in the y direction, of the velocity of the fluid film boundaries (journal and sleeve surfaces) are

$$V_1 = 0$$
$$V_2 = R\omega_j \frac{\partial h}{\partial x} \tag{6-19}$$

For the general case of a combined rolling and sliding, the expression on the right-hand side of Reynolds equation is obtained by substituting the preceding components of the surface velocity:

$$6(U_1 - U_2)\frac{\partial h}{\partial x} + 12(V_2 - V_1) = 6(\xi U - U)\frac{\partial h}{\partial x} + 12\left(U\frac{\partial h}{\partial x} - 0\right)$$
$$= 6U(1+\xi)\frac{\partial h}{\partial x} \tag{6-20}$$

Here, $U = R\omega_j$ is the journal surface velocity. Finally, the Reynolds equation for a combined rolling and sliding of a journal bearing is as follows:

$$\frac{\partial}{\partial x}\left(\frac{h^3}{\mu}\frac{\partial p}{\partial x}\right) + \frac{\partial}{\partial z}\left(\frac{h^3}{\mu}\frac{\partial p}{\partial z}\right) = 6R\omega_j(1+\xi)\frac{\partial h}{\partial x} \tag{6-21a}$$

For $\xi = 0$ and $\xi = 1$, the right-hand-side of the Reynolds equation is in agreement with the previous derivations for pure sliding and pure rolling, respectively. The right-hand side of Eq. (6-21a) indicates that pure rolling action doubles the pressure wave in comparison to a pure sliding.

Equation (6-21a) is often written in the basic form:

$$\frac{\partial}{\partial x}\left(\frac{h^3}{\mu}\frac{\partial p}{\partial x}\right) + \frac{\partial}{\partial z}\left(\frac{h^3}{\mu}\frac{\partial p}{\partial z}\right) = 6R(\omega_j + \omega_b)\frac{\partial h}{\partial x} \tag{6-21b}$$

In journal bearings, the difference between the journal radius and the bearing radius is small and we can assume that $R_1 \approx R$.

6.5 PRESSURE WAVE IN A LONG JOURNAL BEARING

For a common long journal bearing with a stationary sleeve, the pressure wave is derived by a double integration of Eq. (6.12). After the first integration, the following explicit expression for the pressure gradient is obtained:

$$\frac{dp}{dx} = 6U\mu\frac{h + C_1}{h^3} \tag{6-22}$$

Here, C_1 is a constant of integration. In this equation, a regular derivative replaces the partial one, because in a long bearing, the pressure is a function of one variable, x, only. The constant, C_1, can be replaced by h_0, which is the film thickness at the point of peak pressure. At the point of peak pressure,

$$\frac{dp}{dx} = 0 \quad \text{at} \quad h = h_0 \tag{6-23}$$

Substituting condition (6-23), in Eq. (6-22) results in $C_1 = -h_0$, and Eq. (6-22) becomes

$$\frac{dp}{dx} = 6U\mu\frac{h - h_0}{h^3} \tag{6-24}$$

Equation (6-24) has one unknown, h_0, which is determined later from additional information about the pressure wave.

The expression for the pressure distribution (pressure wave) around a journal bearing, along the x direction, is derived by integration of Eq. (6-24), and there will be an additional unknown—the constant of integration. By using the two boundary conditions of the pressure wave, we solve the two unknowns, h_0, and the second integration constant.

The pressures at the start and at the end of the pressure wave are usually used as boundary conditions. However, in certain cases the locations of the start and the end of the pressure wave are not obvious. For example, the fluid film of a practical journal bearing involves a fluid cavitation, and other boundary conditions of the pressure wave are used for solving the two unknowns. These boundary conditions are discussed in this chapter. The solution method of two unknowns for these boundary conditions is more complex and requires computer iterations.

Replacing x by an angular coordinate θ, we get

$$x = R\theta \tag{6-25}$$

Equation (6-24) takes the form

$$\frac{dp}{d\theta} = 6UR\mu \frac{h - h_0}{h^3} \tag{6-26}$$

For the integration of Eq. (6-26), the boundary condition at the start of the pressure wave is required. The pressure wave starts at $\theta = 0$, and the magnitude of pressure at $\theta = 0$ is p_0. The pressure, p_0, can be very close to atmospheric pressure, or much higher if the oil is fed into the journal bearing by an external pump.

The film thickness, h, as a function of θ for a journal bearing is given in Eq. (6-3), $h(\theta) = C(1 + \varepsilon \cos \theta)$. After substitution of this expression into Eq. (6-26), the pressure wave is given by

$$\frac{C^2}{6\mu UR}(p - p_0) = \int_0^\theta \frac{d\theta}{(1 + \varepsilon \cos \theta)^2} - \frac{h_0}{C} \int_0^\theta \frac{d\theta}{(1 + \varepsilon \cos \theta)^3} \tag{6-27}$$

The pressure, p_0, is determined by the oil supply pressure (inlet pressure). In a common journal bearing, the oil is supplied through a hole in the sleeve, at $\theta = 0$. The oil can be supplied by gravitation from an oil container or by a high-pressure pump. In the first case, p_0 is only slightly above atmospheric pressure and can be approximated as $p_0 = 0$. However, if an external pump supplies the oil, the pump pressure (at the bearing inlet point) determines the value of p_0. In industry, there are often central oil circulation systems that provide oil under pressure for the lubrication of many bearings.

The two integrals on the right-hand side of Eq. (6-27) are functions of the eccentricity ratio, ε. These integrals can be solved by numerical or analytical integration. Sommerfeld (1904) analytically solved these integrals for a full (360°) journal bearing. Analytical solutions of these two integrals are included in integral tables of most calculus textbooks. For a full bearing, it is possible to solve for the load capacity even in cases where p_0 can't be determined. This is because p_0 is an extra constant hydrostatic pressure around the bearing, and, similar to atmospheric pressure, it does not contribute to the load capacity.

In a journal bearing, the pressure that is predicted by integrating Eq. (6-27) increases (above the inlet pressure, p_0) in the region of converging clearance, $0 < \theta < \pi$. However, in the region of a diverging clearance, $\pi < \theta < 2\pi$, the pressure wave reduces below p_0. If the oil is fed at atmospheric pressure, the pressure wave that is predicted by integration of Eq. (6-27) is negative in the divergent region, $\pi < \theta < 2\pi$. This analytical solution is not always valid, because a negative pressure would result in a "fluid cavitation." The analysis may predict negative pressures below the absolute zero pressure, and that, of course, is physically impossible.

A continuous fluid film cannot be maintained at low negative pressure (relative to the atmospheric pressure) due to fluid cavitation. This effect occurs whenever the pressure reduces below the vapor pressure of the oil, resulting in an oil film rupture. At low pressures, the boiling process can take place at room temperature; this phenomenon is referred to as *fluid cavitation*. Moreover, at low pressure, the oil releases its dissolved air, and the oil foams in many tiny air bubbles. Antifoaming agents are usually added into the oil to minimize this effect.

As a result of cavitation and foaming, the fluid is not continuous in the divergent clearance region, and the actual pressure wave cannot be predicted anymore by Eq. (6-27). In fact, the pressure wave is maintained only in the converging region, while the pressure in most of the diverging region is close to ambient pressure. Cole and Hughes (1956) conducted experiments using a transparent sleeve. Their photographs show clearly that in the divergent region, the fluid film ruptures into filaments separated by air and lubricant vapor.

Under light loads or if the supply pressure, p_0, is high, the minimum predicted pressure in the divergent clearance region is not low enough to generate a fluid cavitation. In such cases, Eq. (6-27) can be applied around the complete journal bearing. The following solution referred to as the *Sommerfeld solution*, is limited to cases where there is a full fluid film around the bearing.

6.6 SOMMERFELD SOLUTION OF THE PRESSURE WAVE

Sommerfeld (1904) solved Eq. (6-27) for the pressure wave and load capacity of a full hydrodynamic journal bearing (360°) where a fluid film is maintained around the bearing without any cavitation. This example is of a special interest because this was the first analytical solution of a hydrodynamic journal bearing based on the Reynolds equation. In practice, a full hydrodynamic lubrication around the bearing is maintained whenever at least one of the following two conditions are met:

a. The feed pressure, p_0, (from an external oil pump), into the bearing is quite high in order to maintain positive pressures around the bearing and thus prevent cavitation.
b. The journal bearing is lightly loaded. In this case, the minimum pressure is above the critical value of cavitation.

Sommerfeld assumed a periodic pressure wave around the bearing; namely, the pressure is the same at $\theta = 0$ and $\theta = 2\pi$:

$$p_{(\theta=0)} = p_{(\theta=2\pi)} \tag{6-28}$$

The unknown, h_0, that represents the film thickness at the point of a peak pressure can be solved from Eq. (6-27) and the Sommerfeld boundary condition in Eq. (6-28). After substituting $p - p_0 = 0$, at $\theta = 2\pi$, Eq. (6-27) yields

$$\int_0^{2\pi} \frac{d\theta}{(1+\varepsilon\cos\theta)^2} - \frac{h_0}{C}\int_0^{2\pi} \frac{d\theta}{(1+\varepsilon\cos\theta)^3} = 0 \tag{6-29}$$

This equation can be solved for the unknown, h_0. The following substitutions for the values of the integrals can simplify the analysis of hydrodynamic journal bearings:

$$J_n = \int_0^{2\pi} \frac{d\theta}{(1+\varepsilon\cos\theta)^n} \tag{6-30}$$

$$I_n = \int_0^{2\pi} \frac{\cos\theta\, d\theta}{(1+\varepsilon\cos\theta)^n} \tag{6-31}$$

Equation (6-29) is solved for the unknown, h_0, in terms of the integrals J_n:

$$\frac{h_0}{C} = \frac{J_2}{J_3} \tag{6-32}$$

Here, the solutions for the integrals J_n are:

$$J_1 = \frac{2\pi}{(1-\varepsilon^2)^{1/2}} \tag{6-33}$$

$$J_2 = \frac{2\pi}{(1-\varepsilon^2)^{3/2}} \tag{6-34}$$

$$J_3 = \left(1 + \frac{1}{2}\varepsilon^2\right)\frac{2\pi}{(1-\varepsilon^2)^{5/2}} \tag{6-35}$$

$$J_4 = \left(1 + \frac{3}{2}\varepsilon^2\right)\frac{2\pi}{(1-\varepsilon^2)^{7/2}} \tag{6-36}$$

The solutions of the integrals I_n are required later for the derivation of the expression for the load capacity. The integrals I_n can be obtained from J_n by the following equation:

$$I_n = \frac{J_{n-1} - J_n}{\varepsilon} \tag{6-37}$$

Sommerfeld solved for the integrals in Eq. (6-27), and obtained the following equation for the pressure wave around an infinitely long journal bearing with a full film around the journal bearing:

$$p - p_0 = \frac{6\mu U R}{C^2}\frac{\varepsilon(2+\varepsilon\cos\theta)\sin\theta}{(2+\varepsilon^2)(1+\varepsilon\cos\theta)^2} \tag{6-38}$$

The curves in Fig. 6-4 are dimensionless pressure waves, relative to the inlet pressure, for various eccentricity ratios, ε. The pressure wave is an antisymmetrical function on both sides of $\theta = \pi$. The curves indicate that the peak pressure considerably increases with the eccentricity ratio, ε. According to Eq. (6-38), the peak pressure approaches infinity when ε approaches 1. However, this is not possible in practice because the surface asperities prevent a complete contact between the sliding surfaces.

6.7 JOURNAL BEARING LOAD CAPACITY

Figure 6-5 shows the load capacity, W, of a journal bearing and its two components, W_x and W_y. The direction of W_x is along the bearing symmetry line $O-O_1$. This direction is inclined at an attitude angle, ϕ, from the direction of the external force and load capacity, W. In Fig. 6-5 the external force is in a vertical direction. The direction of the second component, W_y is normal to the W_x direction.

The elementary load capacity, dW, acts in the direction normal to the journal surface. It is the product of the fluid pressure, p, and an elementary area,

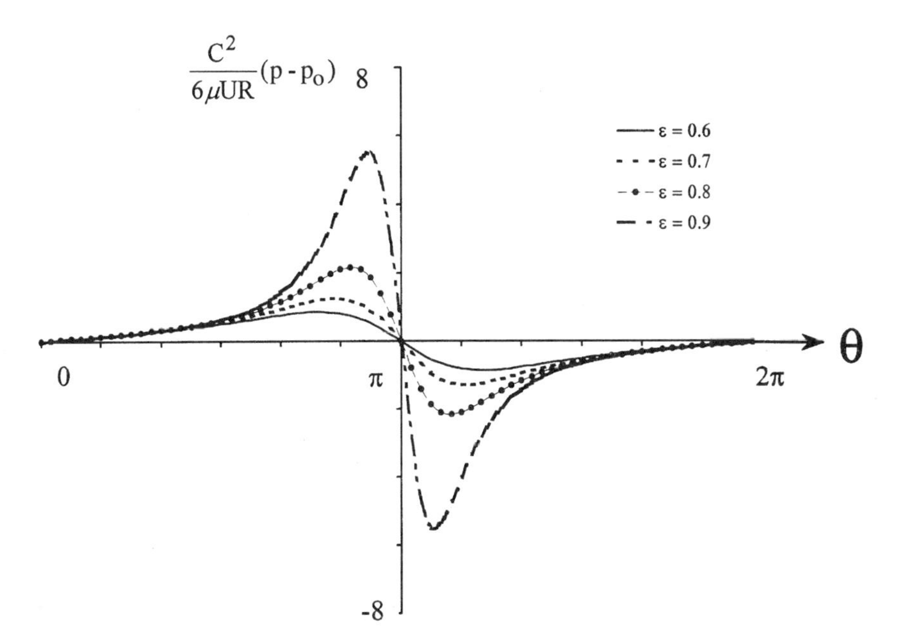

FIG. 6-4 Pressure waves in an infinitely long, full bearing according to the Sommerfeld solution.

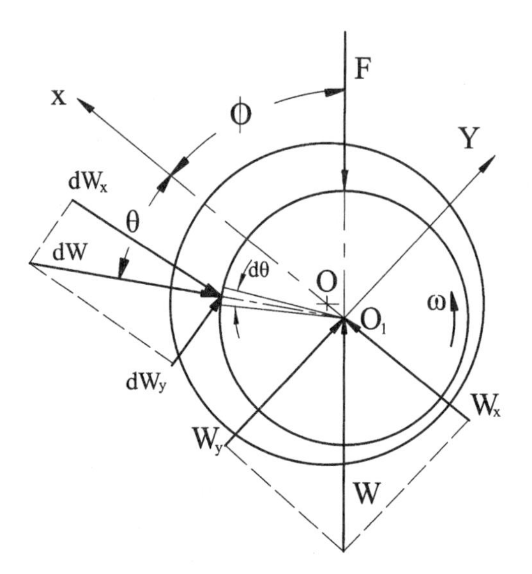

FIG. 6-5 Hydrodynamic bearing force components.

$dA = LR\,d\theta$, of the journal surface bounded by a small journal angle, $d\theta$. An elementary fluid force ($dW = p\,dA$) is given by,

$$dW = pLR\,d\theta \tag{6-39}$$

The pressure is acting in the direction normal to the journal surface, and dW is a radial elementary force vector directed toward the journal center, as shown in Fig. 6-5.

In a plane-slider, the pressure is acting in one direction and the load capacity has been derived by integration of the pressure wave. However, in a journal bearing, the direction of the pressure varies around the bearing, and simple summation of the elementary forces, dW, will not yield the resultant force. In order to allow summation, the elementary force, dW, is divided into two components, dW_x and dW_y, in the directions of W_x and W_y, respectively. By using force components, it is possible to have a summation (by integration) of each component around the bearing. The elementary force components are:

$$dW_x = -pLR\cos\theta\,d\theta \tag{6-40}$$

$$dW_y = pLR\sin\theta\,d\theta \tag{6-41}$$

The two components of the load capacity, W_x and W_y, are in the X and Y directions, respectively, as indicated in Fig. 6-5. Note the negative sign in Eq. (6-40), since dW_x is opposite to the W_x direction. The load components are:

$$W_x = -LR \int_0^{2\pi} p \cos\theta \, d\theta \tag{6-42}$$

$$W_y = LR \int_0^{2\pi} p \sin\theta \, d\theta \tag{6-43}$$

The attitude angle ϕ in Fig. 6-5 is determined by the ratio of the force components:

$$\tan\phi = \frac{W_y}{W_x} \tag{6-44}$$

6.8 LOAD CAPACITY BASED ON SOMMERFELD CONDITIONS

The load capacity components can be solved by integration of Eqs. (6-42 and (6-43), where p is substituted from Eq. (6-38). However, the derivation can be simplified if the load capacity components are derived directly from the basic Eq. (6-24) of the pressure gradient. In this way, there is no need to integrate the complex Eq. (6-38) of the pressure wave. This can be accomplished by employing the following identity for product derivation:

$$(uv)' = uv' + vu' \tag{6-45}$$

Integrating and rearranging Eq. (6-45) results in

$$\int uv' = uv - \int u'v \tag{6-46}$$

In order to simplify the integration of Eq. (6-42) for the load capacity component W_x, the substitutions $u = p$ and $v' = \cos\theta$ are made. This substitution allows the use of the product rule in Eq. (6-46), and the integral in Eq. (6-42) results in the following terms:

$$\int p\cos\theta \, d\theta = p\sin\theta - \int \frac{dp}{d\theta} \sin\theta \, d\theta \tag{6-47}$$

In a similar way, for the load capacity component, W_y, in Eq. (6-43), the substitutions $u = p$ and $v' = \sin\theta$ result in

$$\int p\sin\theta \, d\theta = -p\cos\theta + \int \frac{dp}{d\theta} \cos\theta \, d\theta \tag{6-48}$$

Equations (6-47) and (6-48) indicate that the load capacity components in Eqs. (6-42) and (6-43) can be solved directly from the pressure gradient. By using this method, it is not necessary to solve for the pressure wave in order to find the load capacity components (it offers the considerable simplification of one simple integration instead of a complex double integration). The first term, on the right-hand side in Eqs. (6-47) and (6-48) is zero, when integrated around a full bearing, because the pressure, p, is the same at $\theta = 0$ and $\theta = 2\pi$.

Integration of the last term in Eq. (6-47), in the boundaries $\theta = 0$ and $\theta = 2\pi$, indicates that the load component, W_x, is zero. This is because it is an integration of the antisymmetrical function around the bearing (the function is antisymmetric on the two sides of the centerline O–O_1, which cancel each other). Therefore:

$$W_x = 0 \tag{6-49}$$

Integration of Eq. (6-43) with the aid of identity (6-48), and using the value of h_0 in Eq. 6-32 results in

$$W_y = \frac{6\mu U R^2 L}{C^2}\left(I_2 - \frac{J_2}{J_3} I_3\right) \tag{6-50}$$

Substituting for the values of I_n and J_n as a function of ε yields the following expression for the load capacity component, W_y. The other component is $W_x = 0$; therefore, for the Sommerfeld conditions, W_y is equal to the total load capacity, $W = W_y$:

$$W = \frac{12\pi\mu U R^2 L}{C^2} \frac{\varepsilon}{(2 + \varepsilon^2)(1 - \varepsilon^2)^{1/2}} \tag{6-51}$$

The attitude angle, ϕ, is derived from Eq. (6-44). For Sommerfeld's conditions, $W_x = 0$ and $\tan\phi \to \infty$; therefore,

$$\phi = \frac{\pi}{2} \tag{6-52}$$

Equation (6-52) indicates that in this case, the symmetry line O–O_1 is normal to the direction of the load capacity W.

6.9 FRICTION IN A LONG JOURNAL BEARING

The bearing friction force, F_f, is the viscous resistance force to the rotation of the journal due to high shear rates in the fluid film. This force is acting in the tangential direction of the journal surface and results in a resistance torque to the rotation of the journal. The friction force is defined as the ratio of the friction torque, T_f, to the journal radius, R:

$$F_f = \frac{T_f}{R} \tag{6-53}$$

The force is derived by integration of the shear stresses over the area of the journal surface, at $y = h$, around the bearing. The shear stress distribution at the journal surface (shear at the wall, τ_w) around the bearing, is derived from the velocity gradient, as follows:

$$\tau_w = \mu \left.\frac{du}{dy}\right|_{(y=h)} \tag{6-54}$$

The friction force is obtained by integration:

$$F_f = \int_A \tau_{(y=h)}\, dA \tag{6-55}$$

After substituting $dA = RL\, d\theta$ in Eq. (6-55), the friction force becomes

$$F_f = \mu RL \int_0^{2\pi} \tau_{(y=h)} d\theta \tag{6-56}$$

Substitution of the value of the shear stress, Eq. (6-56) becomes

$$F_f = \mu URL \int_0^{2\pi} \left(\frac{4}{h} - \frac{3h_0}{h^2}\right) d\theta \tag{6-57}$$

If we apply the integral definitions in Eq. (6-30), the expression for the friction force becomes

$$F_f = \frac{\mu RL}{C}\left(4J_1 - 3\frac{J_2^2}{J_3}\right) \tag{6-58}$$

The integrals J_n are functions of the eccentricity ratio. Substituting the solution of the integrals, J_n, in Eqs. (6.33) to (6.37) results in the following expression for the friction force:

$$F_f = \frac{\mu URL}{C} \frac{4\pi(1 + 2\varepsilon^2)}{(2 + \varepsilon^2)(1 - \varepsilon^2)^{1/2}} \tag{6-59}$$

Let us recall that the bearing friction coefficient, f, is defined as

$$f = \frac{F_f}{W} \tag{6-60}$$

Substitution of the values of the friction force and load in Eq. (6-60) results in a relatively simple expression for the coefficient of friction of a long hydrodynamic journal bearing:

$$f = \frac{C}{R}\frac{1 + 2\varepsilon^2}{3\varepsilon} \tag{6-61}$$

Comment: The preceding equation for the friction force in the fluid film is based on the shear at $y = h$. The viscous friction force around the bearing bore surface, at $y = 0$, is not equal to that around the journal surface, at $y = h$. The viscous friction torque on the journal surface is unequal to that on the bore surface because the external load is eccentric to the bore center, and it is an additional torque. However, the friction torque on the journal surface is the actual total resistance to the journal rotation, and it is used for calculating the friction energy losses in the bearing.

6.10 POWER LOSS ON VISCOUS FRICTION

The energy loss, per unit of time (power loss) $\dot{E}_f$, is determined from the friction torque, or friction force, by the following equations:

$$\dot{E}_f = T_f\omega = F_f U \tag{6-62}$$

where ω (rad/s) is the angular velocity of the journal. Substituting Eq. (6-59) into Eq. (6-62) yields

$$\dot{E}_f = \frac{\mu U^2 RL}{C}\frac{4\pi(1 + 2\varepsilon^2)}{(2 + \varepsilon^2)(1 - \varepsilon^2)^{1/2}} \tag{6-63}$$

The friction energy losses are dissipated in the lubricant as heat. Knowledge of the amount of friction energy that is dissipated in the bearing is very important for ensuring that the lubricant does not overheat. The heat must be transferred from the bearing by adequate circulation of lubricant through the bearing as well as by conduction of heat from the fluid film through the sleeve and journal.

6.11 SOMMERFELD NUMBER

Equation (6-51) is the expression for the bearing load capacity in a long bearing operating at steady conditions with the Sommerfeld boundary conditions for the pressure wave. This result was obtained for a full film bearing without any cavitation around the bearing. In most practical cases, this is not a realistic expression for the load capacity, since there is fluid cavitation in the diverging clearance region of negative pressure. The Sommerfeld solution for the load capacity has been improved by applying a more accurate analysis with realistic boundary conditions of the pressure wave. This solution requires iterations performed with the aid of a computer.

In order to simplify the design of hydrodynamic journal bearings, the realistic results for the load capacity are provided in dimensionless form in tables or graphs. For this purpose, a widely used dimensionless number is the Sommerfeld number, S.

Equation (6-51) can be converted to dimensionless form if all the variables with dimensions are placed on the left-hand side of the equations and the dimensionless function of ε on the right-hand side. Also, it is the tradition in this discipline to have the Sommerfeld dimensionless group as a function of journal speed, n, in revolutions per second, and the average bearing pressure, P, according to the following substitutions:

$$U = 2\pi R n \tag{6-64}$$

$$W = 2RLP \tag{6-65}$$

By substituting Eqs. (6.64) and (6.65) into Eq. (6.51), we obtain the following dimensionless form of the Sommerfeld number for an infinitely long journal bearing where cavitation is disregarded:

$$S = \frac{\mu n}{P}\left(\frac{R}{C}\right)^2 = \frac{(2+\varepsilon^2)(1-\varepsilon^2)^{1/2}}{12\pi^2\varepsilon} \tag{6-66}$$

The dimensionless Sommerfeld number is a function of ε only. For an infinitely long bearing and the Sommerfeld boundary conditions, the number S can be calculated via Eq. (6-66). However, for design purposes, we use the realistic conditions for the pressure wave. The values of S for various ratios of length and diameter, L/D, have been computed; the results are available in Chapter 8 in the form of graphs and tables for design purposes.

6.12 PRACTICAL PRESSURE BOUNDARY CONDITIONS

The previous discussion indicates that for most practical applications in machinery the pressures are very high and cavitation occurs in the diverging region of the clearance. For such cases, the Sommerfeld boundary conditions do not apply and the pressure distribution can be solved for more realistic conditions. In an actual journal bearing, there is no full pressure wave around the complete bearing, and there is a positive pressure wave between θ_1 and θ_2. Outside this region, there is cavitation and low negative pressure that can be ignored for the purpose of calculating the load capacity. The value of θ_2 is unknown, and an additional condition must be applied for the solution. The boundary condition commonly used is that the pressure gradient is zero at the end of the pressure wave, at θ_2.

This is equivalent to the assumption that there is no flow, u, in the x direction at θ_2. When the lubricant is supplied at θ_1 at a pressure p_0, the following boundary conditions can be applied:

$$\begin{aligned} p &= p_0 \quad \text{at} \quad \theta = \theta_1 \\ \frac{dp}{d\theta} &= 0 \quad \text{at} \quad \theta = \theta_2 \\ p &= 0 \quad \text{at} \quad \theta = \theta_2 \end{aligned} \tag{6-67}$$

In most cases, the feed pressure is close to ambient pressure at $\theta = 0$, and the first boundary condition is $p = 0$ at $\theta = 0$. The pressure wave with the foregoing boundary conditions is shown in Fig. 6-6.

The location of the end of the pressure wave, θ_2, is solved by iterations. The solution is performed by guessing a value for θ_2 that is larger then 180°, then integrating Eq. (6-27) at the boundaries from 0 to θ_2. The solution is obtained when the pressure at θ_2 is very close to zero.

Figure 6-6 indicates that the angle θ_2 and the angle of the peak pressure are symmetrical on both sides of $\theta = \pi$; therefore both have an equal film thickness h (or clearance thickness). The constant, h_0, representing the film thickness at the maximum pressure is

$$h_0 = C(1 + \cos\theta_2) \tag{6-68}$$

After the integration is completed, the previous guess for θ_2 is corrected until a satisfactory solution is obtained (the pressure at θ_2 is very close to zero). Preparation of a small computer program to solve for θ_2 by iterations is recommended as a beneficial exercise for the reader. After h_0 is solved the pressure wave can be plotted.

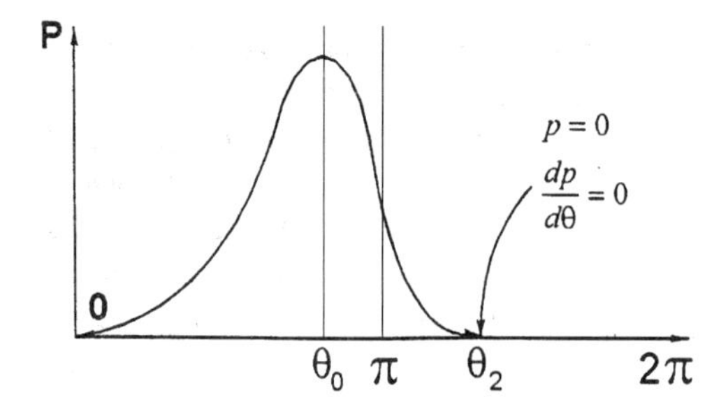

FIG. 6-6 Pressure wave plot under realistic boundary conditions.

Example Problem 6-1

Ice Sled

Two smooth cylindrical sections support a sled as shown in Fig. 6-7. The sled is running over ice on a thin layer of water film. The total load (weight of the sled and the person) is 1000 N. This load is acting at equal distances between the two blades. The sled velocity is 15 km/h, the radius of the blade is 30 cm, and its width is $L = 90$ cm. The viscosity of water $\mu = 1.792 \times 10^{-3}$ N-s/m^2.

a. Find the pressure distribution at the entrance region of the fluid film under the ski blade as it runs over the ice. Derive and plot the pressure wave in dimensionless form.
b. Find the expression for the load capacity of one blade.
c. Find the minimum film thickness ($h_n = h_{min}$) of the thin water layer.

Solution

a. Pressure Distribution and Pressure Wave

For $h_n/R \ll 1$, the pressure wave is generated only near the minimum-film region, where $x \ll R$, or $x/R \ll 1$.

For a small value of x/R, the equation for the clearance, between the quarter-cylinder and the ice, $h = h(x)$, can be approximated by a parabolic wedge. The following expression is obtained by expanding the equation for the variable clearance, $h(x)$, which is equal to the fluid-film thickness, into a Taylor series and truncating powers higher than x^2 (see Chapter 4).

If the minimum thickness of the film is $h_n = h_{min}$, the equation for the variable clearance becomes

$$h(x) = h_n + \frac{x^2}{2R}$$

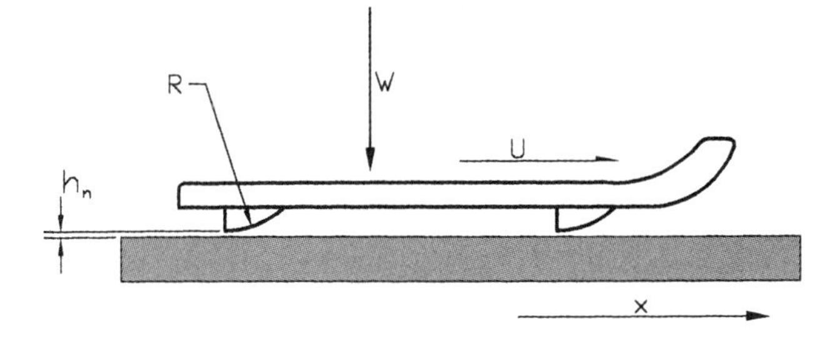

FIG. 6-7 Sled made of two quarter-cylinder.

For dimensionless analysis, the clearance function is written as

$$h(x) = h_n\left(1 + \frac{x^2}{2Rh_n}\right)$$

The width, L (in the direction normal to the sled speed), is very large in comparison to the film length in the x direction. Therefore, it can be considered an infinitely long bearing, and the pressure gradient is

$$\frac{dp}{dx} = 6\mu U \frac{h - h_0}{h^3}$$

The unknown, h_0, is the fluid film thickness at the point of peak pressure, where $dp/dx = 0$.

Conversion to Dimensionless Terms. The conversion to dimensionless terms is similar to that presented in Chapter 4. The length, x, is normalized by $\sqrt{2Rh_n}$, and the dimensionless terms are defined as

$$\bar{x} = \frac{x}{\sqrt{2Rh_n}} \qquad \text{and} \qquad \bar{h} = \frac{h}{h_n}$$

The dimensionless clearance gets a simple form:

$$\bar{h}(\bar{x}) = 1 + \bar{x}^2$$

Using the preceding substitutions, the pressure gradient equation takes the form

$$\frac{dp}{dx} = \frac{6\mu U}{h_n^2} \frac{\bar{h} - \bar{h}_0}{(1 + \bar{x}^2)^3}$$

Here the unknown, h_0, is replaced by

$$\bar{h}_0 = 1 + \bar{x}_0^2$$

The unknown, h_0, is replaced by unknown x_0, which describes the location of the peak pressure. Using the foregoing substitutions, the pressure gradient equation is reduced to

$$\frac{dp}{dx} = \frac{6\mu U}{h_n^2} \frac{\bar{x}^2 - \bar{x}_0^2}{(1 + \bar{x}^2)^3}$$

Converting dx into dimensionless form produces

$$dx = \sqrt{2Rh_n}\, d\bar{x}$$

The dimensionless differential equation for the pressure takes the form

$$\frac{h_n^2}{\sqrt{2Rh_n}\, 6\mu U} dp = \frac{\bar{x}^2 - \bar{x}_0^2}{(1 + \bar{x}^2)^3} d\bar{x}$$

Here, the dimensionless pressure is

$$\bar{p} = \frac{h_n^2}{6U\mu\sqrt{2Rh_n}} p$$

The final equation for integration of the dimensionless pressure is

$$\bar{p} = \frac{h_n^2}{\sqrt{2Rh_n}} \frac{1}{6\mu U} \int_0^p dp = \int_\infty^x \frac{\bar{x}^2 - \bar{x}_0^2}{(1+\bar{x}^2)^3} d\bar{x} + p_0$$

Here p_0 is a constant of integration, which is atmospheric pressure far from the minimum clearance. In this equation, p_0 and x_0 are two unknowns that can be solved for by the following boundary conditions of the pressure wave:

$$\text{at } x = \infty, \qquad p = 0$$
$$\text{at } x = 0, \qquad p = 0$$

The first boundary condition, $p = 0$ at $x = \infty$, yields $p_0 = 0$. The second unknown, x_0, is solved for by iterations (trial and error). The value of x_0 is varied until the second boundary condition of the pressure wave, $p = 0$ at $x = 0$, is satisfied.

Numerical Solution by Iterations. For numerical integration, the boundary $\bar{x} = \infty$ is replaced by a relatively large finite dimensionless value where pressure is small and can be disregarded, such as $\bar{x} = 4$.

The numerical solution involves iterations for solving for x_0. Each one of the iterations involves integration in the boundaries from 4 to 0. The solution for x_0 is obtained when the pressure at $x = 0$ is sufficiently close to zero. Using trial and error, we select each time a value for x_0 and integrate. This is repeated until the value of the dimensionless pressure is nearly zero at $x = 0$. The solution of the dimensionless pressure wave is plotted in Fig. 6-8. The maximum dimensionless pressure occurs at a dimensionless distance $\bar{x} = 0.55$, and the maximum dimensionless pressure is $\bar{p} = 0.109$.

$$\text{Dimensionless Pressure}, \bar{p} = \frac{h_n^2}{\sqrt{2Rh_n}} \frac{1}{6\mu U} p$$

Discussion of Numerical Iterations The method of iteration is often referred to as the *shooting method*. In order to run iteration, we guess a certain value of $\bar{x}_0$, and, using this value, we integrate and attempt to hit the target point $\bar{x} = 0, \bar{p} = 0$.

For the purpose of illustrating the shooting method, three iterations are shown in Fig. 6-9. The iteration for $\bar{x}_0 = 0.25$ results in a pressure that is too high at $\bar{x} = 0$; for $\bar{x}_0 = 0.85$, the pressure is too low at $\bar{x} = 0$. When the final iteration of $\bar{x}_0 = 0.55$ is made, the target point $\bar{x} = 0, \bar{p} = 0$ is reached with sufficient accuracy. The solution requires a small computer program.

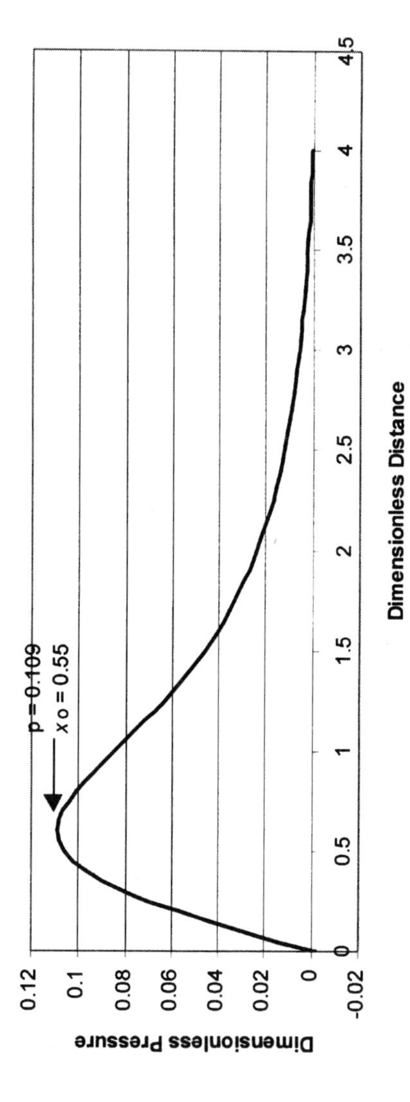

FIG. 6-8 Dimensionless pressure wave.

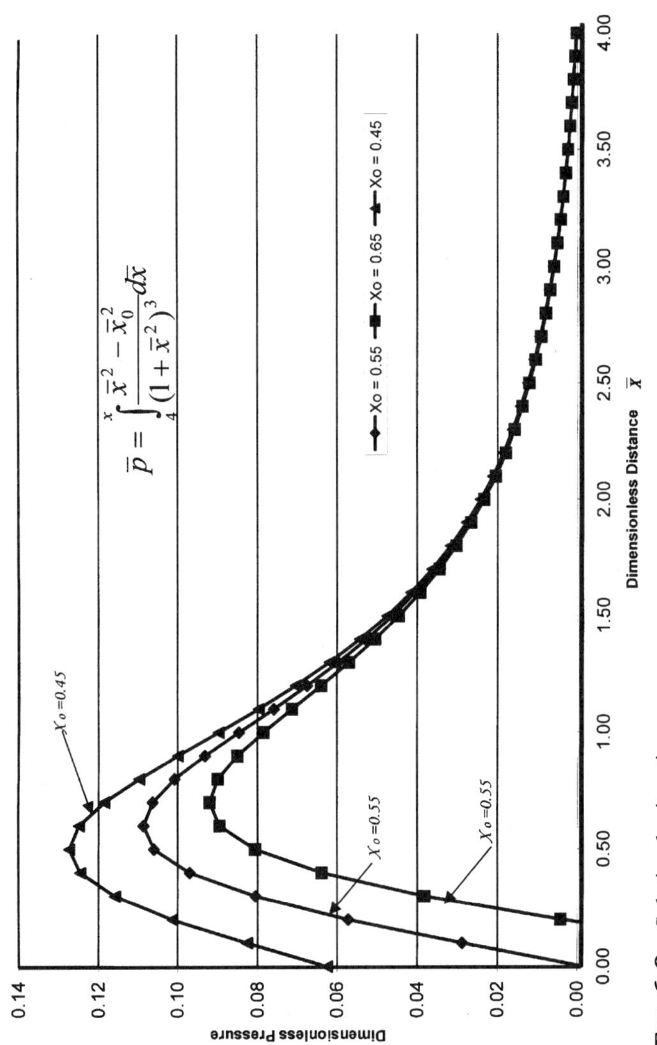

FIG. 6-9 Solution by iterations.

b. Load Capacity

The load capacity for one cylindrical section is solved by the equation

$$W = L\int_{(A)} p\,dx$$

Converting to dimensionless terms, the equation for the load capacity becomes

$$W = L\frac{\sqrt{2Rh_n}}{h_n^2}6\mu U\sqrt{2Rh_n}\int_{(A)} \bar{p}\,d\bar{x}$$

For the practical boundaries of the pressure wave, the equation takes the form

$$W = L\frac{2R}{h_n}6\mu U\int_4^0 \bar{p}\,d\bar{x}$$

c. Minimum Film Thickness

Solving for h_n, the minimum thickness between the ice and sled blade is

$$h_n = \frac{12LR\mu U}{W}\int_4^0 \bar{p}\,d\bar{x}$$

Numerical integration is performed based on the previous results of the dimensionless pressure in Fig. 6-8. The result for the load capacity is obtained by numerical integration, in the boundaries 4 to 0:

$$\int_4^0 \bar{p}\,d\bar{x} \approx \sum_4^0 \bar{p}_i\,\Delta x_i = 0.147$$

The minimum film thickness is

$$h_n = \frac{12 \times 0.9\text{ m} \times 0.3\text{ m} \times 1.792 \times 10^{-3}(\text{N-s/m}^2) \times 4.2\text{ m/s}}{500\text{ N}}(0.147)$$

$$h_n = 7.27 \times 10^{-6}\text{ m} = 7.27 \times 10^{-3}\text{ mm}$$

The result is: $h_n = 7.27\ \mu\text{m}$.

Example Problem 6-2

Cylinder on a Flat Plate

A combination of a long cylinder of radius R and a flat plate surface are shown in Fig. 5-4. The cylinder rotates and slides on a plane (there is a combination of rolling and sliding), such as in the case of gears where there is a theoretical line contact with a combination of rolling and sliding. However, due to the hydrodynamic action, there is a small minimum clearance, h_n. The viscosity, μ, of the lubricant is constant, and the surfaces of the cylinder and flat plate are rigid.

Assume practical pressure boundary conditions [Eqs. (6-67)] and solve the pressure wave (use numerical iterations). Plot the dimensionless pressure-wave for various rolling and sliding ratios ξ.

Solution

In Chapter 4, the equation for the pressure gradient was derived. The following is the integration for the pressure wave. The clearance between a cylinder of radius R and a flat plate is discussed in Chapter 4; see Eq. (4-33). For a fluid film near the minimum clearance, the approximation for the clearance is

$$h(x) = h_{\min} + \frac{x^2}{2R}$$

The case of rolling and sliding is similar to that of Eq. (6-20), (see Section 6.4). The Reynolds equation is in the form

$$\frac{\partial}{\partial x}\left(\frac{h^3}{\mu}\frac{\partial p}{\partial x}\right) + \frac{\partial}{\partial z}\left(\frac{h^3}{\mu}\frac{\partial p}{\partial z}\right) = 6U(1+\xi)\frac{\partial h}{\partial x}$$

Here, the coefficient ξ is the ratio of rolling and sliding. In terms of the velocities of the two surfaces, the ratio is

$$\xi - \frac{\omega R}{U}$$

The fluid film is much wider in the z direction in comparison to the length in the x direction. Therefore, the pressure gradient in the axial direction can be neglected in comparison to that in the x direction. The Reynolds equation is simplified to the form

$$\frac{\partial}{\partial x}\left(\frac{h^3}{\mu}\frac{\partial p}{\partial x}\right) = 6U(1+\xi)\frac{\partial h}{\partial x}$$

The preceding equation is converted into dimensionless terms (see Example Problem 6-1):

$$\bar{x} = \frac{x}{\sqrt{2Rh_n}}, \qquad \bar{h} = \frac{h}{h_n}, \qquad \text{and} \qquad \overline{h_0} = \frac{h_0}{h_n}$$

$$\frac{dp}{dx} = \frac{6\mu U(1+\xi)}{h_n^2} \frac{\bar{x}^2 - \bar{x}_0^2}{(1+\bar{x}^2)^3}$$

Converting the pressure gradient to dimensionless form yields

$$\frac{h_n^2}{\sqrt{2Rh_n}6\mu U} dp = (1+\xi)\frac{\bar{x}^2 - \bar{x}_0^2}{(1+\bar{x}^2)^3} d\bar{x}$$

The left hand side of the equation is the dimensionless pressure:

$$\bar{p} = (1+\xi)\frac{h_n^2}{\sqrt{2RH_n}} \frac{1}{6\mu U} \int_0^p dp = (1+\xi)\int_{-\infty}^{x} \frac{\bar{x}^2 - \bar{x}_0^2}{(1+\bar{x}^2)^3} d\bar{x} + p_0$$

Here, p_0 is a constant of integration, which is atmospheric pressure far from the minimum clearance. In this equation, p_0 and x_0 are two unknowns that can be solved for by the practical boundary conditions of the pressure wave; compare to Eqs. (6-67):

$$p = p_0 \quad \text{at} \quad x = x_1$$

$$\frac{dp}{dx} = 0 \quad \text{at} \quad x = x_2$$

$$p = 0 \quad \text{at} \quad x = x_2$$

Atmospheric pressure is zero, and the first boundary condition results in $p_0 = 0$. The location of the end of the pressure wave, x_2, is solved by iterations. The solution is performed by guessing a value for $x = x_2$; then x_0 is taken as x_2, because at that point the pressure gradient is zero.

The solution requires iterations in order to find $x_0 = x_2$, which satisfies the boundary conditions. For each iteration, integration is performed in the boundaries from 0 to x_2, and the solution is obtained when the pressure at x_2 is very close to zero.

For numerical integration, the boundary $\bar{x}_1$, where the pressure is zero, is taken as a small value, such as $\bar{x}_1 = -4$. The solution is presented in Fig. 6-10. The curves indicate that the pressure wave is higher for higher rolling ratios. This means that the rolling plays a stronger role in hydrodynamic pressure generation in comparison to sliding.

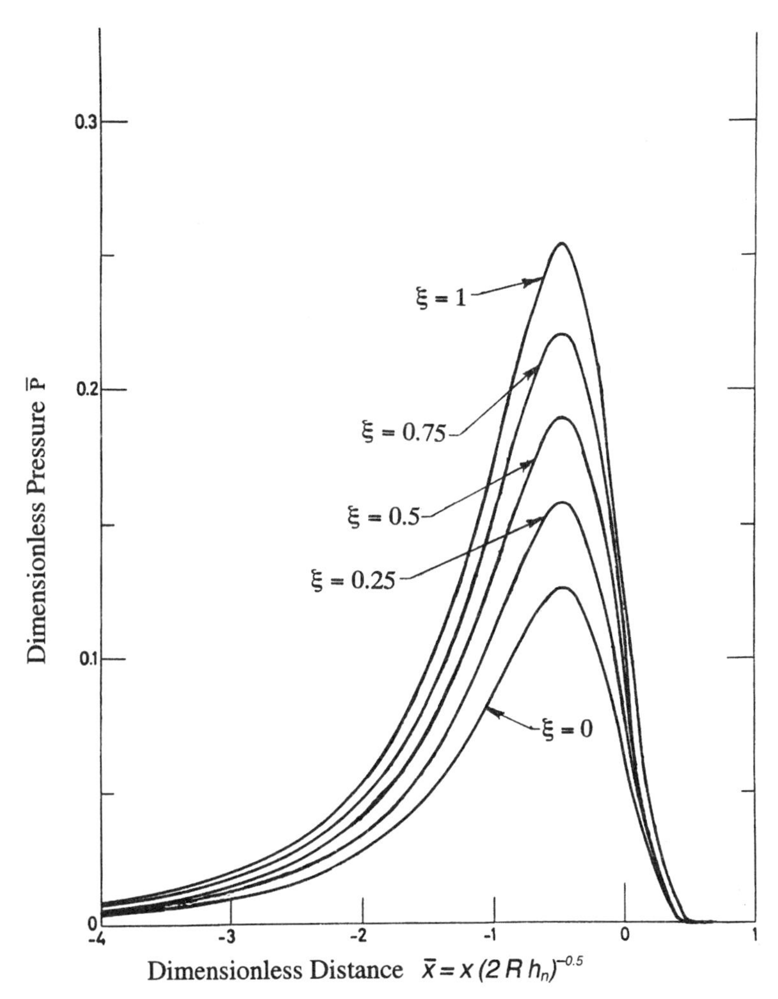

FIG. 6-10 Pressure wave along a fluid film between a cylinder and a flat plate for various rolling-to-sliding ratios.

Problems

6-1 A journal bearing is fed a by high-pressure external pump. The pump pressure is sufficient to avoid cavitation. The bearing length $L = 2D$. The diameter $D = 100$ mm, the shaft speed is 6000 RPM, and the clearance ratio is $C/R = 0.001$.

Assume that the infinitely-long-bearing analysis can be approximated for this bearing, and find the maximum load capacity for lubricant SAE 10 at average fluid film temperature of 80°C, if the maximum allowed eccentricity ratio $\varepsilon = 0.7$.

6-2 A flat plate slides on a lubricated cylinder as shown in Fig. 4-7. The cylinder radius is R, the lubricant viscosity is μ, and the minimum clearance between the stationary cylinder and plate is h_n. The elastic deformation of the cylinder and plate is negligible.

1. Apply numerical iterations, and plot the dimensionless pressure wave. Assume practical boundary conditions of the pressure, according to Eq. 6.67.
2. Find the expression for the load capacity by numerical integration.

6-3 In problem 6-2, the cylinder diameter is 250 mm, the plate slides at $U = 0.5$ m/s, and the minimum clearance is 1 μm (0.001 mm). The lubricant viscosity is constant, $\mu = 10^{-4}$ N-s/m^2. Find the hydrodynamic load capacity.

6-4 Oil is fed into a journal bearing by a pump. The supply pressure is sufficiently high to avoid cavitation. The bearing operates at an eccentricity ratio of $\varepsilon = 0.85$, and the shaft speed is 60 RPM. The bearing length is $L = 3D$, the journal diameter is $D = 80$ mm, and the clearance ratio is $C/R = 0.002$. Assume that the pressure is constant along the bearing axis and there is no axial flow (long-bearing theory).

a. Find the maximum load capacity for a lubricant SAE 20 operating at an average fluid film temperature of 60°C.
b. Find the bearing angle θ where there is a peak pressure.
c. What is the minimum supply pressure from the pump in order to avoid cavitation and to have only positive pressure around the bearing?

6-5 An air bearing operates inside a pressure vessel that has sufficiently high pressure to avoid cavitation in the bearing. The average viscosity of the air inside the bearing is $\mu = 2 \times 10^{-4}$ N-s/m^2. The bearing operates at an eccentricity ratio of $\varepsilon = 0.85$. The bearing length is $L = 2D$, the journal diameter is $D = 30$ mm, and the clearance ratio is $C/R = 8 \times 10^{-4}$. Assume that the pressure is constant along the bearing axis and there is no axial flow (long-bearing theory).

a. Find the journal speed in RPM that is required for a bearing load capacity of 200 N. Find the bearing angle θ where there is a peak pressure.
b. What is the minimum ambient pressure around the bearing (inside the pressure vessel) in order to avoid cavitation and to have only positive pressure around the bearing?

7

Short Journal Bearings

7.1 INTRODUCTION

The term *short journal bearing* refers to a bearing of a short length, L, in comparison to the diameter, D, ($L \ll D$). The bearing geometry and coordinates are shown in Fig. 7-1. Short bearings are widely used and perform successfully in various machines, particularly in automotive engines. Although the load capacity, per unit length, of a short bearing is lower than that of a long bearing, it has the following important advantages.

1. In comparison to a long bearing, a short bearing exhibits improved heat transfer, due to faster oil circulation through the bearing clearance. The flow rate of lubricant in the axial direction through the bearing clearance of a short bearing is much faster than that of a long bearing. This relatively high flow rate improves the cooling by continually replacing the lubricant that is heated by viscous shear. Overheating is a major cause for bearing failure; therefore, operating temperature is a very important consideration in bearing design.
2. A short bearing is less sensitive to misalignment errors. It is obvious that short bearings reduce the risk of damage to the journal and the bearing edge resulting from misalignment of journal and bearing bore centrelines.
3. Wear is reduced, because abrasive wear particles and dust are washed away by the oil more easily in short bearings.

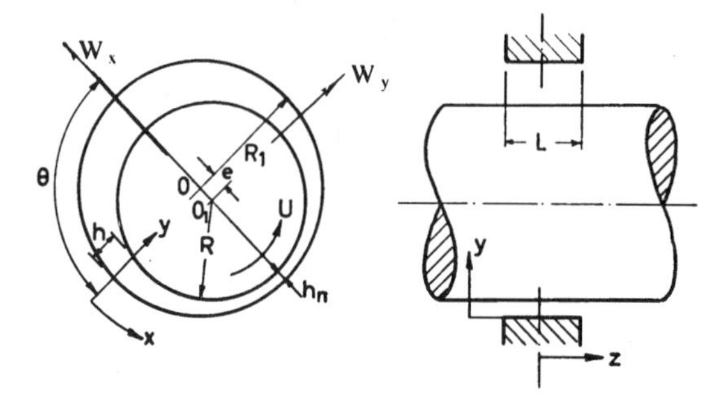

FIG. 7-1 Short hydrodynamic journal bearing.

4. Short bearings require less space and result in a more compact design. The trend in machine design is to reduce the size of machines. Short hydrodynamic journal bearings compete with rolling-element bearings, which are usually short relative to their diameter.

For all these reasons, short bearings are commonly used today. Long bearings were widely used a few decades ago and are still operating, mostly in old machines or in special applications where high load capacity is required.

The analysis of infinitely short hydrodynamic bearings has been introduced by Dubois and Ocvirk (1953). Unlike the analysis of a long bearing with realistic boundary conditions or of a finite bearing, short-bearing analysis results in a closed-form equation for the pressure wave and load capacity and does not require complex numerical analysis. Also, the closed-form expression of the bearing load capacity simplifies the analysis of a short bearing under dynamic conditions, such as unsteady load and speed.

Dubois and Ocvirk assumed that the pressure gradient around the bearing (in the x direction in Fig. 7-1) is small and can be neglected in comparison to the pressure gradient in the axial direction (z direction). In short bearings, the oil film pressure gradient in the axial direction, dp/dz, is larger by an order of magnitude or more in comparison to the pressure gradient, dp/dx, around the bearing:

$$\frac{dp}{dz} \gg \frac{dp}{dx} \tag{7-1}$$

By disregarding dp/dx in comparison to dp/dz, it is possible to simplify the Reynolds equation. This allows a closed-form analytical solution for the fluid film pressure distribution and load capacity. The resulting closed-form expression of the load capacity can be conveniently applied in bearing design as well as in dynamical analysis. One should keep in mind that when the bearing is not very short and the bearing length, L, approaches the diameter size, D, the true load capacity is, in fact, higher than that predicted by the short-bearing analysis. If we use the short-bearing equation to calculate finite-length bearings, the bearing design will be on the safe side, with high safety coefficient. For this reason, the short-journal equations are widely used by engineers for the design of hydrodynamic journal bearings, even for bearings that are not very short, if there is no justification to spend too much time on elaborate calculations to optimize the bearing design.

7.2 SHORT-BEARING ANALYSIS

The starting point of the derivation is the Reynolds equation, which was discussed in Chapter 5. Let us recall that the Reynolds equation for incompressible Newtonian fluids is

$$\frac{\partial}{\partial x}\left(\frac{h^3}{\mu}\frac{\partial p}{\partial x}\right) + \frac{\partial}{\partial z}\left(\frac{h^3}{\mu}\frac{\partial p}{\partial z}\right) = 6(U_1 - U_2)\frac{\partial h}{\partial x} + 12(V_2 - V_1) \tag{7-2}$$

Based on the assumption that the pressure gradient in the x direction can be disregarded, we have

$$\frac{\partial}{\partial x}\left(\frac{h^3}{\mu}\frac{\partial p}{\partial x}\right) \approx 0 \tag{7-3}$$

Thus, the Reynolds equation (7-2) reduces to the following simplified form:

$$\frac{\partial}{\partial z}\left(\frac{h^3}{\mu}\frac{\partial p}{\partial z}\right) = 6(U_1 - U_2)\frac{\partial h}{\partial x} + 12(V_2 - V_1) \tag{7-4}$$

In a journal bearing, the surface velocity of the journal is not parallel to the x direction along the bore surface, and it has a normal component V_2 (see diagram of velocity components in Fig. 6-2). The surface velocity components of the journal surface are

$$U_2 \approx U; \qquad V_2 \approx U\frac{\partial h}{\partial x} \tag{7-5}$$

On the stationary sleeve, the surface velocity components are zero:

$$U_1 = 0; \qquad V_1 = 0 \tag{7-6}$$

After substituting Eqs. (7-5) and (7-6) into the right-hand side of Eq. (7-4), it becomes

$$6(U_1 - U_2)\frac{\partial h}{\partial x} + 12(V_2 - V_1) = 6(0 - U)\frac{\partial h}{\partial x} + 12U\frac{\partial h}{\partial x} = 6U\frac{\partial h}{\partial x} \tag{7-7}$$

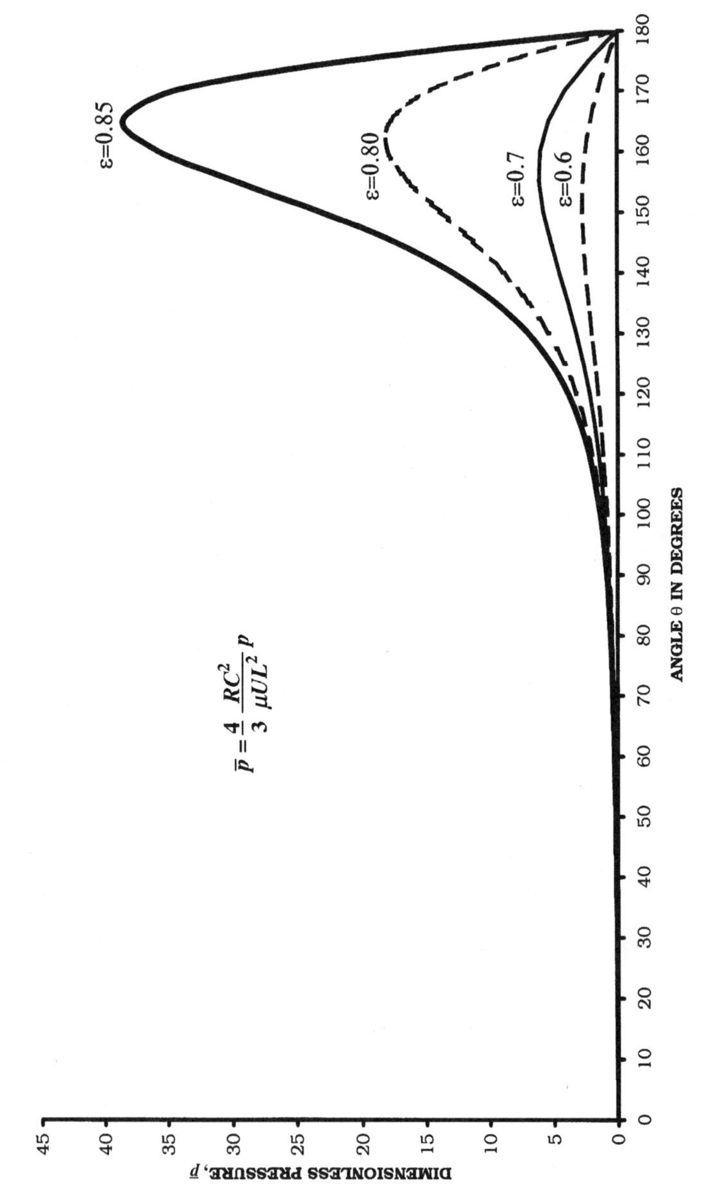

FIG. 7-2 Pressure wave at $z = 0$ in a short bearing for various values of ε.

The Reynolds equation for a short journal bearing is finally simplified to the form

$$\frac{\partial}{\partial z}\left(\frac{h^3}{\mu}\frac{\partial p}{\partial z}\right) = 6U\frac{\partial h}{\partial x} \tag{7-8}$$

The film thickness h is solely a function of x and is constant for the purpose of integration in the z direction. Double integration results in the following parabolic pressure distribution, in the z direction, with two constants, which can be obtained from the boundary conditions of the pressure wave:

$$p = -\frac{6\mu U}{h^3}\frac{dh}{dx}\frac{z^2}{2} + C_1 z + C_2 \tag{7-9}$$

At the two ends of the bearing, the pressure is equal to the atmospheric pressure, $p = 0$. These boundary conditions can be written as

$$p = 0 \quad \text{at } z = \pm\frac{L}{2} \tag{7-10}$$

Solving for the integration constants, and substituting the function for h in a journal bearing, $h(\theta) = C(1 + \varepsilon\cos\theta)$, the following expression for the pressure distribution in a short bearing (a function of θ and z) is obtained:

$$p(\theta, z) = \frac{3\mu U}{RC^2}\left(\frac{L^2}{4} - z^2\right)\frac{\varepsilon\sin\theta}{(1 + \varepsilon\cos)^3} \tag{7-11}$$

In a short journal bearing, the film thickness h is converging (decreasing h vs. θ) in the region ($0 < \theta < \pi$), resulting in a viscous wedge and a positive pressure wave. At the same time, in the region ($\pi < \theta < 2\pi$), the film thickness h is diverging (increasing h vs. θ). In the diverging region ($\pi < \theta < 2\pi$), Eq. (7-11) predicts a negative pressure wave (because sin θ is negative). The pressure according to Eq. (7-11) is an antisymmetrical function on the two sides of $\theta = \pi$.

In an actual bearing, in the region of negative pressure ($\pi < \theta < 2\pi$), there is fluid cavitation and the fluid continuity is breaking down. There is fluid cavitation whenever the negative pressure is lower than the vapor pressure. Therefore, Eq. (7-11) is no longer valid in the diverging region. In practice, the contribution of the negative pressure to the load capacity can be disregarded. Therefore in a short bearing, only the converging region with positive pressure ($0 < \theta < \pi$) is considered for the load capacity of the oil film (see Fig. 7-2).

Similar to a long bearing, the load capacity is solved by integration of the pressure wave around the bearing. But in the case of a short bearing, the pressure is a function of z and θ. The following are the two equations for the integration for the load capacity components in the directions of W_x and W_y of the bearing centerline and the normal to it:

$$W_x = -2\int_0^{\pi}\int_0^{L/2} p\cos\theta\, R\, d\theta\, dz = -\frac{\mu U L^3}{2C^2}\int_0^{\pi}\frac{\varepsilon\sin\theta\cos\theta}{(1+\varepsilon\cos\theta)^3}d\theta \tag{7-12a}$$

$$W_y = 2\int_0^{\pi}\int_0^{L/2} p\sin\theta\, d\theta\, dz = \frac{\mu U L^3}{2C^2}\int_0^{\pi}\frac{\varepsilon\sin^2\theta}{(1+\varepsilon\cos\theta)^3}d\theta \tag{7-12b}$$

The following list of integrals is useful for short journal bearings.

$$J_{11} = \int_0^{\pi}\frac{\sin^2\theta}{(1+\varepsilon\cos\theta)^3}d\theta = \frac{\pi}{2(1-\varepsilon^2)^{3/2}} \tag{7-13a}$$

$$J_{12} = \int_0^{\pi}\frac{\sin\theta\cos\theta}{(1+\varepsilon\cos\theta)^3}d\theta = \frac{-2\varepsilon}{(1-\varepsilon^2)^2} \tag{7-13b}$$

$$J_{22} = \int_0^{\pi}\frac{\cos^2\theta}{(1+\varepsilon\cos\theta)^3}d\theta = \frac{\pi(1+2\varepsilon^2)}{2(1-\varepsilon^2)^{5/2}} \tag{7-13c}$$

The load capacity components are functions of the preceding integrals:

$$W_x = -\frac{\mu U L^3}{2C^2}\varepsilon J_{12} \tag{7-14}$$

$$W_y = \frac{\mu U L^3}{2C^2}\varepsilon J_{11} \tag{7-15}$$

Using Eqs. (7-13) and substitution of the values of the integrals results in the following expressions for the two load components:

$$W_x = \frac{\mu U L^3}{C^2}\frac{\varepsilon^2}{(1-\varepsilon^2)^2} \tag{7-16a}$$

$$W_y = \frac{\mu U L^3}{4C^2}\frac{\pi\varepsilon}{(1-\varepsilon^2)^{3/2}} \tag{7-16b}$$

Equations (7-16) for the two load components yield the resultant load capacity of the bearing, W:

$$W = \frac{\mu U L^3}{4C^2}\frac{\varepsilon}{(1-\varepsilon^2)^2}[\pi^2(1-\varepsilon^2)+16\varepsilon^2]^{1/2} \tag{7-17}$$

The attitude angle, ϕ, is determined from the two load components:

$$\tan\phi = \frac{W_y}{W_x} \tag{7-18}$$

Via substitution of the values of the load capacity components, the expression for the attitude angle of a short bearing becomes

$$\tan\phi = \frac{\pi}{4}\frac{(1-\varepsilon^2)^{1/2}}{\varepsilon} \tag{7-19}$$

7.3 FLOW IN THE AXIAL DIRECTION

The velocity distribution of the fluid in the axial, z, direction is

$$w = 3U\frac{zh'}{Rh^3}(y^2 - hy) \tag{7-20}$$

Here,

$$h' = \frac{dh}{dx} \tag{7-21}$$

The gradient h' is the clearance slope (wedge angle), which is equal to the fluid film thickness slope in the direction of $x = R\theta$ (around the bearing). This gradient must be negative in order to result in a positive pressure wave as well as positive flow, w, in the z direction.

Positive axial flow is directed outside the bearing (outlet flow from the bearing). There is positive axial flow where h is converging (decreasing h vs. x) in the region ($0 < \theta < \pi$). At the same time, there is inlet flow, directed from outside into the bearing, where h is diverging (increasing h vs. θ) in the region ($\pi < \theta < 2\pi$). In the diverging region, $h' > 0$, there is fluid cavitation that is causing deviation from the theoretical axial flow predicted in Eq. (7-20). However, in principle, the lubricant enters into the bearing in the diverging region and leaves the bearing in the converging region. In a short bearing, there is much faster lubricant circulation relative to that in a long bearing. Fast lubricant circulation reduces the peak temperature of the lubricant. This is a significant advantage of the short bearing, because high peak temperature can cause bearing failure.

7.4 SOMMERFELD NUMBER OF A SHORT BEARING

The definition of the dimensionless Sommerfeld number for a short bearing is identical to that for a long bearing; however, for a short journal bearing, the expression of the Sommerfeld number is given as,

$$S = \frac{\mu n}{P}\left(\frac{R}{C}\right)^2 = \left(\frac{D}{L}\right)^2 \frac{(1-\varepsilon^2)^2}{\pi\varepsilon[\pi^2(1-\varepsilon^2) + 16\varepsilon^2]^{1/2}} \tag{7-22}$$

Let us recall that the Sommerfeld number of a long bearing is only a function of ε. However, for a short bearing, the Sommerfeld number is a function of ε as well as the ratio L/D.

7.5 VISCOUS FRICTION

The friction force around a bearing is obtained by integration of the shear stresses. The shear stress in a short bearing is

$$\tau = \mu \frac{U}{h} \tag{7-23}$$

For the purpose of computing the friction force, the shear stresses are integrated around the complete bearing. The fluid is present in the diverging region ($\pi < \theta < 2\pi$), and it is contributing to the viscous friction, although its contribution to the load capacity has been neglected. The friction force, F_f, is obtained by integration of the shear stress, τ, over the complete surface area of the journal:

$$F_f = \int_{(A)} \tau \, dA \tag{7-24}$$

Substituting $dA = LR\, d\theta$, the friction force becomes

$$F_f = RL \int_0^{2\pi} \tau \, d\theta \tag{7-25}$$

To solve for the friction force, we substitute the expression for h into Eq. (7-23) and substitute the resulting equation of τ into Eq. (7-25). For solving the integral, the following integral equation is useful:

$$J_1 = \int_0^{2\pi} \frac{1}{1 + \cos\theta} d\theta = \frac{2\pi}{(1 - \varepsilon^2)^{1/2}} \tag{7-26}$$

Note that J_1 has the limits of integration $0\text{–}2\pi$, while for the first three integrals in Eqs. (7-13), the limits are $0\text{–}\pi$. The final expression for the friction force is

$$F_f = \frac{\mu LRU}{C} \int_0^{2\pi} \frac{d\theta}{1 + \varepsilon\cos\theta} = \frac{\mu LRU}{C} \frac{2\pi}{(1 - \varepsilon^2)^{1/2}} \tag{7-27}$$

The bearing friction coefficient f is defined as

$$f = \frac{F_f}{W} \tag{7-28}$$

The friction torque T_f is

$$T_f = F_f R; \qquad T_f = \frac{\mu L R^2 U}{C} \frac{2\pi}{(1 - \varepsilon^2)^{1/2}} \tag{7-29}$$

The energy loss per unit of time, $\dot{E}_f$ is determined by the following:

$$\dot{E}_f = F_f U \tag{7-30a}$$

Substituting Eq. (7-27) into Eq. (7-30a) yields the following expression for the power loss on viscous friction:

$$\dot{E}_f = \frac{\mu LRU^2}{C} \frac{2\pi}{(1-\varepsilon^2)^{1/2}} \tag{7-30b}$$

7.6 JOURNAL BEARING STIFFNESS

Journal bearing stiffness, k, is the rate of increase of load W with displacement e in the same direction, dW/de (similar to that of a spring constant). High stiffness is particularly important in machine tools, where any displacement of the spindle centerline during machining would result in machining errors. Hydrodynamic journal bearings have low stiffness at low eccentricity (under light load).

The displacement of a hydrodynamic bearing is not in the same direction as the force W. In such cases, the journal bearing has *cross-stiffness* components. The stiffness components are presented as four components related to the force components W_x and W_y and the displacement components in these directions.

In a journal bearing, the load is divided into two components, W_x and W_y, and the displacement of the bearing center, e, is divided into two components, e_x and e_y. The two components of the journal bearing stiffness are

$$k_x = \frac{dW_x}{de_x}; \qquad k_y = \frac{dW_y}{de_y} \tag{7-31}$$

and the two components of the cross-stiffness are defined as

$$k_{xy} = \frac{dW_x}{de_y}; \qquad k_y = \frac{dW_y}{de_x} \tag{7-32}$$

Cross-stiffness components cause instability, in the form of an oil whirl in journal bearings.

Example Problem 7-1

A short bearing is designed to operate with an eccentricity ratio $\varepsilon = 0.8$. The journal diameter is 60 mm, and its speed is 1500 RPM. The journal is supported by a short hydrodynamic bearing of length $L/D = 0.5$, and clearance ratio $C/R = 10^{-3}$. The radial load on the bearing is 1 metric ton (1 metric ton = 9800 [N]).

a. Assume that infinitely-short-bearing theory applies to this bearing, and find the Sommerfeld number.

b. Find the minimum viscosity of the lubricant for operating at $\varepsilon = 0.8$.
c. Select a lubricant if the average bearing operating temperature is 80°C.

Solution

a. Sommerfeld Number

The Sommerfeld number is

$$S = \frac{\mu n}{P}\left(\frac{R}{C}\right)^2 = \left(\frac{D}{L}\right)^2 \frac{(1-\varepsilon^2)^2}{\pi\varepsilon[\pi^2(1-\varepsilon^2)+16\varepsilon^2]^{1/2}}$$

From the right-hand side of the equation,

$$S = (2)^2 \frac{(1-0.8^2)^2}{\pi 0.8[\pi^2(1-0.8^2)+16\times 0.8^2]^{1/2}} = \frac{0.36^2}{\pi \times 0.8 \times 3.71} = 0.0139$$

b. Minimum Viscosity

The average pressure is

$$P = \frac{W}{LD} = \frac{9800}{0.06 \times 0.03} = 5.44 \times 10^6 \text{ Pa}$$

The load in SI units (newtons) is 9800 [N], and the speed is $n = 1500/60 = 25$ RPS.

The viscosity is determined by equating:

$$S = \frac{\mu n}{P}\left(\frac{R}{C}\right)^2 = 0.0139$$

$$\mu = \frac{5.44 \times 10^6 \times 0.0139}{25 \times 10^6} = 0.0030 \text{ [N-s/m}^2\text{]}$$

c. Lubricant

For lubricant operating temperature of 80°C, mineral oil SAE 10 has suitable viscosity of 0.003 [N-s/m^2] (Fig. 2-3). This is the minimum required viscosity for the operation of this bearing with an eccentricity ratio no higher than $\varepsilon = 0.8$.

Example Problem 7-2

A journal of 75-mm diameter rotates at 3800 RPM. The journal is supported by a short hydrodynamic bearing of length $L = D/4$ and a clearance ratio $C/R = 10^{-3}$. The radial load on the bearing is 0.5 metric ton, (1 metric ton = 9800 [N]). The lubricant is SAE 40, and the operating temperature of the lubricant in the bearing is 80°C.

a. Assume infinitely-short-bearing theory, and find the eccentricity ratio, ε, of the bearing (use a graphic method to solve for ε) and the minimum film thickness, h_n.
b. Derive the equation for the pressure wave around the bearing, at the center of the width (at $z = 0$).
c. Find the hydrodynamic friction torque and the friction power losses (in watts).

Solution

The following conversion is required for calculation in SI units:

Speed of shaft: $n = 3800/60 = 63.3$ [RPS]
Radial load: $W = 0.5 \times 9800 = 4900$ [N]
Axial length of shaft: $L = D/4 = 0.075/4 = 0.01875$ [m]
$C/R = 10^{-3}$; hence $R/C = 10^3$.

a. Eccentricity Ratio and Minimum Film Thickness

For an operating temperature of $T = 80°\text{C}$, the viscosity of SAE-40 oil is obtained from the viscosity–temperature chart: $\mu = 0.0185$ [N-s/m^2]. The equation for the load capacity is applied to solve for the eccentricity ratio, ε, the only unknown in the following equation:

$$W = \frac{\mu U L^3}{4C^2} \frac{\varepsilon}{(1-\varepsilon^2)^2} [\pi^2(1-\varepsilon^2) + 16\varepsilon^2]^{1/2} \tag{7-32}$$

To simplify the mathematical derivation of ε, the following substitution is helpful:

$$f(\varepsilon) = \frac{\varepsilon}{(1-\varepsilon^2)^2} [\pi^2(1-\varepsilon^2) + 16\varepsilon^2]^{1/2} \tag{7-33}$$

First, we can solve for $f(\varepsilon)$, and later we can obtain the value of ε from the graph of $f(\varepsilon)$ vs. ε (Fig. 7-3).

Equations (7-32) and (7-33) yield

$$f(\varepsilon) = \frac{4WC^2}{\mu U L^3} = \frac{4W(10^{-3} \times D/2)^2}{\mu \pi n D (D/4)^3} = \frac{4 \times 16 \times 4900 \times 10^{-6}}{0.0185\pi \times 63.3(0.075)^2} = 15.15$$

According to the curve of $f(\varepsilon)$ vs. ε, for $f(\varepsilon) = 15.15$, the eccentricity of the bearing is $\varepsilon = 0.75$.

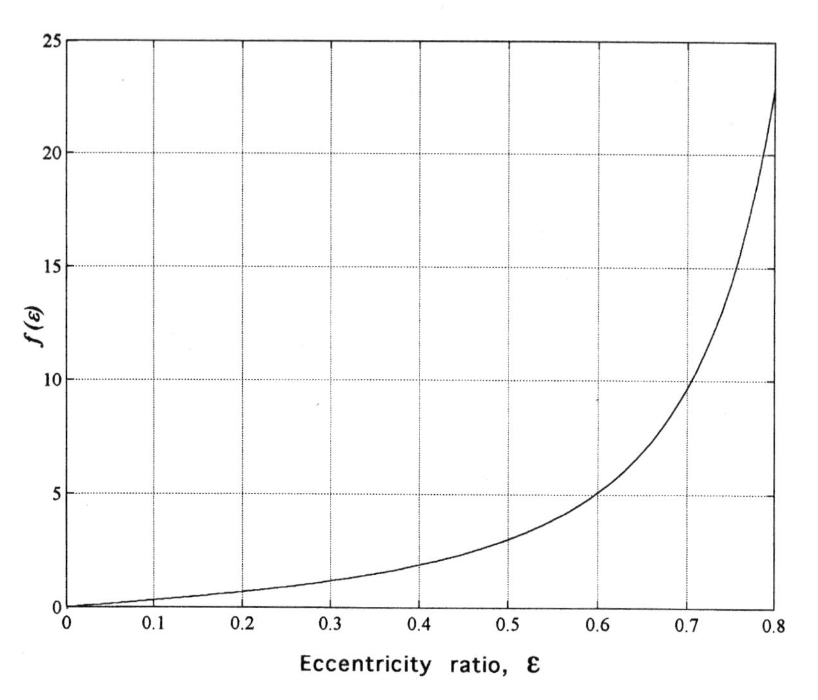

FIG. 7-3 Graph of $f(\varepsilon)$ vs. ε, describing Eq. (7-33).

We find the minimum film thickness, $h_{\min}$, as follows:

$$h_{\min} = C(1-\varepsilon) = 10^{-3} \times R(1-0.75) = 10^{-3}\left(\frac{0.075}{2}\right)0.25$$
$$= 9.4 \times 10^{-6}\ \text{m}$$
$$h_{\min} = 9.4\ \mu\text{m}$$

b. Pressure Wave

The pressure is a function of θ and z, according to the equation

$$p = \frac{3\mu U}{h^3}\frac{dh}{dx}\left(Z^2 - \frac{L^2}{4}\right)$$

where $h = C(1+\varepsilon\cos\theta)$, $x = R\theta$, and $dx = Rd\theta$. The clearance slope is

$$\frac{dh}{dx} = \frac{dh}{d\theta}\frac{d\theta}{dx} = -\frac{C\varepsilon\sin\theta}{R}$$

The pressure wave at the width center, $z = 0$, is

$$p(\theta) = \frac{3\mu UL^2}{4h^3}\frac{C\varepsilon\sin\theta}{R}$$

$$p(\theta)\ (\text{at } z = 0) = \frac{3 \times 0.0185 \times \pi \times 63.3 \times 0.075 \times 0.075^2 \times 0.75}{64 \times 10^3} \times \frac{\sin\theta}{(1 + 0.75\cos\theta)^3}$$

$$p(\theta)\ (\text{at } z = 0) = \frac{5.5 \times 10^{-8}\sin\theta}{(1 + 0.75\cos\theta)^3}$$

c. *Friction Torque and Friction Power Loss*

We find the friction torque as follows:

$$T_f = F_f R = \frac{\mu LR^2 U}{C}\frac{2\pi}{(1-\varepsilon^2)^{1/2}}$$

$$= \frac{0.0185 \times 0.01875 \times (0.075/2) \times \pi \times 0.075 \times 63.3}{10^{-3}}\frac{2\pi}{(1-0.75^2)^{1/2}}$$

$$T_f = 1.80\ \text{N-m}$$

The friction power loss is found as follows:

$$\dot{E}_f = T_f\omega$$

$$\dot{E}_f = \frac{\mu LRU^2}{C}\frac{2\pi}{(1-\varepsilon^2)^{1/2}}$$

$$\dot{E}_f = \frac{0.0185 \times 0.01875 \times (\pi \times 63.3 \times 0.075)^2}{10^{-3}}\frac{2\pi}{(1-0.75^2)^{1/2}} = 733\ [\text{W}]$$

Problems

7-1 A short bearing is designed to operate with an eccentricity ratio of $\varepsilon = 0.7$. Find the journal diameter if the speed is 30,000 RPM and the radial load on the bearing is 8000 N. The bearing length ratio $L/D = 0.6$, and the clearance ratio is $C/R = 10^{-3}$. The lubricant is SAE 30 and the average operating temperature in the bearing is 70°C. Assume that infinitely-short-bearing theory applies.

7-2 Plot the dimensionless pressure distribution (function of θ) at the bearing center, $z = 0$, in Example Problem 7-2.

7-3 A short bearing is designed to operate with an eccentricity ratio of $\varepsilon = 0.75$. The journal is 80 mm in diameter, and its speed is

3500 RPM. The journal is supported by a short hydrodynamic bearing of length $D/L = 4$ and a clearance ratio of $C/R = 10^{-3}$. The radial load on the bearing is 1000 N.

a. Assume that infinitely-short-bearing theory applies to this bearing, and find the minimum viscosity of the lubricant.
b. Select a lubricant for an average operating temperature in the bearing of 60°C.

7-4 The journal speed of a 100 mm diameter journal is 2500 RPM. The journal is supported by a short hydrodynamic bearing of length $L = 0.6D$ and a clearance ratio of $C/R = 10^{-3}$. The radial load on the bearing is 10,000 [N]. The lubricant is SAE 30, and the operating temperature of the lubricant in the bearing is 70°C.

a. Assume infinitely-short-bearing theory, and find the eccentricity ratio, ε, of the bearing and the minimum film thickness, h_n (use a graphic method to solve for ε).
b. Derive the equation and plot the pressure distribution around the bearing, at the center of the width (at $z = 0$).
c. Find the hydrodynamic friction torque and the friction power losses (in watts) for each bearing.

8

Design Charts for Finite-Length Journal Bearings

8.1 INTRODUCTION

In the preceding chapters, the analysis of infinitely long and short journal bearings have been presented. In comparison, the solution of a finite-length journal bearing (e.g., $L/D = 1$) is more complex and requires a computer program for a numerical solution of the Reynolds equation. The first numerical solution of the Reynolds equation for a finite-length bearing was performed by Raimondi and Boyd (1958). The results were presented in the form of dimensionless charts and tables, which are required for journal bearing design. The presentation of the results in the form of dimensionless charts and tables is convenient for design purposes because one does not need to repeat the numerical solution for each bearing design. The charts and tables present various dimensionless *performance parameters*, such as minimum film thickness, friction, and temperature rise of the lubricant as a function of the Sommerfeld number, S. Let us recall that the dimensionless Sommerfeld number is defined as

$$S = \left(\frac{R}{C}\right)^2 \frac{\mu n}{P} \tag{8-1}$$

where n is the speed of the journal in revolutions per second (RPS), R is the journal radius, C is the radial clearance, and P is the average bearing pressure (load, F, per unit of projected contact area of journal and bearing), given by

$$P = \frac{F}{2RL} = \frac{F}{DL} \tag{8-2}$$

Note that S is a dimensionless number, and any system of units can be applied for its calculation as long as one is consistent with the units. For instance, if the Imperial unit system is applied, length should be in inches, force in lbf, and μ in reynolds [lbf-s/in.^{2}]. In SI units, length is in meters, force in newtons, and the viscosity, μ, in [N-s/m^2]. The journal speed, n, should always be in revolutions per second (RPS), irrespective of the system of units used, and the viscosity, μ, must always include seconds as the unit of time.

8.2 DESIGN PROCEDURE

The design procedure starts with the selection of the bearing dimensions: the journal diameter D, the bearing length L, and the radial clearance between the bearing and the journal C. At this stage of the design, the shaft diameter should already have been computed according to strength-of-materials considerations. However, in certain cases the designer may decide, after preliminary calculations, to increase the journal diameter in order to improve the bearing hydrodynamic load capacity.

One important design decision is the selection of the L/D ratio. It is obvious from hydrodynamic theory of lubrication that a long bearing has a higher load capacity (per unit of length) in comparison to a shorter bearing. On the other hand, a long bearing increases the risk of bearing failure due to misalignment errors. In addition, a long bearing reduces the amount of oil circulating in the bearing, resulting in a higher peak temperature inside the lubrication film and the bearing surface. Therefore, short bearings (L/D ratios between 0.5 and 0.7) are recommended in many cases. Of course, there are many unique circumstances where different ratios are selected.

The bearing clearance, C, is also an important design factor, because the load capacity in a long bearing is proportional to $(R/C)^2$. Experience over the years has resulted in an empirical rule used by most designers. They commonly select a ratio R/C of about 1000. The ratio R/C is equal to the ratio $D/\Delta D$ between the diameter and the diameter clearance; i.e., a journal of 50-mm diameter should have a 50-μm (fifty-thousandth of a millimeter)-diameter clearance. The designer should keep in mind that there are manufacturing tolerances of bearing bore and journal diameters, resulting in significant tolerances in the journal bearing clearance, ΔD. The clearance can be somewhat smaller or larger, and thus the bearing should be designed for the worst possible scenario. In general, high-precision manufacturing is required for journal bearings, to minimize the clearance tolerances as well as to achieve good surface finish and optimal alignment.

For bearings subjected to high dynamic impacts, or very high speeds, somewhat larger bearing clearances are chosen. The following is an empirical equation that is recommended for high-speed journal bearings having an L/D ratio of about 0.6:

$$\frac{C}{D} = (0.0009 + \frac{n}{83,000}) \tag{8-3}$$

where n is the journal speed (RPS). This equation is widely used to determine the radial clearance in motor vehicle engines.

8.3 MINIMUM FILM THICKNESS

One of the most critical design decisions concerns the minimum film thickness, h_n. Of course, the minimum fluid film thickness must be much higher than the surface roughness, particularly in the presence of vibrations. Even for statically loaded bearings, there are always unexpected disturbances and dynamic loads, due to vibrations in the machine, and a higher value of the minimum film thickness, h_n, is required to prevent bearing wear. In critical applications, where the replacement of bearings is not easy, such as bearings located inside an engine, more care is required to ensure that the minimum film thickness will never be reduced below a critical value at which wear can initiate.

Another consideration is the fluid film temperature, which can increase under unexpected conditions, such as disturbances in the operation of the machine. The temperature rise reduces the lubricant viscosity; in turn, the oil film thickness is reduced. For this reason, designers are very careful to select h_n much larger than the surface roughness. The common design practice for hydrodynamic bearings is to select a minimum film thickness in the range of 10–100 times the average surface finish (in RMS). For instance, if the journal and the bearing are both machined by fine turning, having a surface finish specified by an RMS value of 0.5 μm (0.5 thousandths of a millimeter), the minimum film thickness can be within the limits of 5–50 μm. High h_n values are chosen in the presence of high dynamic disturbances, whereas low values of h_n are chosen for steady operation that involves minimal vibrations and disturbances.

Moreover, if it is expected that dust particles would contaminate the lubricant, a higher minimum film thickness, h_n, should be selected. Also, for critical applications, where there are safety considerations, or where bearing failure can result in expensive machine downtime, a coefficient of safety is applied in the form of higher values of h_n.

The surface finish of the two surfaces (bearing and journal) must be considered. A dimensionless film parameter, Λ, relating h_n to the average surface finish, has been introduced; see Hamrock (1994):

$$\Lambda = \frac{h_n}{(R_{s,j}^2 + R_{s,b}^2)^{1/2}} \tag{8-4}$$

where $R_{s,j}$ = surface finish of the journal surface (RMS) and $R_{s,b}$ = surface finish of the bearing surface (RMS).

As discussed earlier, the range of values assigned to Λ depends on the operating conditions and varies from 5 to 100. The minimum film thickness is not the only limitation encountered in the design of a journal bearing. Other limitations, which depend on the bearing material, determine in many cases the maximum allowable bearing load. The most important limitations are as follows.

1. Maximum allowed *PV* value (depending on the bearing material) to avoid bearing overheating during the start-up of boundary lubrication. This is particularly important in bearing materials that are not good heat conductors, such as plastics materials.
2. Maximum allowed peak pressure to prevent local failure of the bearing material.
3. Maximum allowed peak temperature, to prevent melting or softening of the bearing material.

In most applications, the inner bearing surface is made of a thin layer of a soft white metal (babbitt), which has a low melting temperature. The design procedure must ensure that the allowed values are not exceeded, for otherwise it can result in bearing failure. If the preliminary calculations indicate that these limitations are exceeded, it is necessary to introduce design modifications. In most cases, the design of hydrodynamic bearing requires trial-and-error calculations to verify that all the requirements are satisfied.

8.4 RAIMONDI AND BOYD CHARTS AND TABLES

8.4.1 Partial Bearings

A partial journal bearing has a bearing arc, β, of less than 360°, and only part of the bearing circumference supports the journal. A full bearing is where the bearing arc $\beta = 360°$; in a partial bearing, the bearing arc is less than 360°, such as $\beta = 60°,\ 120°$, and 180°. A partial bearing has two important advantages in comparison to a full bearing. First, there is a reduction of the viscous friction coefficient; second, in a partial bearing there is a faster circulation of the lubricant, resulting in better heat transfer from the bearing. The two advantages result in a lower bearing temperature as well as lower energy losses from viscous friction. In high-speed journal bearings, the friction coefficient can be relatively high, and partial bearings are often used to mitigate this problem. At the same time, the load capacity of a partial bearing is only slightly below that of a full bearing, which make the merits of using a partial bearing quite obvious.

8.4.2 Dimensionless Performance Parameters

Using numerical analysis, Raimondi and Boyd solved the Reynolds equation. They presented the results in dimensionless terms via graphs and tables. Dimensionless performance parameters of a finite-length bearing were presented as a function of the Sommerfeld number, *S*. The Raimondi and Boyd performance parameters are presented here by charts for journal bearings with the ratio $L/D = 1$; see Figs. 8-1 to 8-10. For bearings having different L/D ratios, the performance parameters are given in tables; see Tables 8-1 to 8-4.

The charts and tables of Raimondi and Boyd have been presented for both partial and full journal bearings, and for various L/D ratios. Partial journal bearings include multi-lobe bearings that are formed by several eccentric arcs.

The following ten dimensionless performance parameters are presented in charts and tables.

1. Minimum film thickness ratio, h_n/C. Graphs of minimum film thickness ratio versus Sommerfeld number, *S*, are presented in Fig. 8-1.
2. Attitude angle, ϕ, i.e., the angle at which minimum film thickness is attained. The angle is measured from the line along the load direction as shown in Fig. 8-2.
3. Friction coefficient variable, $(R/C)f$. Curves of the dimensionless friction coefficient variable versus *S* are presented in Fig. 8-3.
4. In Fig. 8-4, curves are plotted of the dimensionless total bearing flow rate variable, $Q/nRCL$, against the Sommerfeld number.
5. The ratio of the side flow rate (in the *z* direction) to the total flow rate, Q_s/Q, as a function of the Sommerfeld number is shown in Fig. 8-5. The side flow rate, Q_s, is required for determining the end leakage, since the bearing is no longer assumed to be infinite. The side flow rate is important for cooling of the bearing.
6. The dimensionless temperature rise variable, $c\rho\Delta T/P$, is presented in Fig. 8-6. It is required for determining the temperature rise of the lubricant due to friction. The temperature rise, ΔT, is of the lubricant from the point of entry into the bearing to the point of discharge from the bearing. The estimation of the temperature rise is discussed in greater detail later.

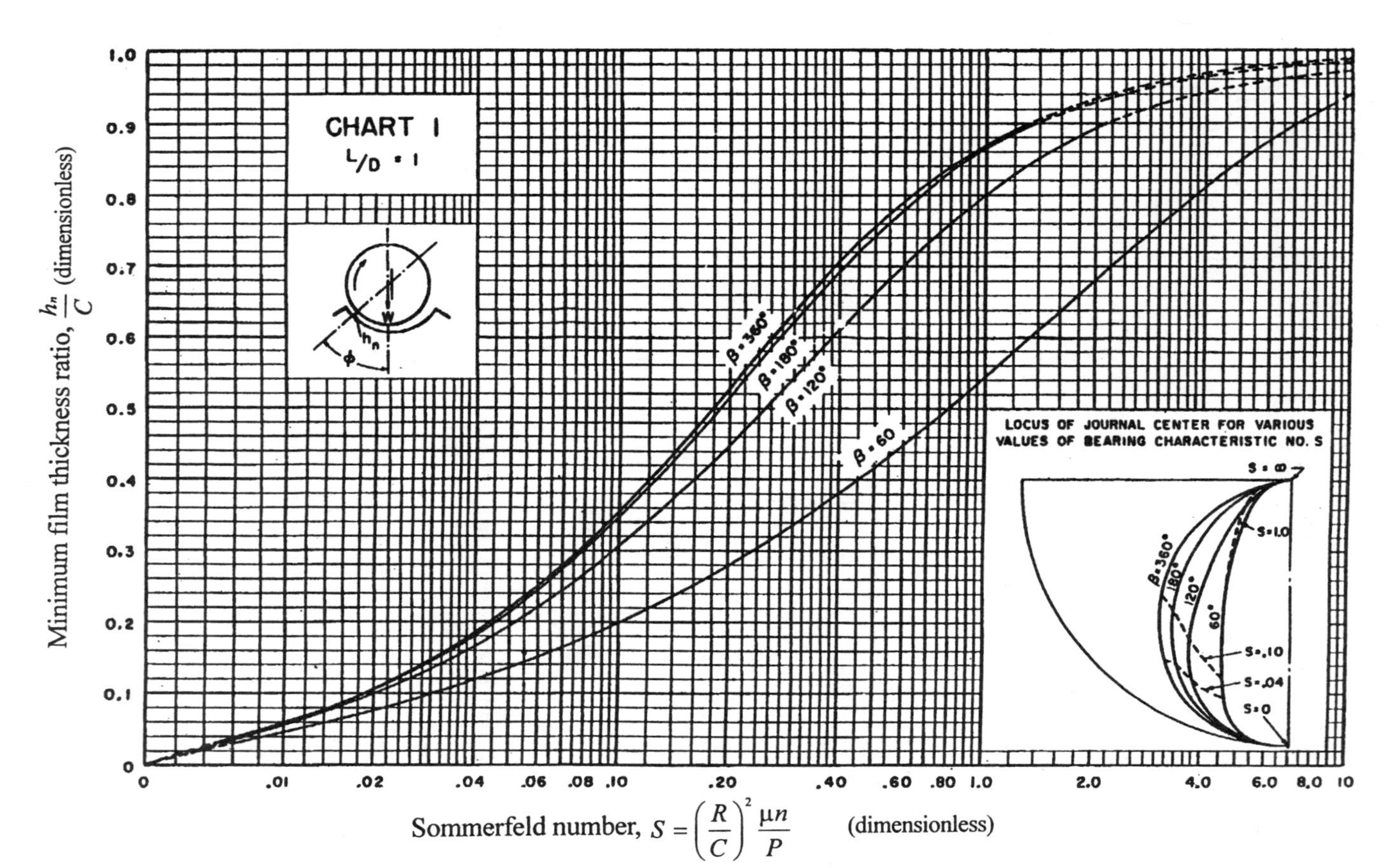

FIG. 8-1 Minimum film thickness ratio versus Sommerfeld number for variable bearing arc β, $L/D = 1$. (From Raimondi and Boyd, 1958, with permission of STLE.)

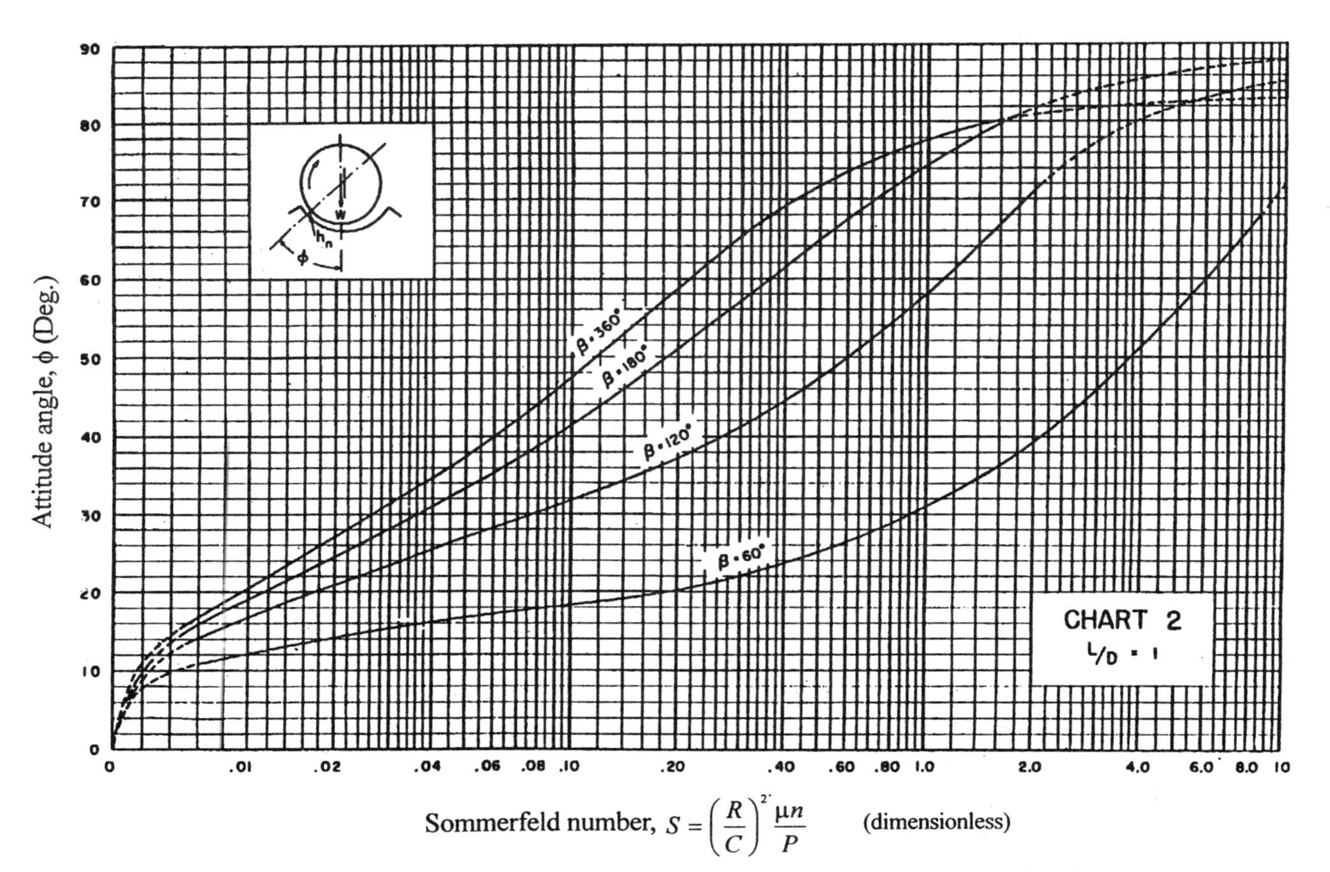

FIG. 8-2 Attitude angle versus Sommerfeld number for variable bearing arc β, $L/D = 1$. (From Raimondi and Boyd, 1958, with permission of STLE.)

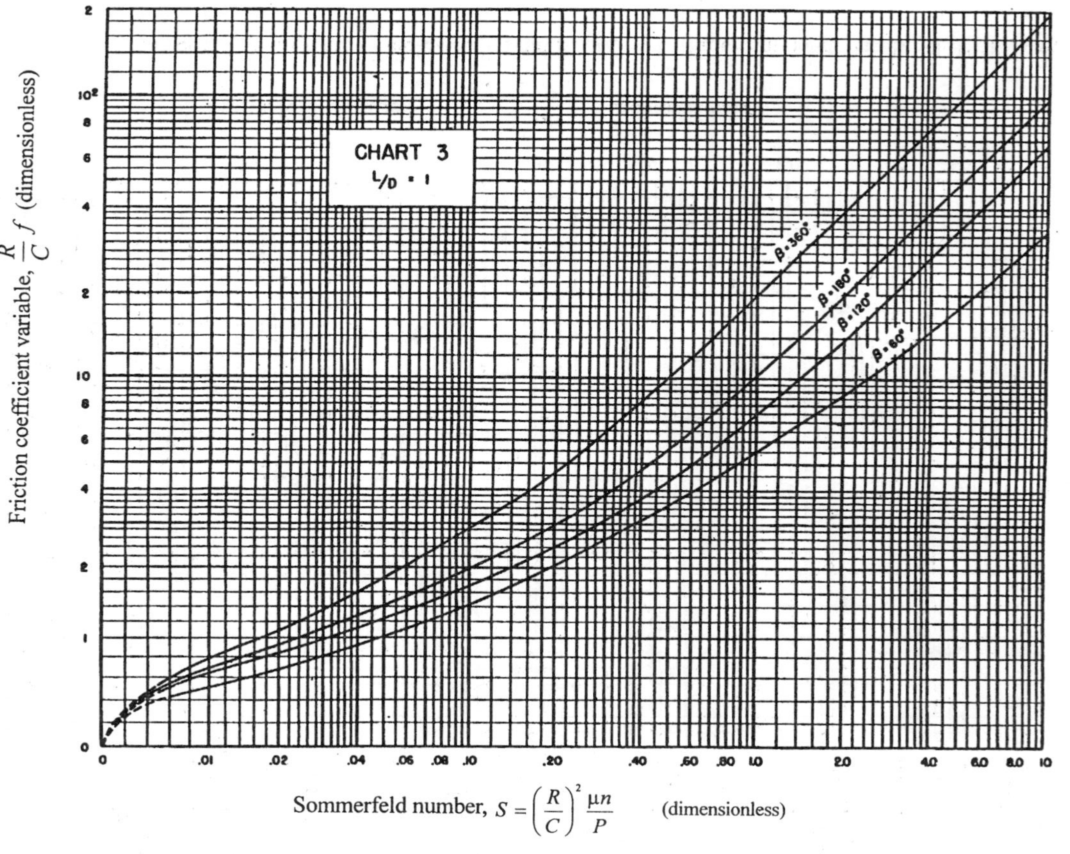

FIG. 8-3 Friction coefficient versus Sommerfeld number for variable bearing arc β, $L/D = 1$. (From Raimondi and Boyd, 1958, with permission of STLE.)

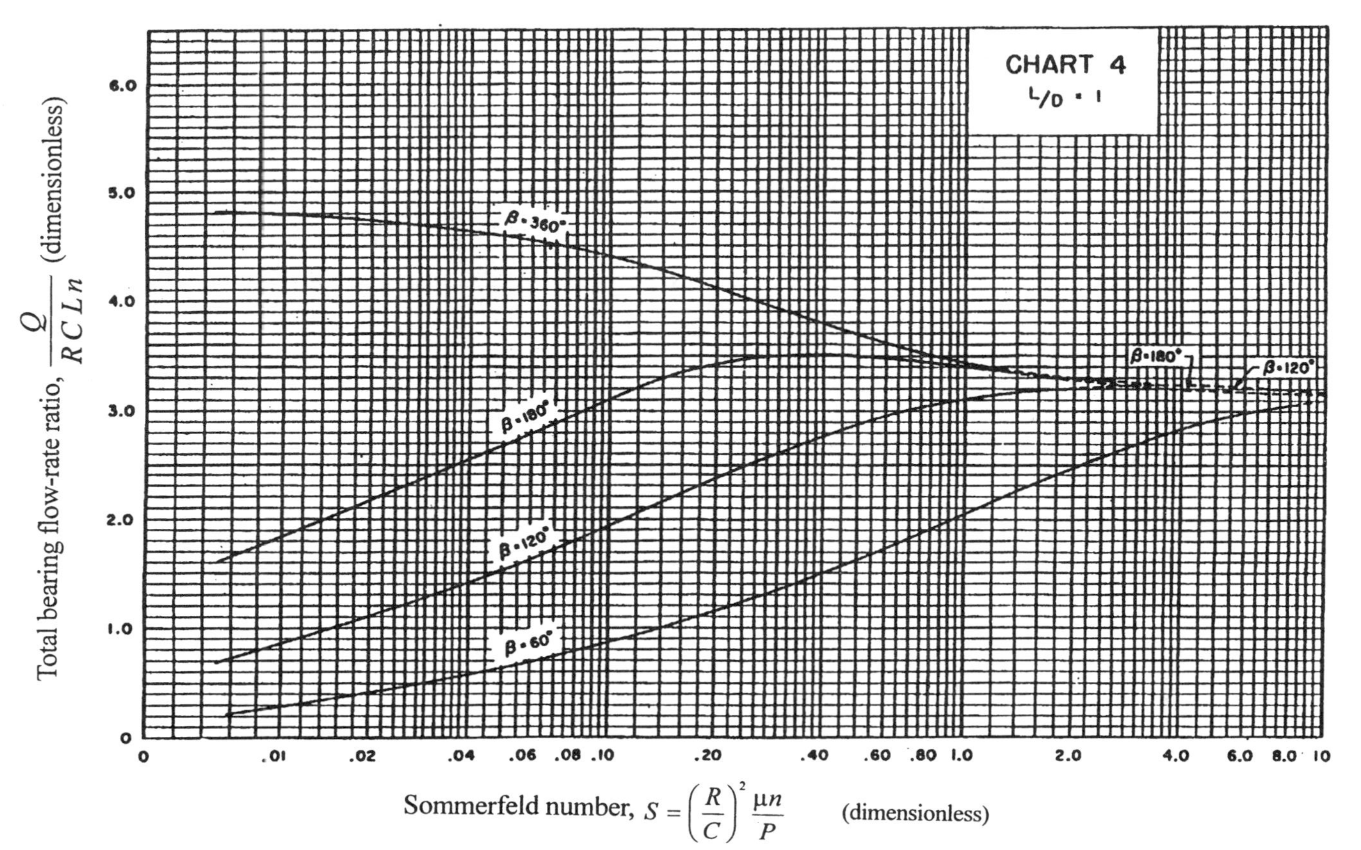

FIG. 8-4 Total bearing flow rate variable versus Sommerfeld number, $(L/D = 1)$. (From Raimondi and Boyd, 1958, with permission of STLE.)

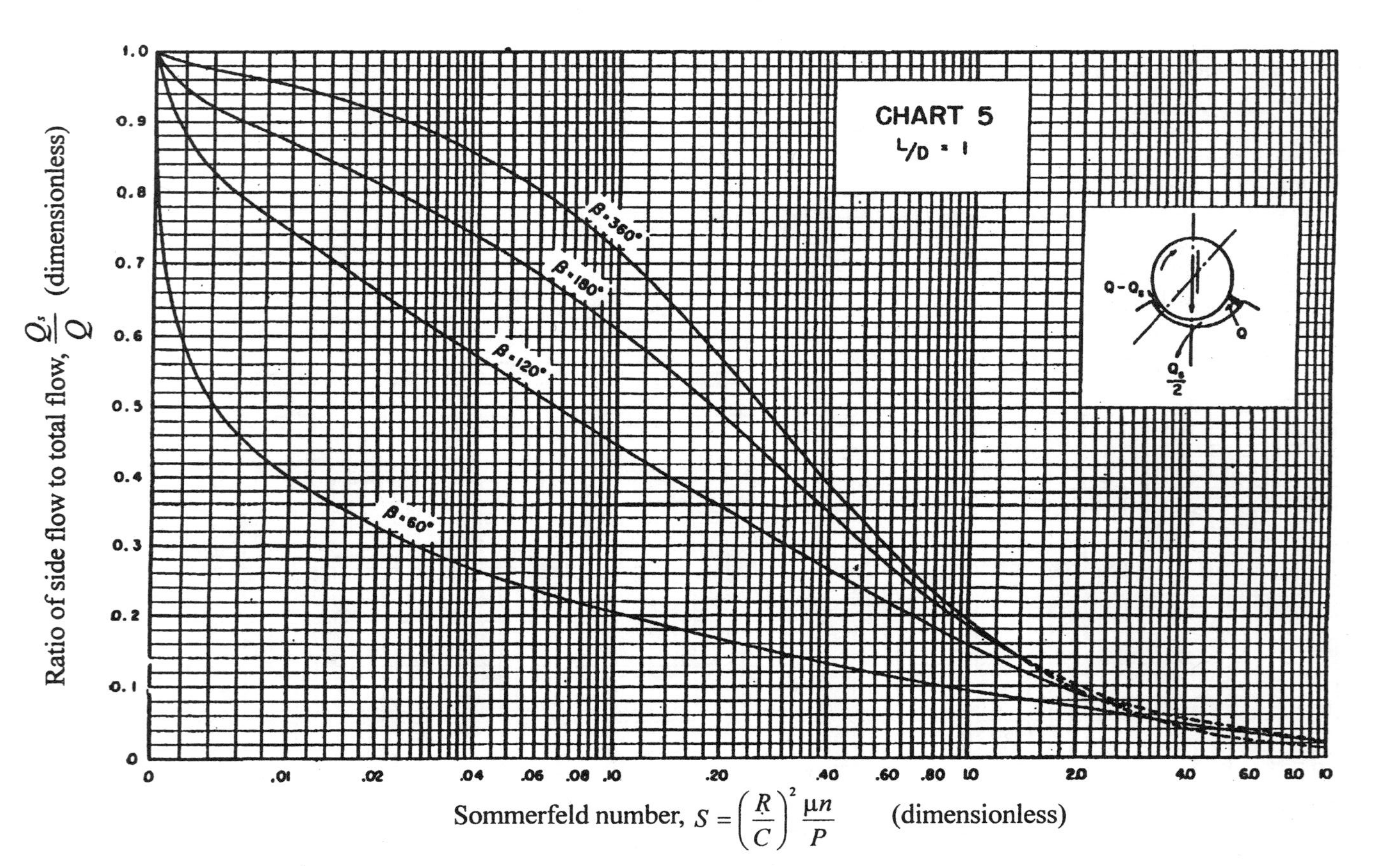

FIG. 8-5 Ratio of side flow (axial direction) to total flow versus Sommerfeld number ($L/D = 1$). (From Raimondi and Boyd, 1958, with permission of STLE.)

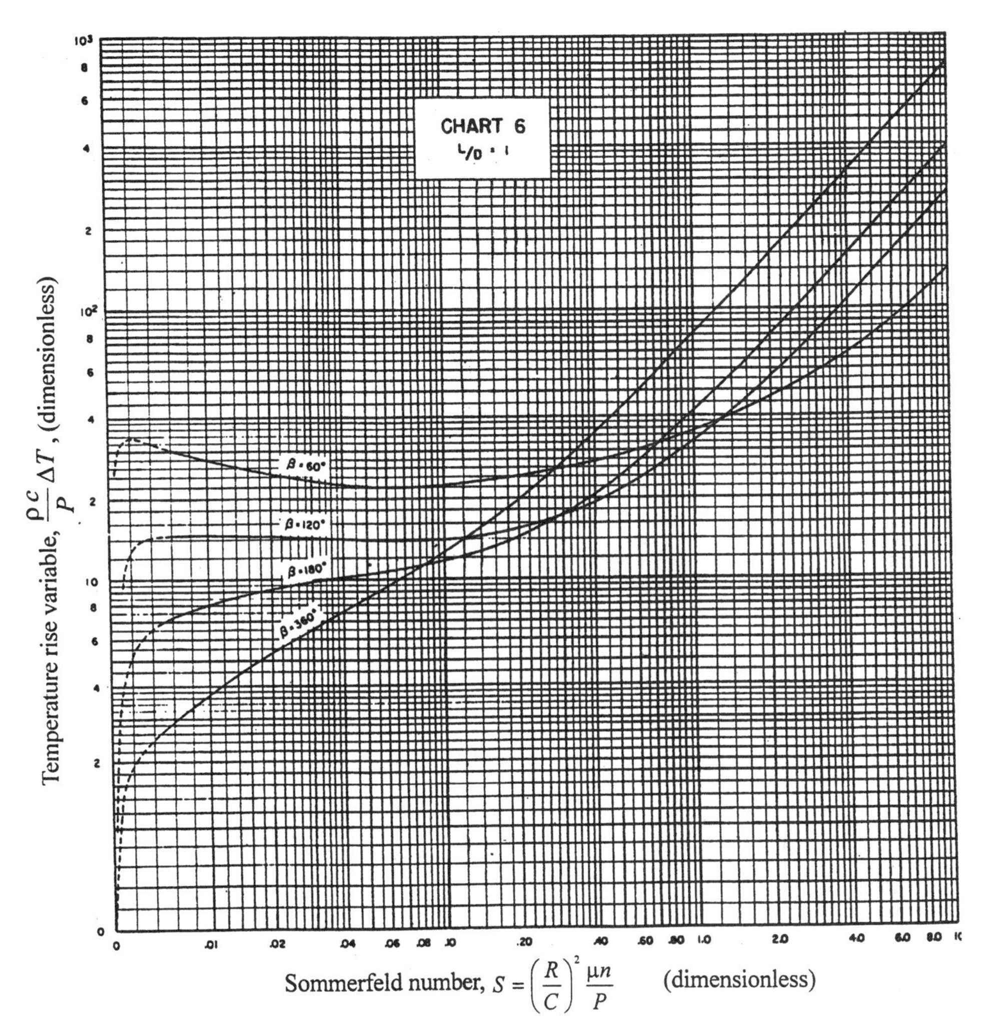

FIG. 8-6 Temperature rise variable versus Sommerfeld number ($L/D = 1$). (From Raimondi and Boyd, 1958, with permission of STLE.)

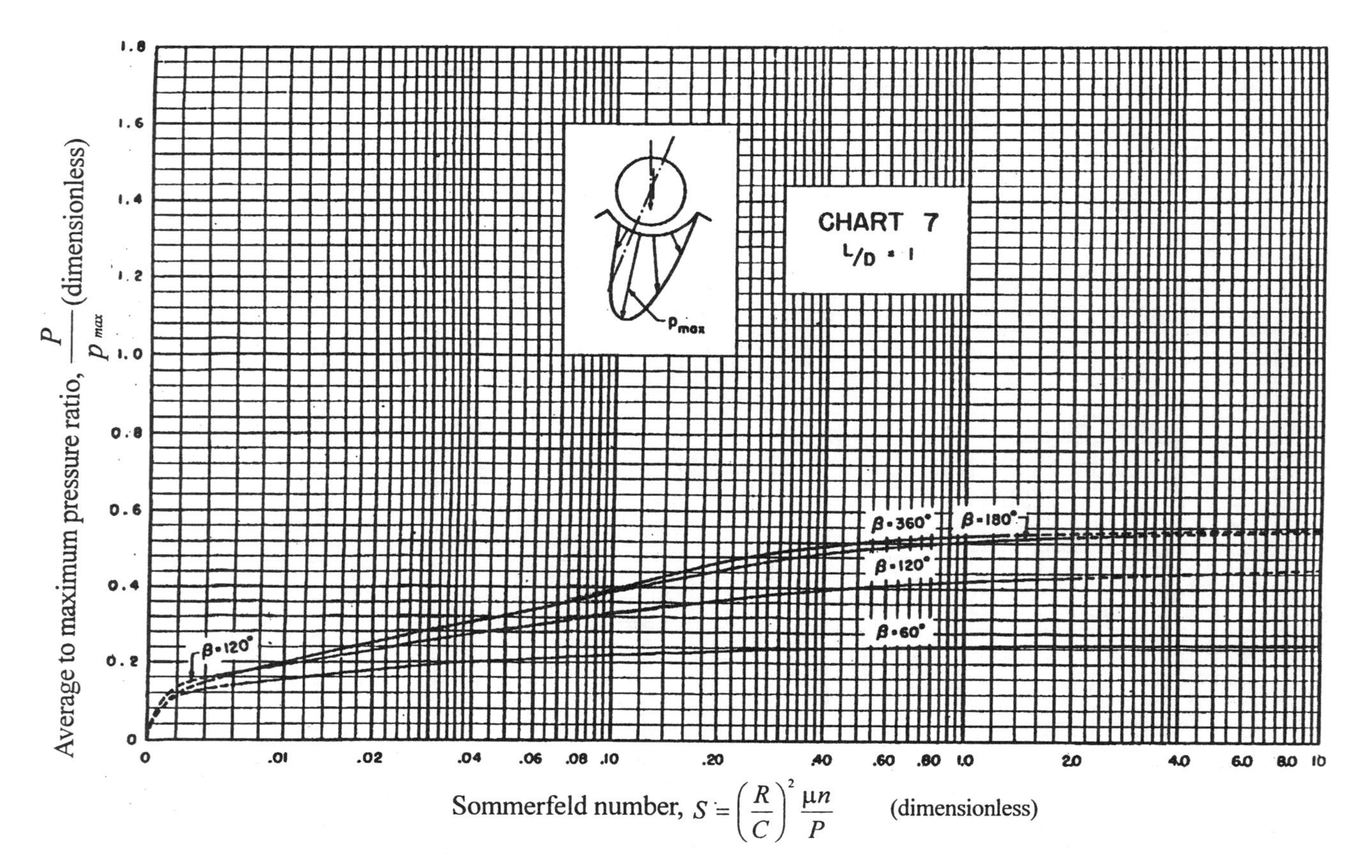

FIG. 8-7 Average to maximum pressure ratio versus Sommerfeld number ($L/D = 1$). (From Raimondi and Boyd, 1958, with permission of STLE.)

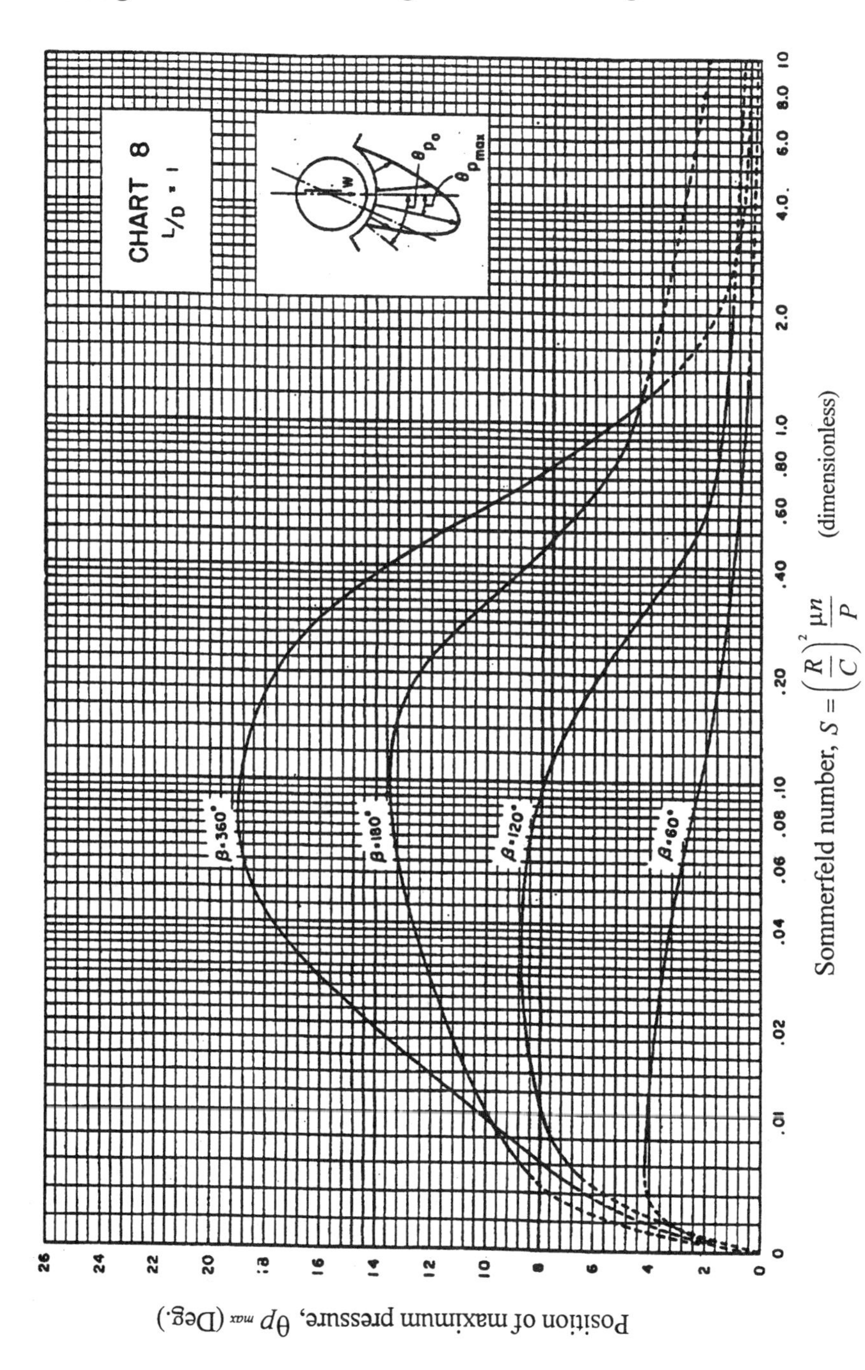

FIG. 8-8 Position of maximum pressure versus Sommerfeld number ($L/D = 1$). (From Raimondi and Boyd, 1958, with permission of STLE.)

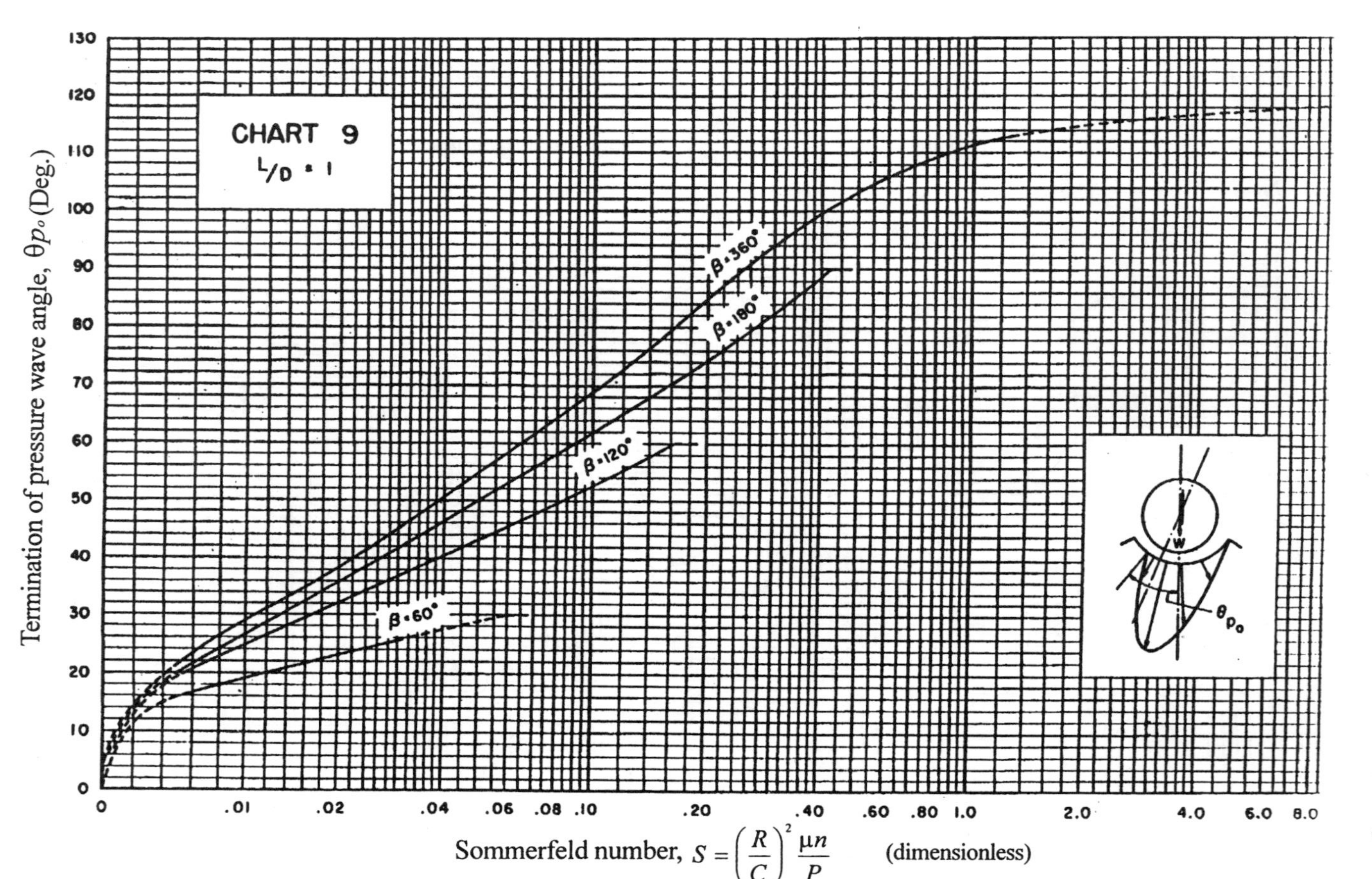

FIG. 8-9 Termination of pressure wave angle versus Sommerfeld number ($L/D = 1$). (From Raimondi and Boyd, 1958, with permission of STLE.)

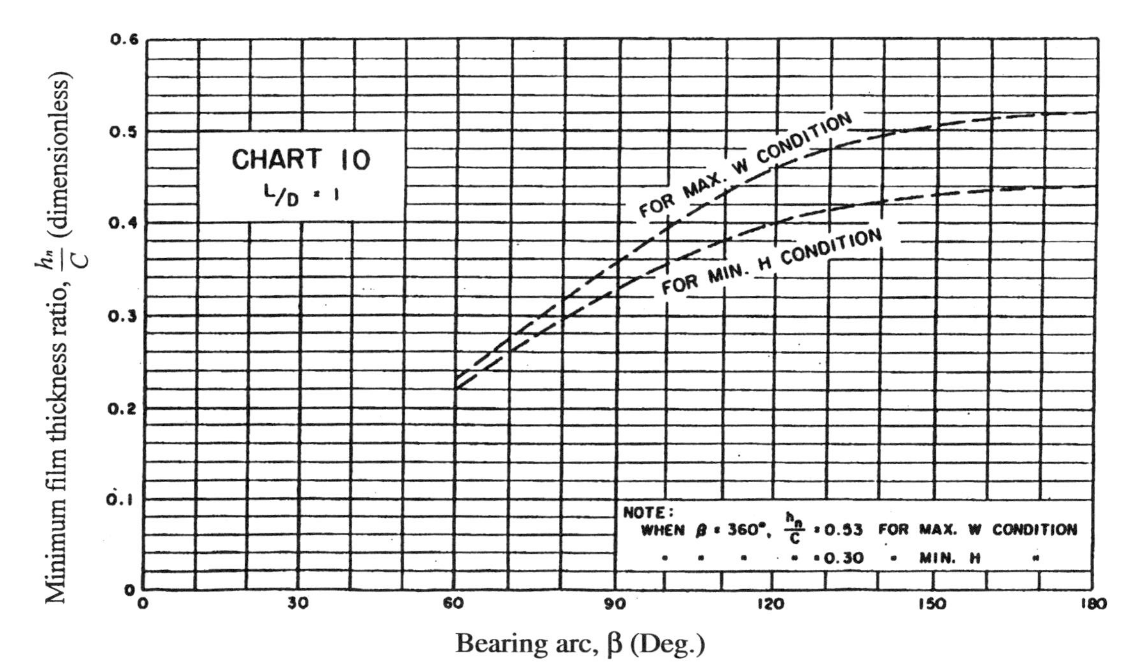

FIG. 8-10 Chart for determining the value of the minimum film thickness versus bearing arc for maximum load, and minimum power loss ($L/D = 1$). (From Raimondi and Boyd, 1958, with permission of STLE.)

TABLE 8-1 Performance Characteristics for a Centrally Loaded 360° Bearing

$\frac{L}{D}$	ε	$\frac{h_n}{C}$	θ_A	$\frac{\alpha}{\beta}$	S	ϕ	$\frac{R}{C}f$	$\frac{Q}{nRCL}$	$\frac{Q_s}{Q}$	$\frac{\rho c}{P}\Delta T$	$\frac{P}{p_{max}}$	$\theta_{p_{max}}$	θ_{p_0}
∞	0	1	0	---	∞	70.92	∞	π	0	∞	---	0	149.38
	0.1	0.9	0	0.308	0.24	69.1	4.8	3.03	0	19.9	0.826	0	137
	0.2	0.8	0	0.314	0.123	67.26	2.57	2.83	0	11.4	0.814	5.6	128
	0.4	0.6	0	0.328	0.0626	61.94	1.52	2.26	0	8.47	0.764	14.4	107
	0.6	0.4	0	0.349	0.0389	54.31	1.2	1.56	0	9.73	0.667	20.8	86
	0.8	0.2	0	0.383	0.021	42.22	0.961	0.76	0	15.9	0.495	21.5	58.8
	0.9	0.1	0	0.412	0.0115	31.62	0.756	0.411	0	23.1	0.358	19	44
	0.97	0.03	0	---	---	---	---	---	0	---	---	---	---
	1	0	0	---	0	0	0	0	0	□□	0	0	0
1	0	1	0	---	∞	85	∞	π	0	∞	---	0	119
	0.1	0.9	0	0.279	1.33	79.5	26.4	3.37	0.15	106	0.54	3.5	113
	0.2	0.8	0	0.294	0.631	74.02	12.8	3.59	0.28	52.1	0.529	9.2	106
	0.4	0.6	0	0.325	0.264	63.1	5.79	3.99	0.497	24.3	0.484	16.5	91.2
	0.6	0.4	0	0.36	0.121	50.58	3.22	4.33	0.68	14.2	0.415	18.7	72.9
	0.8	0.2	0	0.399	0.0446	36.24	1.7	4.62	0.842	8	0.313	18.2	52.3
	0.9	0.1	0	0.426	0.0188	26.45	1.05	4.74	0.919	5.16	0.247	13.8	37.3
	0.97	0.03	0	0.457	0.00474	15.47	0.514	4.82	0.973	2.61	0.152	7.1	20.5
	1	0	0	---	0	0	0	---	1	0	0	0	0
0.5000	0	1	0	---	∞	88.5	∞	π	0	∞	---	0	107
	0.1	0.9	0	0.273	4.31	81.62	85.6	3.43	0.173	343	0.523	5.8	99.2
	0.2	0.8	0	0.292	2.03	74.97	40.9	3.72	0.318	164	0.506	11.9	92.5
	0.4	0.6	0	0.329	0.779	61.45	17	4.29	0.552	68.6	0.441	16.9	78.8
	0.6	0.4	0	0.366	0.319	48.14	8.1	4.85	0.73	33	0.365	17.1	64.3
	0.8	0.2	0	0.408	0.0923	33.31	3.26	5.41	0.874	13.4	0.267	15.3	44.2
	0.9	0.1	0	0.434	0.0313	23.66	1.6	5.69	0.939	6.66	0.206	11	33.8
	0.97	0.03	0	0.462	0.00609	13.75	0.61	5.88	0.98	2.56	0.126	3.8	19.1
	1	0	0	---	0	0	0	---	1	0	0	0	0
0.2500	0	1	0	---	∞	89.5	∞	π	0	∞	---	0	99
	0.1	0.9	0	0.271	16.2	82.31	322	3.45	0.18	1287	0.515	7.4	98.9
	0.2	0.8	0	0.291	7.57	75.18	153	3.76	0.33	611	0.489	13.5	85
	0.4	0.6	0	0.331	2.83	60.86	61.2	4.37	0.567	245	0.415	17.4	70
	0.6	0.4	0	0.37	1.07	46.72	26.7	4.99	0.746	107	0.334	16.4	55.5
	0.8	0.2	0	0.414	0.261	31.04	8.8	5.6	0.884	35.4	0.24	11.5	39.7
	0.9	0.1	0	0.439	0.0736	21.85	3.5	5.91	0.945	14.1	0.18	8.6	27.8
	0.97	0.03	0	0.466	0.0101	12.22	0.922	6.12	0.984	3.73	0.108	4	17.7
	1	0	0	---	0	0	0	---	1	0	0	0	0

TABLE 8-2 Performance Characteristics for a Centrally Loaded 180° Bearing

$\frac{L}{D}$	ε	$\frac{h_n}{C}$	θ_A	$\frac{\alpha}{\beta}$	S	ϕ	$\frac{R}{C}f$	$\frac{Q}{nRCL}$	$\frac{Q_s}{Q}$	$\frac{\rho c}{P}\Delta T$	$\frac{P}{p_{max}}$	$\theta_{p_{max}}$	θ_{p_0}
∞	0	1	0	0.5	∞	90.00	∞	π	0	∞	----	0.0	90.0
	0.10	0.90	17.00	0.5	0.347	72.90	3.550	3.040	0	14.70	0.778	1.6	90.0
	0.20	0.80	28.60	0.5	0.179	61.32	2.010	2.800	0	8.99	0.759	3.4	90.0
	0.40	0.60	40.00	0.5	0.0898	49.99	1.290	2.200	0	7.34	0.700	7.7	90.0
	0.60	0.40	46.90	0.5	0.0523	43.15	1.060	1.520	0	8.71	0.607	12.3	71.8
	0.80	0.20	56.70	0.5	0.0253	33.35	0.859	0.767	0	14.10	0.459	13.7	51.2
	0.90	0.10	64.20	0.501	0.0128	25.57	0.681	0.380	0	22.50	0.337	13.3	36.8
	0.97	0.03	74.65	0.5	0.00384	15.43	0.416	0.119	0	44.00	0.190	9.1	20.7
	1	0	90	1	0	0	0	0	0	∞	0	0	0
1	0	1	0	0.500	∞	90.00	∞	π	0	∞	----	0.0	90.0
	0.10	0.90	11.500	0.500	1.40000	78.50	14.10	3.34	0.139	57.00	0.525	4.5	90.0
	0.20	0.80	21.000	0.500	0.67000	67.93	7.15	3.46	0.252	29.70	0.513	5.9	90.0
	0.40	0.60	34.167	0.500	0.27800	58.86	3.61	3.49	0.425	16.50	0.466	10.8	80.8
	0.60	0.40	45.000	0.502	0.12800	44.67	2.28	3.25	0.572	12.40	0.403	13.4	66.8
	0.80	0.20	58.000	0.498	0.04630	32.33	1.39	2.63	0.721	10.40	0.313	12.9	48.5
	0.90	0.10	66.000	0.499	0.01930	24.14	0.92	2.14	0.818	9.13	0.244	11.3	35.2
	0.97	0.03	75.584	0.499	0.00483	14.57	0.48	1.60	0.915	6.96	0.157	8.2	20.0
	1	0	90	0.5	0	0	0	----	1	0	0	0	0
0.5000	0	1	0	0.500	∞	90.00	∞	π	0	∞	----	0.0	90.0
	0.10	0.90	10.0	0.500	4.38000	79.97	44.00	3.41	0.167	177.00	0.518	5.4	90.0
	0.20	0.80	17.8	0.500	2.06000	72.14	21.60	3.64	0.302	87.80	0.499	9.1	90.0
	0.40	0.60	32.0	0.500	0.79400	58.01	9.96	3.93	0.506	42.70	0.438	13.7	71.9
	0.60	0.40	45.0	0.500	0.32100	45.01	5.41	3.93	0.665	25.90	0.365	14.1	59.1
	0.80	0.20	59.0	0.498	0.09210	31.29	2.54	3.56	0.806	15.00	0.273	12.1	43.9
	0.90	0.10	67.2	0.500	0.03140	22.80	1.38	3.17	0.886	9.80	0.208	10.4	31.5
	0.97	0.03	76.5	0.499	0.00625	13.63	0.58	2.62	0.951	5.30	0.132	6.9	18.1
	1	0	90	0.5	0	0	0	0	1	0	0	0	0
0.2500	0	1	0	0.5	∞	90.00	∞	π	0	∞	---	0.0	90.0
	0.10	0.9251	9.0	0.498	16.3000	81.40	163.00	3.44	0.176	653.0	0.513	6.3	90.0
	0.20	0.8242	16.3	0.500	7.6000	73.70	79.40	3.71	0.32	320.0	0.489	12.4	80.9
	0.40	0.6074	31.0	0.500	2.8400	58.99	35.10	4.11	0.534	146.0	0.417	15.8	70.5
	0.60	0.4000	45.0	0.500	1.0800	44.96	17.60	4.25	0.698	79.8	0.336	14.8	53.6
	0.80	0.2000	59.3	0.502	0.2630	30.43	6.88	4.07	0.837	36.5	0.241	11.4	39.0
	0.90	0.1000	68.9	0.498	0.0736	21.43	2.99	3.72	0.905	18.4	0.180	9.3	27.3
	0.97	0.0300	77.7	0.500	0.0104	12.28	0.88	3.29	0.961	6.5	0.110	6.3	17.5
	1	0	90.0	0.500	0	0	0	---	1	0	0	0	0

TABLE 8-3 Performance Characteristics for a Centrally Loaded 120° Bearing

$\frac{L}{D}$	ε	$\frac{h_n}{C}$	θ_A	$\frac{\alpha}{\beta}$	S	ϕ	$\frac{R}{C}f$	$\frac{Q}{nRCL}$	$\frac{Q_s}{Q}$	$\frac{\rho c}{P}\Delta T$	$\frac{P}{p_{max}}$	$\theta_{p\,max}$	$\theta_{p\,0}$
∞	0.0000	1.0000	30.0000	0.5000	∞	90.0000	∞	π	0.0000	∞	---	0.0000	60.0000
	0.1000	0.9007	53.3000	0.5000	0.8770	66.6900	6.0200	3.0200	0.0000	25.1000	0.6100	0.4000	60.0000
	0.2000	0.8000	67.4000	0.5000	0.4310	52.6000	3.2600	2.7500	0.0000	14.9000	0.5990	0.9000	60.0000
	0.4000	0.6000	81.0000	0.5000	0.1810	39.0200	1.7800	2.1300	0.0000	10.5000	0.5660	2.4000	60.0000
	0.6000	0.4000	87.3000	0.5000	0.0845	32.6700	1.2100	1.4700	0.0000	10.3000	0.5090	5.1000	60.0000
	0.8000	0.2000	93.2000	0.5000	0.0328	26.8000	0.8530	0.7590	0.0000	14.1000	0.4050	8.2000	44.2000
	0.9000	0.1000	98.5000	0.5000	0.0147	21.5100	0.6530	0.3880	0.0000	21.2000	0.3110	8.7000	32.9000
	0.9700	0.0300	106.1500	0.5000	0.0041	13.8600	0.3990	0.1180	0.0000	42.4000	0.1990	6.6000	19.6000
	1.0000	0.0000	120.0000	0.5000	0.0000	0.0000	0.0000	0.0000	0.0000		0.0000	0.0000	0.0000
1.0000	0.0000	1.0000	30.0000	0.5000	∞	90.0000	∞	π	0.0000	∞	---	0.0000	60.0000
	0.1000	0.9024	47.5000	0.5000	2.1400	74.9900	14.5000	3.2000	0.0876	59.5000	0.4270	1.1000	60.0000
	0.2000	0.8000	62.0000	0.4980	1.0100	63.3800	7.4400	3.1100	0.1570	32.6000	0.4200	1.3000	60.0000
	0.4000	0.6000	76.0000	0.5000	0.3850	48.0700	3.6000	2.7500	0.2720	19.0000	0.3960	3.2000	60.0000
	0.6000	0.4000	84.5000	0.4990	0.1620	38.5000	2.1600	2.2400	0.3840	15.0000	0.3560	6.5000	60.0000
	0.8000	0.2000	92.6000	0.5000	0.0531	28.0200	1.2700	1.5700	0.5350	13.9000	0.2900	8.6000	43.6000
	0.9000	0.1000	98.6670	0.5000	0.0208	21.0200	0.8550	1.1100	0.6570	14.4000	0.2330	8.5000	32.5000
	0.9700	0.0300	106.5000	0.5000	0.0050	13.0000	0.4610	0.6940	0.8120	14.0000	0.1620	6.3000	19.3000
	1.0000	0.0000	120.0000	0.5000	0.0000	0.0000	0.0000		1.0000	0.0000	0.0000	0.0000	0.0000
0.5000	0.0000	1.0000	30.0000	0.5000	∞	90.0000	∞	π	0.0000	∞	---	0.0000	60.0000
	0.1000	0.9034	45.0000	0.5000	5.4200	74.9900	36.6000	3.2900	0.1240	149.0000	0.4310	1.2000	60.0000
	0.2000	0.8003	56.6500	0.5000	2.5100	63.3800	18.1000	3.3200	0.2250	77.2000	0.4240	2.4000	60.0000
	0.4000	0.6000	72.0000	0.5000	0.9140	48.0700	8.2000	3.1500	0.3860	40.5000	0.3890	4.8000	60.0000
	0.6000	0.4000	81.5000	0.5000	0.3540	38.5000	4.4300	2.8000	0.5300	27.0000	0.3360	8.1000	53.4000
	0.8000	0.2000	92.0000	0.5000	0.0973	28.0200	2.1700	2.1800	0.6840	19.0000	0.2610	9.0000	40.5000
	0.9000	0.1000	99.0000	0.5000	0.0324	21.0200	1.2400	1.7000	0.7870	15.1000	0.2030	8.2000	30.4000
	0.9700	0.0300	107.0000	0.5000	0.0063	13.0000	0.5500	1.1900	0.8990	10.6000	0.1360	6.0000	18.2000
	1.0000	0.0000	120.0000	0.5000	0.0000	0.0000	0.0000	---	1.0000	0.0000	0.0000	0.0000	0.0000
0.2500	0.0000	1.0000	30.0000	0.5000	∞	90.0000	∞	π	0.0000	∞	---	0.0000	60.0000
	0.1000	0.9044	43.0000	0.5000	18.4000	76.9700	124.0000	3.3400	0.1430	502.0000	0.4560	3.0000	60.0000
	0.2000	0.8011	54.0000	0.5000	8.4500	65.9700	60.4000	3.4400	0.2600	254.0000	0.4380	4.8000	60.0000
	0.4000	0.6000	68.8330	0.5000	3.0400	51.2300	26.6000	3.4200	0.4420	125.0000	0.3890	8.4000	60.0000
	0.6000	0.4000	79.6000	0.5000	1.1200	40.4200	13.5000	3.2000	0.5990	75.8000	0.3210	10.4000	48.3000
	0.8000	0.2000	91.5600	0.5000	0.2680	28.3800	5.6500	2.6700	0.7530	42.7000	0.2370	9.4000	35.8000
	0.9000	0.1000	99.4000	0.5000	0.0743	20.5500	2.6300	2.2100	0.8460	25.9000	0.1780	7.8000	26.9000
	0.9700	0.0300	108.0000	0.4990	0.0105	12.1100	0.8320	1.6900	0.9310	11.6000	0.1120	5.5000	17.4000
	1.0000	0.0000	120.0000	0.5000	0.0000	0.0000	0.0000	---	1.0000	0.0000	0.0000	0.0000	0.0000

TABLE 8-4 Performance Characteristics for a Centrally Loaded 60° Bearing

$\frac{L}{D}$	ε	$\frac{h_n}{C}$	θ_A	$\frac{\alpha}{\beta}$	S	ϕ	$\frac{Q}{nRCL}$	$\frac{R}{C}f$	$\frac{Q_s}{Q}$	$\frac{\rho c}{P}\Delta T$	$\frac{P}{p_{max}}$	$\theta_{p_{max}}$	θ_{p_0}
∞	0.0000	1.0000	60.0000	0.5000	∞	90.0000	∞	π	0.0000	∞	----	0.0000	30.0000
	0.1000	0.9191	84.0000	0.5020	5.7500	65.9100	19.7000	3.0100	0.0000	82.3000	0.3370	0.1600	30.0000
	0.2000	0.8109	101.0000	0.5020	2.6600	48.9100	10.1000	2.7300	0.0000	46.5000	0.3360	0.1800	30.0000
	0.4000	0.6002	118.0000	0.5010	0.9310	31.9600	4.6700	2.0700	0.0000	28.4000	0.3290	0.2500	30.0000
	0.6000	0.4000	126.8000	0.5000	0.3220	23.2100	2.4000	1.4000	0.0000	21.5000	0.3170	0.5400	30.0000
	0.8000	0.2000	132.6000	0.5000	0.0755	17.3900	1.1000	0.7220	0.0000	19.2000	0.2870	1.7000	30.0000
	0.9000	0.1000	135.0600	0.5000	0.0241	14.9400	0.6670	0.3720	0.0000	22.5000	0.2430	3.2000	25.5000
	0.9700	0.0300	139.1400	0.5000	0.0050	10.8800	0.3720	0.1150	0.0000	40.7000	0.1630	4.2000	16.9000
	1.0000	0.0000	150.0000	0.5000	0.0000	0.0000	0.0000	0.0000	0.0000	∞	0.0000	0.0000	0.0000
1	0.0000	1.0000	60.0000	0.5000	∞	90.0000	∞	π	0.0000	∞	----	0.0000	30.0000
	0.1000	0.9212	82.0000	0.5010	8.5200	67.9200	29.1000	3.0700	0.0267	121.0000	0.2520	0.3000	30.0000
	0.2000	0.8133	99.0000	0.5010	3.9200	50.9600	14.8000	2.8200	0.4810	67.4000	0.2510	0.3000	30.0000
	0.4000	0.6010	116.0000	0.5000	1.3400	33.9900	6.6100	2.2200	0.0849	39.1000	0.2470	0.5400	30.0000
	0.6000	0.4000	125.5000	0.4990	0.4500	24.5600	3.2900	1.5600	0.1270	28.2000	0.2390	0.9500	30.0000
	0.8000	0.2000	131.6000	0.5010	0.1010	18.3300	1.4200	0.8830	0.2000	22.5000	0.2200	2.2000	30.0000
	0.9000	0.1000	134.6700	0.5000	0.0309	15.3300	0.8220	0.5190	0.2870	23.2000	0.1920	3.5000	25.9000
	0.9700	0.0300	139.1000	0.5000	0.0058	10.8800	0.4220	0.2260	0.4650	30.5000	0.1390	4.2000	16.9000
	1.0000	0.0000	150.0000	0.5000	0.0000	0.0000	0.0000	----	1.0000	0.0000	0.0000	0.0000	0.0000
0.5000	0.0000	1.0000	60.0000	0.5000	∞	90.0000	∞	π	0.0000	∞	---	0.0000	30.0000
	0.1000	0.9223	81.0000	0.5000	14.2000	69.0000	48.6000	3.1100	0.0488	201.0000	0.2390	0.0000	30.0000
	0.2000	0.8152	97.5000	0.4980	6.4700	52.6000	24.2000	2.9100	0.0883	109.0000	0.2390	0.0300	30.0000
	0.4000	0.6039	113.0000	0.5000	2.1400	37.0000	10.3000	2.3800	0.1600	59.4000	0.2330	0.4500	30.0000
	0.6000	0.4000	123.0000	0.5000	0.6950	26.9800	4.9300	1.7400	0.2360	40.3000	0.2250	1.0000	30.0000
	0.8000	0.2000	130.4000	0.5000	0.1490	19.5700	2.0200	1.0500	0.3500	29.4000	0.2010	2.2000	30.0000
	0.9000	0.1000	134.0900	0.5000	0.0422	15.9100	1.0800	0.6640	0.4640	26.5000	0.1720	3.8000	25.4000
	0.9700	0.0300	139.2200	0.4990	0.0070	10.8500	0.4900	0.3290	0.6500	27.8000	0.1220	4.2000	16.6000
	1.0000	0.0000	150.0000	0.5000	0.0000	0.0000	0.0000	----	1.0000	0.0000	0.0000	0.0000	0.0000
0.2500	0.0000	1.0000	60.0000	0.5000	∞	90.0000	∞	π	0.0000	∞	---	0.0000	30.0000
	0.1000	0.9251	78.5000	0.4990	35.8000	71.5500	121.0000	3.1600	0.0666	499.0000	0.2510	0.0000	30.0000
	0.2000	0.8242	91.5000	0.5000	16.0000	58.5100	58.7000	3.0400	0.1310	260.0000	0.2490	0.1000	30.0000
	0.4000	0.6074	109.0000	0.5000	5.2000	41.0100	24.5000	2.5700	0.2360	136.0000	0.2420	0.5000	30.0000
	0.6000	0.4000	119.8000	0.5010	1.6500	30.1400	11.2000	1.9800	0.3460	86.1000	0.2280	1.5000	30.0000
	0.8000	0.2000	128.3000	0.5000	0.3330	21.7000	4.2700	1.3000	0.4960	54.9000	0.1950	3.2000	30.0000
	0.9000	0.1000	133.1000	0.5000	0.0844	16.8700	2.0100	0.8940	0.6200	41.0000	0.1590	4.3000	23.7000
	0.9700	0.0300	139.2000	0.5000	0.0110	10.8100	0.7130	0.5070	0.7860	29.1000	0.1070	4.1000	15.9000
	1.0000	0.0000	150.0000	0.5000	0.0000	0.0000	0.0000	---	1.0000	0.0000	0.0000	0.0000	0.0000

7. The ratio of average pressure to maximum pressure, P/p_{max}, in the fluid film as a function of Sommerfeld number is given in Fig. 8-7.
8. The location of the point of maximum pressure is given in Fig. 8-8. It is measured in degrees from the line along the load direction as shown in Fig. 8-8.
9. The location of the point of the end of the pressure wave is given in Fig. 8-9. It is measured in degrees from the line along the load direction as shown in Fig. 8-9. This is the angle θ_2 in this text, and it is referred to as θ_{p° in the chart of Raimondi and Boyd.
10. Curves of the minimum film thickness ratio, h_n/C as a function of the bearing arc, β (Deg.), are presented in Fig. 8-10 for two cases: a. maximum load capacity, b. minimum power losses due to friction. These curves are useful for the design engineer for selecting the optimum bearing arc, β, based on the requirement of maximum load capacity, or minimum power loss due to viscous friction.

Note that the preceding performance parameters are presented by graphs only for journal bearings with the ratio $L/D = 1$. For bearings having different L/D ratios, the performance parameters are listed in tables.

In Fig. 8-1 (chart 1), curves are presented of the film thickness ratio, h_n/C, versus the Sommerfeld number, S for various bearing arcs β. The curves for $\beta = 180^\circ$ and $\beta = 360^\circ$ nearly coincide. This means that for an identical bearing load, a full bearing ($\beta = 360^\circ$) does not result in a significantly higher value of h_n in comparison to a partial bearing of $\beta = 180^\circ$. This means that for an identical h_n, a full bearing ($\beta = 360^\circ$) does not have a much higher load capacity than a partial bearing. At the same time, it is clear from Fig. 8-3 (chart 3) that lowering the bearing arc, β, results in a noticeable reduction in the bearing friction (viscous friction force is reduced because of the reduction in oil film area).

In conclusion, the advantage of a partial bearing is that it can reduce the friction coefficient of the bearing without any significant reduction in load capacity (this advantage is for identical geometry and viscosity in the two bearings). In fact, the advantage of a partial bearing is more than indicated by the two figures, because it has a lower fluid film temperature due to a faster oil circulation. This improvement in the thermal characteristics of a partial bearing in comparison to a full bearing is considered an important advantage, and designers tend to select this type for many applications.

8.5 FLUID FILM TEMPERATURE

8.5.1 Estimation of Temperature Rise

After making the basic decisions concerning the bearing dimensions, bearing arc, and determination of the minimum film thickness h_n, the lubricant is selected. At this stage the bearing temperature is unknown, and it should be estimated. We assume an average bearing temperature and select a lubricant that would provide the required bearing load capacity (equal to the external load). The next step is to determine the flow rate of the lubricant in the bearing, Q, in the axial direction. Knowledge of this flow rate allows one to determine the temperature rise inside the fluid film from the charts. This will allow one to check and correct the initial assumptions made earlier concerning the average oil film temperature. Later, it is possible to select another lubricant for the desired average viscosity, based on the newly calculated temperature. A few iterations are required for estimation of the average temperature.

The temperature inside the fluid film increases as it flows inside the bearing, due to high shear rate flow of viscous fluid. The energy loss from viscous friction is dissipated in the oil film in the form of heat. There is an energy balance, and a large part of this heat is removed from the bearing by continuous convection as the hot oil flows out and is replaced by a cooler oil that flows into the bearing clearance. In addition, the heat is transferred by conduction through the sleeve into the bearing housing. The heat is transferred from the housing partly by convection to the atmosphere and partly by conduction through the base of the housing to the other parts of the machine. In most cases, precise heat transfer calculations are not practical, because they are too complex and because many parameters, such as contact resistance between the machine parts, are unknown.

For design purposes it is sufficient to estimate the temperature rise of the fluid ΔT, from the point of entry into the bearing clearance (at temperature T_{in}) to the point of discharge from the bearing (at temperature T_{max}). This estimation is based on the simplified assumptions that it is possible to neglect the heat conduction through the bearing material in comparison to the heat removed by the continuous replacement of fluid. In fact, the heat conduction reduces the temperature rise; therefore, this assumption results in a design that is on the safe side, because the estimated temperature rise is somewhat higher than in the actual bearing. The following equation for the temperature rise of oil in a journal bearing, ΔT, was presented by Shigley and Mitchell (1983):

$$\Delta T = \frac{8.3P(f\,R/C)}{10^6\left(\dfrac{Q}{nRCL}\right)(1 - 0.5Q_s/Q)} \tag{8-5}$$

where ΔT is the temperature rise [°C], $P = F/2RL$ [Pa]. All the other parameters required for calculation of the temperature rise are dimensionless parameters.

They can be obtained directly from the charts or tables of Raimondi and Boyd as a function of the Sommerfeld number and L/D ratio. The average temperature in the fluid film is determined from the temperature rise by the equation

$$T_{av} = \frac{T_{in} + T_{max}}{2} = T_i + \frac{\Delta T}{2} \qquad (8\text{-}6)$$

Equation (8-5) is derived by assuming that all the heat that is generated by viscous shear in the fluid film is dissipated only in the fluid (no heat conduction through the boundaries). This heat increases the fluid temperature. In a partial bearing, the maximum temperature is at the outlet at the end of the bearing arc. In a full bearing, the maximum temperature is after the minimum film thickness at the end of the pressure wave (angle θ_2). The mean temperature of the fluid flowing out, in the axial direction, Q, has been assumed as T_{av}, the average of the inlet and outlet temperatures.

Example Problem 8-1

Calculation of Temperature Rise

A partial journal bearing ($\beta = 180^\circ$) has a radial load $F = 10{,}000$ N. The speed of the journal is $N = 6000$ RPM, and the viscosity of the lubricant is 0.006 N-s/m^2. The geometry of the bearing is as follows:

Journal diameter: $D = 40$ mm
Bearing length: $L = 10$ mm
Bearing clearance: $C = 30 \times 10^{-3}$ mm

a. Find the following performance parameters:

Minimum film thickness h_n
Friction coefficient f
Flow rate Q
Axial side leakage Q_s
Rise in temperature ΔT if you ignore the heat conduction through the sleeve and journal

b. Given an inlet temperature of the oil into the bearing of 20°C, find the maximum and average temperature of the oil.

Solution

This example is calculated from Eq. (8-5) in SI units.

The bearing data is given by:

$$\beta = 180^\circ$$
$$\frac{L}{D} = \frac{1}{4}$$
$$P = \frac{F}{LD} = 25 \times 10^6 \text{ Pa}$$
$$n = \frac{6000}{60} = 100 \text{ rPS}$$

The Sommerfeld number [using Eq. (8-1)] is:

$$S = \left(\frac{2 \times 10^{-2}}{30 \times 10^{-6}}\right)^2 \frac{0.006 \times 100}{25 \times 10^6} = 0.0106$$

a. Performance Parameters

From the table for a $\beta = 180^\circ$ bearing) and $L/D = 1/4$, the following operating parameters can be obtained for $S = 0.0106$, the calculated Sommerfeld number.

Minimum Film Thickness:

$$\frac{h_n}{C} = 0.03 \qquad h_n = 0.9 \times 10^{-3} \text{ mm}$$

If the minimum film thickness obtained is less than the design value, the design has to modified.

Coefficient of Friction: The coefficient of friction is obtained from the table:

$$\frac{R}{C} f = 0.877 \qquad f = 0.0013$$

Flow rate:

$$\frac{Q}{nRCL} = 3.29 \qquad Q = 1.974 \times 10^{-6} \text{ m}^3/\text{s}$$

Side Leakage:

$$\frac{Q_s}{Q} = 0.961 \qquad Q_s = 1.897 \times 10^{-6} \text{ m}^3/\text{s}$$

Temperature Rise ΔT: The estimation of the temperature rise is based on Eq. (8-5) in SI units. The dimensionless operating parameters, from the appropriate table of Raimondi and Boyd, are substituted:

$$P = 25 \times 10^6 \text{ Pa} \qquad \left(1 - 0.5\frac{Q_s}{Q}\right) = 0.5195 \qquad \frac{R}{C}f = 0.877$$

$$\Delta T_m = \frac{8.3P[R/C(f)]}{10^6\left(\frac{Q}{nRCL}\right)[1 - (0.5)Q_s/Q]} = \frac{8.3 \times 25 \times 0.877}{3.29 \times 0.5195} = 106^\circ\text{C}$$

b. Maximum and Average Oil Temperatures:

Maximum temperature:

$$T_{\text{max}} = T_{\text{in}} + \Delta T = 20 + 106 = 123^\circ\text{C}$$

Average temperature:

$$T_{\text{av}} = T_{\text{in}} + \frac{\Delta T}{2} = 20 + \frac{106}{2} = 73^\circ\text{C}$$

Since the bearing material is subjected to the maximum temperature of 123°C, the bearing material that is in contact with the lubricant should be resistant to this temperature. Bearing materials are selected to have a temperature limit well above the maximum temperature in the fluid film.

For bearing design, the Sommerfeld number, S, is determined based on lubricant viscosity at the average temperature of 73°C.

8.5.2 Temperature Rise Based on the Tables of Raimondi and Boyd

The specific heat and density of the lubricant affect the rate of heat transfer and the resulting temperature rise of the fluid film. However, Eq. (8-5) does not consider the properties of the lubricant, and it is an approximation for the properties of mineral oils. For other fluids, such as synthetic lubricants, the temperature rise can be determined more accurately from a table of Raimondi and Boyd. The advantage of the second method is that it can accommodate various fluid properties. The charts and tables include a temperature-rise variable as a function of the Sommerfeld number. The temperature-rise variable is a dimensionless ratio that includes the two properties of the fluid: the specific heat, c (Joule/kg-°C), and the density, ρ (kg/m^3). Table 8-5 lists these properties for engine oil as a function of temperature.

The following two problems illustrate the calculation of the temperature rise, based on the charts or tables of Raimondi and Boyd. The two examples involve calculations in SI units and Imperial units.* We have to keep in mind that

* The original charts of Raimondi and Boyd were prepared for use with Imperial units (the conversion of energy from BTU to lbf-inch units is included in the temperature-rise variable). In this text, the temperature-rise variable is applicable for any unit system.

TABLE 8-5 Specific Heat and Density of Engine Oil

Temperature, T		Specific heat, c		Density, ρ	
°F	°C	J/kg-°C	BTU/lb_m-°F	kg/m^3	lbm/ft^3
32	0	1796	0.429	899.1	56.13
68	20	1880	0.449	888.2	55.45
104	40	1964	0.469	876.1	54.69
140	60	2047	0.489	864.0	53.94
176	80	2131	0.509	852.0	53.19
212	100	2219	0.529	840.0	52.44
248	120	2307	0.551	829.0	51.75
284	140	2395	0.572	816.9	50.99
320	160	2483	0.593	805.9	50.31

both solutions are adiabatic, in the sense that the surfaces of the journal and the bearing are assumed to be ideal insulation. In practice, it means that conduction of heat through the sleeve and journal is disregarded in comparison to the heat taken away by the fluid. In this way, the solution is on the safe side, because it predicts a higher temperature than in the actual bearing.

Example Problem 8-2

Calculation of Transformation Rise in SI Units

Solve for the temperature rise ΔT for the journal bearing in Example Problem 8-1. Use the temperature-rise variable according to the Raimondi and Boyd tables and solve in SI units. Use Table 8-5 for the oil properties. Assume that the properties can be taken as for engine oil at 80°C

Solution

The temperature rise is solved in SI units based on the tables of Raimondi and Boyd. The properties of engine oil at 80°C are:

Specific heat (from Table 8-5): $c = 2131$ [Joule/kg-°C]
Density of oil (from Table 8-5): $\rho = 852$ [kg/m^3]
Bearing average pressure (see Example Problem 8-1): $P = 25 \times 10^6$ [N/m^2]
Temperature-rise variable (from Table 8-2 for $\beta = 180°$) is 6.46.

The equation is

$$\frac{c\rho}{P}\Delta T = 6.46$$

The properties and P are given, and the preceding equation can be solved for the temperature rise:

$$\Delta T = 6.46 \frac{P}{c\rho} = 6.46 \frac{25 \times 10^6}{2131 \times 852} = 88.9°\text{C}$$

This temperature rise is considerably lower than that obtained by the equation of Shigley and Mitchell (1983) in Example Problem 8-1.

Example Problem 8-3

Calculation of Temperature Rise in Imperial Units

Solve for the temperature rise ΔT of the journal bearing in Example Problem 8-1. Use the temperature-rise variable according to the Raimondi and Boyd tables and solve in Imperial units. Use Table 8-5 for the oil properties. Assume that the properties can be taken as for engine oil at 176°F (equal to 80°C in Example Problem 8-2).

Solution

The second method is to calculate ΔT from the tables of Raimondi and Boyd in Imperial units. The following values are used:

Density of engine oil (at 176°F, from Table 8-1): $\rho = 53.19$ [lbm/ft^3] = $53.19/12^3 = 0.031$ [lbm/in^3.]
Specific heat of oil (from Table 8-5): $c = 0.509$ [BTU/lbm-F°]
Mechanical equivalent of heat: $J = 778$ [lbf-ft/BTU] = 778×12 [lbf-inch/BTU]

This factor converts the thermal unit BTU into the mechanical unit lbf-ft:

$$c = 0.509\ [\text{BTU/lbm} - °\text{F}] \times 778 \times 12\ [\text{lbf-inch/BTU}]$$
$$= 4752\ [\text{lbf-inch/lbm} - \text{F}°]$$

The bearing average pressure (from Example Problem 8-1):

$$P = 2.5 \times 10^6\ \text{Pa} = (25 \times 10^6)/6895 = 3626\ [\text{lbf/in}^2.]$$

The data in Imperial units results in a dimensionless temperature-rise variable where the temperature rise is in °F.

Based on the table of Raimondi and Boyd, the same equation is applied as in Example Problem 8-2:

$$\frac{c\rho}{P} \Delta T = 6.46$$

Solving for the temperature rise:

$$\Delta T = 6.46\frac{P}{c\rho} = 6.46\frac{3626}{4752 \times 0.031} = 159^{\circ}\text{F}$$

$$\Delta T = 147.7^{\circ}\text{F} \times \frac{5}{9}\ (^{\circ}\text{C}/^{\circ}\text{F}) = 88.3^{\circ}\text{C}$$

(close to the previous solution in SI units)

Note: The reference 32°F does not play a role here because we solve for the temperature difference, ΔT.

8.5.3 Journal Bearing Design

Assuming an initial value for viscosity, the rise in temperature, ΔT, is calculated and an average temperature of the fluid film is corrected. Accordingly, after using the calculated average temperature, the viscosity of the oil can be corrected. The new viscosity is determined from the viscosity–temperature chart (Fig. 2-3). The inlet oil temperature to the bearing can be at the ambient temperature or at a higher temperature in central circulating systems.

If required, the selection of the lubricant may be modified to account for the new temperature. In the next step, the Sommerfeld number is modified for the corrected viscosity of the previous oil, but based on the new temperature. Let us recall that the Sommerfeld number is a function of the viscosity, according to Eq. (8-1). If another oil grade is selected, the viscosity of the new oil grade is used for the new Sommerfeld number. Based on the new Sommerfeld number S, the calculation of Q and the temperature rise estimation ΔT are repeated. These iterations are repeated until there is no significant change in the average temperature between consecutive iterations. If the temperature rise is too high, the designer can modify the bearing geometry.

After the average fluid film temperature is estimated, it is necessary to select the bearing material. Knowledge of the material properties allows one to test whether the allowable limits are exceeded. At this stage, it is necessary to calculate both the peak pressure and the peak temperature and to compare those values with the limits for the bearing material that is used. The values of the maximum pressure and temperature rise in the fluid film are easy to determine from the charts or tables of Raimondi and Boyd.

8.5.4 Accurate Solutions

For design purposes, the average temperature of the fluid-film can be estimated as described in the preceding section. Temperature estimation is suitable for most practical cases. However, in certain critical applications, more accurate analysis is required. The following is a general survey and references that the reader can use for advanced study of this complex heat transfer problem.

In a fluid film bearing, a considerable amount of heat is generated by viscous friction, which is dissipated in the oil film and raises its temperature. The fluid film has a non-uniform temperature distribution along the direction of motion (x direction) and across the film (z direction). The peak fluid film temperature is near the point of minimum film thickness. The rise in the oil temperature results in a reduction of the lubricant viscosity; in turn, there is a significant reduction of the hydrodynamic pressure wave and load carrying capacity. Accurate solution of the temperature distribution in the fluid film includes heat conduction through the bearing material and heat convection by the oil. This solution requires a numerical analysis, and it is referred to as a full *thermohydrodynamic* (THD) analysis. This analysis is outside the scope of this text, and the reader is referred to available surveys, such as by Pinkus (1990) and by Khonsari (1987). The results are in the form of isotherms mapping the temperature distribution in the sleeve. An example is included in Chap. 18.

8.6 PEAK TEMPERATURE IN LARGE, HEAVILY LOADED BEARINGS

The maximum oil film temperature of large, heavily loaded bearings is higher than the outlet temperature. Heavily loaded bearings have a high eccentricity ratio, and at high speed they are subjected to high shear rates and much heat dissipation near the minimum film thickness. For example, in high-speed turbines having journals of the order of magnitude of 10 in. (250 mm) and higher, it has been recognized that the maximum temperature near the minimum film thickness, h_n, is considerably higher than $T_{in} + \Delta T$, which has been calculated in the previous section. In bearings made of white metal (babbitt), it is very important to limit the maximum temperature to prevent bearing failure.

In a bearing with a white metal layer on its surface, creep of this layer can initiate at temperatures above 260°F. The risk of bearing failure due to local softening of the white metal is high for large bearings operating at high speeds and small minimum film thickness. Plastic bearings can also fail due to local softening of the plastic at elevated temperatures. The peak temperature along the bearing surface is near the minimum film thickness, where there is the highest shear rate and maximum heat dissipation by viscous shear. This is exacerbated by the combination of local high oil film pressure and high temperature at the same point, which initiates an undesirable creep process of the white metal. Therefore, it is important to include in the bearing design an estimation of the peak temperature near the minimum film thickness (in addition to the temperature rise, ΔT).

The yield point of white metals reduces significantly with temperature. The designer must ensure that the maximum pressure does not exceed its limit. If the temperature is too high, the designer can use bearing material with a higher melting point. Another alternative is to improve the cooling by providing faster oil circulation by means of several oil grooves. An example is the three-lobe bearing that will be described in Chapter 9.

Adiabatic solutions were developed by Booser et al. (1970) for calculating the maximum temperature, based on the assumption that the heat conduction through the bearing can be neglected in comparison to the heat removed by the flow of the lubricant. This assumption is justified in a finite-length journal bearing, where the axial flow rate has the most significant role in heat removal.

The derivation of the maximum temperature considers the following viscosity–temperature relation:

$$\mu = kT^{-n} \tag{8-7}$$

where the constants k and n are obtained from the viscosity–temperature charts. The viscosity is in units of lb-s/in^2. and the temperature is in deg. F.

The maximum temperatures obtained according to Eq. (8-8) were experimentally verified, and the computation results are in good agreement with the measured temperatures. The equation for the maximum temperature, T_{max}, is (Booser et al., 1970):

$$T_{max}^{n+1} - T_1^{n+1} = \frac{4\pi k(n+1)N}{60\rho c_p}\left(\frac{R}{C}\right)^2 \Delta G_j \tag{8-8}$$

Here, ρ is the lubricant density and c_p is its specific heat at constant pressure. The temperatures T_m and T_1 are the maximum and inlet temperatures, respectively. The temperatures, in deg. F, have an exponent of $(n+1)$ from the viscosity–temperature equation (8-7). The journal speed N is in revolutions per minute. The coefficient ΔG_j is a temperature-rise multiplier. It can be obtained from Fig. 8-11. It shows the rapid increase of ΔG_j at high eccentricity ratios ($\varepsilon = 0.8$–0.9), indicating that the maximum temperature is highly dependent on the film thickness, particularly under high loads.

For turbulent fluid films, the equation is

$$T_{max} - T_1 = \frac{f\pi^2 N^2 D^3}{2gc_p(1-\varepsilon^2)}(\pi - \theta_1) \tag{8-9}$$

where f is the friction coefficient, D is the journal diameter, and g is gravitational acceleration, 386 in./s^2. The angle θ_1 is the oil inlet angle (in radians). The

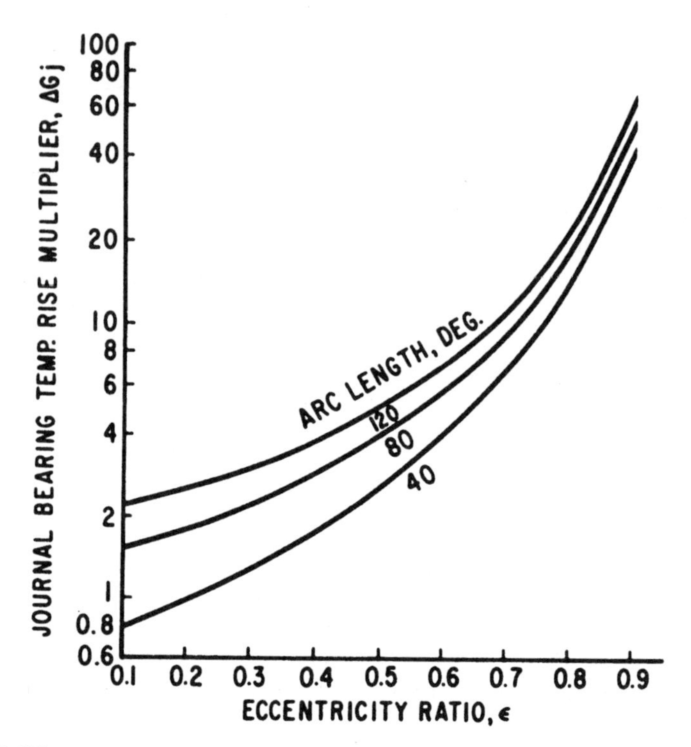

FIG. 8-11 Journal bearing temperature-rise multiplier for Eq. (8-8). (From Booser et al., 1970, with permission of STLE.)

friction coefficient is determined by experiment or taken from the literature for a similar bearing.

8.7 DESIGN BASED ON EXPERIMENTAL CURVES

In the preceding discussion, it was shown that the complete design of hydrodynamic journal bearings relies on important decisions: determination of the value of the minimum film thickness, h_n, and the upper limit of bearing operating temperature. The minimum value of h_n is determined by the surface finish of the bearing and the journal as well as other operating conditions that have been discussed in this chapter. However, the surface finish may vary after running the machine, particularly for the soft white metal that is widely used as bearing material. In addition to the charts of Raimondi and Boyd, which are based on hydrodynamic analysis, bearing design engineers need design tools that are based on previous experience and experiments. In particular for bearing design for critical applications, there is a merit in also relying on experimental curves for determining the limits of safe hydrodynamic performance.

In certain machines, there are design constraints that make it necessary to have highly loaded bearings operating with very low minimum film thickness. Design based on hydrodynamic theory is not very accurate for highly loaded bearings at very thin h_n. The reason is that in such cases, it is difficult to predict the temperature rise, ΔT, and the h_n that secure hydrodynamic performance. In such cases, the limits of hydrodynamic bearing operation can be established only by experiments or experience with similar bearings. There are many examples of machines that are working successfully with hydrodynamic bearings having much lower film thickness than usually recommended.

For journal bearings operating in the full hydrodynamic region, the friction coefficient, f, is an increasing function of the Sommerfeld number. Analytical curves of $(R/C)f$ versus the Sommerfeld number are presented in the charts of Raimondi and Boyd; see Fig. 8-3. These curves are for partial and full journal bearings, for various bearing arcs, β. Of course, the designer would like to operate the bearing at minimum friction coefficient. However, these charts are only for the hydrodynamic region and do not include the boundary and mixed lubrication regions. These curves do not show the lowest limit of the Sommerfeld number for maintaining a full hydrodynamic film. A complete curve of $(R/C)f$ versus the Sommerfeld number over the complete range of boundary, mixed, and hydrodynamic regions can be obtained by testing the bearing friction against variable speed or variable load. These experimental curves are very helpful for bearing design. Description of several friction testing systems is included in Chapter 14.

8.7.1 Friction Curves

The friction curve in the boundary and mixed lubrication regions depends on the material as well as on the surface finish. For a bearing with constant C/R ratio, the curves of $(R/C)f$ versus the Sommerfeld number, S, can be reduced to dimensionless, experimental curves of the friction coefficient, f, versus the dimensionless ratio, $\mu n/P$. These experimental curves are very useful for design purposes. In the early literature, the notation for viscosity is z, and the variable zN/P has been widely used. In this text, the ratio $\mu n/P$ is preferred, because it is dimensionless and any unit system can be used as long as the units are consistent. In addition, this ratio is consistent with the definition of the Sommerfeld number.

The variable zN/P is still widely used, because it is included in many experimental curves that are provided by manufacturers of bearing materials. Curves of f versus zN/P are often used to describe the performance of a specific bearing of constant geometry and material combination. This ratio is referred to as the Hersey* number. The variable zN/P is not completely dimensionless, because it is used as a combination of Imperial units with metric units for the viscosity. The average pressure is in Imperial units [psi], the journal speed, N, is in revolutions per minute [RPM], and the viscosity, z, is in centipoise. In order to have dimensionless variables, the journal speed, n, must always be in revolutions per second (RPS), irrespective of the system of units used, and the viscosity, μ, must always include seconds as the unit of time. The variable zN/P is proportional and can be converted to the dimensionless variable $\mu n/P$.

Transition from Mixed to Hydrodynamic Lubrication

A typical experimental curve of the friction coefficient, f, versus the dimensionless variable, $\mu n/P$, is shown in Fig. 8-12. The curve shows the region of hydrodynamic lubrication, at high values of $\mu n/P$, and the region of mixed

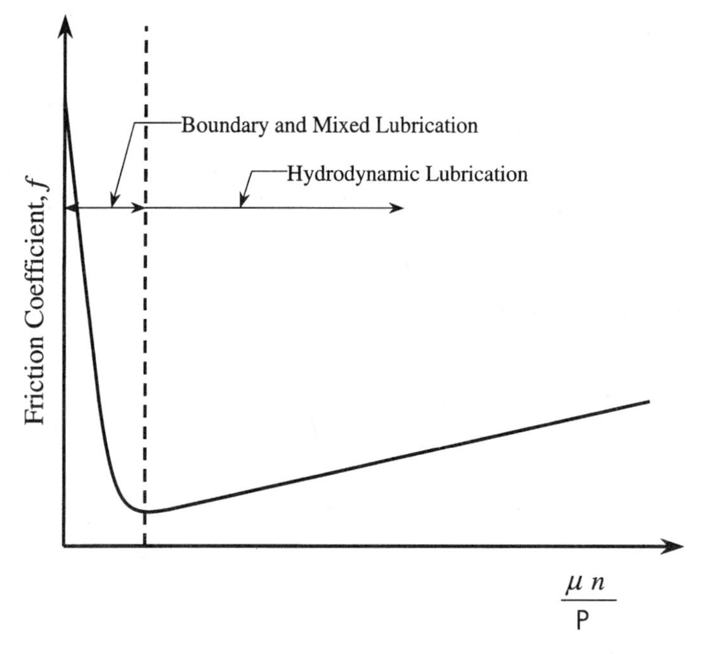

FIG. 8-12 Friction coefficient, f, versus variable $\mu n/P$ in a journal bearing,

* After Mayo D. Hersey, for his contribution to the lubrication field.

lubrication, at lower values of $\mu n/P$. The transition point $(\mu n/P)_{tr}$ from mixed to hydrodynamic lubrication is at the point of minimum friction coefficient.

Hydrodynamic theory indicates that minimum film thickness increases with the variable $\mu n/P$. Full hydrodynamic lubrication is where $\mu n/P$ is above a certain transition value $(\mu n/P)_{tr}$. At the transition point, the minimum film thickness is equal to the size of surface asperities. However, in the region of full hydrodynamic lubrication, the minimum film thickness is higher than the size of surface asperities, and there is no direct contact between the sliding surfaces. Therefore, there is only viscous friction, which is much lower in comparison to direct contact friction. In the hydrodynamic region, viscous friction increases with $\mu n/P$, because the shear rates and shear stresses in the fluid film are increasing with the product of viscosity and speed.

Below the critical value $(\mu n/P)_{tr}$, there is mixed lubrication where the thickness of the lubrication film is less than the size of the surface asperities. Under load, there is direct contact between the surfaces, resulting in elastic as well as plastic deformation of the asperities. In the mixed region, the external load is carried partly by the pressure of the hydrodynamic fluid film and partly by the mechanical elastic reaction of the deformed asperities. The film thickness increases with $\mu n/P$; therefore, as the velocity increases, a larger portion of the load is carried by the fluid film. In turn, the friction decreases with $\mu n/P$ in the mixed region, because the fluid viscous friction is lower than the mechanical friction due to direct contact between the asperities. The transition value, $(\mu N/P)_{tr}$, is at the minimum friction, where there is a transition in the trend of the friction slope.

Design engineers are often tempted to design the bearing at the transition point $(\mu n/P)_{tr}$ in order to minimize friction-energy losses as well as to minimize the temperature rise in the bearing. However, a close examination of bearing operation indicates that it is undesirable to design at this point. The purpose of the following discussion is to explain that this point does not have the desired operation stability. The term *stability* is used here in the sense that the hydrodynamic operation would recover and return to normal operation after any disturbance, such as overload for a short period or unexpected large vibration of the machine. In contrast, unstable operation is where any such disturbance would result in deterioration in bearing operation that may eventually result in bearing failure.

Although it is important to minimize friction-energy losses, if the bearing operates at the point $(\mu N/P)_{tr}$, where the friction is minimal, any disturbance would result in a short period of higher friction. This would cause a chain of events that may result in overheating and even bearing failure. The higher friction would result in a sudden temperature rise of the lubricant film, even if the disturbance discontinues. Temperature rise would immediately reduce the fluid viscosity, and the magnitude of the variable $\mu n/P$ would decrease with the viscosity. In turn, the bearing would operate in the mixed region, resulting in higher friction. The higher friction causes further temperature rise and further reduction in the value of $\mu n/P$. This can lead to an unstable chain reaction that may result in bearing failure, particularly for high-speed hydrodynamic bearings.

In contrast, if the bearing is designed to operate on the right side of the transition point, $\mu n/P > (\mu N/P)_{tr}$, any unexpected temperature rise would also reduce the fluid viscosity and the value of the variable $\mu n/P$. However in that case, it would shift the point in the curve to a lower friction coefficient. The lower friction would help to restore the operation by lowering the fluid film temperature. The result is that a bearing designed to operate at somewhat higher value of $\mu n/P$ has the important advantage of stable operation.

The decision concerning h_n relies in many cases on previous experience with bearings operating under similar conditions. In fact, very few machines are designed without any previous experience as a first prototype, and most designs represent an improvement on previous models. In order to gain from previous experience, engineers should follow several important dimensionless design parameters of the bearings in each machine. As a minimum, engineers should keep a record of the value of $\mu n/P$ and the resulting analytical minimum film thickness, h_n, for each bearing. Experience concerning the relationship of these variables to successful bearing operation, or early failure, is essential for future designs of similar bearings or improvement of bearings in existing machinery. However, for important applications, where early bearing failure is critical, bearing tests should be conducted *before* testing the machine in service. This is essential in order to prevent unexpected expensive failures. Testing machines will be discussed in Chapter 14.

Problems

8-1 Select the lubricant for a full hydrodynamic journal bearing ($\beta = 360°$) under a radial load of 1 ton. The design requirement is that the minimum film thickness, h_n, during steady operation, not be less than 16×10^{-3} mm. The inlet oil temperature is 40°C, and the journal speed is 3600 RPM. Select the oil type that would result in the required performance. The bearing dimensions are: $D = 100$ mm; $L = 50$ mm, $C = 80 \times 10^{-3}$ mm.

Directions: First, determine the required Sommerfeld number, based on the minimum film thickness, and find the required viscosity. Second evaluate the temperature rise Δt and the average temperature, and select the oil type (use Fig. 2-2).

8-2 Use the Raimondi and Boyd charts to find the maximum load capacity of a full hydrodynamic journal bearing ($\beta = 360°$). The lubrication is SAE 10. The bearing dimensions are: $D = 50\,\text{mm}$, $L = 50\,\text{mm}$, $C = 50 \times 10^{-3}\,\text{mm}$. The minimum film thickness, h_n, during steady operation, should not be below $10 \times 10^{-3}\,\text{mm}$. The inlet oil temperature is 30°C and the journal speed is 6000 RPM.

Directions: Trial-and-error calculations are required for solving the temperature rise. Assume a temperature rise and average temperature. Find the viscosity for SAE 10 as a function of temperature, and use the chart to find the Sommerfeld number and the resulting load capacity. Use the new average pressure to recalculate the temperature rise. Repeat iterations until the temperature rise is equal to that in the previous iteration.

8-3 The dimensions of a partial hydrodynamic journal bearing, $\beta = 180°$, are: $D = 60\,\text{mm}$, $L = 60\,\text{mm}$, $C = 30 \times 10^{-3}\,\text{mm}$. During steady operation, the minimum film thickness, h_n, should not go below $10 \times 10^{-3}\,\text{mm}$. The maximum inlet oil temperature (in the summer) is 40°C, and the journal speed is 7200 RPM. Given a lubricant of SAE 10, use the chart to find the maximum load capacity and the maximum fluid film pressure, p_{max}.

8-4 A short journal bearing is loaded by 500 N. The journal diameter is 25 mm, the L/D ratio is 0.6, and $C/R = 0.002$. The bearing has a speed of 600 RPM. An experimental curve of friction coefficient, f, versus variable $\mu n/P$ of this bearing is shown in Fig. 8-12. The minimum friction is at $\mu n/P = 3 \times 10^{-8}$.

a. Find the lubrication viscosity for which the bearing would operate at a minimum friction coefficient.
b. Use infinitely-short-bearing theory and find the minimum film thickness at the minimum-friction point.
c. Use the charts of Raimondi and Boyd to find the minimum film thickness at the minimum-friction point.
d. For stable bearing operation, increase the variable $\mu n/P$ by 20% and find the minimum film thickness and new friction coefficient. Use the short bearing equations.

9

Practical Applications of Journal Bearings

9.1 INTRODUCTION

A hydrodynamic journal bearing operates effectively when it has a full fluid film without any contact between the asperities of the journal and bearing surfaces. However, under certain operating conditions, this bearing has limitations, and unique designs are used to extend its application beyond these limits.

The first limitation of hydrodynamic bearings is that a certain minimum speed is required to generate a full fluid film of sufficient thickness for complete separation of the sliding surfaces. When the bearing operates below that speed, there is only mixed or boundary lubrication, with direct contact between the asperities. Even if the bearing is well designed and successfully operating at the high-rated speed, it can be subjected to excessive friction and wear at low speed, during starting and stopping of the machine. In particular, hydrodynamic bearings undergo severe wear during start-up, when the journal accelerates from zero speed, because static friction is higher than dynamic friction. In addition, there is a limitation on the application of hydrodynamic bearings in machinery operating at variable speed, because the bearing has high wear rate when the machine operates in the low-speed range.

The second important limitation of hydrodynamic journal bearings is the low stiffness to radial displacement of the journal, particularly under light loads and high speed, when the eccentricity ratio, ε, is low. Low stiffness rules out the application of hydrodynamic bearings for precision applications, such as machine tools and measurement machines. In addition, under dynamic loads, the low stiffness of the hydrodynamic bearings can result in dynamic instability, referred to as *bearing whirl*. It is important to prevent bearing whirl, which often causes bearing failure. It is possible to demonstrate bearing whirl in a variable-speed testing machine for journal bearings. When the speed is increased, it reaches the critical whirl speed, where noise and severe vibrations are generated.

In a rotating system of a rotor supported by two hydrodynamic journal bearings, the stiffness of the shaft combines with that of the hydrodynamic journal bearings (similar to the stiffness of two springs in series). This stiffness and the distributed mass of the rotor determine the *natural frequencies*, also referred to as the *critical speeds* of the rotor system. Whenever the force on the bearing oscillates at a frequency close to one of the critical speeds, bearing instability results (similar to resonance in dynamic systems), which often causes bearing failure. An example of an oscillating force is the centrifugal force due to imbalance in the rotor and shaft unit.

9.2 HYDRODYNAMIC BEARING WHIRL

In addition to resonance near the critical speeds of the rotor system, there is a failure of the oil film in hydrodynamic journal bearings under certain dynamic conditions. The stiffness of long hydrodynamic bearings is not similar to that of a spring support. The bearing reaction force increases with the radial displacement, o–o_1, of the journal center (or eccentricity, e). However, the reaction force is not in the same direction as the displacement. There is a component of cross-stiffness, namely, a reaction-force component in a direction perpendicular to that of the displacement. In fact, the bearing force based on the Sommerfeld solution is only in the normal direction to the radial displacement of the journal center.

The cross-stiffness of hydrodynamic bearings causes the effect of the *half-frequency whirl*; namely, the journal bearing loses its load capacity when the external load oscillates at a frequency equal to about half of the journal rotation speed. It is possible to demonstrate this effect by computer simulation of the trajectory of the journal center of a long bearing under external oscillating force. If the frequency of the dynamic force is half of that of the journal speed, the eccentricity increases very fast, until there is contact of the bearing and journal surfaces. In practice, hydrodynamic bearing whirl is induced at relatively high speed under light, steady loads superimposed on oscillating loads. In actual machinery, oscillating loads at various frequencies are always present, due to imbalance in the various rotating parts of the machine.

Several designs have been used to eliminate the undesired half-frequency whirl. Since the bearing whirl takes place under light loads, it is possible to prevent it by introducing internal preload in the bearing. This is done by using a

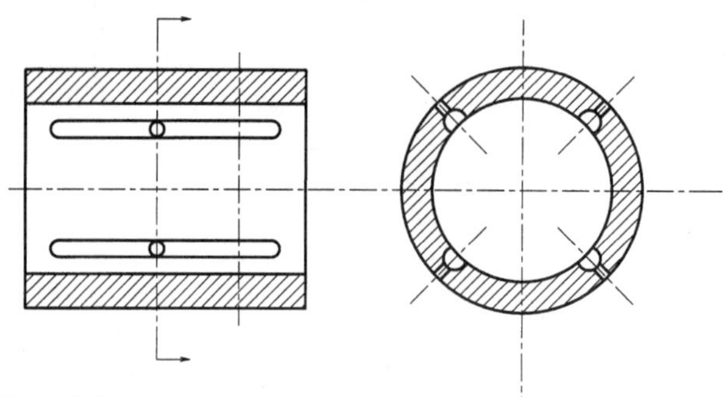

FIG. 9-1 Bearing with axial oil grooves.

bearing made of several segments; each segment is a partial hydrodynamic bearing. In this way, each segment has hydrodynamic force, in the direction of the bearing center, that is larger than the external load. The partial bearings can be rigid or made of tilting pads. Elliptical bearings are used that consist of only two opposing partial pads. However, for most applications, at least three partial pads are desirable. An additional advantage is improved oil circulation, which reduces the bearing operating temperature.

Some resistance to oil whirl is obtained by introducing several oil grooves, in the axial direction of the internal cylindrical bore of the bearing, as shown in Fig. 9-1. The oil grooves are along the bearing length, but they are not completely open at the two ends, as indicated in the drawing. It is important that the oil grooves not be placed at the region of minimum film thickness, where it would disturb the pressure wave. Better resistance to oil whirl is achieved by designs that are described in the following sections.

9.3 ELLIPTICAL BEARINGS

The geometry of the basic elliptical bearing is shown in Fig. 9-2a. The bore is made of two arcs of larger radius than for a circular bearing. It forms two pads with opposing forces. In order to simplify the manufacturing process, the bearing bore is machined after two shims are placed at a split between two halves of a round sleeve. After round machining, the two shims are removed. In fact, the shape is not precisely elliptical, but the bearing has larger clearances on the two horizontal sides and smaller clearance in the upper and bottom sides. In this way, the bearing operates as a two-pad bearing, with action and reaction forces in opposite directions.

The additional design shown in Fig. 9-2b is made by shifting the upper half of the bearing, relative to the lower half, in the horizontal direction. In this way,

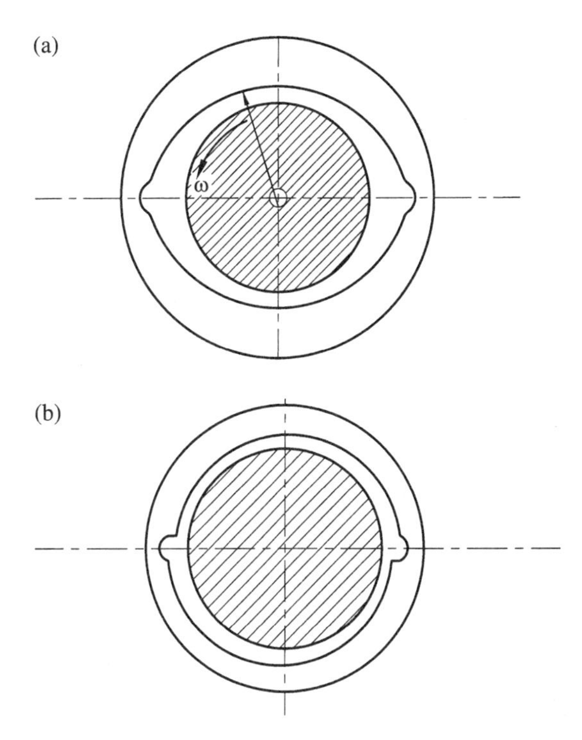

FIG. 9-2 Elliptical bearing: (a) basic bearing design; (b) shifted bearing.

each half has a converging fluid film of hydrodynamic action and reaction forces in opposite directions.

Elliptical and shifted bearings offer improved resistance to oil whirl at a reasonable cost. They are widely used in high-speed turbines and generators and other applications where the external force is in the vertical direction. The circulation of oil is higher in comparison to a full circular bearing with equivalent minimum clearance.

9.4 THREE-LOBE BEARINGS

Various designs have been developed to prevent the undesired effect of bearing whirl. An example of a successful design is the *three-lobe journal bearing* shown in Fig. 9-3. It has three curved segments that are referred to as *lobes*. During operation, the geometry of the three lobes introduces preload inside the bearing. This design improves the stability because it increases the bearing stiffness and reduces the magnitude of the cross-stiffness components. The preferred design

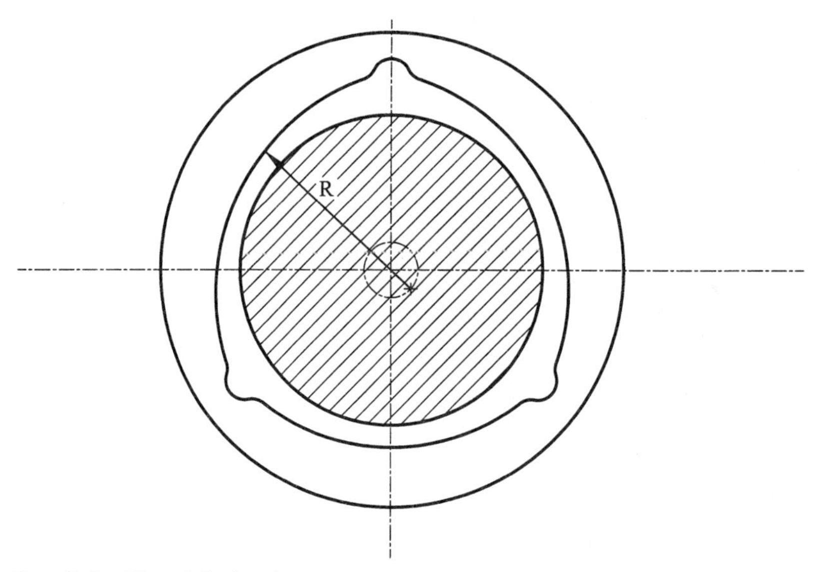

FIG. 9-3 Three-lobe bearing.

for optimum stability is achieved if the center of curvature of each lobe lies on the journal center trajectory. This trajectory is the small circle generated by the journal center when the journal is rolling in contact with the bearing surface around the bearing. According to this design, the journal center is below the center of each of the three lobes, and the load capacity of each lobe is directed to the bearing center.

The calculation of the load capacity of each lobe is based on a simplifying assumption that the journal is running centrally in the bearing. This assumption is justified because this type of bearing is commonly used at low loads and high speeds, where the shaft eccentricity is very small.

An additional advantage of the three-lobe bearing is that it has oil grooves between the lobes. The oil circulation is obviously better than for a regular journal bearing (360°). This bearing can carry higher loads when the journal center is over an oil groove rather than over the center of a lobe.

9.5 PIVOTED-PAD JOURNAL BEARING

Figure 9-4 shows a *pivoted-pad bearing*, also referred to as *tilt-pad bearing*, where a number of tilting pads are placed around the circumference of the journal. The best design is a universal self-aligning pad; namely, each pad is free to align in both the tangential and axial directions. These two degrees of freedom

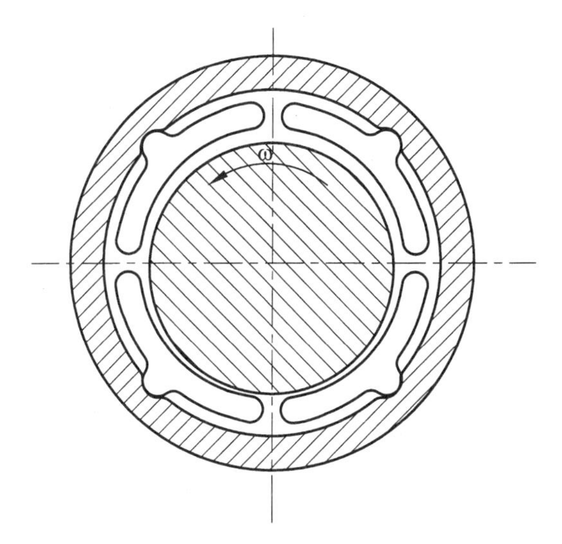

FIG. 9-4 Pivoted-pad journal bearing.

allow a full adjustment for any misalignment between the journal and bearing. This design has clear advantages in comparison to a rigid bearing, and it is used in critical applications where continuous operation of the machine, without failure, is essential. This is particularly important in large bearings that have large tolerances due to manufacturing errors, such as a propeller shaft of a ship. In many cases, a pivoted-pad journal bearing is used for this application. During a storm, the ship experiences a large elastic deformation, resulting in considerable misalignment between the bearings that are attached to the body of the ship and the propeller shaft. Self-aligning of pivoted-pad journal bearings can prevent excessive wear due to such misalignment.

Self-aligning pivoted-pad journal bearings are widely used in high-speed machines that have a relatively low radial load, resulting in low eccentricity. For example, the pivoted-pad journal bearing is used in high-speed centrifugal compressors. This design offers stability of operation and resistance to oil whirl by increasing the bearing stiffness.

The pivoted-pad journal bearing has the advantages of high radial stiffness of the bearing and low cross-stiffness. These advantages are important in applications where it is necessary to resist bearing whirl or where high precision is required. Better precision results from higher stiffness that results in lower radial run-out (eccentricity) of the journal center. The reaction force of each pad is in the radial direction. The forces of all the pads preload the journal at the bearing center and tend to increase the stiffness and minimize the eccentricity.

9.6 BEARINGS MADE OF COMPLIANT MATERIALS

Pivoted pad bearings are relatively expensive. For many applications, where only a small alignment is required, low cost bearings that are made of elastic materials such as an elastomer (rubber) can align the contact surface to the journal. Of course, the alignment is much less in comparison to that of the tilting pad. Rubber-to-metal bonding techniques have been developed with reference to compliant surface bearings; see a report by Rightmire (1967). Water-lubricated rubber bearings can be used in boats, see Orndorff and Tiedman (1977).

Bearings made of plastic materials are also compliant, although to a lesser degree than elastomer materials. Plastics have higher elasticity than metals, since their modulus of elasticity, E, is much lower.

In rolling-element bearings or gears there is a theoretical point or line contact resulting in very high maximum contact pressure. When the gears or rolling elements are made of soft compliant materials, such as plastics, the maximum pressure is reduced because there is a larger contact area due to more elastic deformation. Even steel has a certain elastic deformation (compliance) that plays an important role in the performance of elastohydrodynamic lubrication in gears and rollers.

Similar effects take place in journal bearings. The journal has a smaller diameter than the bearing bore, and for a rigid surface under load there is a theoretical line contact resulting in a peak contact pressure that is much higher than the average pressure. Engineers realized that in a similar way to gears and rollers, it is possible to reduce the high peak pressure in rigid bearings by using compliant bearing materials. Although the initial application of hydrodynamic bearings involved only rigid materials, the later introduction of a wide range of plastic materials has motivated engineers to test them as alternative materials that would result in a more uniform pressure distribution. In fact, plastic materials demonstrated successful performance in light-duty applications under low load and speed (relatively low PV). The explanations for the improved performance are the self-lubricating properties and compliant surfaces of plastic materials. In fact, biological joints, such as the human hip joint, have soft compliant surfaces that are lubricated by synovial fluid. The superior performance of the biological bearings suggested that bearings in machinery could be designed with compliant surfaces with considerable advantages.

Plastic materials have a low dry-friction coefficient against steel. In addition, experiments indicated that bearings made of rubber or plastic materials have a low friction coefficient at the boundary or mixed lubrication region. This is explained by the surface compliance near the minimum film thickness, where the high pressure forms a depression in the elastic material. In the presence of lubricant, the depression is a puddle of lubricant under pressure.

Another important advantage of compliant materials is that they have a certain degree of elastic self-aligning. The elastic deformation compensates for misaligning or other manufacturing errors of the bearing or sleeve. In contrast, metal bearings are very sensitive to any deviation from a perfect roundness of the bearing and journal. For hydrodynamic metal bearings, high precision as well as perfect surface finish is essential for successful performance with minimum contact between the surface asperities. In comparison, plastic bearings can be manufactured with lower precision due to their compliance characteristic. The advantage of surface compliance is that it relaxes the requirement for high precision, which involves high cost.

Moreover, compliant surfaces usually have better wear resistance. Elastic deformation prevents removal of material due to rubbing of rough and hard surfaces. Compliant materials allow the rough asperities to pass through without tearing. In addition, it has better wear resistance in the presence of abrasive particles in the lubricant, such as dust, sand, and metal wear derbies. Rubber sleeves are often used with slurry lubricant in pumps. Embedding of the abrasive particles in the sleeve is possible by means of elastic deformation. Later, elastic deformation allows the abrasive particles to roll out and leave the bearing.

For all these advantages, bearings made of plastic material are widely used. However, their application is limited to light loads and moderate speeds. For heavy-duty applications, metal bearings are mostly used, because they have better heat conduction. Plastic bearings would fail very fast at elevated temperatures.

9.7 FOIL BEARINGS

The foil bearing has the ultimate bending compliance, and its principle is shown in Fig. 9-5. The foil is thin and lacks any resistance to bending. The flexible foil stretches around the journal. In the presence of lubricant, a thin fluid film is formed, which separates the foil from the journal. At high speeds, air can perform as a lubricant, and a thin air film prevents direct contact between the rotating shaft and the foil. An air film foil bearing has considerable advantages at high speeds. Air film can operate at much higher temperatures in comparison to oils, and, of course, air lubrication is much simpler and less expensive than oil lubrication.

Lubricant flow within the foil bearing has a converging region, which generates bearing pressure, and a parallel region, of constant clearance, h_0, supports the load. Foil bearings have important applications wherever there is a requirement for surface compliance at elevated temperature. Mineral oils or synthetic oils deteriorate very fast at high temperature; therefore, several designs have already been developed for foil bearings that operate as air bearings.

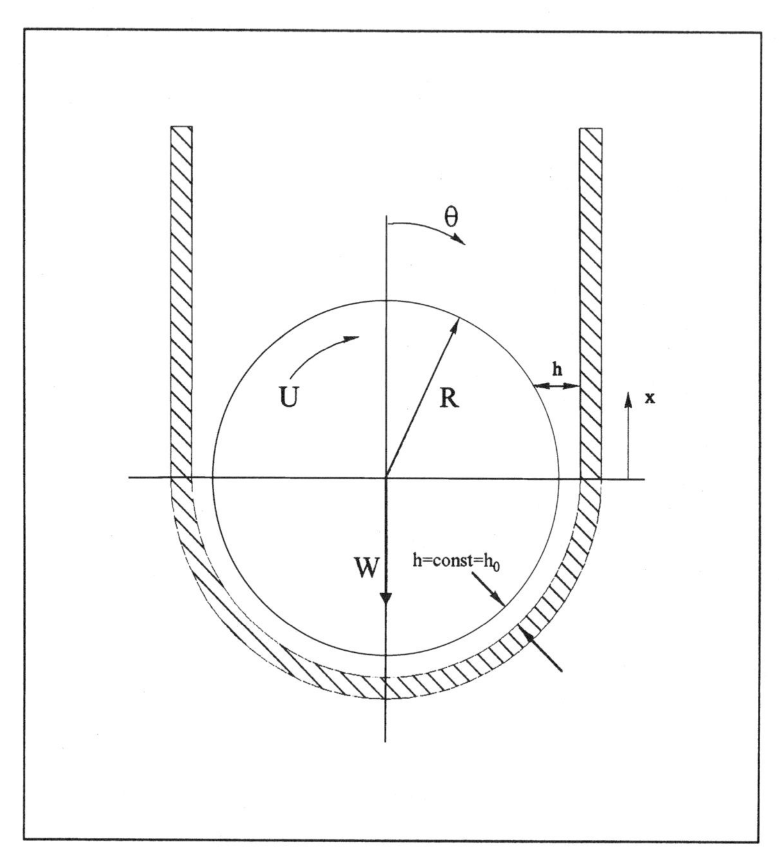

FIG. 9-5 Foil bearing.

9.8 ANALYSIS OF A FOIL BEARING

The following analysis is presented to illustrate the concept of the hydrodynamic foil bearing. The foil is stretched around a journal of radius R rotating at constant speed, and resulting in a tangential velocity U at the journal surface. A vertical external load F is applied to the journal that is equal to the load capacity W of the fluid film.

The following assumptions are made for solving the load capacity W of the foil bearing.

1. The foil sleeve is parallel to the journal surface (constant clearance) along the lower half of the journal, due to its high flexibility. In turn, there is no localized pressure concentration.

2. The contributions to load-carrying capacity of the converging and diverging ends are neglected.
3. The simplified equation for an infinitely long bearing can be applied.

This problem is similar to that of the sled of Example problem 4-4. In the sled problem, there is also a converging clearance between a flat plate and a cylinder. Let us recall that the film thickness in the converging region between a flat plate and a cylinder is given by the following equation [see equation (4-24)]:

$$h(\theta) = h_0 + R(1 - \cos\theta) \tag{9-1}$$

Here, θ is measured from the line $x = 0$. The expression for the film thickness h is approximated by (see Sec. 4.10.1)

$$h(x) = h_0 + \frac{x^2}{2R} \tag{9-2}$$

This approximation is valid within the relevant range of the converging region. The boundary conditions of lubricant pressure is $p = 0$ at $x = \infty$. The Reynolds equation for an infinitely long bearing is

$$\frac{dp}{dx} = 6\mu U \frac{h_0 - h}{h^3} \tag{9-3}$$

In the parallel region, $h = h_0$. So it follows from Eq. (9-3) that $dp/dx = 0$ (parallel region). This means that the pressure is constant within the parallel region. The pressure acting over the entire lower parallel region is constant and solely responsible for carrying the journal load.

The force dW on an elementary area $dA = LR\,d\theta$ is

$$dW = p_0 LR\,d\theta \tag{9-4}$$

This force is in the direction normal to the journal surface. The vertical component of this elementary force is

$$dW = LRp_0 \sin\theta\,d\theta \tag{9-5}$$

Integration of the vertical elementary force component over the parallel region would result in the load capacity W, according to the equation

$$W = LR\int_{\pi/2}^{3\pi/2} p_0 \sin\theta\,d\theta \tag{9-6}$$

Here, L is the bearing length (in the axial direction, z) and p_0 is the constant pressure in the parallel region.

Upon integration, this equation reduces to

$$W = 2LRp_0 \tag{9-7}$$

Similar to the case of the sled, the fluid flow in the converging region generates the pressure, which is ultimately responsible for supporting the journal load. The viscous fluid is dragged into the foil–journal convergence, due to the journal rotation, generating the pressure supplied to the parallel region.

In a similar way to the parallel sled problem in Chapter 4, by substituting the value of h into Eq. (9-3) and integrating, the equation of pressure distribution becomes

$$p(x) = 24\mu UR^2 \int_{\infty}^{x} \frac{x^2}{(2Rh_0 + x^2)^3} dx \tag{9-8}$$

The following example is a numerical solution of the pressure wave at the inlet to the foil bearing. This solution is in dimensionless form. The advantage of a dimensionless solution is that it is universal, in the sense that it can be applied to any design parameters.

Example Problem 9-1

Find the expression of the film thickness, h_0, as a function of the load capacity, W.

Solution

For conversion of Eq. (9-8) to a dimensionless form, the following dimensionless terms are introduced:

$$\bar{x} = \frac{x}{\sqrt{2Rh_0}} \qquad \bar{h} = \frac{h}{h_0} \qquad \bar{h} = 1 + \bar{x}^2$$

Similar to the sled problem in Chapter 4, the expression for the dimensionless pressure becomes

$$\bar{p} = \frac{h_0^2}{\sqrt{2Rh_0}} \frac{1}{6\mu U} \int_0^p dp = \int_{\infty}^{x} \frac{\bar{x}^2}{(1 + \bar{x}^2)^3} d\bar{x}$$

Here, the dimensionless pressure is defined as

$$\bar{p} = \frac{h_0^2}{\sqrt{2Rh_0}} \frac{1}{6\mu U} p$$

The dimensionless pressure can be integrated analytically or numerically. The pressure becomes significant only near $x = 0$. Therefore, for numerical integration, the infinity is replaced by a finite number, such as 4. Numerical integration results in a dimensionless pressure distribution, as shown in Fig. 4-9.

A comparison of numerical and analytical solutions, in Chapter 4, resulted in the following same solution for the constant pressure in the parallel clearance:

$$\bar{p} = \int_0^{p_0} dp = \int_\infty^0 \frac{\bar{x}^2}{(1+\bar{x}^2)^3} d\bar{x}; \qquad \bar{p}_0 = 0.196$$

Minimum Film Clearance: Neglecting the pressure in the entrance and outlet regions, the equation for the load capacity as a function of the uniform pressure, p_0, in the clearance, h_0, is given in Eq. (9-7)

$$W = p_0 DL$$

Using the value of dimensionless pressure, $\bar{p}_0 = 0.196$, and converting to pressure p_0, the following equation is obtained for the clearance, h_0, as a function of the pressure, p_0:

$$\frac{h_0}{R} = 4.78\left(\frac{\mu n}{p_0}\right)^{2/3}$$

Here, n is the journal speed, in RPS, and the pressure, p_0, is determined by the load:

$$p_0 = \frac{W}{LD}$$

9.9 FOIL BEARINGS IN HIGH-SPEED TURBINES

It has been realized that a compliant air bearing can offer considerable advantages in high-speed gas turbines that operate at very high temperatures. Unlike compliant bearings made of polymers or elastomer materials, foil air bearings made of flexible metal foils can operate at relatively high temperature because metals that resist high temperature can be selected. In addition, the viscosity of air increases with temperature and it does not deteriorate at elevated temperature like oils. In addition, a foil bearing does not require precision manufacturing with close tolerances, as does a conventional rigid journal bearing.

Hydrodynamic bearings have a risk of catastrophic failure by seizure, because, in the case of overheating due to direct contact, the journal has larger thermal expansion than the sleeve. This problem is eliminated in compliant foil because the clearance can adjust itself. An additional advantage is that the foil bearing has very good resistance to whirl instability.

The foil air bearing concept has already been applied in high-speed gas turbines, with a few variations. Two types of compliant air bearings made of

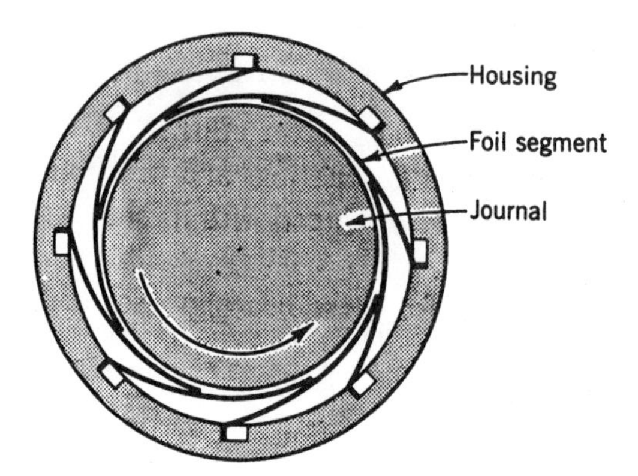

FIG. 9-6 Compliant journal bearing A. (From Suriano et al., 1983.)

flexible metal have been extensively investigated and tested. One problem is the high friction and wear during the start-up or occasional contact. The thin flexible metal strips and journal undergo severe wear during the start-up. Several coatings, such as titanium nitride, have been tested in an attempt to reduce the wear. Experimental investigation of the dynamic characteristics of a turborotor simulator of gas-lubricated foil bearings is described by Licht (1972).

The two designs are shown in Figs. 9-6 and 9-7. The first design [see Suriano et al. (1983)] is shown in Fig. 9-6, and a second design [see Heshmat et

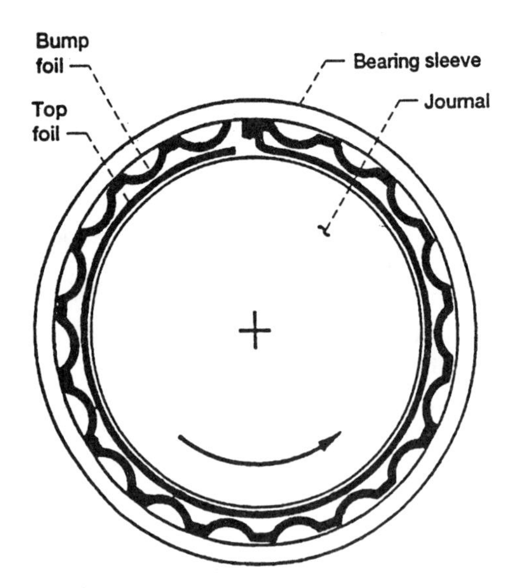

FIG. 9-7 Compliant journal bearing B. (From Heshmat et al., 1982.)

al. (1982)] is shown in Fig. 9-7. The tests indicated that foil bearings have adequate load capacity for gas turbines. In addition, foil bearings demonstrated very good whirl stability at very high speeds. Foil bearings have already been tested successfully in various applications. Additional information about research and development of foil bearings for turbomachines is included in a report by the Air Force Aero Propulsion Laboratory (1977).

Extensive research and development are still conducted in order to improve the performance of the compliant air bearing for potential use in critical applications such as aircraft turbines.

9.10 DESIGN EXAMPLE OF A COMPLIANT BEARING

In Sec. 9.5, the advantages of the pivoted-pad bearing were discussed. In addition to self-aligning, it offers high stiffness, which is essential for high-speed operation. However, in many cases, compliance can replace the tilting action. A compliant-pad bearing has been suggested by the KMC Company [see Earles et al. (1989)]. A compliant-pad journal bearing is shown in Fig. 9-8. The most important advantage is that it is made of one piece, in comparison to the pivoted pad, which consists of many parts (Fig. 9-9). In turn, the compliant pad is easier

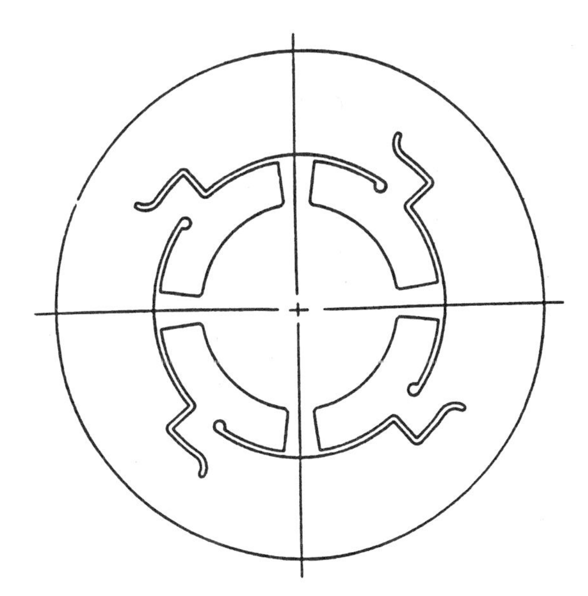

FIG. 9-8 Compliant-pad journal bearing. (With permission from KMC Co.)

FIG. 9-9 Comparison of a compliant and a pivoted pad. (With permission from KMC Co.)

to manufacture and assemble. The bearing is made from one solid bronze cylinder by an electrical discharge machining process. In addition, tests indicated that the flexural compliance of the pads improves the bearing characteristics at high speed, such as in centrifugal compressors.

Problems

9-1 A foil bearing operates as shown in Fig. 9-5. Find the uniform film thickness, h_0, around the journal of a foil bearing. The bearing has the following design parameters:

$R = 50$ mm $\quad \mu = 0.015$ N-s/m^2

$L = 100$ mm $\quad W = 10{,}000$ N

$N = 5000$ RPM

9-2 Find the uniform film thickness around the journal of a foil bearing, as shown in Fig. 9-5, that is floating on a hydrodynamic air film. The temperature of the air is 20°C. The bearing has the following design parameters:

$$D = 120 \text{ mm} \qquad \mu_{air} = 184 \times 10^{-7} \text{ N-s/m}^2$$

$$L = 60 \text{ mm} \qquad N = 25{,}000 \text{ RPM}$$

a. Find the load capacity, W, for a fluid film thickness around the bearing of 50 μm.
b. Use the infinitely-short-bearing equation and compare the load capacity with that of a short bearing having the given design parameters, air viscosity, and minimum film thickness. The radial clearance is $C/R = 0.001$ ($R = 60$ mm).

10

Hydrostatic Bearings

10.1 INTRODUCTION

The design concept for hydrostatic bearings is the generation of a high-pressure fluid film by using an external pump. The hydrostatic system of a journal bearing is shown in Fig. 1-4. The fluid is fed from a pump into several recesses around the bore of the bearing. From the recesses, the fluid flows out through a thin clearance, h_0, between the journal and bearing surfaces, at the *lands* outside the recesses. Previous literature on hydrostatic bearings includes Opitz (1967), Rowe (1989), Bassani and Piccigallo (1992), and Decker and Shapiro (1968).

The fluid film in the clearance separates the two surfaces of the journal and the bearing and thus reduces significantly the friction and wear. At the same time, the thin clearance at the land forms a resistance to the outlet flow from each recess. This flow resistance is essential for maintaining high pressure in the recess. The hydrodynamic load capacity that carries the external load is the resultant force of the pressure around the bearing.

There is also a fluid film in a hydrodynamic bearing. However, unlike the hydrodynamic bearing, where the pressure wave is generated by the hydrodynamic action of the rotation of the journal, hydrostatic bearing pressure is generated by an external pump.

There are certain designs of hydrodynamic bearings where the oil is also supplied under pressure from an external oil pump. However, the difference is that in hydrostatic bearings the design entails recesses, and the operation does not depend on the rotation of the journal for generating the pressure wave that supports the load. For example, the hydrodynamic journal bearing does not generate hydrodynamic pressure and load capacity when the journal and sleeve are stationary. In contrast, the hydrostatic bearing maintains pressure and load capacity when it is stationary; this characteristic is important for preventing wear during the bearing start-up. In fact most hydrostatic journal bearings are hybrid, in the sense that they combine hydrostatic and hydrodynamic action.

An important advantage of a hydrostatic bearing, in comparison to the hydrodynamic bearing, is that it maintains complete separation of the sliding surfaces at low velocities, including zero velocity. The hydrostatic bearing requires a high-pressure hydraulic system to pump and circulate the lubricant. The hydraulic system involves higher initial cost; in addition, there is an extra-long-term cost for the power to pump the fluid through the bearing clearances. Although hydrostatic bearings are more expensive, they have important advantages. For many applications, the extra expenses are justified, because these bearings have improved performance characteristics, and, in addition, the life of the machine is significantly extended. The following is a summary of the most important advantages of hydrostatic bearings in comparison to other bearings.

Advantages of Hydrostatic Bearings

1. The journal and bearing surfaces are completely separated by a fluid film at all times and over the complete range of speeds, including zero speed. Therefore, there is no wear due to direct contact between the surfaces during start-up. In addition, there is a very low sliding friction, particularly at low sliding speeds.
2. Hydrostatic bearings have high stiffness in comparison to hydrodynamic bearings. High stiffness is important for the reduction of journal radial and axial displacements. High stiffness is important in high-speed applications in order to minimize the level of vibrations. Also, high stiffness is essential for precise operation in machine tools and measurement machines. Unlike in hydrodynamic bearings, the high stiffness of hydrostatic bearings is maintained at low and high loads and at all speeds. This is a desirable characteristic for precision machines (Rowe, 1989) and high-speed machinery such as turbines.
3. Hydrostatic bearings operate with a thicker fluid film, which reduces the requirement for high-precision manufacture of the bearing (in comparison to hydrodynamic bearings). This means that it is possible to get more precision (lower journal run-out) in comparison to other bearings made with comparable manufacturing.
4. The continuous oil circulation prevents overheating of the bearing.
5. The oil pumped into the bearing passes through an oil filter and then through the bearing clearance. In this way, dust and other abrasive

particles are removed and do not damage the bearing surface. This is an important advantage in a dusty environment.

A hydrostatic bearing has one or several recesses where a uniformly high pressure is maintained by an external pump. Hydrostatic bearings are used for radial and thrust loads. Various types of recess geometry are used, such as circular or rectangular recesses in hydrostatic pads. The following is an example of a hydrostatic circular pad for a thrust load.

10.2 HYDROSTATIC CIRCULAR PADS

A circular pad is a hydrostatic thrust bearing that has a load-carrying capacity in the axial (vertical) direction, as shown in Fig. 10-1. This hydrostatic thrust bearing comprises two parallel concentric disks having a small clearance, h_0, between them. The radius of the disk is R, and the radius of the round recess is R_0. The bearing under a thrust load. Fluid is fed from an external pump into the round recess. The recess pressure is p_r above atmospheric pressure. The circular pad has load capacity W while maintaining complete separation between the surfaces. The external pump pressurizes the fluid in the recess, at a uniform pressure p_r. The clearance outside the recess, h_0, is thin relative to that of the recess. The thin clearance forms resistance to the outlet flow from the recess, in the radial direction. In this way, high pressure in the circular recess is continually maintained.

In the area of thin clearance, h_0 (often referred to as the *land*) the pressure reduces gradually in the radial direction, due to viscous friction. The uniform pressure in the recess combined with the radial pressure distribution in the land carry the external load. The recess pressure, p_r, increases with the flow rate, Q, that the pump feeds into the bearing. In turn, the thin clearance, h_0, adjusts its thickness to maintain the pressure and load capacity to counterbalance the thrust load. The clearance h_0 increases with the flow rate, Q, and the designer can control the desired clearance by adjusting the flow rate.

10.3 RADIAL PRESSURE DISTRIBUTION AND LOAD CAPACITY

Example Problem 10-1

Derive the equation for the radial pressure distribution, p, at the land and the load capacity, W, for a stationary circular pad under steady load as shown in Fig. 10-1. The fluid is incompressible, and the inlet flow rate, Q (equal to the flow in the radial direction), is constant. The clearance is h_0, and the radius of the circular pad and of the recess are R and R_0, respectively.

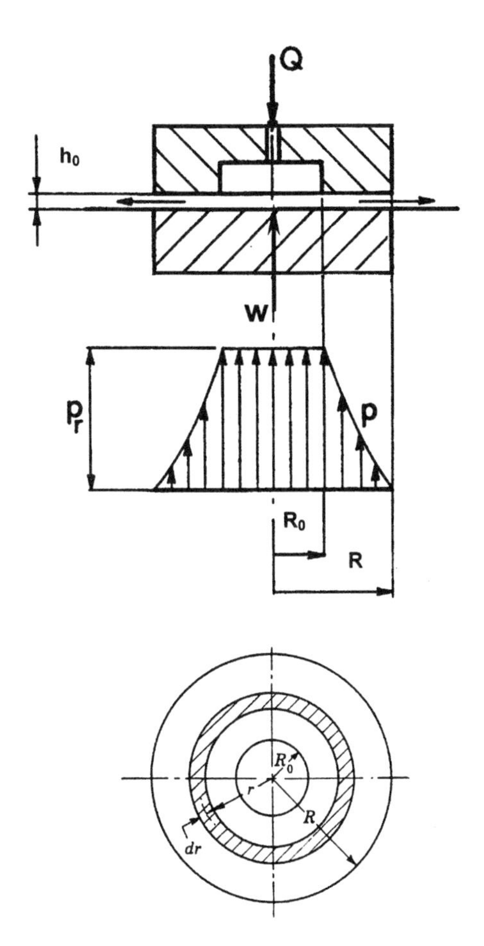

FIG. 10-1 Hydrostatic circular pad made up of two parallel disks and a round recess.

Solution

In Sec. 4.9, Example Problem 4-3 is solved for the pressure gradient in a parallel flow between two stationary parallel plates. The pressure gradient for a constant clearance, h_0, and rate of flow Q is

$$\frac{dp}{dx} = -\frac{12\mu}{bh_0^3}Q \tag{10-1}$$

The flow is in the direction of x, and b is the width (perpendicular to the flow direction). In a circular pad, there are radial flow lines, having radial symmetry.

The flow in the radial direction is considered between the round recess, at $r = R_0$, to the pad exit at $r = R$.

Pressure Distribution

In order to solve the present problem, the flow is considered in a thin ring, of radial thickness dr, as shown in Fig. 10-1. Since dr is small in comparison to the radius, r, the curvature can be disregarded. In that case, the flow along dr is assumed to be equal to a unidirectional flow between parallel plates of width $b = 2\pi r$. The pressure gradient dp is derived from Eq. (10-1):

$$\frac{dp}{dr} = -\frac{12\mu Q}{2\pi r h_0^3} \qquad \text{or} \qquad dp = \frac{12\mu Q}{2\pi h_0^3}\frac{dr}{r} \tag{10-2}$$

Integrating of Eq. (10-2) yields

$$p = -\frac{6\mu Q}{\pi h_0^3}\ln r + C \tag{10-3}$$

The constant of integration, C, is determined by the following boundary condition:

$$p = 0 \qquad \text{at} \qquad r = R \tag{10-4}$$

After solving for C, the expression for the radial pressure distribution in the radial clearance as a function of r is

$$p = \frac{6\mu Q}{\pi h_0^3}\ln\frac{R}{r} \tag{10-5}$$

The expression for the pressure at the recess, p_r, at $r = R_0$ is

$$p_r = \frac{6\mu Q}{\pi h_0^3}\ln\frac{R}{R_0} \tag{10-6}$$

Load Capacity

The load capacity is the integration of the pressure over the complete area according to the following equation:

$$W = \int_{(A)} p\,dA \tag{10-7}$$

The area dA of a ring thickness dr is (see Fig. 10-1)

$$dA = 2\pi r\,dr \tag{10-8}$$

The pressure in the recess, p_r, is constant, and the load capacity of the recess is derived by integration:

$$W = \int_0^{R_0} 2\pi r p_r dr \tag{10-9}$$

For the total load capacity, the pressure is integrated in the recess and in the thin clearance (land) according to the following equation:

$$W = \int_0^{R_0} p_r(2\pi r\, dr) + \int_{R_0}^{R} p(2\pi r dr) \tag{10-10}$$

After substituting the pressure equation into Eq. (10-10) and integrating, the following expression for the load capacity of a circular hydrostatic pad is obtained:

$$W = \frac{\pi}{2} \frac{R^2 - R_0^2}{\ln(R/R_0)} p_r \qquad \text{(10-11}$$

Equation (10-11) can be rearranged as a function of the recess ratio, R_0/R, and the expression for load capacity of a hydrostatic pad is

$$W = \frac{\pi R^2}{2} \frac{1 - (R_0/R)^2}{\ln(R/R_0)} p_r \tag{10-12}$$

The expression for the flow rate Q is obtained by rearranging Eq. (10-6) as follows:

$$Q = \frac{\pi}{6\mu} \frac{h_0^3}{\ln(R/R_0)} p_r \tag{10-13}$$

Equations (10-12) and (10-13) are for two stationary parallel disks. These equations can be extended to a hydrostatic pad where one disk is rotating, because according to the assumptions of classical lubrication theory, the centrifugal forces of the fluid due to rotation can be disregarded. In fact, the centrifugal forces are negligible (in comparison to the viscous forces) if the clearance h_0 is small and the viscosity is high (low Reynolds number). Equations (10-12) and (10-13) are used for the design of this thrust bearing. Whenever the fluid is supplied at constant flow rate Q, Eq. (10-13) is used to determine the flow rate. It is necessary to calculate the flow rate Q, which results in the desired clearance, h_0.

Unlike hydrodynamic bearings, the desired clearance, h_0, is based not only on the surface finish and vibrations, but also on the minimization of the power losses for pumping the fluid and for rotation of the pad. In general, hydrostatic bearings operate with larger clearances in comparison to their hydrodynamic counterparts. Larger clearance has the advantage that it does not require the high-precision machining that is needed in hydrodynamic bearings.

10.4 POWER LOSSES IN THE HYDROSTATIC PAD

The fluid flows through the clearance, h_0, in the radial direction, from the recess to the bearing exit. The pressure, p_r, is equal to the pressure loss due to flow resistance in the clearance (pressure loss p_r results from viscous friction loss in the thin clearance). In the clearance, the pressure has a negative slope in the radial directions, as shown in Fig. 10-1. The bearing is loaded by a thrust force in the vertical direction (direction of the centerline of the disks). Under steady conditions, the resultant load capacity, W, of the pressure distribution is equal to the external thrust load.

The upper disk rotates at angular speed ω, driven by an electrical motor, and power is required to overcome the viscous shear in the clearance. All bearings require power to overcome friction; however, hydrostatic bearings require extra power in order to circulate the fluid. An important task in hydrostatic bearing design is to minimize the power losses.

The following terms are introduced for the various components of power consumption in the hydrostatic bearing system.

$\dot{E}_h$—is the hydraulic power required to pump the fluid through the bearing and piping system. The flow resistance is in the clearance (land) and in the pipes. In certain designs there are flow restrictors at the inlet to each recess that increase the hydraulic power. The hydraulic power is dissipated as heat in the fluid.

$\dot{E}_f$—is the mechanical power provided by the drive (electrical motor) to overcome the friction torque resulting from viscous shear in the clearance due to relative rotation of the disks. This power is also dissipated as heat in the oil. This part of the power of viscous friction is present in hydrodynamic bearings without an external pump.

$\dot{E}_t$—is the total hydraulic power and mechanical power required to maintain the operation of the hydrostatic bearing, $\dot{E}_t = \dot{E}_h + \dot{E}_f$.

The mechanical torque of the motor, T_f, overcomes the viscous friction of rotation at angular speed ω. The equation for the motor-driving torque is (see Problem 2-1)

$$T_f = \frac{\pi}{2}\mu\frac{R^4}{h_0}\left(1 - \frac{R_0^4}{R^4}\right)\omega \tag{10-14}$$

The mechanical power of the motor, $\dot{E}_f$, that is required to overcome the friction losses in the pad clearance is

$$\dot{E}_f = T_f\omega = \frac{\pi}{2}\mu\frac{R^4}{h_0}\left(1 - \frac{R_0^4}{R^4}\right)\omega^2 \tag{10-15}$$

The power of a hydraulic pump, such as a gear pump, is discussed in Sec. (10.14). For hydrostatic pads where each recess is fed by a constant flow rate Q, there is no need for flow restrictors at the inlet to the recess. In that case, it is possible to simplify the calculations, since the pressure loss in the piping system is small and can be disregarded in comparison to the pressure loss in the bearing clearance:

$$\dot{E}_h \approx Qp_r \tag{10-16}$$

Here, Q is the flow rate through the pad and p_r is the recess pressure. The total power consumption $\dot{E}_t$ is the sum of the power of the drive for turning the bearing and the power of the pump for circulating the fluid through the bearing resistance. The following equation is for net power consumption. In fact, the pump and motor have power losses, and their efficiency should be considered for the calculation of the actual power consumption:

$$\dot{E}_{t(\text{net})} = Qp_r + \frac{\pi}{2}\mu\frac{R^4}{h_0}\left(1 - \frac{R_0^4}{R^4}\right)\omega^2 \tag{10-17}$$

Substituting the value of Q and dividing by the efficiency of the motor and drive η_1 and the efficiency of the pump η_2, the following equation is obtained for the total power consumption (in the form of electricity consumed by the hydrostatic system) for the operation of a bearing:

$$\dot{E}_t = \frac{1}{\eta_2}\frac{1}{6}\frac{\pi h_0^3}{\mu\ln(R/R_0)}p_r^2 + \frac{1}{\eta_1}\frac{\pi}{2}\mu\frac{R^4}{h_0}\left(1 - \frac{R_0^4}{R^4}\right)\omega^2 \tag{10-18}$$

The efficiency η_2 of a gear pump is typically low, about 0.6–0.7. The motor-drive system has a higher efficiency η_1 of about 0.8–0.9.

10.5 OPTIMIZATION FOR MINIMUM POWER LOSS

The total power consumption, $\dot{E}_t$, is a function of the clearance h_0. In general, hydrostatic bearings operate with larger clearance in comparison to hydrodynamic bearings. In order to optimize the clearance for minimum power consumption, it is convenient to rewrite Eq. (10-18) as a function of the clearance while all the other terms are included in a constant coefficient in the following form:

$$\dot{E}_t = C_1 h_0^3 + \frac{C_2}{h_0} \tag{10-19}$$

where

$$C_1 = \frac{1}{6\eta_2} \frac{\pi}{\mu \ln(R/R_0)} p_r^2 \qquad \text{and} \qquad C_2 = \frac{1}{2\eta_1} \pi\mu R^4 \left(1 - \frac{R_0^4}{R^4}\right)\omega^2 \tag{10-20}$$

Plotting the curve of power consumption versus clearance according to Eq. (10-19) allows optimization of the clearance, h_0, for minimum power loss in the bearing system. Hydrostatic bearings should be designed to operate at this optimal clearance.

Example Problem 10-2

Optimization of a Circular Hydrostatic Pad

A circular hydrostatic pad, as shown in Fig. 10-1, is supporting a load of $W = 1000$ N, and the upper disk has rotational speed of 5000 RPM. The disk diameter is 200 mm, and the diameter of the circular recess is 100 mm. The oil is SAE 10 at an operating temperature of 70°C, having a viscosity of $\mu = 0.01$ N-s/m^2. The efficiency of the hydraulic pump system is 0.6 and that of the motor and drive system is 0.9. Optimize the clearance, h_0, for minimum total power consumption.

Solution

The radius of the circular pad is $R = 100$ mm. The recess ratio is $R/R_0 = 2$. The angular speed is

$$N = 5000 \text{ RPM} \Rightarrow \omega \frac{2\pi N}{60} = \frac{2\pi 5000}{60} = 523.6 \text{ rad/s}$$

The first step is to find p_r by using the following load capacity equation:

$$W = R^2 \left(\frac{\pi}{2} \frac{1 - R_0^2/R^2}{\ln(R/R_0)} \right) p_r$$

Substituting the known values, the following equation is obtained, with p_r as unknown:

$$1000 = 0.1^2 \left(\frac{\pi}{2} \frac{1 - 0.5^2}{\ln 2} \right) p_r$$

Solving gives

$$p_r = 58{,}824 \text{ N/m}^2$$

Substituting p_r in Eq. (10-20), the constant C_1, which is associated with the pump, is solved for:

$$C_1 = \frac{1}{0.6} \times \frac{1}{6} \times \frac{\pi}{0.01 \ln(1/0.5)} (58{,}824)^2 = 4.35 \times 10^{11} N/\text{s-m}^2$$

In a similar way, the second constant, C_2, which is associated with the motor, is calculated from Eq. (10-20):

$$C_2 = \frac{1}{0.9} \frac{\pi}{2} (0.01)(0.1^4)(1 - 0.5^4) 523.6^2 = 0.448 \text{ N-m}^2/\text{s}$$

Substituting these values of C_1 and C_2 into Eq. (10-19), the total power as a function of the clearance becomes

$$\dot{E}_t = (4.35 \times 10^{11}) h_0^3 + \frac{0.448}{h_0}$$

In this equation, the power of the pump for circulating the fluid is the first term, which is proportional to h_0^3, while the second term, which is proportional to h_0^{-1}, is the power of the motor for rotating the disk.

The powers of the pump and of the drive are the two power components required to maintain the operation of the hydrostatic bearing. In Fig. 10-2, the curves of the hydraulic power and mechanical power are plotted as a function of the clearance, h_0. The curve of the mechanical power, $\dot{E}_f$, that is provided by the motor points down with increasing clearance, h_0. The power, $\dot{E}_f$, is for rotating one disk relative to the other and overcoming the viscous friction in the clearance between the two disks. The second curve is of the hydraulic power, $\dot{E}_h$, which is provided by the pump to maintain hydrostatic pressure in the recess. This power is rising with increasing clearance, h_0. The hydraulic power, $\dot{E}_h$, is for overcoming the flow resistance in the thin clearance at the outlet from the recess. The total power, $\dot{E}_t$, is the sum of these curves. For the hydrostatic pad in this problem, the optimal point (minimum power) is for a clearance of about 0.75 mm, and the total power consumption of the bearing in this problem is below 0.8 kW.

The result of 0.8 kW is too high for a hydrostatic bearing. It is possible to reduce the power consumption by using lower-viscosity oil.

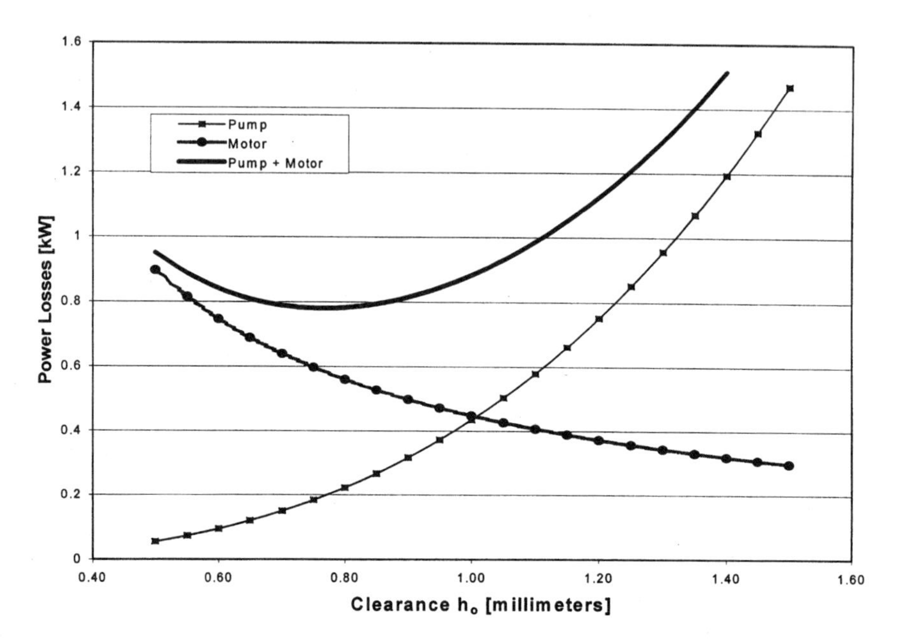

FIG. 10-2 Optimization of clearance for minimum power consumption.

10.6 LONG RECTANGULAR HYDROSTATIC BEARINGS

There are several applications where a long rectangular hydrostatic pad is used. An important example is the hydrostatic slideway in machine tools, as well as other machines where slideways are applied.

A long rectangular pad is shown in Fig. 10-3. The pad, as well as the recess, is long in comparison to its width, $L \gg B$ and $l \gg b$. Therefore, the flow through the clearance of width b is negligible relative to that through length l. The flow is considered one dimensional, because it is mostly in the x direction, while the flow in the y direction is negligible. For solving the pressure distribution, unidirectional flow can be assumed.

The pressure in the recess is constant, and it is linearly decreasing along the land of clearance, h_0, in the x direction. Integration of the pressure distribution results in the fluid load capacity W. At the same time, the flow rate in the two directions of the two sides of the recess is a flow between two parallel plates according to Eq. (10-1). The pressure gradient is linear, $dp/dx =$ constant. The total flow rate Q in the two directions is

$$Q = 2\frac{h_0^3}{12}\frac{l}{\mu}\frac{dp}{dx} = 2\frac{h_0^3}{12}\frac{l}{\mu}\frac{p_r}{(B-b)/2} = \frac{l\,h_0^3}{3\mu(B-b)}p_r \tag{10-21}$$

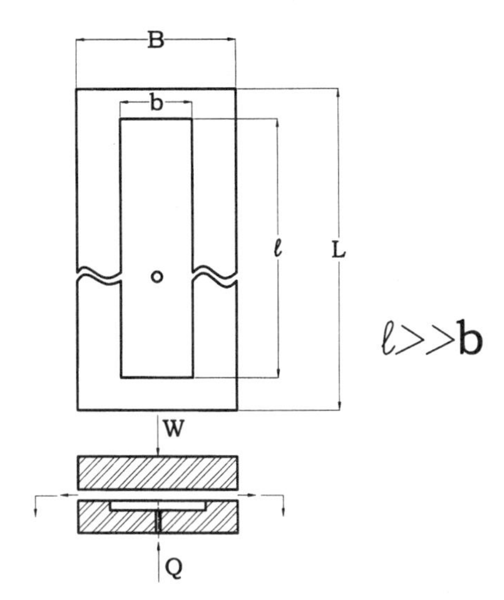

FIG. 10-3 Long rectangular pad.

Here, $B - b$ is the clearance (land) dimension along the direction of flow and l is the dimension of clearance (land) normal to the flow direction.

10.7 MULTIDIRECTIONAL HYDROSTATIC SUPPORT

Slideways are used in machinery for accurate linear motion. Pressurized fluid is fed into several recesses located at all contact surfaces, to prevent any direct metal contact between the sliding surfaces. Engineers already recognized that friction has an adverse effect on precision, and it is important to minimize friction by providing hydrostatic pads.

In machine tools, multirecess hydrostatic bearings are used for supporting the slideways as well as the rotating spindle. Hydrostatic slideways make the positioning of the table much more accurate, because it reduces friction that limits the precision of sliding motion.

A slideway supported by constant-flow-rate pads is shown in Fig. 10-4. Whenever there is a requirement for high stiffness in the vertical direction, the preferred design is of at least two hydrostatic pads with vertical load capacity in two opposite directions. Bidirectional support is also necessary when the load is changing its direction during operation. Additional pads are often included with horizontal load capacity in opposite directions for preventing any possible direct contact.

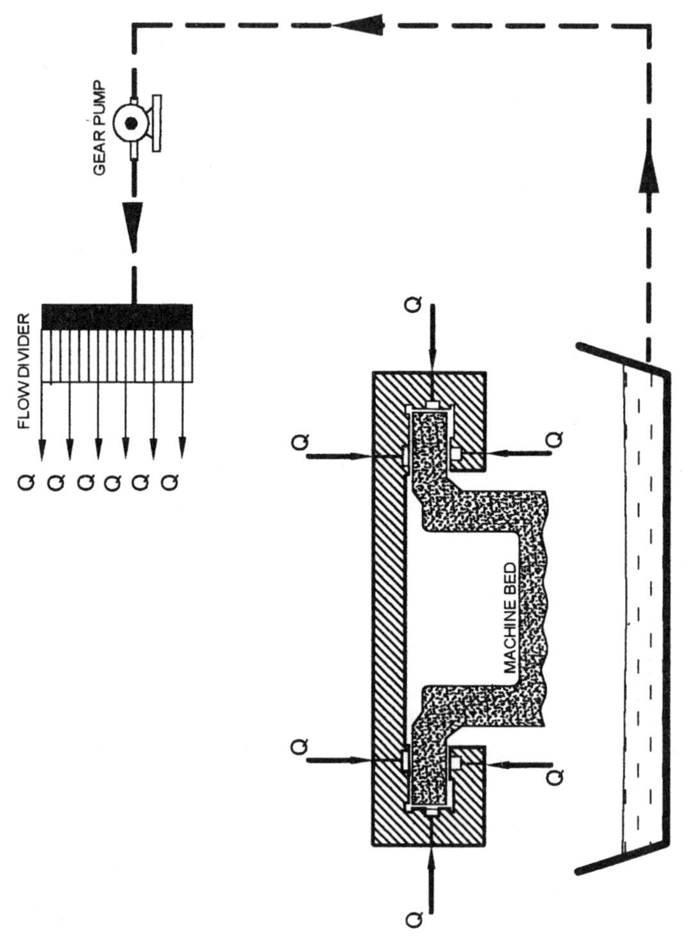

Fig. 10-4 Slideway supported by constant-flow-rate pads.

In a multipad support, one of the following two methods for feeding the oil into each recess is used.

1. Constant-flow-rate system, where each recess is fed by a constant flow rate Q.
2. Constant pressure supply, where each recess is fed by a constant pressure supply p_s. The oil flows into each recess through a flow restrictor (such as a capillary tube). The flow restrictor causes a pressure drop, and the recess pressure is reduced to a lower level, $p_r < p_s$. The flow restrictor makes the bearing stiff to displacement due to variable load.

In the case of the constant-flow-rate system, the fluid is fed from a pump to a flow divider that divides the flow rate between the various recesses. The flow divider is essential for the operation because it ensures that the flow will be evenly distributed to each recess and not fed only into the recesses having the least resistance.

High stiffness is obtained whenever each pad is fed by a constant flow rate Q. The explanation for the high stiffness lies in the relation between the clearance and recess pressure. For a bearing with given geometry, the constant flow rate Q is proportional to

$$Q \propto \left(\frac{h_0^3}{\mu} p_r\right) \tag{10-22}$$

A vertical displacement, Δh, of the slide will increase and decrease the clearance h_0 at the lands of the opposing hydrostatic pads. For constant flow rate Q and viscosity, Eq. (10-22) indicates that increase and decrease in the clearance h_0 would result in decrease and increase, respectively, of the recess pressure (the recess pressure is inversely proportional to h_0^3). High stiffness means that only a very small vertical displacement of the slide is sufficient to generate a large difference of pressure between opposing recesses. The force resulting from these pressure differences acts in the direction opposite to any occasional additional load on the thrust bearing.

Theoretically, the bearing stiffness can be very high for a hydrostatic pad with a constant flow rate to each recess; but in practice, the stiffness is limited by the hydraulic power of the motor and its maximum flow rate and pressure. This theoretical explanation is limited in practice because there is a maximum limit to the recess pressure, p_r. The hydraulic power of the pump and the strength of the complete system limit the recess pressure. A safety relief valve is installed to protect the system from exceeding its allowable maximum pressure. In addition, the fluid viscosity, μ, is not completely constant. When the clearance, h_0, reduces, the viscous friction increases and the temperature rises. In turn, the viscosity is lower in comparison to the opposing side, where the clearance, h_0, increases.

10.8 HYDROSTATIC PAD STIFFNESS FOR CONSTANT FLOW RATE

In this system, each recess is fed by a constant flow rate, Q. This system is also referred to as *direct supply system*. For this purpose, each recess is fed from a separate positive-displacement pump of constant flow rate. Another possibility, which is preferred where there are many hydrostatic pads, is to use a *flow divider*. A flow divider is designed to divide the constant flow rate received from one pump into several constant flow rates that are distributed to several recesses. Each recess is fed by constant flow rate directly from the divider. (The design of a flow divider is discussed in this chapter.) The advantage of using flow dividers is that only one pump is used. If properly designed, the constant-flow-rate system would result in high stiffness.

The advantage of this system, in comparison to the constant pressure supply with restrictors, is that there are lower viscous friction losses. In the flow restrictors there is considerable resistance to the flow (pressure loss), resulting in high power losses. In turn, the system with flow restrictors requires a pump and motor of higher power. However, the flow divider is an additional component, which also increases the initial cost of the system.

An example of a constant-flow-rate system is the machine tool slideway shown in Fig. 10-4. The areas of the two opposing recesses, in the vertical direction, are not equal. The purpose of the larger recess area is to support the weight of the slide, while the small pad recess is for ensuring noncontact sliding and adequate stiffness.

10.8.1 Constant-Flow-Rate Pad Stiffness

The bearing stiffness, k, is the rate of rise of the load capacity, W, as a function of incremental reduction of the clearance, h_0, by a small increment dh_0. It is equivalent to the rise of the load capacity with a small downward vertical displacement dh_0 of the upper surface in Fig. 10-1, resulting in lower clearance. The bearing stiffness is similar to a spring constant:

$$k = -\frac{dW}{dh_0} \tag{10-23}$$

The meaning of the negative sign is that the load increases with a reduction of the clearance. High stiffness is particularly important in machine tools where any displacement of the slide or spindle during machining would result in machining errors. The advantage of the high-stiffness bearing is that it supports any additional load with minimal displacement.

For the computation of the stiffness with a constant flow rate, it is convenient to define the bearing clearance resistance, R_c, at the land (resistance to flow through the bearing clearance) and the effective bearing area, A_e. The flow resistance to flow through the bearing clearance, R_c, is defined as

$$Q = \frac{p_r}{R_c} \qquad \text{or} \qquad R_c = \frac{p_r}{Q} \tag{10-24}$$

The effective bearing area, A_e, is defined by the relation

$$A_e p_r = W \qquad \text{or} \qquad A_e = \frac{W}{p_r} \tag{10-25}$$

For a constant flow rate, the load capacity, in terms of the effective area and bearing resistance, is

$$W = A_e R_c Q \tag{10-26}$$

Comparison with the equations for the circular pad indicates that the resistance is proportional to h_0^{-3} or

$$R_c = \kappa h_0^{-3}; \qquad \text{and} \qquad W = \kappa A_e Q \frac{1}{h_0^3} \tag{10-27}$$

Here, κ is a constant that depends on bearing geometry, flow rate, and fluid viscosity

$$\text{Stiffness } k = -\frac{dW}{dh_0} = K \frac{1}{h_0^4} \qquad \text{where } K = 3\kappa A_e Q \tag{10-28a}$$

$$\text{Stiffness } k = -\frac{dW}{dh_0} = 3\kappa A_e Q \frac{1}{h_0^4} \tag{10-28b}$$

Equation (10-28b) indicates that stiffness increases very fast with reduction in the bearing clearance. This equation can be applied as long as the flow rate Q to the recess is constant. As discussed earlier, deviation from this can occur in practice if the pressure limit is reached and the relief valve of the hydraulic system is opened. In that case, the flow rate is no longer constant.

Equation (10-28a) can be used for any hydrostatic bearing, after the value of K is determined. For a circular pad:

$$K = 9\mu Q (R^2 - R_0^2) \tag{10-29}$$

The expression for the stiffness of a circular pad becomes

$$k = \frac{9\mu Q(R^2 - R_0^2)}{h_0^4} \tag{10-30}$$

Whenever there are two hydrostatic pads in series (bidirectional hydrostatic support), the stiffnesses of the two pads are added for the total stiffness.

Example Problem 10-3

Stiffness of a Constant Flow Rate Pad

A circular hydrostatic pad, as shown in Fig. 10-1, has a constant flow rate Q. The circular pad is supporting a load of $W = 5000$ N. The outside disk diameter is 200 mm, and the diameter of the circular recess is 100 mm. The oil viscosity is $\mu = 0.005$ N-s/m^2. The pad is operating with a clearance of 120 μm.

a. Find the recess pressure, p_r.
b. Calculate the constant flow rate Q of the oil through the bearing to maintain the clearance.
c. Find the effective area of this pad.
d. Find the stiffness of the circular pad operating under the conditions in this problem.

Solution

Given:

$$W = 5000 \text{ N}$$
$$R = 0.1 \text{ m}$$
$$R_0 = 0.05 \text{ m}$$
$$\mu = 0.005 \text{ N-s/m}^2$$
$$h_0 = 120 \text{ μm}$$

a. Recess Pressure

In order to solve for the flow rate, the first step is to determine the recess pressure. The recess pressure is calculated from Eq. (10-12) for the load capacity:

$$W = R^2\left(\frac{\pi}{2}\frac{1 - R_0^2/R^2}{\ln(R/R_0)}\right)p_r$$

After substitution, the recess pressure is an unknown in the following equation:

$$5000 = 0.1^2\left(\frac{\pi}{2} \times \frac{1 - 0.25}{\ln(2)}\right)p_r$$

Solving for the recess pressure p_r yields:

$$p_r = 294.12 \text{ kPa}$$

b. Flow Rate

The flow rate Q can now be determined from the recess pressure. It is derived from Eq. (10-13):

$$Q = \frac{\pi}{6\mu} \frac{h_0^3}{\ln(R/R_0)} p_r; \qquad Q = \left(\frac{1}{6} \times \frac{\pi(120 \times 10^{-6})^3}{0.005 \times \ln(0.1/0.05)} \times 294{,}120\right)$$

The result for the flow rate is

$$Q = 76.8 \times 10^{-6} \text{ m}^3/\text{s}$$

c. Pad Effective Area

The effective area is defined by

$$W = A_e p_r$$

Solving for A_e as the ratio of the load and the recess pressure, we get

$$A_e = \frac{5000}{294{,}120}$$

$$A_e = 0.017 \text{ m}^2$$

d. Bearing Stiffness

Finally, the stiffness of the circular pad fed by a constant flow rate can be determined from Eq. (10-30):

$$k = \frac{9\mu Q(R^2 - R_0^2)}{h_0^4}$$

Substituting the values in this stiffness equation yields

$$k = \frac{9 \times 0.005 \times 76.8 \times 10^{-6} \times (0.1^2 - 0.05^2)}{(120 \times 10^{-6})^4}$$

$$k = 125 \times 10^6 \text{ N/m}$$

This result indicates that the stiffness of a constant-flow-rate pad is quite high. This stiffness is high in comparison to other bearings, such as hydrodynamic bearings and rolling-element bearings. This fact is important for designers of machine tools and high-speed machinery. The high stiffness is not obvious. The bearing is supported by a fluid film, and in many cases this bearing is not selected because it is mistakenly perceived as having low stiffness.

Example Problem 10-4

Bidirectional Hydrostatic Pads

We have a machine tool with four hydrostatic bearings, each consisting of two bidirectional circular pads that support a slider plate. Each recess is fed by a constant flow rate, Q, by means of a flow divider. Each bidirectional bearing is as shown in Fig. 10-4 (of circular pads). The weight of the slider is 20,000 N, divided evenly on the four bearings (5000-N load on each bidirectional bearing). The total manufactured clearance of the two bidirectional pads is $(h_1 + h_2) = 0.4$ mm. Each circular pad is of 100-mm diameter and recess diameter of 50 mm. The oil viscosity is 0.01 N-s/m^2.

In order to minimize vertical displacement under load, the slider plate is prestressed. The pads are designed to have 5000 N reaction from the top, and the reaction from the bottom is 10,000 N (equivalent to the top pad reaction plus weight).

a. Find the flow rates Q_1 and Q_2 in order that the top and bottom clearances will be equal, $(h_1 = h_2)$.
b. Given that the same flow rate applies to the bottom and top pads, $Q_1 = Q_2$, find the magnitude of the two clearances, h_1 and h_2. What is the equal flow rate, Q, into the two pads?
c. For the first case of equal clearances, find the stiffness of each bidirectional bearing.
d. For the first case of equal clearances, if an extra vertical load of 120 N is placed on the slider (30 N on each pad), find the downward vertical displacement of the slider.

Solution

a. Flow rates Q_1 and Q_2

Given that $h_1 = h_2 = 0.2$ mm, the flow rate Q can be obtained via Eq. (10-13):

$$Q = \frac{1}{6} \frac{\pi h_0^3}{\mu \ln(R/R_0)} p_r$$

The load capacity of a circular hydrostatic pad is obtained from Eq. (10-12):

$$W = \frac{\pi R^2}{2} \frac{1 - (R_0/R)^2}{\ln(R/R_0)} p_r$$

The first step is to find p_r by using the load capacity equation (for a top pad). Substituting the known values, the recess pressure is the only unknown:

$$5000 = 0.05^2 \frac{\pi}{2} \frac{1 - (0.025^2/0.05^2)}{\ln(0.05/0.025)} p_{r1}$$

The result for the recess pressure at the upper pad is

$$p_{r1} = 1.176 \times 10^6 \text{ Pa}$$

Substituting this recess pressure in Eq. (10-13), the following flow rate, Q_1, is obtained:

$$Q_1 = \frac{1}{6} \frac{\pi\, 0.0002^3}{0.01 \times \ln(0.05/0.025)} \times 1.176 \times 10^6 = 7.1 \times 10^{-4} \text{ m}^3/\text{s}$$

The second step is to find p_{r2} by using the load capacity equation (for the bottom pad), substituting the known values; the following equation is obtained, with P_{r2} as unknown:

$$10{,}000 = 0.05^2 \frac{\pi}{2} \frac{1 - (0.025^2/0.05^2)}{\ln(0.05/0.025)} p_{r2}$$

The recess pressure at the lower pad is

$$p_{r2} = 2.352 \times 10^6 \text{ Pa}$$

Substituting the known values, in Eq. (10-13) the flow rate Q_2 is:

$$Q_2 = \frac{1}{6} \frac{\pi(0.0002^3)}{0.01 \ln(0.05/0.025)} \times 2.352 \times 10^6 = 14.2 \times 10^{-4} \text{ m}^3/\text{s}$$

b. Upper and Lower Clearances h_1 and h_2, for $Q_1 = Q_2$

The flow rate equation (10-13) is

$$Q = \frac{\pi}{6\mu} \frac{h_0^3}{\ln(R/R_0)} p_r$$

For $Q_1 = Q_2$ the following two equations with two unknowns, h_1 and h_2, are obtained:

$$Q = \frac{\pi}{6\mu} \frac{h_1^3}{\ln R/R_0)} p_{r1} = \frac{\pi}{6\mu} \frac{h_2^3}{\ln(R/R_0)} p_{r2}$$
$$h_1 + h_2 = 0.0004 \text{ m}$$

Substituting yields

$$\frac{1}{6} \frac{\pi h_1^3}{0.01 \ln 2} 1.176 \times 10^6 = \frac{1}{6} \frac{\pi(0.0004 - h_1)^3}{0.01 \ln 2} 2.352 \times 10^6$$

The equation can be simplified to the following:

$$1.176 \times h_1^3 = 2.352 \times (0.0004 - h_1)^3$$

Converting to millimeters, the solution for h_1 and h_2 is

$$h_1 = 0.223 \text{ mm} \qquad \text{and} \qquad h_2 = 0.177 \text{ mm}$$

c. *Stiffness of Each Pad*

Equation (10-30) yields the stiffness of a constant-flow rate circular pad:

$$k = \frac{9\mu Q(R^2 - R_0^2)}{h_0^4}$$

Substitute in Eq. (10-30) (for the top pad):

$$k \text{ (top pad)} = \frac{9\,(0.01)(7.07 \times 10^{-4})(0.05^2 - 0.025^2)}{0.0002^4} = 74.56 \times 10^6 \text{ N/m}$$

Substitute in Eq. (10-30) (for the bottom pad):

$$k \text{ (lower pad)} = \frac{9\,(0.01)(0.00142((0.05^2 - 0.025^2)}{0.0002} = 149.76 \times 10^6 \text{ N/m}$$

The total bidirectional bearing stiffness is obtained by adding the top and bottom stiffnesses, as follows:

$$k \text{ (bearing)} = k \text{ (top pad)} + k \text{ (lower pad)} = 224.32 \times 10^6 \text{ N/m}$$

d. Vertical Downward Displacement Δh *of the Slider*

$$k = \frac{\Delta W}{\Delta h} \qquad \text{where } \Delta W = 30\ \text{N}$$

$$\Delta h = \frac{\Delta W}{k}$$

$$\Delta h = \frac{30\ \text{N}}{224.32 \times 10^{6}} = 1.33 \times 10^{-7}\ \text{m} = 0.133\ \mu\text{m}$$

This example shows that under extra force, the displacement is very small.

10.9 CONSTANT-PRESSURE-SUPPLY PADS WITH RESTRICTORS

Hydrostatic pads with a constant flow rate have the desirable characteristic of high stiffness, which is important in machine tools as well as many other applications. However, it is not always practical to supply a constant flow rate to each of the many recesses, because each recess must be fed from a separate positive-displacement pump or from a flow divider. For example, in designs involving many recesses, such as machine tool spindles, a constant flow rate to each of the many recesses requires an expensive hydraulic system that may not be practical.

An alternative arrangement is to use only one pump that supplies a constant pressure to all the recesses. This system is simpler, because it does not require many pumps or flow dividers. Unlike in the constant-flow-rate system, in this system each recess is fed from a *constant supply pressure*, p_s. The oil flows into each recess through a flow restrictor (such as a capillary tube). The flow restrictor causes a pressure drop, and the recess pressure is reduced to a lower level, p_r. The important feature of the flow restrictor is that it is making the bearing stiff to displacement under variable load.

Although hydraulic pumps are usually of the positive-displacement type, such as a gear pump or a piston pump, and have a constant flow rate, the system can be converted to a constant pressure supply by installing a relief valve that returns the surplus flow into the oil sump. The relief valve makes the system one of constant pressure supply. The preferred arrangement is to have an adjustable relief valve so that the supply pressure, p_s, can be adjusted for optimizing the bearing performance. In order to have the desired high bearing stiffness, constant-pressure-supply systems operate with flow restrictors at the inlet to each recess.

10.9.1 Flow Restrictors and Bearing Stiffness

A system of bidirectional hydrostatic pads with a constant pressure supply is presented in Fig. 10-5. The oil flows from a pump, through a flow restrictor, and into each recess on the two sides of this thrust bearing. From the recesses, the fluid flows out, in the radial direction, through the thin clearances, h_1 and h_2 along the lands (outside the recesses). This thin clearance forms a resistance to the outlet flow from each recess. This resistance at the outlet is subject to variations resulting from any small vertical displacement of the slider due to load variations. The purpose of feeding the fluid to the recesses through flow restrictors is to make the bearing stiffer under thrust force; namely, it reduces vertical displacement of the slider when extra load is applied.

When the vertical load on the slider rises, the slider is displaced downward in the vertical direction, and under constant pressure supply a very small displacement results in a considerable reaction force to compensate for the load rise. After a small vertical displacement of the slider, the clearances at the lands of the opposing pads are no longer equal. In turn, the resistances to the outlet flow from the opposing recesses decrease and increase, respectively. It results in unequal flow rates in the opposing recesses. The flow increases and decreases, respectively (the flow is inversely proportional to h_0^3). An important

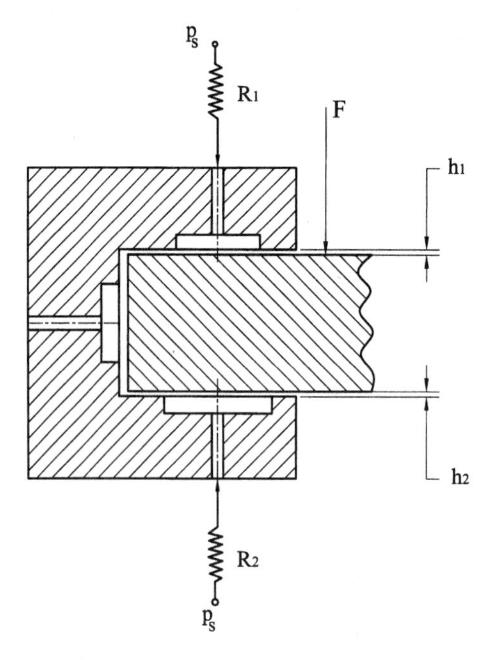

FIG. 10-5 Bidirectional hydrostatic pads with flow restrictors.

characteristic of a flow restrictor, such as a capillary tube, is that its pressure drop increases with the flow rate. In turn, this causes the pressures in the opposing recesses to decrease and increase, respectively. The bearing load capacity resulting from these pressure differences acts in the direction opposite to the vertical load on the slider. In this way, the bearing supports the slider with minimal vertical displacement, Δh. In conclusion, the introduction of inlet flow restrictors increases the bearing stiffness, because only a very small vertical displacement of the slider is sufficient to generate a large difference of pressure between opposing recesses.

10.10 ANALYSIS OF STIFFNESS FOR A CONSTANT PRESSURE SUPPLY

Where the fluid is fed to each recess through a flow restrictor, the fluid in the recesses is bounded between the inlet and outlet flow resistance. The following equations are for derivation of the expression for the stiffness of one hydrostatic pad with a constant pressure supply.

In general, flow resistance causes a pressure drop. Flow resistance R_f is defined as the ratio of pressure loss, Δp (along the resistance), to the flow rate, Q. Flow resistance is defined, similar to Ohm's law in electricity, as

$$R_f = \frac{\Delta p}{Q} \tag{10-31}$$

For a given resistance, the flow rate is determined by the pressure difference:

$$Q = \frac{\Delta p}{R_f} \tag{10-32}$$

The resistance of the inlet flow restrictor is R_{in}, and the resistance to outlet flow through the bearing clearance is R_c (resistance at the clearance). The pressure at the recess, p_r, is bounded between the inlet and outlet resistances; see a schematic representation in Fig. 10-6. The supply pressure, p_s, is constant; therefore, any change in the inlet or outlet resistance would affect the recess pressure.

From Eq. (10-32), the flow rate into the recess is

$$Q = \frac{p_s - p_r}{R_{in}} \tag{10-33}$$

The flow rate through the clearance resistance is given by

$$Q = \frac{p_r}{R_c} \tag{10-34}$$

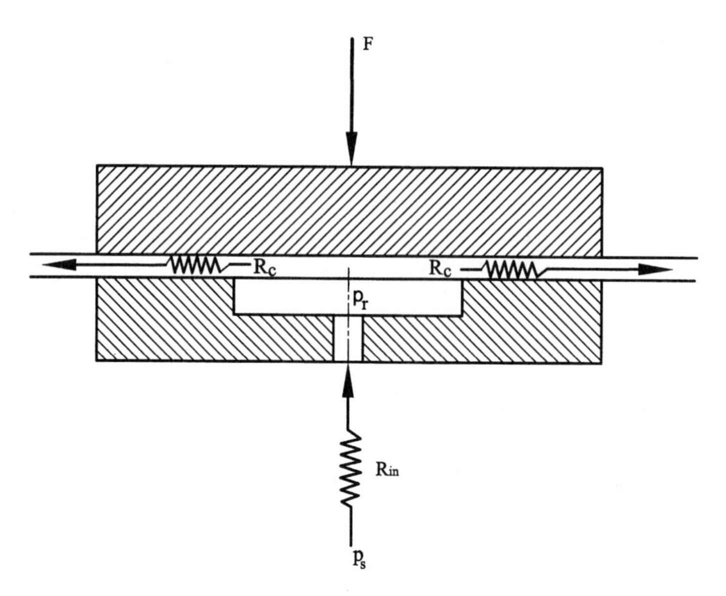

FIG. 10-6 Recess pressure bounded between inlet and outlet resistances.

The fluid is incompressible, and the flow rate Q is equal through the inlet resistance and clearance resistance (continuity). Equating the preceding two flow rate expressions yields

$$\frac{p_r}{R_c} = \frac{p_s - p_r}{R_{\text{in}}} \tag{10-35}$$

The recess pressure is solved for as a function of the supply pressure and resistances:

$$p_r = \frac{1}{(1 + R_{\text{in}}/R_c)} p_s \tag{10-36}$$

The load capacity is [see Eq. (10-25)]

$$W = A_e p_r \tag{10-37}$$

In terms of the supply pressure and the effective pad area, the load capacity is

$$W = A_e \frac{1}{(1 + R_{\text{in}}/R_c)} p_s \tag{10-38}$$

The inlet resistance of laminar flow through a capillary tube is constant, and the pressure drop is proportional to the flow rate. However, the flow resistance through the variable pad clearance is proportional to h_0^{-3} see Eq. (10-27). The clearance resistance can be written as

$$R_c = \kappa h_0^{-3} \tag{10-39}$$

Here, κ is a constant that depends on the pad geometry and fluid viscosity. Equation (10-38) can be written in the form,

$$W = A_e \frac{1}{(1 + K_1 h_0^3)} p_s \tag{10-40}$$

where K is defined as

$$K_1 = \frac{R_{in}}{\kappa} \tag{10-41}$$

In Eq. (10-40) for the load capacity, all the terms are constant except the clearance thickness. Let us recall that the expression for the stiffness is

$$\text{Stiffness } k = -\frac{dW}{dh_0} \tag{10-42}$$

Differentiating Eq. (10-40) for the load capacity W by h_0 results in

$$\text{Stiffness } k = -\frac{dW}{dh_0} = A_e \frac{3K_1 h_0^2}{(1 + K_1 h_0^3)^2} p_s \tag{10-43}$$

Equation (10-43) is for the stiffness of a hydrostatic pad having a constant supply pressure p_s. If the inlet flow is through a capillary tube, the pressure drop is

$$\Delta p = \frac{64 \mu l_c}{\pi d_i^4} Q \tag{10-44}$$

Here, d_i is the inside diameter of the tube and l_c is the tube length. The inlet resistance by a capillary tube is

$$R_{in} = \frac{64 \mu l_c}{\pi d_i^4} \tag{10-45}$$

For calculating the pad stiffness in Eq. (10-43), the inlet resistance is calculated from Eq. (10-45), and the value of κ is determined from the pad equations.

Equation (10-43) can be simplified by writing it as a function of the ratio of the recess pressure to the supply pressure, β, which is defined as

$$\beta = \frac{p_r}{p_s} \tag{10-46}$$

Equations (10-43) and (10-46) yield a simplified expression for the stiffness as a function of β:

$$k = -\frac{dW}{dh_0} = \frac{3}{h_0} A_e(\beta - \beta^2)p_s \tag{10-47}$$

Equation (10-47) indicates that the maximum stiffness is when $\beta = 0.5$, or

$$\frac{p_r}{p_s} = 0.5 \tag{10-48}$$

For maximum stiffness, the supply pressure should be twice the recess pressure. This can be obtained if the inlet resistance were equal to the recess resistance. This requirement will double the power of the pump that is required to overcome viscous friction losses. The conclusion is that the requirement for high stiffness in constant-supply-pressure systems would considerably increase the friction losses and the cost of power for operating the hydrostatic bearings.

Example Problem 10-5

Stiffness of a Circular Pad with Constant Supply Pressure

A circular hydrostatic pad as shown in Fig. 10-1 has a constant supply pressure, p_s. The circular pad is supporting a load of $W = 5000$ N. The outside disk diameter is 200 mm, and the diameter of the circular recess is 100 mm. The oil viscosity is $\mu = 0.005$ N-s/m^2. The pad is operating with a clearance of 120 μm.

a. Find the recess pressure, p_r.
b. Calculate the flow rate Q of the oil through the bearing to maintain the clearance.
c. Find the effective area of the pad.
d. If the supply pressure is twice the recess pressure, $p_s = 2p_r$, find the stiffness of the circular pad.
e. Compare with the stiffness obtained in Example Problem 10-3 for a constant flow rate.
f. Find the hydraulic power required for circulating the oil through the bearing. Compare to the hydraulic power in a constant-flow-rate pad.

Solution

a. Recess Pressure

Similar to Example Problem 10-3 for calculating the flow rate Q, the first step is to solve for the recess pressure. This pressure is derived from the equation of the load:

$$W = R^2\left(\frac{\pi}{2}\frac{1 - R_0^2/R^2}{\ln(R/R_0)}\right)p_r$$

After substitution, the recess pressure is only unknown in the following equation:

$$5000 = 0.1^2\left(\frac{\pi}{2} \times \frac{1 - 0.25}{\ln(2)}\right)p_r$$

Solving for the recess pressure, p_r yields

$$p_r = 294.18 \text{ kPa}$$

b. Flow Rate

The flow rate Q can now be determined. It is derived from the following expression [see Eq. (10-13)] for Q as a function of the clearance pressure:

$$Q = \frac{\pi}{6\mu}\frac{h_0^3}{\ln(R/R_0)}p_r$$

Similar to Example Problem 10-3, after substituting the values, the flow rate is

$$Q = 76.8 \times 10^{-6} \text{ m}^3/\text{s}$$

c. Pad Effective Area

The effective area is defined by

$$W = A_e p_r$$

Solving for A_e as the ratio of the load and the recess pressure, we get

$$A_e = \frac{5000}{294.180}A_e = 0.017 \text{ m}^2$$

d. Pad Stiffness

Supply Pressure: Now the supply pressure can be solved for as well as the stiffness for constant supply pressure:

$$p_s = 2p_r$$
$$= 2 \times 294.18 \text{ kPa}$$
$$p_s = 588.36 \text{ kPa}$$

Pad Stiffness of Constant Pressure Supply: The stiffness is calculated according to Eq. (10-47):

$$k = \frac{3}{h_0} A_e(\beta - \beta^2)p_s \quad \text{where } \beta = 0.5$$

$$k = \frac{3}{120 \times 10^{-6}} \times 0.017(0.5 - 0.5^2)588.36 \times 10^3,$$

and the result is

$$k = 62.5 \times 10^6 \text{ N/m} \quad \text{(for constant pressure supply)}$$

e. Stiffness Comparison

In comparison, for a constant flow rate (see Example Problem 10-3) the stiffness is

$$k = 125 \times 10^6 \text{ N/m} \quad \text{(for constant flow rate}$$

For the bearing with a constant pressure supply in this problem, the stiffness is about half of the constant-flow-rate pad in Example Problem 10-3.

f. Hydraulic Power

The power for circulating the oil through the bearing for constant pressure supply is twice of that for constant flow rate. Neglecting the friction losses in the pipes, the equation for the net hydraulic power for circulating the oil through the bearing in a constant-flow-rate pad is

$$\dot{E}_h \approx Qp_r \quad \text{(for constant-flow-rate pad)}$$
$$= 76.8 \times 10^{-6} \times 294.18 \times 10^3 = 22.6 \text{ W} \quad \text{(constant-flow-rate pad).}$$

In comparison, the equation for the net hydraulic power for constant pressure supply is

$$\dot{E}_h \approx Qp_s \quad \text{(For a constant pressure supply pad).}$$

Since $p_s = 2p_r$, the hydraulic power is double for constant pressure supply:

$$\dot{E}_h \approx Qp_r = 76.8 \times 10^{-6} \times 588.36 \times 10^3 = 45.20 \text{ W}$$
$$\text{(for a constant-pressure-supply pad)}$$

Example Problem 10-6

Constant-Supply-Pressure Bidirectional Pads

A bidirectional hydrostatic bearing (see Fig. 10-5) consists of two circular pads, a constant supply pressure, p_s, and flow restrictors. If there is no external load, the two bidirectional circular pads are prestressed by an equal reaction force, $W = 21{,}000$ N, at each side.

The clearance at each side is equal, $h_1 = h_2 = 0.1$ mm. The upper and lower circular pads are each of 140-mm diameter and circular recess of 70-mm diameter. The oil is SAE 10, and the operation temperature of the oil in the clearance is 70°C. The supply pressure is twice the recess pressure, $p_s = 2p_r$.

a. Find the recess pressure, p_r, and the supply pressure, p_s, at each side to maintain the required prestress.
b. Calculate the flow rate Q of the oil through each pad.
c. Find the stiffness of the bidirectional hydrostatic bearing.
d. The flow restrictor at each side is a capillary tube of inside diameter $d_i = 1$ mm. Find the length of the capillary tube.
e. If there is no external load, find the hydraulic power required for circulating the oil through the bidirectional hydrostatic bearing.

Solution

a. Recess Pressure and Supply Pressure

The recess pressure is derived from the equation of the load capacity:

$$W = R^2 \frac{\pi}{2} \cdot \frac{1 - (R_0/R)^2}{\ln(R/R_0)} \cdot p_r$$

After substitution, the recess pressure is the only unknown in the following equation:

$$21{,}000 \text{ N} = 0.07^2 \frac{\pi}{2} \frac{1 - (0.035/0.07)^2}{\ln(0.07/0.035)} p_r$$

The solution for the recess pressure yields

$$p_r = 2.52 \text{ MPa}$$

The supply pressure is

$$p_s = 2p_r = 5.04 \text{ MPa}$$

b. Flow Rate Through Each Pad

The flow rate Q can now be derived from the following expression as a function of the recess pressure:

$$Q = \frac{\pi}{6\mu} \cdot \frac{h_0^3}{\ln(R/R_0)} \cdot p_r$$

Substituting the known values gives

$$Q = \frac{\pi}{6 \times 0.01} \times \frac{10^{-12}}{\ln 2} \times 2.52 \times 10^6 = 190.4 \times 10^{-6} \frac{m^3}{s}$$

c. Stiffness of the Bidirectional Hydrostatic Pad

In order to find the stiffness of the pad, it is necessary to find the effective area:

$$W = A_e p_r$$
$$21{,}000 \text{ N} = A_e \times 2.52 \text{ MPa}$$

Solving for A_e as the ratio of the load capacity and the recess pressure yields

$$A_e = \frac{21{,}000 \text{ N}}{2.52 \text{ MPa}} = 0.0083 \text{ m}^2$$

The ratio of the pressure to the supply pressure, β, is

$$\beta = \frac{p_r}{p_s} = 0.5$$

The stiffness of the one circular hydrostatic pad is

$$k = \frac{3}{h_0} A_e(\beta - \beta^2)p_s$$

$$k = \frac{3}{0.1 \times 10^{-3}} \times 0.0083 \times (0.5 - 0.5^2) \times 5.04 \times 10^6 = 315 \times 10^6 \text{ N/m}$$

and the stiffness of the bidirectional bearing is

$$K = 2 \times 315 \times 10^6 = 630 \times 10^6 \text{ N/m}$$

d. Length of the Capillary Tube

The internal diameter of the tube is $d_i = 1$ mm.

The equation for the flow rate in the recess is

$$Q = \frac{p_s - p_r}{R_{\text{in}}}$$

After substituting the known values for p_s, p_r, and Q, the inlet resistance becomes

$$R_{\text{in}} = \frac{p_s - p_r}{Q} = \frac{5.04 \times 10^6 - 2.52 \times 10^6}{190.4 \times 10^{-6}} = 1.32 \times 10^{10} \text{ N-s/m}^4$$

The inlet resistance of capillary tube is given by the following tube equation:

$$R_{\text{in}} = \frac{64\, \mu l_c}{\pi d_i^4}$$

Here, d_i is the inside diameter of the tube and l_c is the tube length. The tube length is

$$l_c = \frac{R_{\text{in}} \pi d_i^4}{64\mu}$$

$$= \frac{1.32 \times 10^{10} \times \pi 0.001^4}{64 \times 0.01} = 65 \times 10^{-3} \text{ m},$$

$$l_c = 65 \text{ mm}$$

e. Hydraulic Power for Circulating Oil Through the Bidirectional Hydrostatic Bearing

Neglecting the friction losses in the pipes, the equation for the net hydraulic power for one pad is

$$\dot{E}_h \approx Q p_s$$

Substituting the values for Q and P_s results in

$$\dot{E}_h \approx 190.4 \times 10^{-6} \times 5.04 \times 10^6 = 960 \text{ W}$$

Hydraulic power for bidirectional bearing is

$$\dot{E}_h \text{ (bidirectional bearing)} = 2 \times 960 = 1920\ W$$

10.11 JOURNAL BEARING CROSS-STIFFNESS

The hydrodynamic thrust pad has its load capacity and the stiffness in the same direction. However, for journal bearings the stiffness is more complex and involves four components. For most designs, the hydrostatic journal bearing has hydrodynamic as well as hydrostatic effects, and it is referred to as a hybrid bearing. The hydrodynamic effects are at the lands around the recesses. The displacement is not in the same direction as the force W. In such cases, the journal bearing has cross-stiffness (see Chapter 7). The stiffness components are presented as four components related to the force components and the displacement component.

Similar to the hydrodynamic journal bearing, the load of the hydrostatic journal bearing is also divided into two components, W_x and W_y, and the displacement of the bearing center e is divided into two components, e_x and e_y. In Chapter 7, the two components of the journal bearing stiffness are defined [Eq. (7-31)], and the cross-stiffness components are defined in Eq. (7-32). Cross-stiffness components can result in bearing instability, which was discussed in Chapter 9.

10.12 APPLICATIONS

An interesting application is the hydrostatic pad in machine tool screw drives (Rumberger and Wertwijn, 1968). For high-precision applications, it is important to prevent direct metal contact, which results in stick-slip friction and limits the machining precision. Figure 10-7 shows a noncontact design that includes hydrostatic pads for complete separation of the sliding surfaces of screw drive. Another important application is in a friction testing machine, which will be described in Chapter 14.

10.13 HYDRAULIC PUMPS

An example of a positive-displacement pump that is widely used for lubrication is the gear pump. The use of gear pumps is well known in the lubrication system of automotive engines. Gear pumps, as well as piston pumps, are positive-displacement pumps; i.e., the pumps deliver, under ideal conditions, a fixed quantity of liquid per cycle, irrespective of the flow resistance (head losses in the system). However, it is possible to convert the discharge at a constant flow rate to discharge at a constant pressure by installing a pressure relief valve that maintains a constant pressure and returns the surplus flow.

A cross section of a simple gear pump is shown in Fig. 10-8a. A gear pump consists of two spur gears (or helix gears) meshed inside a pump casing, with one of the gears driven by a constant-speed electric motor. The liquid at the suction side is trapped between the gear teeth, forcing the liquid around the casing and finally expelling it through the discharge. The quantity of liquid discharged per revolution of the gear is known as *displacement*, theoretically equal to the sum of the volumes of all the spaces between the gear teeth and the casing. However, there are always tolerances and small clearance for a free fit between the gears and casing. The presence of clearance in pump construction makes it practically impossible to attain the theoretical displacement.

The advantages of the gear pump, in comparison to other pumps, are as follows.

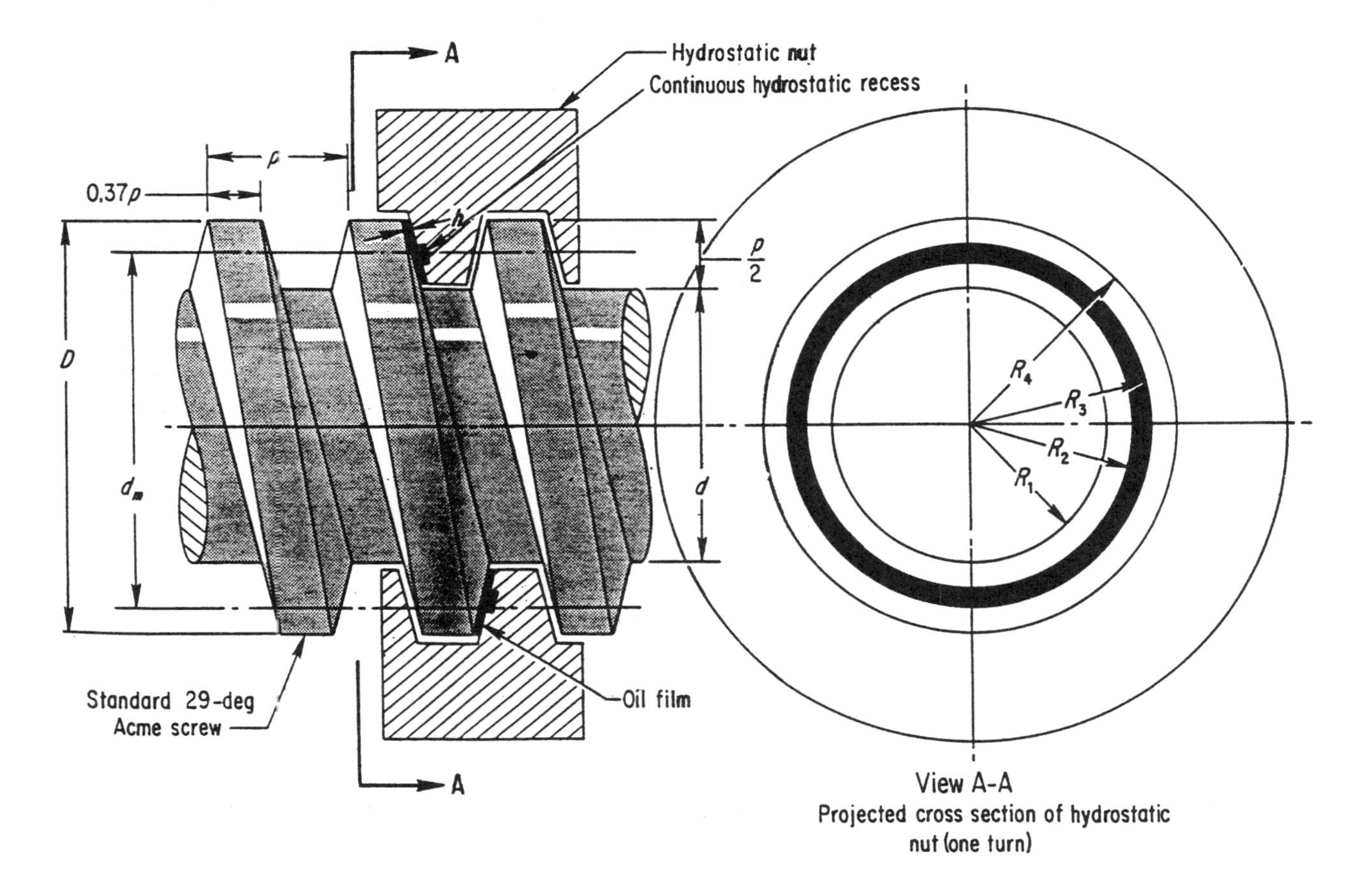

FIG. 10-7 Noncontact screw drive with hydrostatic pads. From Rumberger and Wertwijn (1968), reprinted with permission from *Machine Design*.

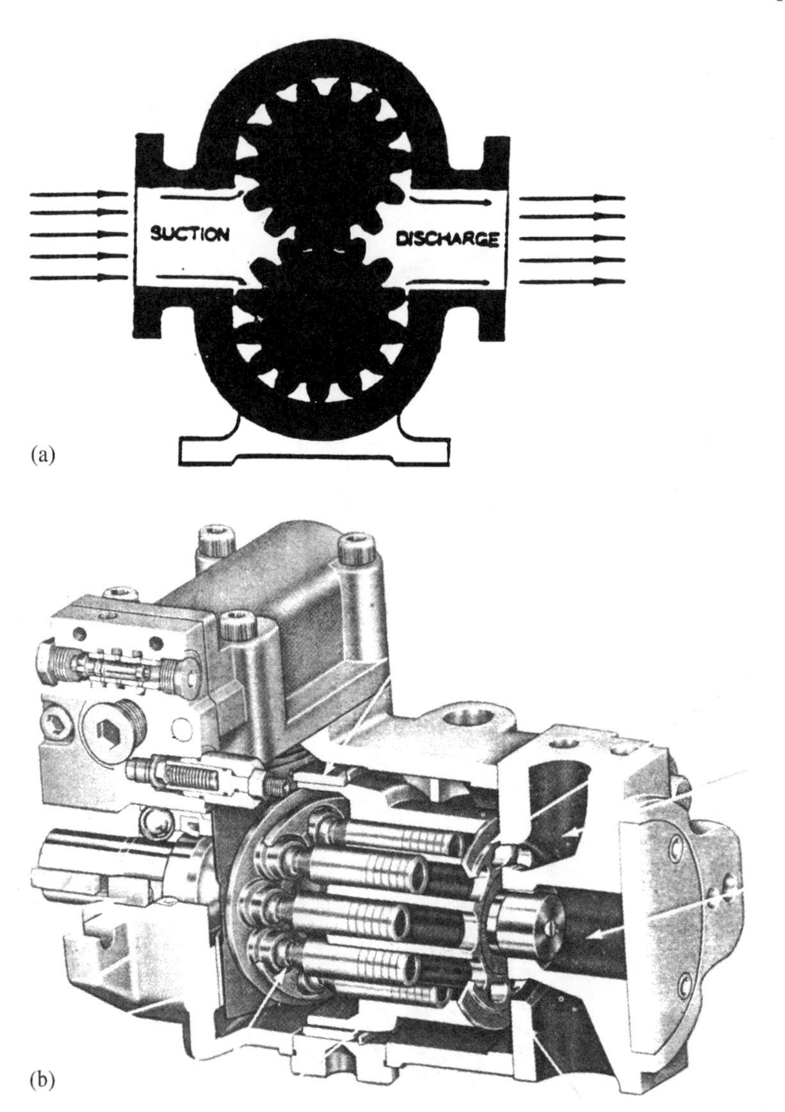

(a)

(b)

FIG. 10-8 (a) Cross section of a gear pump. (b) Multipiston hydraulic pump. (Reprinted with permission from The Oilgear Company.)

1. It is a simple and compact pump, and does not need inlet and outlet valves, such as in the piston pump. However, gear pumps require close running clearances.
2. It involves continuous flow (unlike positive-displacement reciprocating pumps).

3. It can handle very high-viscosity fluid.
4. It can generate very high heads (or outlet pressure) in comparison to centrifugal pumps.
5. It is self-priming (unlike the centrifugal pump). It acts like a compressor and pumps out trapped air or vapors.
6. It has good efficiency at very high heads.
7. It has good efficiency over a wide speed range.
8. It requires relatively low suction heads.

The flow rate of a gear pump is approximately constant, irrespective of its head losses. If we accidentally close the discharge valve, the discharge pressure would rise until the weakest part of the system fails. To avoid this, a relief valve should be installed in parallel to the discharge valve.

When a small amount of liquid escapes backward from the discharge side to the suction side through the gear pump clearances, this is referred to as *slip*. The capacity (flow rate) lost due to slip in the clearances increases dramatically with the clearance, h_0, between the housing and the gears (proportional to h_0^3) and is inversely proportional to the fluid viscosity. An idea about the amount of liquid lost in slippage can be obtained via the equation for laminar flow between two parallel plates having a thin clearance, h_0, between them:

$$Q = \frac{lh_0^3}{12\mu b}\Delta p \tag{10-49}$$

where

Q = flow rate of flow in the clearance (slip flow rate)
Δp = differential pressure (between discharge and suction)
b = width of fluid path (normal to fluid path)
h_0 = clearance between the two plates
μ = fluid viscosity
l = length along the fluid path

This equation is helpful in understanding the parameters affecting the magnitude of slip. It shows that slip is mostly dependent on clearance, since it is proportional to the cube of clearance. Also, slip is proportional to the pressure differential Δp and inversely proportional to the viscosity μ of the liquid. Gear pumps are suitable for fluids of higher viscosity, for minimizing slip, and are widely used for lubrication, since lubricants have relatively high viscosity (in comparison to water). Fluids with low viscosity, such as water and air, are not suitable for gear pumps.

Piston pumps are also widely used as high-pressure positive-displacement (constant-flow-rate) hydraulic pumps. An example of the multipiston pump is shown in Fig. 10-8b. The advantage of the piston pump is that it is better sealed and the slip is minimized. In turn, the efficiency of the piston pump is higher, compared to that of a gear pump, but the piston pump requires valves, and it is more expensive.

10.14 GEAR PUMP CHARACTERISTICS

The actual capacity (flow rate) and theoretical displacement versus pump head are shown in Fig. 10-9. The constant theoretical displacement is a straight horizontal line. The actual capacity (flow rate) reduces with the head because the "slip" is proportional to the head of the pump (discharge head minus suction head). When the head approaches zero, the capacity is equivalent to the theoretical displacement.

10.14.1 Hydraulic Power and Pump Efficiency

The SI unit of power $\dot{E}$ is the watt. Another widely used unit is the Imperial unit, horse power [HP]. Brake power, $\dot{E}_b$ (BHP in horsepower units), is the mechanical shaft power required to drive the pump by means of electric motor. In the pump, this power is converted into two components: the useful hydraulic power, $\dot{E}_h$, and the frictional losses, $\dot{E}_f$. The useful hydraulic power can be converted back to work done by the fluid. A piston or hydraulic motor can do this energy conversion. In the pump, the friction losses are dissipated as heat. Friction

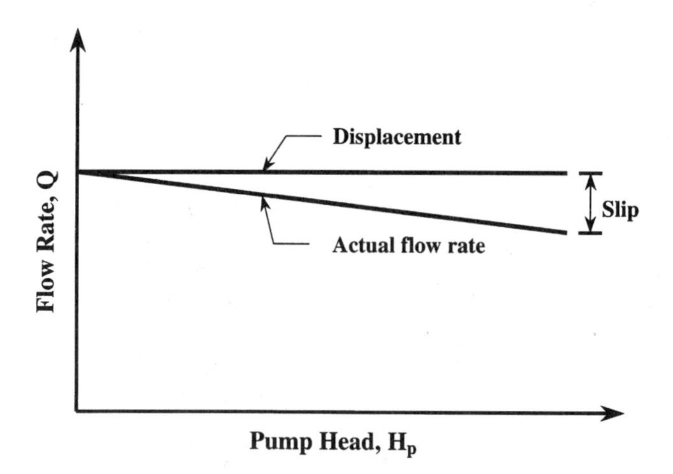

FIG. 10-9 Gear pump Q–H characteristics.

losses result from friction in the bearings, the stuffing box (or mechanical seal), and the viscous shear of the fluid in the clearances.

In Fig. 10-10, the curves of the various power components $\dot{E}$ versus pump head H_p are presented in horsepower [HP] units. The frictional horsepower [FHP] does not vary appreciably with increased head; it is the horizontal line in Fig. 10-10. The other useful component is the hydraulic horsepower [HHP]. This power component is directly proportional to the pump head and is shown as a straight line with a positive slope. This component is added to constant FHP, resulting in the total brake horsepower [BHP].

The BHP curve in Fig. 10-10 is a straight line, and at zero pump head there is still a definite brake horsepower required, due to friction in the pump. In a gear pump, the friction horsepower, FHP, is a function of the speed and the viscosity of the fluid, but not of the head of the pump. Because FHP is nearly constant versus the head, it is a straight horizontal line in Fig. 10-10. On the other hand, HHP is an increasing linear function of H_p (see equation for hydraulic power). (This is approximation, since Q is not constant because it is reduced by the slip.) The sum of the friction power and the hydraulic power is the brake horsepower. The brake horsepower increases nearly linearly versus H_p, as shown in Fig. 10-10. Since FHP is constant, the efficiency η is an increasing function versus H_p. The result is that gear pumps have a higher efficiency at high heads.

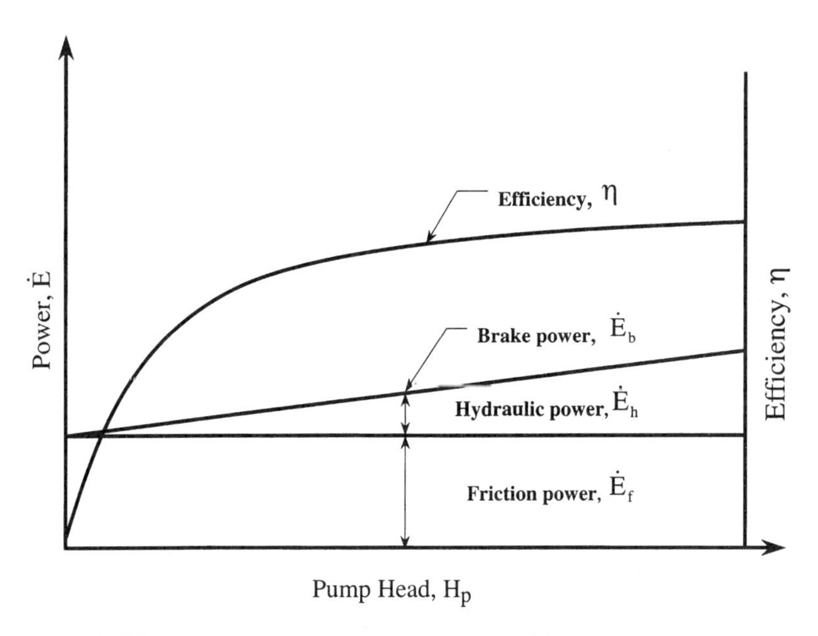

FIG. 10-10 Power and efficiency characteristics of the gear pump.

The head of the pump, H_p, generated by the pump is equal to the head losses in a closed-loop piping system, such as in the hydrostatic bearing system. If the fluid is transferred from one tank to another at higher elevation, the head of the pump is equal to the head losses in the piping system plus the height difference ΔZ.

The head of the pump, H_p, is the difference of the heads between the two points of discharge and suction:

$$H_p = H_d - H_s \tag{10-50}$$

Pump head units are of length (m, ft). Head is calculated from the Bernoulli equation. The expression for discharge and suction heads are:

$$H_d = \frac{p_d}{\gamma} + \frac{V_d^2}{2g} + Z_d \tag{10-51}$$

$$H_s = \frac{p_s}{\gamma} + \frac{V_s^2}{2g} + Z_s \tag{10-52}$$

where

H_d = head at discharge side of pump (outlet)
H_s = head at suction side of pump (inlet)
p_d = pressure measured at discharge side of pump (outlet)
p_s = pressure measured at suction side of pump (inlet)
γ = specific weight of fluid (for water, $\gamma = 9.8 \times 10^3$ [N/m^3])
V = fluid velocity
g = gravitational acceleration
Z = height

The pump head, H_p, is the difference between the discharge head and suction head. In a closed loop, H_p is equal to the head loss in the loop. The expression for the pump head is

$$H_p = \frac{p_d - p_s}{\gamma} + \frac{V_d^2 - V_s^2}{2g} + (Z_d - Z_s) \tag{10-53}$$

The velocity of the fluid in the discharge and suction can be determined from the rate of flow and the inside diameter of the pipes. In most gear pumps, the pipe inside the diameters on the discharge and suction sides are equal. In turn, the discharge velocity is equal to that of the suction. Also, there is no significant difference in height between the discharge and suction. In such cases, the last two terms can be omitted, and the pump is determined by a simplified equation that considers only the pressure difference:

$$H_p = \frac{p_d - p_s}{\gamma} \tag{10-54}$$

10.14.2 Hydraulic Power

The hydraulic power of a pump, $\dot{E}_h$, is proportional to the pump head, H_p, according to the following equation:

$$\dot{E}_h = Q\gamma H_p \tag{10-55}$$

The SI units for Eq. (10-53), (10-54), and (10-55) are

$\dot{E}_h[w]$
Q [m^3/s]
γ [N/m^3]
H [m]
p [N/m^2 or pascals]

The pump efficiency is the ratio of hydraulic power to break power:

$$\eta = \frac{\dot{E}_h}{\dot{E}_b} \tag{10-56}$$

The conversion to horsepower units is 1 HP = 745.7 W. In most gear pumps, the inlet and outlet pipes have the same diameter and the inlet and outlet velocities are equal.

In Imperial units, the hydraulic horsepower (HHP) is given by

$$HHP = \frac{Q\Delta p}{1714} \tag{10-57}$$

Here, the units are as follows:

Δp [psi] $= \gamma(H_p - H_s)$ and Q [GPM]

In imperial units, the efficiency of the pump is:

$$\eta = \frac{HHP}{BHP} \tag{10-58}$$

The BHP can be measured by means of a motor dynamometer. If we are interested in the efficiency of the complete system of motor and pump, the input power is measured by the electrical power, consumed by the electric motor that drives the pump (using a wattmeter). The horsepower lost on friction in the pump, FHP, cannot be measured but can be determined from

$$FHP = BHP - HHP \tag{10-59}$$

10.15 FLOW DIVIDERS

Using many hydraulic pumps for feeding the large number of recesses of hydrostatic pads in machines is not practical. A simple solution of this problem is to use constant pressure and flow restrictors. However, flow restrictors increase the power losses in the system. Therefore, this method can be applied only with small machines or machines that are operating for short periods, and the saving in the initial cost of the machine is more important than the long-term power losses.

Another solution to this problem is to use flow dividers. Flow dividers are also used for distributing small, constant flow rates of lubricant to rolling-element bearings. It is designed to divide the constant flow rate of one hydraulic pump into several constant flow rates. In hydrostatic pads, each recess is fed at a constant flow rate from the flow divider. The advantage of using flow dividers is that only one hydraulic pump is needed for many pads and a large number of recesses.

The design concept of a flow divider is to use the hydraulic power of the main pump to activate many small pistons that act as positive-displacement pumps (constant-flow-rate pumps), and thus the flow of one hydraulic pump is divided into many constant flow rates. A photo of a flow divider is shown in Fig. 10-11a. Figure 10-11b presents a cross section of a flow divider made up of many rectangular blocks connected together for dividing the flow for feeding a large number of bearings. The contact between the blocks is sealed by O-rings. The intricate path of the inlet and outlet of one piston is shown in this drawing.

For a large number of bearings, the flow divider outlets are divided again. An example of such a combination is shown in Fig. 10-12.

10.16 CASE STUDY: HYDROSTATIC SHOE PADS IN LARGE ROTARY MILLS

Size reduction is an important part of the process of the enrichment of ores. Ball-and-rod rotary mills are widely used for grinding ores before the enrichment process in the mines. Additional applications include the reduction of raw-material particle size in cement plants and pulverizing coal in power stations.

In rotary mills, friction and centrifugal forces lift the material and heavy balls against the rotating cylindrical internal shell and liners of the mill, until they fall down by gravity. The heavy balls fall on the material, and reduce the particle size by impact. For this operation, the rotation speed of the mills must be slow,

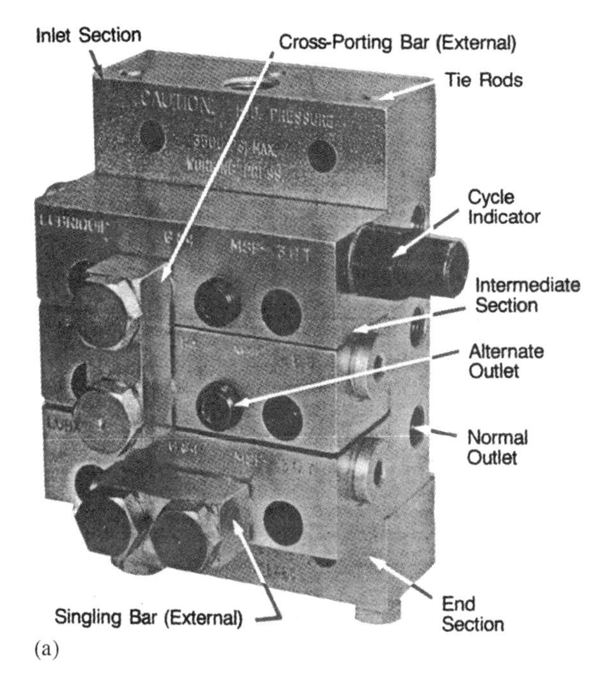

FIG. 10-11a Flow divider. (Reprinted with permission from Lubriquip, Inc.)

about 12–20 RPM. In most cases, the low speed of rotation is not sufficient for adequate fluid film thickness in hydrodynamic bearings.

There has been continual trend to increase the diameter, D, of rotary mills, because milling output is proportional approximately to $D^{2.7}$ and only linearly proportional to the length. In general, large rotary mills are more economical for the large-scale production of ores. Therefore, the outside diameter of a rotary mill, D, is quite large; many designs are of about 5-m diameter, and some rotary mills are as large as 10 m in diameter.

Two bearings on the two sides support the rotary mill, which is rotating slowly in these bearings. Although each bearing diameter is much less than the rotary mill diameter, it is still very large in comparison to common bearings in machinery. The *trunnion* on each side of the rotary mill is a hollow shaft (large-diameter sleeve) that is turning in the bearings; at the same time, it is used for feeding the raw material and as an outlet for the reduced-size processed material. The internal diameter of the trunnion must be large enough to accommodate the high feed rate of ores. The trunnion outside diameter is usually more than 1.2 m.

In the past, as long as the trunnion outside diameter was below 1.2 m, large rolling-element bearings were used to support the trunnion. However, rolling

SERIES PROGRESSIVE DIVIDER VALVE OPERATION

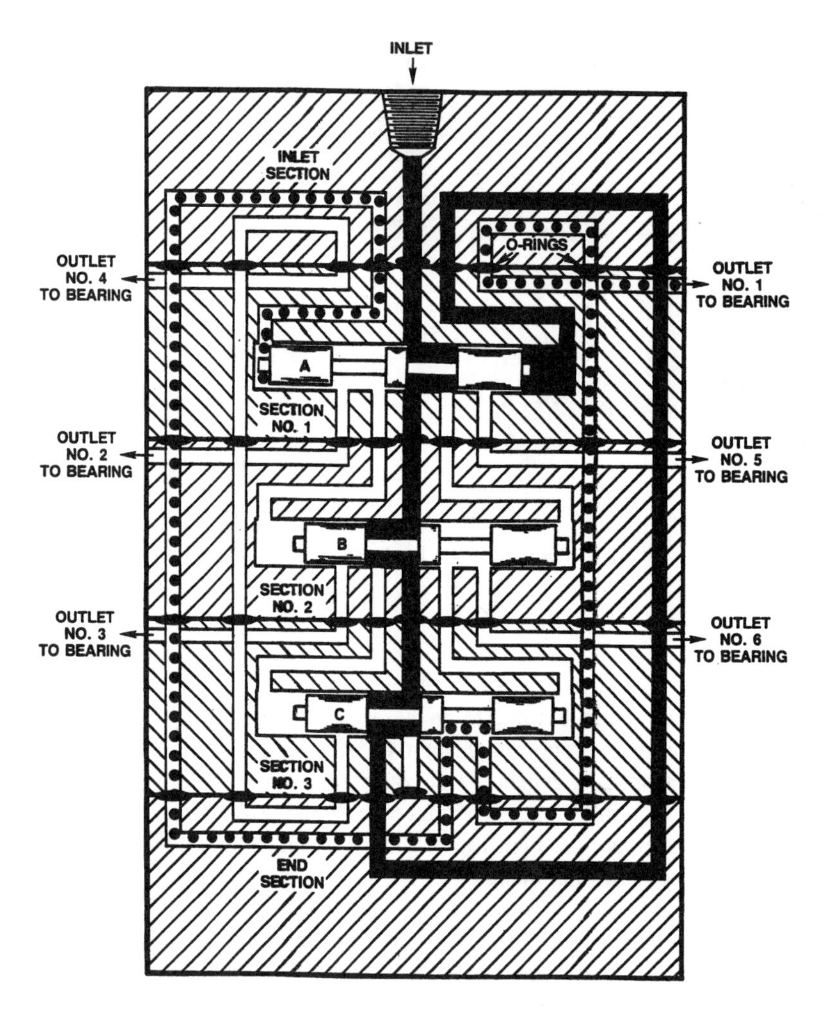

POSITION NO. 1 — LUBE PRESSURE TO INLET MOVES PISTON A TO LEFT FORCING A MEASURED AMOUNT OF LUBRICANT TO OUTLET NO. 1 BEARING.

(b)

FIG. 10-11b Cross section of a flow divider. (Reprinted with permission from Lubriquip, Inc.)

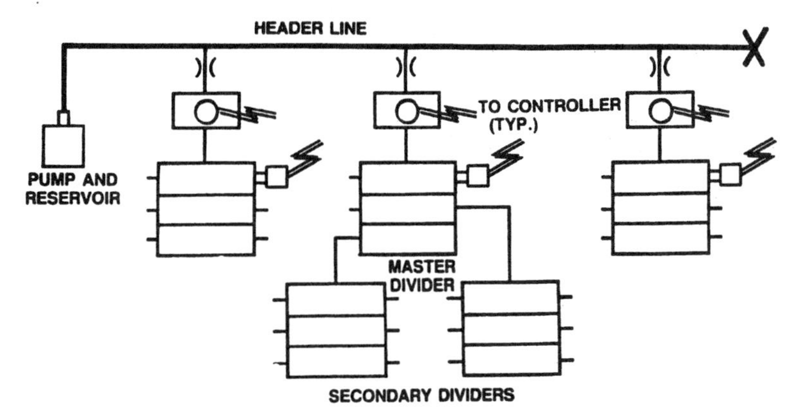

FIG. 10-12 Combination of flow dividers. (Reprinted with permission from Lubriquip, Inc.)

bearings are not practical any more for the larger trunnion diameters currently in use in rotary mills. Several designs of self-aligning hydrodynamic bearings are still in use in many rotary mills. These designs include a hydrostatic lift, of high hydrostatic pressure from an external pump, only during start-up. This hydrostatic lift prevents the wear and high friction torque during start-up. Due to the low speed of rotation, these hydrodynamic bearings operate with very low minimum film thickness. Nevertheless, these hydrodynamic bearings have been operating successfully for many years in various rotary mills. The hydrodynamic bearings are designed with a thick layer of white metal bearing material (babbitt), and a cooling arrangement is included in the bearing. However, ever-increasing trunnion size makes the use of continuous hydrostatic bearings the preferred choice.

Large-diameter bearings require special design considerations. A major problem is the lack of manufacturing precision in large bearings. A large-diameter trunnion is less accurate in comparison to a regular, small-size journal, for the following reasons.

1. Machining errors of round parts, in the form of out-of-roundness, are usually proportional to the diameter.
2. The trunnion supports the heavy load of the mill, and elastic deformation of the hollow trunnion causes it to deviate from its ideal round geometry.
3. Many processes require continuous flow of hot air into the rotary mill through the trunnion, to dry the ores. This would result in thermal distortion of the trunnion; in turn, it would cause additional out-of-roundness errors.

For successful operation, rolling bearings as well as hydrodynamic bearings require precision machining. For rolling-element bearings, any out-of-roundness of the trunnion or the bearing housing would deform the inner or outer rings of the rolling bearing. This undesired deformation would adversely affect the performance of the bearing and significantly reduce its life. Moreover, large-diameter rolling-element bearings are expensive in comparison to other alternatives. For a hydrodynamic bearing, the bearing and journal must be accurately round and fitted together for sustained performance of a full hydrodynamic fluid film. Any out-of-roundness in the bearing or journal results in a direct contact and excessive wear. In addition, rotary mills rotate at relatively low speed, which is insufficient for building up a fluid film of sufficient thickness to support the large trunnion.

An alternative that is often selected is the hydrostatic bearing system. As mentioned earlier, hydrostatic bearings can operate with a thicker fluid film and therefore are less sensitive to manufacturing errors and elastic deformation.

10.16.1 Self-Aligning and Self-Adjusted Hydrostatic Shoe Pads

A working solution to the aforementioned problems of large bearings in rotary mills has been in practice for many years, patented by Arsenius, from SKF (see Arsenius and Goran, 1973) and Trygg and McIntyre (1982). It is in the form of self-aligning hydrostatic shoe pads that support the trunnion as shown in Fig. 10-13. These shoe pads can pivot to compensate for aligning errors, in all directions. Hydrostatic pads that pivot on a sphere for universal self-aligning are also used in small bearings.

When two hydrostatic shoe pads support a circular trunnion (Fig. 10-13a), the load is distributed evenly between these two pads. In fact, the location of the two pads determines the location of the trunnion center. However, whenever three or more pads are supporting the trunnion, the load is no longer distributed evenly, and the design must include radial adjustment of the pads, as shown in Fig. 10-13b.

The load capacity is inversely proportional to h_0^3, where h_0 is the radial clearance between the face of the hydrostatic pad and the trunnion running surface. Due to limitations in precision in the mounting of the pads, the clearance h_0 is never equal in all the pads. Therefore, the design must include adjustment of the pad height to ensure that the load is distributed evenly among all the pads. Adjustment is required only for the extra pads above the first two pads, which do not need adjustment. Therefore, each of the extra hydrostatic pads must be designed to move automatically in the radial direction of the trunnion until the load is divided evenly among all pads. This way, the pads always keep a constant clearance from the trunnion surface.

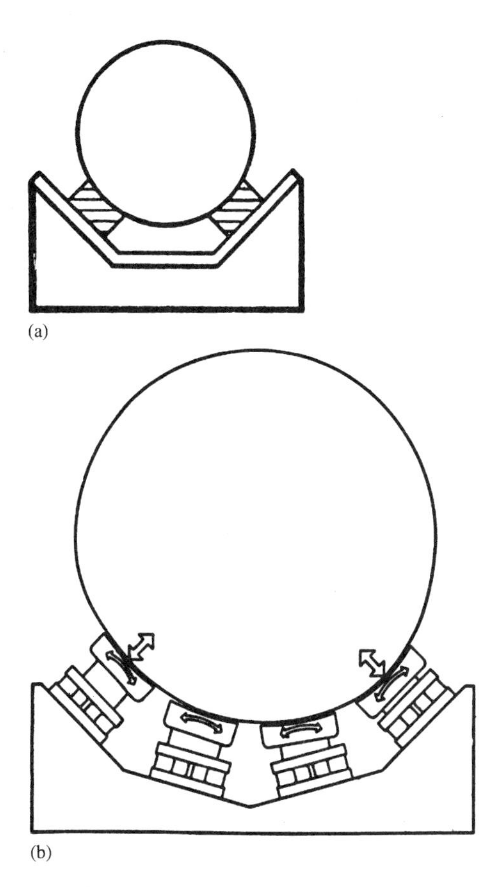

FIG. 10-13 Hydrostatic shoe pads: (a) Two-pad support. (b) Four-pad support: All pads are self-aligning, two have radial adjustment. From Trygg and McIntyre (1982), reprinted with permission from CIM Bulletin.

In addition, the large trunnion becomes slightly oval under the heavy load of the mill. For a large trunnion, out-of-roundness errors due to elastic deformation are of the order of 6 mm (one-quarter inch). In addition, mounting errors, deflection of the mill axis, and out-of-roundness errors in the machining of the trunnion surface all add up to quite significant errors that require continuous clearance adjustment by means of radial motion of the pads. Also, it is impossible to construct the hydrostatic system precisely so that all pads will have equal clearance, h_0, for equally sharing the load among the pads. Therefore, the hydrostatic pads must be designed to be self-adjusting; namely they must move automatically in the radial direction of the trunnion until the load is equalized among all the pads.

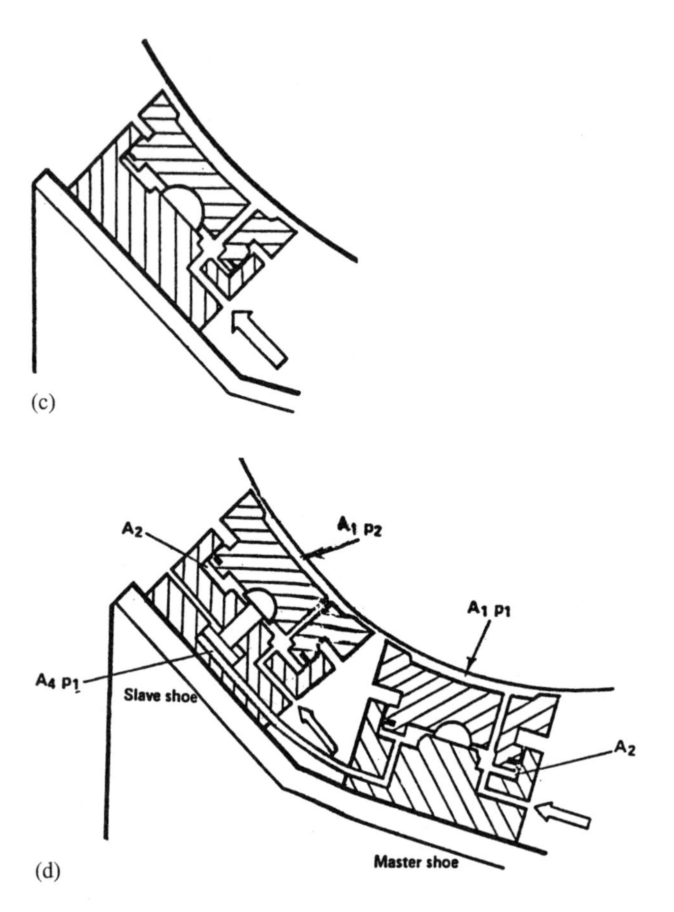

FIG. 10-13 Hydrostatic shoe pads: (c) Self-aligning ball support with pressure relief. (d) Master and slave shoe pads. (From Trygg and McIntyre (1982), reprinted with permission from CIM Bulletin.)

Since the pads are self-aligning and self-adjusting, the foundation's construction does not have to be precise, and a relatively low-cost welded structure can be used as a bed to support the set of hydrostatic pads.

The design concept is as follows: The surface of the pads is designed with the same radius of curvature as the trunnion outside surface. The clearance is adjusted, by pad radial displacement, which requires additional lower piston and hydraulic oil pressure for radial displacement. Explanation of the control of the pad radial motion will follow shortly.

If sufficient constant-flow-rate of oil is fed into each pad from external pumps, it is then possible to build up appropriate pressure in each of the pad recesses for separating completely the mating surfaces by means of a thin oil film. A major advantage of hydrostatic pads is that the fluid film thickness is independent of the trunnion speed. The fluid film is formed when the trunnion is stationary or rotating, and the mating surfaces are completely separated by oil film during start-up as well as during steady operation.

All pads have universal angular self-aligning (see Fig. 10-13b). This is achieved by supporting each pad on a sphere (hard metal ball), as shown in Fig. 10-13c, where the pad has a spheroid recess with its center coinciding with the sphere center. In this way, it can tilt in all directions, and errors in alignment with the trunnion outside surface are compensated.

However, the spheroid pivot arrangement under high load has a relatively high friction torque. This friction torque, combined with the inertia of the pad, would result in slow movement and slow reaction to misalignment. In fact, in large hydrostatic pads the reaction is too slow to adequately compensate the variable misalignment during the rotation of the trunnion.

To improve the self-alignment performance, part of the load on the metal ball is relieved by hydrostatic pressure. The bottom part of the pad has been designed as a piston and is pressurized by oil pressure. The oil pressure relieves a portion of the load on the metal ball, and in turn the undesired friction torque is significantly reduced, as shown schematically in Fig. 10-13c.

In Fig. 10-13b, the radial positions of two inner pads determine the location of the axis of rotation of the trunnion; therefore, these two pads do not require radial adjustments, and they are referred to as *master shoe pads*. Any additional shoe pads require radial adjustment and are referred to as *slave shoes*. The design of the master and slave shoe pads with the hydraulic connections is shown in Fig. 10-13d.

In the slave shoe, there is radial adjustment of the pad clearance with the trunnion surface. The radial adjustment requires an additional lower piston, as shown in Fig. 10-13d. The radial motion of the lower piston is by means of hydrostatic oil pressure. The oil is connected by an additional duct to the space beneath the lower pad. There is a hydraulic duct connection, and the pressures are equalized in the two spaces below the two pads and in the pad recess (in contact with the trunnion surface) of the master and slave shoes. Since there is a constant flow rate, this equal pressure is a load-dependent pressure. If the area of the lower piston is larger than the effective pad area, the lower piston will push the piston and shoe pad in the radial direction (in the slave shoe) and adjust the radial clearance with the trunnion until equal load capacity is reached in all pads.

The recess pressure is a function of the load and the pad effective area. As the load increases, the film thickness diminishes and the pressure rises. It is desirable to limit the pressure and the size of the pad. This can be achieved by increasing the number of pads.

When the oil pump is turned off, the pads with the pistons return to the initial position, where the pistons rest completely on the metal ball. The pistons of the slave shoes must have sufficient freedom of movement in the radial direction of the trunnion; therefore, only the master shoes carry the load when the hydraulic system is not under pressure. To minimize this load, the master shoes are placed in center positions between the slave shoes when four or six shoes are used.

The combination of a master shoe and slave shoe operates as follows: The effective areas of the two pads are equal. If the clearance is the same in both shoes, the hydrostatic recess pressures must also be equal in the two pads. In this case, the load on both shoe pads is equalized.

There is hydraulic connection between the lower piston cylinders of the master and slave recesses (both are supplied by one pump). In this way, the load-dependent pressure in the piston cylinder of the slave shoe will be the same as that in the master shoe, resulting in equal load capacity of each shoe pad at all times.

This design can operate with certain deviation from roundness of the trunnion. For example, if there is a depression (reduced radius) in the trunnion surface, when this depression passes the pads of the slave shoe, the pressure at the recess of this pad drops. At the same time, the master shoe pad has not yet been affected by the depression and the pad of the master shoe will carry most of the load for a short duration. After this disturbance, the pressure at the slave pad would rise and lift the piston until there are equal recess pressures and load capacity in the two pads. Similarly, if there is a bulge (increased radius) on the surface of the trunnion, the process is reversed. In this way, the radial loads on the master and slave shoes are automatically controlled to be equal (with a minimal delay time). In conclusion, the clearances between the pads and trunnion surface are automatically controlled to be equal even if the trunnion is not perfectly round.

Cross-sectional views of the slave and master shoes and an isometric view of the slave shoe are shown in Fig. 10-14. In the master pad, there is one oil inlet and there is a hydraulic connection to the slave shoe. The pad recess of the slave and master shoes is of a unique design. For stable operation, it is important that the pad angular misalignment be immediately corrected. Each pad has one large circular recess and four additional recesses at each corner, all hydraulically connected. The purpose of this design is to have higher hydrostatic restoring torque for fast correction of any misalignment error. Oil is supplied at equal pressure to all the recesses (see Fig. 10-15).

For the large hydrostatic pads in rotary mills, each recess is fed at a constant flowrate, and flow restrictors are not used. Large hydrostatic pads consume a lot of power for the circulation of oil, and flow restrictors considerably increase power losses. The preferred design is to use one central pump with flow dividers. A standby pump in parallel is usually provided, to prevent loss of production in

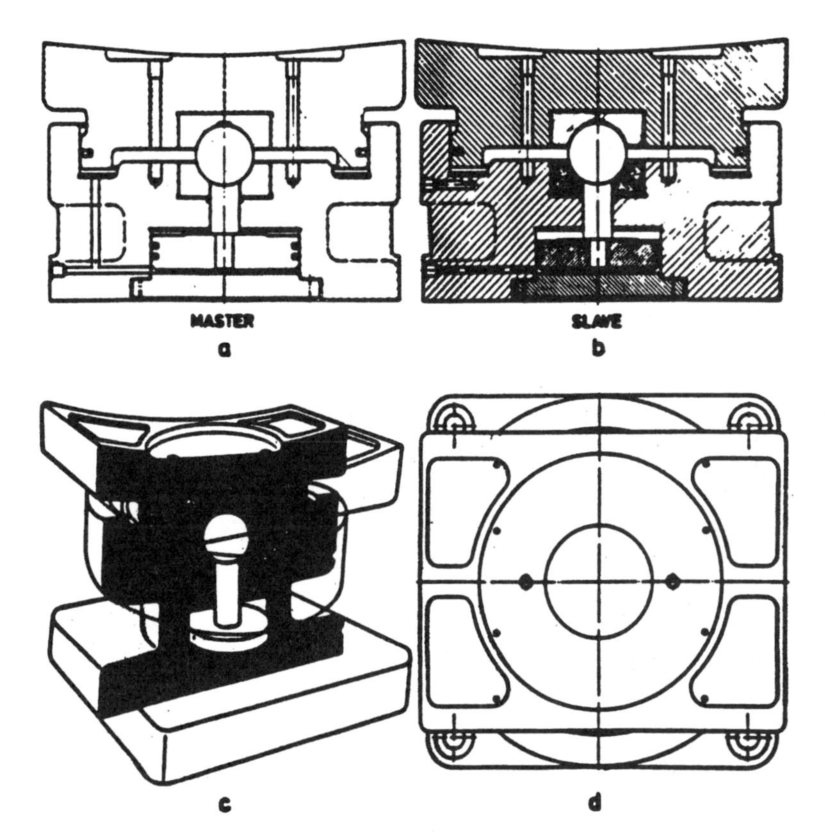

FIG. 10-14 Cross sections and isometric views of master and slave shoe pads. (From Trygg and McIntyre (1982), reprinted with permission from CIM Bulletin.)

case of oil pump failure. The trunnion bearings are continually monitored, to prevent early failure due to unexpected conditions. The bearing temperature and the hydrostatic pressure are continually recorded, and a warning system is set off whenever these values exceed an acceptable limit. In addition, the operation of the mill is automatically cut off when the hydrostatic bearing loses its pressure.

The hydraulic supply system is shown in Fig. 10-15. Each shoe is fed at a constant flow rate by four flow dividers. One positive-displacement pump is used to pump the oil from the oil tank. The oil is fed into the flow dividers through an oil filter, relief valve, and check valve. An accumulator is used to reduce the pressure fluctuations involved in positive-displacement pumps. Four pumps can be used as well. The oil returns to the sump and passes to the oil tank. Additional safety devices are pressure sensors provided to ensure that if the supply pressure

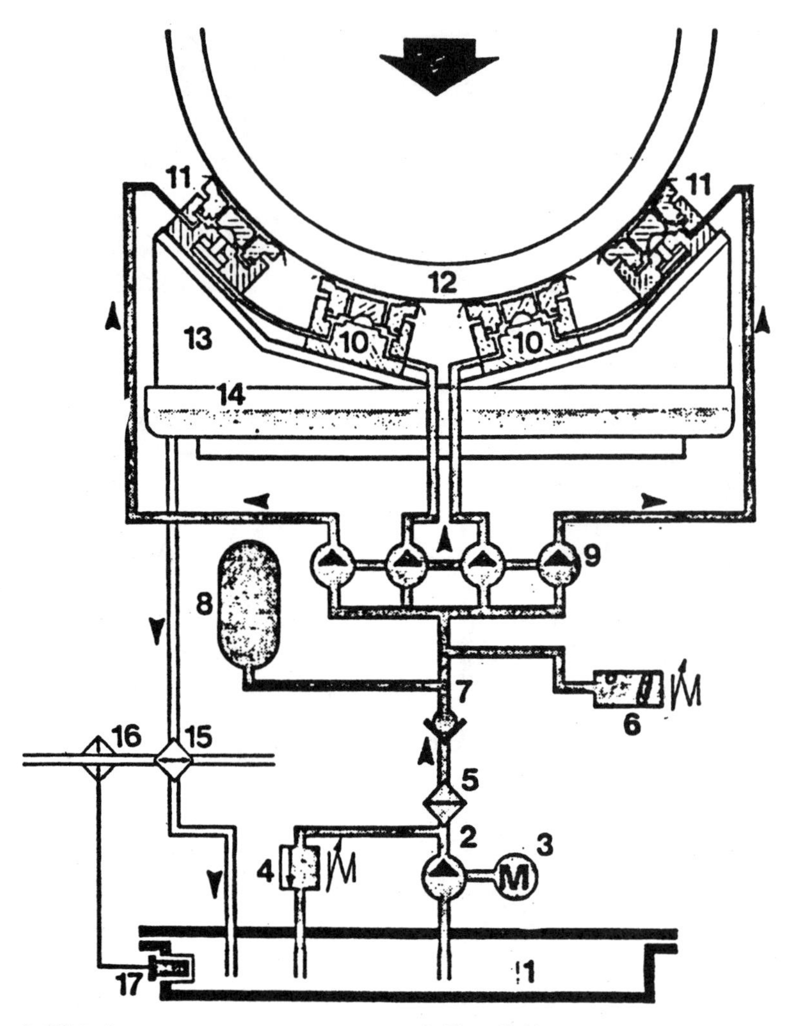

1. Oil tank
2. Pump
3. Electric motor
4. Relief valve
5. Oil filter
6. Pressure switch monitoring the pump motor
7. Check valve
8. Accumulator
9. Flow divider
10. Master shoe
11. Slave shoe
12. Girth ring
13. Pedestal for the shoes
14. Oil sump
15. Oil cooler
16. Valve for cooling water
17. Thermostat

FIG. 10-15 Hydraulic system for hydrostatic pad shoes. (From Trygg and McIntyre (1982), reprinted with permission from CIM Bulletin.)

drops, the mill rotation is stopped. Temperature monitoring is included to protect against overheating of the oil.

10.16.2 Advantages of Self-Aligning Hydrostatic Shoe Pads

Several publications related to the manufacture of these self-adjusting hydrostatic pads claim that there are major advantages in this design: It made it possible to significantly reduce the cost and to reduce the weight of the bearing and trunnion as well as the length of the complete mill in comparison to hydrodynamic bearings. Most important, it improved the bearing performance, namely, it reduced significantly the probability of bearing failure or excessive wear.

The concept of this design is to apply self-aligning hydrostatic shoes, preferably four shoes for each bearing. One important advantage of the design is that the length of the trunnion is much shorter in comparison to that in hydrodynamic bearing design. The shortening of the trunnion results in several advantages.

1. It reduces the weight of the trunnion and thus reduces the total weight of the mill.
2. It simplifies the feed into and from the mill.
3. It reduces the total length of the mill and its weight, resulting in reduced bending moment, and thus the mill can be designed to be lighter. It will, in fact, reduce the cost of the materials and labor for construction of the mill.
4. It reduces the elastic deformation, in the radial direction, of the trunnion.
5. Stiffer trunnion has significant advantages in bearing operation, because it reduces roundness errors; namely, it reduces elastic deformation to an elliptical shape.

In addition to shorter trunnion length, this design eliminates expensive castings followed by expensive precise machining, which are involved in the manufacturing process of the conventional hydrodynamic design. In this case, the casting can be replaced by a relatively low-cost welded construction. The hydrostatic shoes are relatively small, and their machining cost is much lower in comparison to that of large bearings. Moreover, the hydrostatic design operates with a thicker oil film and provides self-aligning bearings in all directions. These improvements prevent unexpected failures due to excessive wear or seizure. This aspect is important because of the high cost involved in rotary mill repair as well as loss of production.

Problems

10-1 A circular hydrostatic pad as shown in Fig. 10-1 has a constant supply pressure, p_s. The circular pad is supporting a load of $W = 1000$ N. The outside disk diameter is 200 mm, and the diameter of the circular recess is 100 mm. The oil is SAE 10 at an operating temperature of 70°C, having a viscosity of $\mu = 0.01$ N-s/m^2. The pad is operating with a clearance of 120 μm.

a. Calculate the flow rate Q of oil through the bearing to maintain the clearance of 120 μm.
b. Find the recess pressure, p_r.
c. Find the effective area of this pad.
d. If the supply pressure is twice the recess pressure, $p_s = 2p_r$, find the stiffness of the circular pad.

10-2 A circular hydrostatic pad, as shown in Fig. 10-1, has a constant flow rate Q. The circular pad is supporting a load of $W = 1000$ N. The outside disk diameter is 200 mm, and the diameter of the circular recess is 100 mm. The oil is SAE 10 at an operating temperature of 70°C, having a viscosity of $\mu = 0.01$ N-s/m^2. The pad is operating with a clearance of 120 μm.

a. Calculate the constant flow rate Q of oil through the bearing to maintain the clearance of 120 μm.
b. Find the recess pressure, p_r.
c. Find the effective area of this pad.
d. For a constant flow rate, find the stiffness of the circular pad operating under the conditions in this problem.

10-3 A long rectangular hydrostatic pad, as shown in Fig. 10-3, has constant flow rate Q. The pad is supporting a load of $W = 10{,}000$ N. The outside dimensions of the rectangular pad are: length is 300 mm and width is 60 mm. The inside dimensions of the central rectangular recess are: length is 200 mm and width is 40 mm. The pad is operating with a clearance of 100 μm. The oil is SAE 20 at an operating temperature of 60°C. Assume that the leakage in the direction of length is negligible in comparison to that in the width direction (the equations for two-dimensional flow of a long pad apply).

a. Calculate the constant flow rate Q of oil through the bearing to maintain the clearance of 100 μm.
b. Find the recess pressure, p_r.

c. Find the effective area of this pad.
d. For a constant flow rate, find the stiffness of the rectangular long pad operating under the conditions in this problem.

10-4 A slider-plate in a machine tool is supported by four bidirectional hydrostatic circular pads. Each recess is fed by a separate pump and has a constant flow rate. Each bidirectional pad is as shown in Fig. 10-3 (but it is a circular and not a rectangular pad). The weight of the slider is 20,000 N, or 5000 N on each pad. The total manufactured clearance between the two pads $(h_1 + h_2)$ is 0.4 mm. Each circular pad is of 100-mm diameter and recess diameter of 50 mm. R = 50 mm and $R_0 = 25$ mm. The oil viscosity is 0.01 N-s/m^2. In order to minimize vertical displacement, the slider plate is prestressed. The reaction force at the top is $W_1 = 5000$ N, and the reaction at the bottom is $W_2 = 10{,}000$ N (reaction to the top bearing reaction plus weight).

a. Find Q_1 and Q_2 in order that the two clearances will be equal $(h_1 = h_2)$.
b. If the flow rate is the same at the bottom and top pads, find the magnitude of the two clearances, h_1 and h_2. What is the equal flow rate, Q, into the two pads?
c. For the first case of equal clearances, find the stiffness of each pad. Add them together for the stiffness of the slider.
d. For the first case of equal clearances, if we place an extra vertical load of 40 N on the slider (10 N on each pad), find the downward vertical displacement of the slider.

10-5 In a machine tool, hydrostatic bearings support the slide plate as shown in Fig. 10-4. The supply pressure reaches each recess through a flow restrictor. The hydrostatic bearings are long rectangular pads. Two bidirectional hydrostatic pads are positioned along the two sides of the slider plate. The weight of the slider is 10,000 N, or 5000 N on each pad. The total manufactured clearance between the two pads $(h_1 + h_2) = 0.4$ mm. The oil viscosity is 0.05 N-s/m^2. For minimizing vertical displacement, the slider plate is prestressed. The reaction of each pad from the top is 5000 N, and the reaction from the bottom of each pad is 10,000 N (reaction to the top bearing reaction plus half slider weight). In order to have equal recess pressure in all pads, the dimensions of the widths of the bottom pad are double those in the top pad. The dimensions of each rectangular pad are as follows.

Top rectangular pad dimensions: 400 mm long and 30 mm wide. A rectangular recess is centered inside the rectangular pad, and its dimensions are 360-mm length and 10-mm width.

Bottom rectangular pad dimensions: 400-mm length and 60-mm width. A rectangular recess is centered inside the rectangular pad and its dimensions are 360-mm length and 20-mm width.

a. Find the recess pressure, p_r, at the bottom and top recesses.
b. The supply pressure from one pump is twice the recess pressure, $p_s = 2p_r$. Find the supply pressure.
c. Find the flow resistance, R_{in}, at the inlet of the bottom and top recesses in order that the two clearances will be equal ($h_1 = h_2$).
d. The flow resistance is made of a capillary tube of 1-mm ID. Find the length, l_c, of the capillary tube at the inlet of the bottom and top recesses.
e. Find the flow rates Q_1 and Q_2 into the bottom and top recesses.
f. For equal clearances, find the stiffness of each pad. Add them together for the stiffness of the bi-directional pad.
g. If we place an extra vertical load of 60 N on the slider (30 N on each bidirectional pad), find the vertical displacement (down) of the slider.

10-6 A long rectangular hydrostatic pad, as shown in Fig. 10-3, has a constant supply pressure, p_s. The pressure is fed into the recess through flow restrictors. The pad supports a load of $W = 20{,}000$ N. The outside dimensions of the rectangular pad are: length is 300 mm and width is 60 mm. The inside dimensions of the central rectangular recess are: length is 200 mm and width is 40 mm. The pad is designed to operate with a minimum clearance of 100 μm. The oil is SAE 30 at an operating temperature of 60°C. Assume that the equations for two-dimensional flow of a long pad apply.

a. Calculate the flow rate Q of oil through the bearing to maintain the clearance of 100 μm.
b. Find the recess pressure, p_r.
c. Find the effective area of this pad.
d. If the supply pressure is twice the recess pressure, $p_s = 2p_r$, find the stiffness of the pad.

11

Bearing Materials

11.1 FUNDAMENTAL PRINCIPLES OF TRIBOLOGY

During the twentieth century, there has been an increasing interest in the friction and wear characteristics of materials. The science of friction and wear of materials has been named *Tribology* (the science of rubbing). A lot of research has been conducted that resulted in significant progress in the understanding of the fundamental principles of friction and wear of various materials. Several journals are dedicated to the publication of original research in this subject, and many reference books have been published where the research findings are presented. The most important objective of the research in tribology is to reduce friction and wear as well as other failure modes in bearings. On the other hand, there are many important applications where it is desirable to maximize friction, such as in brakes and in the friction between tires and road.

The following is a short review of the fundamental principles of tribology that are important to practicing engineers. More detailed coverage of the research work in tribology has been published in several books that are dedicated to this subject. Included in the tribology literature are books by Bowden and Tabor (1956), Rabinowicz (1965), Bowden and Tabor (1986), Blau (1995), and Ludema (1996).

It is well known that sliding surfaces of machine elements have a certain degree of surface roughness. Even highly polished surfaces are not completely smooth, and this roughness can be observed under the microscope or measured by a profilometer. The surface roughness is often compared to a mountainous terrain, where the hills are referred to as *surface asperities*. The root mean square (RMS) of the surface roughness is often used to identify the surface finish, and it can be measured by a profilometer. The RMS roughness value of the best-polished commercial surfaces is about 0.01–250 μm (micrometers). Sliding surfaces are separated by the asperities; therefore, the actual contact area between two surfaces exists only at a few points, where contacts at the tip of the asperities take place. Each contact area is microscopic, and its size is of the order of 10–50 μm. Actual total contact area, A_r, that supports the load is very small relative to the apparent area (by several orders of magnitude).

A very small contact area at the tip of the surface asperities supports the external normal load, F, resulting in very high compression stresses at the contact. The high compression stresses cause elastic as well as plastic deformation that forms the actual contact area, A_r. Experiments indicated that the actual contact area, A_r, is proportional to the load, F, and the actual contact area is not significantly affected by the apparent size of the surface. Moreover, the actual contact area, A_r, is nearly independent of the roughness value of the two surfaces. Under load, the contact area increases by elastic and plastic deformation. The deformation continues until the contact area and compression strength p_h (of the softer material) can support the external load, F. The ultimate compression stress that the softer material can support, p_h, depends on the material hardness, and the equation for the normal load is

$$F = p_h A_r \tag{11-1}$$

The compression strength, p_h, is also referred to as the *penetration hardness*, because the penetration of the hard asperity into the soft one is identical to a hardness test, such as the Vickers test. For elastic materials, p_h is about three times the value of the compression yield stress (Rabinovitz, 1965).

11.1.1 Adhesion Friction

The recent explanation of the friction force is based on the theory of adhesion. Adhesion force is due to intermolecular forces between two rubbing materials. Under high contact pressure, the contact areas adhere together in the form of microscopic junctions. The magnitude of a microscopic junction is about 10–50 μm; in turn, the friction is a continual process of formation and shearing of the microscopic junctions. The tangential friction force, F_f, is the sum of forces required for continual shearing of all the junction points. This process is repeated continually as long as an external tangential force, F_f, is provided to break the adhesion contacts to allow for a relative sliding. The equation for the friction force is

$$F_f = \tau_{av} A_r \qquad (11\text{-}2)$$

Here, τ_{av} is the average shear stress required for shearing the adhesion joints of the actual adhesion contacts of the total area A_r. Equation (11-2) indicates that, in fact, the friction force, F_f, is proportional to the actual total contact area, A_r, and is not affected by the apparent contact area. This explains the Coulomb friction laws, which state that the friction force F_t is proportional to the normal force F_n. The adhesion force is proportional to the actual area A_r, which, in turn, is proportional to the normal force, F, due to the elasticity of the material at the contact. When the normal force is removed, the elastic deformation recovers and there is no longer any friction force.

In many cases, the strength of the adhesion joint is higher than that of the softer material. In such cases, the shear takes place in the softer material, near the junction, because the fracture takes place at the plane of least resistance. In this way, there is a material transfer from one surface to another. The average shear strength, τ_{av}, is in fact the lower value of two: the junction strength and the shear strength of the softer material.

The adhesion and shearing of each junction occurs during a very short time because of its microscopic size. The friction energy is converted into heat, which is dissipated in the two rubbing materials. In turn, the temperature rises, particularly at the tip of the asperities. This results in a certain softening of the material at the contact, and the actual contact areas of adhesion increase, as does the junction strength.

11.1.2 Compatible Metals

A combination of two metals is *compatible* for bearing applications if it results in a low dry friction coefficient and there is a low wear rate. Compatible metals are often referred to as *score resistant*, in the sense that the bearing resists fast scoring, in the form of deep scratches of the surface, which results in bearing failure.

In general, two materials are compatible if they form two separate phases after being melted and mixed together; namely, the two metals have very low solid solubility. In such cases, the adhesion force is a relatively weak bond between the two surfaces of the sliding metals, resulting in low τ_{av}. In turn, there is relatively low friction force between compatible materials. On the other hand, when the two metals have high solubility with each other (can form an alloy), the metals are not compatible, and a high friction coefficient is expected in most cases (Ernst and Merchant, 1940). For example, identical metals are completely soluble; therefore, they are not compatible for bearing applications, such as steel on steel and copper on copper. Aluminum and mild steel are soluble and have a high friction coefficient. On the other hand, white metal (babbitt), which is an alloy of tin, antimony, lead, and copper, is compatible against steel. Steel journal and white metal bearings have low dry friction and demonstrate outstanding score resistance.

Roach, et al. (1956) tested a wide range of metals in order to compare their score resistance (compatibility) against steel. Table 11-1 summarizes the results by classifying the metals into compatibility classes of good, fair, poor, and very poor. Good compatibility means that the metal has good score resistance against steel.

In this table, the atomic number is listed before the element, and the melting point in degrees Celsius is listed after the element. Cadmium has been found to be an intermediate between "good" and "fair" and copper an intermediate between "fair" and "poor." The melting point does not appear to affect the compatibility with steel. Zinc, for example, has a melting point between those of lead and antimony, but has poorer compatibility in comparison to the two.

It should be noted that many metals that are classified as having a good compatibility with steel are the components of white metals (babbitts) that are widely used as bearing material. Roach et al. (1956) suggested an explanation that the shear strength at the junctions determines the score resistance. Metals that are mutually soluble tend to have strong junctions that result in a poor compatibility (poor score resistance).

However, there are exceptions to this rule. For example, magnesium, barium, and calcium are not soluble in steel but do not have a good score resistance against steel. Low friction and score resistance depends on several other factors. Hard metals do not penetrate into each other and do not have a high friction coefficient. Humidity also plays an important role, because the moisture layer acts as a lubricant.

Under light loads, friction results only in a low temperature of the rubbing surfaces. In such cases, the temperature may not be sufficiently high for the metals to diffuse into each other. In turn, there would not be a significant score of the surfaces, although the metals may be mutually soluble. In addition, it has been suggested that these types of bonds between the atoms, in the boundary of the two metals, play an important role in compatibility. Certain atomic bonds are more brittle, and the junctions break easily, resulting in a low friction coefficient.

11.1.3 Coulomb Friction Laws

According to Coulomb (1880), the tangential friction force, F_f, is not dependent on the sliding velocity or on the apparent contact area. However, the friction

TABLE 11-1 Elements Compatible with Steel, from Roach et al. (1956)

Good			Fair			Poor			Very Poor		
Atomic number	Element	Melting Temp. °C	Atomic number	Element	Melting Temp. °C	Atomic number	Element	Melting Temp. °C	Atomic number	Element	Melting Temp. °C
32	Germanium	958	6	Carbon		12	Magnesium	651	4	Beryllium	1280
47	Silver	960	34	Selenium	220	13	Aluminum	660	14	Silicon	1420
49	Indium	155	52	Tellurium	452	30	Zinc	419	20	Calcium	810
50	Tin	232	29	Copper	1083	56	Barium	830	22	Titanium	1800
51	Antimony	630				74	Tungsten	3370	24	Chromium	1615
81	Thallium	303							26	Iron	1535
82	Lead	327							27	Cobalt	1480
83	Bismuth	271							28	Nickel	1455
									40	Zirconium	1900
									41	Niobium (Columbium)	1950
									42	Molybdenum	2620
									45	Rhodium	1985
									46	Palladium	1553
									58	Cerium	640
									73	Tantalum	2850
									77	Iridium	2350
									78	Platinum	1773
									79	Gold	1063
									90	Thorium	1865
									92	Uranium	1130
48	Cadmium	321	29	Copper	1083						

force, F_f, is proportional to the normal load, F. For this reason, the friction coefficient, f, is considered to be constant and it is defined as

$$f = \frac{F_f}{F} \tag{11-3}$$

Equation (11-3) is applicable in most practical problems. However, it is already commonly recognized that the friction laws of Coulomb are only an approximation. In fact, the friction coefficient is also a function of the sliding velocity, the temperature, and the magnitude of the normal load, F.

Substituting Eqs. (11-1) and (11-2) into Eq. (11-3) yields the following expression for the friction coefficient:

$$f = \frac{F_f}{F} = \frac{\tau_{av}}{p_h} \tag{11-4}$$

Equation (11-4) is an indication of the requirements for a low friction coefficient of bearing materials. A desirable combination is of relative high hardness, p_h, and low average shear strength, τ_{av}. High hardness reduces the contact area, while a low shear strength results in easy breaking of the junctions (at the adhesion area or at the softer material). A combination of hard materials and low shear strength usually results in a low friction coefficient.

An example is white metal, which is a multiphase alloy with a low friction coefficient against steel. The hard phase of the white metal has sufficient hardness, or an adequate value of p_h. At the same time, a soft phase forms a thin overlay on the surface. The soft layer on the surface has mild adhesion with steel and can shear easily (low τ_{av}). This combination of a low ratio τ_{av}/p_h results in a low friction coefficient of white metal against steel, which is desirable in bearings. The explanation is similar for the low friction coefficient of cast iron against steel. Cast iron has a thin layer of graphite on the surface of very low τ_{av}. An additional example is porous bronze filled with PTFE, where a thin layer of soft PTFE, which has low τ_{av}, is formed on the surface.

Any reduction of the adhesive energy decreases the friction force. Friction in a vacuum is higher than in air. The reduction of friction in air is due to the adsorption of moisture as well as other molecules from the air on the surfaces. In the absence of lubricant, in most practical cases the friction coefficient varies between 0.2 and 1. However, friction coefficients as low as 0.05 can be achieved by the adsorption of boundary lubricants on metal surfaces in practical applications. All solid lubricants, as well as liquid lubricants, play an important role in forming a thin layer of low τ_{av}, and in turn, the friction coefficient is reduced.

In addition to adhesion friction, there are other types of friction. However, in most cases, adhesion accounts for a significant portion of the friction force. In most practical cases, adhesion is over 90% of the total friction. Additional types of the friction are plowing friction, abrasive friction, and viscous shear friction.

11.2 WEAR MECHANISMS

Unless the sliding surfaces are completely separated by a lubrication film, a certain amount of wear is always present. If the sliding materials are compatible, wear can be mild under appropriate conditions, such as lubrication and moderate stress. However, undesirably severe wear can develop if these conditions are not maintained, such as in the case of overloading the bearing or oil starvation. In addition to the selection of compatible materials and lubrication, the severity of the wear increases with the surface temperature. The bearing temperature increases with the sliding speed, V, because the heat, q, that is generated per unit of time is equal to the mechanical power needed to overcome friction. The power losses are described according to the equation

$$q = f\,FV \tag{11-5}$$

In the absence of liquid lubricant, heat is removed only by conduction through the two rubbing materials. The heat is ultimately removed by convection from the materials to the air. Poor heat conductivity of the bearing material results in elevated surface temperatures. At high surface temperatures, the friction coefficient increases with a further rise of temperature. This chain of events often causes scoring wear and can ultimately cause, under severe conditions, seizure failure of the bearing. The risk of seizure is particularly high where the bearing runs without lubrication. In order to prevent severe wear, compatible materials should always be selected. In addition, the PV value should be limited as well as the magnitudes of P and V separately.

11.2.1 Adhesive Wear

Adhesive wear is associated with adhesion friction, where strong microscopic junctions are formed at the tip of the asperities of the sliding surfaces. This wear can be severe in the absence of lubricant. The junctions must break due to relative sliding. The break of a junction can take place not exactly at the original interface, but near it. In this way, small particles of material are transferred from one surface to another. Some of these particles can become loose, in the form of wear debris. Severe wear can be expected during the sliding of two incompatible materials without lubrication, because the materials have strong adhesion.

For two rubbing metals, high adhesion wear is associated with high solid solubility with each other (such as steel on steel). Adhesion junctions are formed by the high contact pressure at the tip of the surface asperities. However, much stronger junctions are generated when the temperature at the junction points is relatively high. Such strong junctions often cause scoring damage. The source of elevated surface temperature can be the process, such as in engines or turbines, as well as friction energy that generates high-temperature hot spots on the rubbing surfaces. When surface temperature exceeds a certain critical value, wear rate will accelerate. This wear is referred to in the literature as *scuffing* or *scoring*, which can be identified by material removed in the form of lines along the sliding direction. Overheating can also lead to catastrophic bearing failure in the form of seizure.

11.2.2 Abrasion Wear

This type of wear occurs in the presence of hard particles, such as sand dust or metal wear debris between the rubbing surfaces. Also, for rough surfaces, plowing of one surface by the hard asperities of the other results in abrasive wear. In properly designed bearings, with adequate lubrication, it is estimated that 85% of wear is due to abrasion. It is possible to reduce abrasion wear by proper selection of bearing materials. Soft bearing materials, in which the abrasive particles become embedded, protect the shaft as well as the bearing from abrasion.

11.2.3 Fatigue Wear

The damage to the bearing surface often results from fatigue. This wear is in the form of pitting, which can be identified by many shallow pits, where material has been removed from the surface. This type of wear often occurs in line-contact or point-contact friction, such as in rolling-element bearings and gears. The maximum shear stress is below the surface. This often results in fatigue cracks and eventually causes peeling of the surface material.

In rolling-element bearings, gears, and railway wheels, the wear mechanism is different from that in journal bearings, because there is a line or point contact and there are alternating high compression stresses at the contact. In contrast, the surfaces in journal bearings are conformal, and the compression stresses are more evenly distributed over a relatively larger area. Therefore, the maximum compression stress is not as high as in rolling contacts, and adhesive wear is the dominant wear mechanism. In line and point contact, the surfaces are not conformal, and fatigue plays an important role in the wear mechanisms, causing pitting, i.e., shallow pits on the surface. Fatigue failure can start as surface cracks, which extend into the material, and eventually small particles become loose.

11.2.4 Corrosion Wear

Corrosion wear is due to chemical attack on the surface, such as in the presence of acids or water in the lubricant. In particular, a combination of corrosion and fatigue can often cause an early failure of the bearing.

11.3 SELECTION OF BEARING MATERIALS

A large number of publications describe the wear and friction characteristics of various bearing materials.* However, when it comes to the practical design and selection of materials, numerous questions arise concerning the application of this knowledge in practical situations. We must keep in mind that the bearing material is only one aspect of an integrated bearing design, and even the best and most expensive materials would not guarantee successful operation if the other design principles are ignored.

Although hydrodynamic bearings are designed to operate with a full oil film, direct contact of the material surfaces occurs during starting and stopping. Some bearings are designed to operate with boundary or mixed lubrication where there is a direct contact between the asperities of the two surfaces. Proper material combination is required to minimize friction, wear, and scoring damage in all bearing types, including those operating with hydrodynamic or hydrostatic fluid films. In a hydrostatic bearing, the fluid pressure is supplied by an external pump, and a full fluid film is maintained during starting and stopping. The hydrostatic fluid film is much thicker in comparison to the hydrodynamic one. However, previous experience in machinery indicates that the bearing material is important even in hydrostatic bearings. Experience indicates that there are always unexpected vibrations and disturbances as well as other deviations from normal operating conditions. Therefore, the sliding materials are most likely to have a direct contact, even if the bearing is designed to operate as a full hydrodynamic or hydrostatic bearing. For example, in a certain design of a machine tool, the engineers assumed that the hydrostatic bearing maintains full film lubrication at all times, and selected a steel-on-steel combination. However, the machines were recalled after a short operating period due to severe bearing damage.

There is a wide range of bearing materials to select from—metals, plastics, and composite materials—and there is no one ideal bearing material for all cases. The selection depends on the application, which includes type of bearing, speed, load, type of lubrication, and operating conditions, such as temperature and maximum contact pressure.

In general, a bearing metal should have balanced mechanical properties. On the one hand, the metal matrix should be soft, with sufficient plasticity to conform to machining and alignment errors as well as to allow any abrasive particles in the lubricant to be embedded in the bearing metal. On the other hand, the metal should have sufficient hardness and compression strength, even at high operating temperature, to avoid any creep and squeezing flow of the metal under load, as well as having adequate resistance to fatigue and impact. The selection is a

*Examples are Kennedy et al. (1998), Kingsbury (1997), Blau (1992), Booser (1992), Kaufman (1980), and Peterson and Winer (1980).

tradeoff between these contradictory requirements. For the manufacture of the bearing, easy melting and casting properties are required. In addition, the bearing metal must adhere to the steel shell and should not separate from the shell by metal fatigue.

The following is a discussion of the most important performance characteristics that are usually considered in the selection of bearing materials.

11.3.1 Score and Seizure Resistance

Compatibility between two materials refers to their ability to prevent scoring damage and seizure under conditions of friction without adequate lubrication. Compatible materials demonstrate relatively low friction coefficient under dry and boundary lubrication conditions. In metals, junctions at the tip of the surface asperities are formed due to high contact pressure. When these junctions are torn apart by tangential friction force, the surfaces are scored. The high friction energy raises the surface temperature, and in turn stronger junctions are formed, which can result in bearing failure by seizure. Similar metals are not compatible because they tend to have relatively high friction coefficients, e.g., steel on steel. A more compatible combination would be steel on bronze or steel on white metal. Most plastic bearings are compatible with steel shafts.

11.3.2 Embeddability

This is an important characteristic of soft bearing materials, where small hard particles become embedded in the bearing material and thus prevent abrasion damage. Dust, such as silica, and metal particles (wear products) are always present in the oil. These small, hard particles can cause severe damage, in the form of abrasion, particularly when the oil film is very thin, at low speeds under high loads. The abrasion damage is more severe whenever there is overheating of the bearing. When the hard particles are embedded in the soft bearing metal, abrasion damage is minimized.

11.3.3 Corrosion Resistance

Certain bearing metals are subject to corrosion by lubricating oils containing acids or by oils that become acidic through oxidation. Oil oxidation takes place when the oil is exposed to high temperatures for extended periods, such as in engines. Oxidation inhibitors are commonly added to oils to prevent the formation of corrosive organic acids. Corrosion fatigue can develop in the bearing metal in the presence of significant corrosion. Corrosion-resistant materials should be applied in all applications where corrosives may be present in the lubricant or the environment. Improved alloys have been developed that are more corrosion resistant.

11.3.4 Fatigue Resistance

In bearings subjected to oscillating loads, such as in engines, conditions for fatigue failure exist. Bearing failure starts, in most cases, in the form of small cracks on the surface of the bearing, which extend down into the material and tend to separate the bearing material from the housing. When sufficiently large cracks are present in the bearing surface, the oil film deteriorates, and failure by overheating can be initiated. It is impossible to specify the load that results in fatigue, because many operating parameters affect the fatigue process, such as frequency of oscillating load, metal temperature, design of the bearing housing, and the amount of journal flexure. However, materials with high fatigue resistance are usually desirable.

11.3.5 Conformability

This is the ability to deform and to compensate for inaccuracy of the bearing dimensions and its assembly relative to the journal. We have to keep in mind that there are always manufacturing tolerances, and metal deformation can correct for some of these inaccuracies. An example is the ability of a material to conform to misalignment between the bearing and journal. The conformability can be in the form of plastic or elastic deformation of the bearing and its support. The characteristic of having large plastic deformation is also referred to as *deformability*. This property indicates the ability of the material to yield without causing failure. For example, white metal, which is relatively soft, can plastically deform to correct for manufacturing errors.

11.3.6 Friction Coefficient

The friction coefficient is a function of many parameters, such as lubrication, temperature, and speed. Proper selection of the rubbing materials is important, particularly in dry and boundary lubrication. A low coefficient of friction is usually desirable in most applications. In most cases, a combination of hard and soft materials results in a low friction coefficient. A low coefficient of friction is usually related to the compatibility (score-resistance) characteristic, where partial welding of the surface asperities occurring at hot spots on the rubbing surfaces can increase friction.

11.3.7 Porosity

This property indicates the ability of the material to contain fluid or solid lubricants. An example of a porous metal is sintered bronze, which can be impregnated with oil or white metal. Such porous bearings offer a significant advantage, of reduction in maintenance cost in applications of boundary lubrication, where only a small amount of lubricant is required.

11.3.8 Thermal Conductivity

For most applications, a relatively high thermal conductivity improves the performance of the bearing. The friction energy is dissipated in the bearing as heat, and rapid heat transfer reduces the operating temperature at the sliding contact.

11.3.9 Thermal Expansion

The thermal expansion coefficient is an important property in bearing design. It is desirable that the thermal expansion of the bearing be greater than the journal, to reduce the risk of thermal seizure. However, if the expansion coefficient is excessively large in comparison to that of the steel shaft, a very large clearance would result, such as in plastic bearings. Unique designs with elastic flexibility are available to overcome the problem of overexpansion.

11.3.10 Compressive Strength

A high compressive strength is required for most applications. The bearing should be capable of carrying the load at the operating temperature. This characteristic is in conflict with that of conformability and embeddability. Usually for high compressive strength, high-hardness bearing material is required. However, for conformability and embeddability, relatively low hardness values are desired.

11.3.11 Cost

The bearing material should be cost effective for any particular application. To reduce the bearing cost, material should be selected that can be manufactured in a relatively low-cost process. For metals, easy casting and machining properties are desirable to reduce the manufacturing cost. For metal bearings, a bronze bushing is considered a simple low-cost solution, while silver is the most expensive. Plastic bearings are widely used, primarily for their low cost as well as their low-cost manufacturing process. Most plastic bearings are made in mass production by injection molding.

11.3.12 Manufacturing

Consideration should be given to the manufacturing process. Bearing metals must have a relatively low melting point and have good casting properties. Also, they should exhibit good bonding properties, to prevent separation from the backing material during operation.

11.3.13 Classification of Bearing Materials

Bearing materials can be metallic or nonmetallic. Included in the metallic category are several types of white metals (tin and lead-based alloys), bronzes, aluminum alloys, and porous metals. Certain thin metallic coatings are widely used, such as white metals, silver, and indium. The nonmetallic bearing materials include plastics, rubber, carbon-graphite, ceramics, cemented carbides, metal oxides, glass, and composites, such as glass-fiber- and carbon-fiber-reinforced PTFE (Teflon).

11.4 METAL BEARINGS

White Metal: Tin- and Lead-Based Alloys (Babbitts)

Isaac Babbitt invented and in 1839 obtained a U.S. patent on the use of a soft white alloy for a bearing. This was a tin-based alloy with small amounts of added copper, antimony, and lead. These alloys are often referred to as *babbitts*. The term *white metal* is used today for tin- as well as lead-based alloys. A white metal layer is cast as a bearing surface for steel, aluminum, bronze, or cast iron sleeves. White metal can undergo significant plastic deformation, resulting in excellent embeddability and conformability characteristics. Hard crystals are dispersed in the soft matrix and increase the hardness of the alloy, but they do not have a significant adverse effect on the frictional properties because the soft matrix spreads out on the surface during sliding to form a thin lubricating film. This results in a low friction coefficient, since the shearing stress of the soft matrix is relatively low. The limit to the use of white metals is their relatively low melting temperature. Also, there are limits to the magnitude of steady compression pressure—$7\,N/mm^2$ (7 Mpa)—and much lower limits whenever there is fatigue under oscillating loads. Using very thin layers of white metal can extend the limit. Of course, the maximum loads must be reduced at elevated temperature, such as in engines, where the temperature is above 100°C, where white metal loses nearly 50% of its compression strength. Specifications and tables of properties of white metals are included in ASTM B23 (1990).

White metal has considerable advantages as a bearing material, and it is recommended as the first choice for most applications. In order to benefit from these advantages, the design should focus on limiting the peak pressure and maximum operating temperature. Soft sleeve materials can tolerate some misalignment, and dust particles in the oil can be embedded in the soft material, thus preventing excessive abrasion and wear. However, white metal has a relatively low melting point and can creep if the maximum pressure is above its compressive strength. Thin white metal linings offer better resistance to creep and fatigue; see Fig. 11-1.

At the beginning of the twentieth century, white metal linings were much thicker (5 mm and more) in comparison to current applications. The requirement to reduce the size of machines resulted in smaller bearings that have to support higher compressive loads. Also, faster machines require bearings with greater fatigue strength. These requirements were met by reducing the thickness of the white metal lining to 800 μm, and in heavy-duty applications to as low as 50–120 μm. Fatigue strength is increased by decreasing the thickness of the white metal lining. The reduction of thickness is a tradeoff between fatigue resistance and the properties of embeddabilty and conformability of the thicker white metal lining. For certain applications, thick layers of white metal are still applied successfully.

In automotive engines, a very thin lining, of thickness below 800 μm, is commonly applied. Tin-based white metal has been used exclusively in the past, but now has been replaced in many cases by the lower-cost lead-based white metal. The yield point of the lead-based white metal is lower, but when the white metal layer is very thin on a backing lining of good heat-conductor metals, such as aluminum or copper-lead, the lead-based white metal bearings give satisfactory performance.

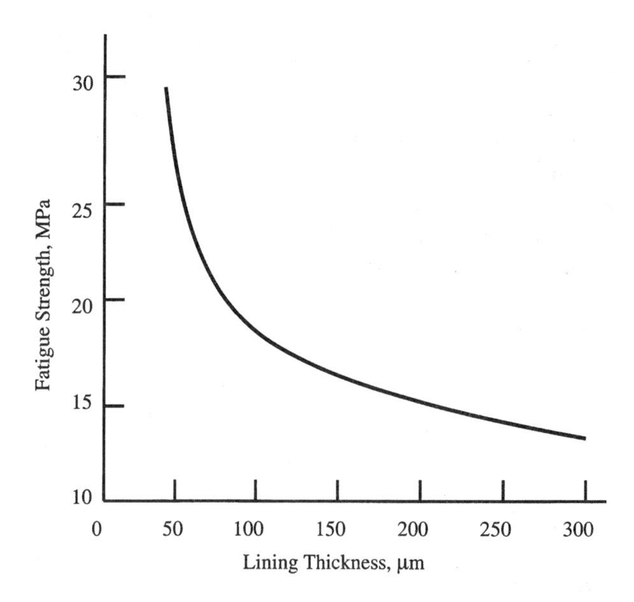

FIG. 11-1 Fatigue resistance as a function of white metal thickness.

One advantage of the white metal is good adhesion to the shell material, such as steel or bronze. Also, it has better seizure resistance in comparison to harder materials, in the case of oil starvation or during starting and stopping. A thick wall lining has the advantage that the sleeve can be replaced (in most cases by centrifugal casting). For large bearings, there is an additional advantage in applying a thick layer, since the white metal can be scraped and fitted to the journal during assembly of the machine. Therefore, a thick white metal layer is still common in large bearings. White metal has been considered the best bearing material, and the quality of other bearing materials can be determined by comparison to it.

11.4.2 Tin-Based Versus Lead-Based White Metals

The advantages of tin-based white metals in comparison to their lead-based counterparts include higher thermal conductivity, higher compression strength, higher fatigue and impact strength, and higher corrosion resistance. On the other hand, lead-based white metals exhibit a lower friction coefficient, better bonding to the shells, and better properties for casting. However, the increase in use of lead-based white metal is attributed mostly to its lower cost.

11.4.3 Copper–Lead Alloys

These alloys contain from 28% to 40% Pb. They are used primarily in the automotive and aircraft industries. They are also used in general engineering applications. They have a higher load capacity and higher fatigue resistance in comparison to white metal. Also, they can operate at higher temperatures. But they have a relatively lower antiseizure characteristic. These alloys are usually cast or sintered to a steel backing strip. The higher-lead-content alloy is used on steel or cast iron–backed bearings. These are commonly used for medium-duty automotive bearings. In order to maintain the soft copper matrix, the tin content in these alloys is restricted to a low level. The higher lead content improves the corrosion and antiseizure properties. However, in most applications, the corrosion and antiseizure properties are improved by a thin lead-tin or lead-indium overlay.

In engine bearings, bare copper-lead bearings are no longer common. Corrosive acids that are formed in the crankcase lubricant attack the lead material. Many of the copper-lead alloys, with lead contents near 25%, are plated with additional overlays. This forms the three-layer bearing—a steel backing covered by a layer of copper-lead alloy and a thin overlay of lead-tin or lead-indium. Such three-layer metal bearings are widely applied in automotive and diesel engines. Sintered and impregnated porous alloys are included in this group, such as SAE 482, 484, and 485.

11.4.4 Bronze

All bronzes can be applied as bearing materials, but the properties of bronzes for bearings are usually improved by adding a considerable amount of lead. Lead improves the bearing performance by forming a foundation for the hard crystals. But lead involves manufacturing difficulties, since it is not easily kept in solution and its alloys require controlled casting.

Bronzes with about 30% lead are referred to as *plastic bronze*. This significant lead content enhances the material's friction properties. However, the strength and hardness are reduced. These bronzes have higher strength than the white metals and are used for heavy mill bearings. A small amount of nickel in bearing bronze helps in keeping the lead in solution. Also, the resistance to compression and shock is improved. Iron content of up to 1% improves the resistance to shock and hardens the bronze. However, at the same time it reduces the grain size and tends to segregate the lead.

11.4.5 Cast Iron

In most applications, the relatively high hardness of cast iron makes it unsuitable as a bearing material. But in certain applications it is useful, particularly for its improved seizure resistance, caused by the graphite film layer formed on its surface. The most important advantages of cast iron are a low friction coefficient, high seizure resistance, high mechanical strength, the formation of a good bond with the shell, and, finally, low cost.

11.4.6 Aluminum Alloys

Aluminum alloys have two important advantages. The major advantage is their high thermal conductivity (236 W/m°C). They readily transfer heat from the bearing, resulting in a lower operating temperature of the bearing surface. The second advantage is their high compressive strength [34 Mpa (5000 psi)]. The aluminum alloys are widely used as a backing material with an overlay of white metal.

Examples of widely used aluminum alloys in automotive engines are an alloy with 4% silicon and 4% cadmium, and alloys containing tin, nickel, copper, and silicon. Also, aluminum-tin alloys are used, containing 20% to 30% tin, for heavily loaded high-speed bearings. These bearings are designed, preferably with steel backings, to conserve tin, which is relatively expensive, as well as to add strength. Addition of 1% copper raises the hardness and improves the physical qualities. The limit for the copper component is usually 3%. The copper is alloyed with the aluminum. However, the tin exists in the form of a continuous crystalline network in the aluminum alloy. These alloys must have a matrix of aluminum through which various elements are dispersed and not dissolved.

11.4.7 Silver

Silver is used only in unique applications in which the use of silver is required. Its high cost prohibits extensive application of this material. One example of an important application is the connecting rod bearings in aircraft engines. The major advantages of silver are its high thermal conductivity and excellent fatigue resistance. However, other mechanical properties of silver are not as good. Therefore, a very thin overlay of lead or lead-indium alloy (thickness of 25–100 μm) is usually applied to improve compatibility and embeddability.

11.4.8 Porous Metal Bearings

Porous metal bearings, such as porous sintered bronze, contain fluid or solid lubricants. The porous material is impregnated with oil or a solid lubricant, such as white metal. A very thin layer of oil or solid lubricant migrates through the openings to the bearing surface. These bearings are selected for applications where boundary lubrication is adequate with only a small amount of lubricant. Porous bearings impregnated with lubricant offer a significant advantage of reduction of maintenance cost.

11.5 NONMETAL BEARING MATERIALS

Nonmetallic materials are widely used for bearings because they offer diversified characteristics that can be applied in a wide range of applications. Generally, they have lower heat conductivity, in comparison to metals; therefore, they are implemented in applications that have a low *PV* (load–speed product) value. Nonmetallic bearings are selected where self-lubrication and low cost are required (plastic materials) and where high temperature stability must be maintained as well as chemical resistance (e.g., carbon graphite). Nonmetallic bearing materials include the following groups:

Plastics: PTFE (Teflon), nylon, phenolics, fiber-reinforced plastics, etc.
Ceramics
Carbon Graphite
Rubber
Other diverse materials, such as wood and glass

11.5.1 Plastic Bearing Materials

Lightly loaded bearings are fabricated mostly from plastic materials. Plastics are increasingly used as bearing materials, not only for the appropriate physical characteristics, but also for their relatively low cost in comparison to metal bearings. Many polymers, such as nylon, can be formed to their final shape by injection molding, where large quantities are manufactured in mass production, resulting in low unit manufacturing cost. Plastic bearings can be used with or without liquid lubrication. If possible, liquid lubrication should be applied, because it reduces friction and wear and plays an important role in cooling the bearing. Whenever liquid lubrication is not applied, solid lubricants can be blended into the base plastics to reduce friction, often referred to as improving the *lubricity* of plastics.

Polymer is synthetically made of a monomer that is a basic unit of chemical composition, such as ethylene or tetrafluorethlene. The monomer molecules always have atoms of carbon in combination with other atoms. For example, ethylene is composed of carbon and hydrogen, and in tetrafluorethlene, the hydrogen is replaced by fluorine. The polymers are made by polymerization; that is, each monomer reacts with many other similar monomers to form a very long-chain molecule of repeating monomer units (see Fig. 11-2). The polymers become stronger as the molecular weight increases. For example, low-molecu-

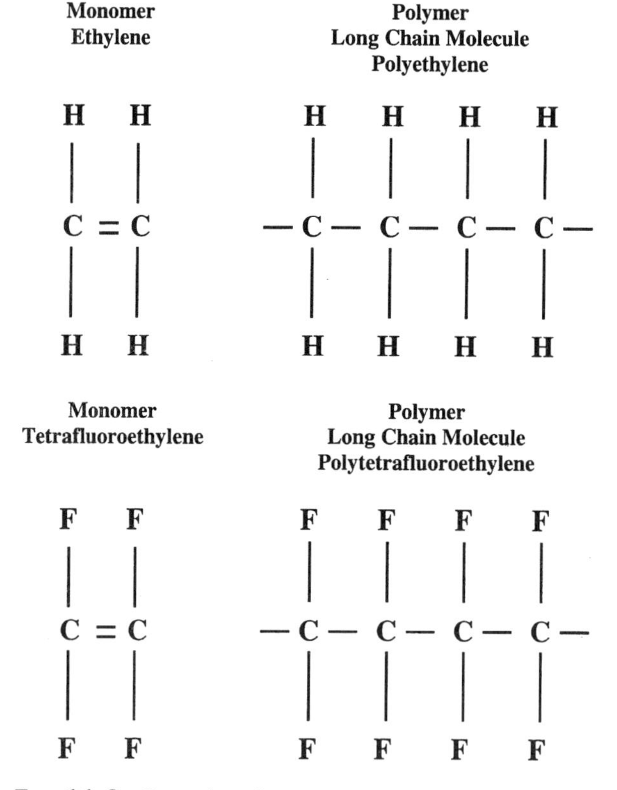

FIG. 11-2 Examples of monomers and their multiunit polymers.

lar-weight polyethylene, which has at least 100 units of CH_2, is a relatively soft material. Increasing the number of units makes the material stronger and tougher. The longest chain is ultrahigh-molecular-weight polyethylene (UHMWPE). It has up to half a million units of CH_2, and it is the toughest polyethylene. This material has an important application as a bearing material in artificial replacement joints, such as hip joints.

Over the last few decades, there has been an increasing requirement for low-cost bearings for various mass-produced machinery and appliances. This resulted in a dramatic rise in the development and application of new plastic materials for bearings. It was realized that plastics are lighter and less expensive than metals, have good surface toughness, can be manufactured by mass production processes such as injection molding, and are available in a greater variety than metallic sleeve bearings. In automotive applications, plastic bearings have steadily replaced bronze bushings for most lightly loaded bearings. The recent rise in the use of plastic bearings can also be attributed to the large volume of research and development that resulted in a better understanding of the properties of various polymers and to the development of improved manufacturing technology for new engineering plastics. An additional reason for the popularity of plastic bearings is the development of the technology of composite materials. Fiber-reinforced plastics improve the bearing strength, and additives of solid lubricants improve wear resistance. Also, significant progress has been made in testing and documenting the properties of various plastics and composites.

Widely used engineering plastics for bearings include phenolics, acetals, polyamides, polyesters, and ultrahigh-molecular-weight polyethylene. For many applications, composites of plastics with various materials have been developed that combine low friction with low wear rates and creep rates and good thermal conductivity. Reinforced plastics offer a wide selection of wear-resistant bearing materials at reasonable cost. Various plastics can be mixed together in the polymer melt phase. Also, they can be combined in layers, interwoven, or impregnated into other porous materials, including porous metals. Bearing materials can be mixed with reinforcement additives, such as glass or carbon fibers combined with additives of solid lubricants. There are so many combinations that it is difficult to document the properties of all of them.

11.5.1.1 Thermoplastics vs. Thermosets

Polymers are classified into two major groups: thermoplastics and thermosets.

11.5.1.1.1 Thermoplastics

The intermolecular forces of thermoplastics, such as nylon and polyethylene, become weaker at elevated temperature, resulting in gradual softening and melting (similar to the melting of wax). Exposure to high temperature degrades the polymer properties because the long molecular chains fracture. Therefore, thermoplastics are usually processed by extrusion or injection molding, where high pressure is used to compress the high-viscosity melt into the mold in order to minimize the process temperature. In this way, very high temperature is not required to lower the melt viscosity.

11.5.1.1.2 Thermosets

Unlike the thermoplastics, the thermosets are set (or cured) by heat. The final stage of polymerization is completed in the mold by a cross-linking reaction between the molecular chains. The thermosets solidify under pressure and heat and will not melt by reheating, so they cannot be remolded. An example of thermosets is the various types of phenolics, which are used for bearings. In the first stage, the phenolics are partially polymerized by reacting phenol with formaldehyde under heat and pressure. This reaction is stopped before the polymer completely cures, and the resin can be processed by molding it to its final shape. In the mold, under pressure and heat, the reaction ends, and the polymer solidifies into its final shape. Although the term *thermoset* means "set by heat", the thermosets include polymers such as epoxy and polyester, which do not require heat and which cure via addition of a curing agent. These thermosets are liquid and can be cast. Two ingredients are mixed together and cast into a mold, where the molecular chains cross-link and solidify. In most cases, heat is supplied to the molds to expedite the curing process, but it can be cured without heating.

11.5.2 Solid Lubricant Additives

Whenever liquid lubrication is not applied, solid lubricants can reduce friction and wear. Solid lubricants are applied only once during installation, but better results can be achieved by blending solid lubricants in the plastic material. Bearings made of thermoplastics can be blended with a variety of solid lubricants, resulting in a significant reduction of the friction and improved wear resistance. Solid lubricant additives include graphite powder and molybdenum disulfide, MoS_2, which are widely used in nylon bearings. Additional solid lubricant additives are PTFE and silicone, separately or in combination, which are blended in most plastics to improve the friction and wear characteristics. The amount of the various additives may vary for each plastic material; however, the following are recommended quantities, as a fraction of the base plastic:

PTFE	15–20%
Silicone	1–5%
Graphite	8–10%
MoS_2	2–5%

These solid lubricants are widely added to nylon and acetal, which are good bearing materials. In certain cases, solid lubricants are blended with base plastics having poor tribological properties but desirable other properties. An example is polycarbonate, which has poor wear resistance but can be manufactured within precise tolerances and has relatively high strength. Bearings and gears are made of polycarbonate blended with dry lubricants.

11.5.1.2 Advantages of Plastic Bearings

Low cost: Plastic materials are less expensive than metals and can be manufactured by mass production processes, such as injection molding. When mass-produced, plastic hearings have a far lower unit manufacturing cost in comparison to metals. In addition, plastics can be easily machined. These advantages are important in mass-produced machines, such as home appliances, where more expensive bearings would not be cost effective. In addition to initial cost, the low maintenance expenses of plastic bearings is a major advantage when operating without liquid lubricant.

Lubricity (self-lubrication): Plastic bearings can operate well with very little or no liquid lubricant, particularly when solid lubricants are blended with the base plastics. This characteristic is beneficial in applications where it is necessary for a bearing to operate without liquid lubrication, such as in the pharmaceutical and food industries, where the lubricant could be a factor in contamination. In vacuum or cryogenic applications it is also necessary to operate without oil lubrication.

Plastic bearings have relatively high compatibility with steel shafts, because they do not weld to steel. This property results in a lower friction coefficient and eliminates the risk of bearing seizure. The friction coefficient of plastic bearings in dry and boundary lubrication is lower than that of metal bearings. Their friction coefficients range from 0.15 to 0.35, and coefficients of friction as low as 0.05 have been obtained for certain plastics.

Conformability: This is the ability to deform in order to compensate for inaccuracy of the bearing dimensions. Plastics are less rigid in comparison to metals, and therefore they have superior conformability. Plastic materials have a relatively low elastic modulus and have the ability to deform to compensate for inaccuracy of the bearing-journal assembly. Tolerances are less critical for plastics than for metals because they conform readily to mating parts.

Vibration absorption: Plastic bearings are significantly better at damping vibrations. This is an important characteristic, since undesirable vibrations are always generated in rotating machinery. Also, most plastics can absorb relatively high-impact loads without permanent deformation. In many applications, plastic bearings are essential for quiet operation.

Embeddability: Contaminating particles, such as dust, tend to be embedded into the plastic material rather than scoring, which occurs in metal bearings. Also, plastics are far less likely to attract dust when running dry, compared with oil- or grease-lubricated bearings.

Low density: Plastics have low density in comparison to metals. Lightweight materials reduce the weight of the machine. This is an important advantage in automotives and particularly in aviation.

Corrosion resistance: An important property of plastics is their ability to operate in adverse chemical environments, such as acids, without appreciable corrosion. In certain applications, sterility is an additional important characteristic associated with the chemical stability of plastics.

Low wear rate: Plastics, particularly reinforced plastics, have relatively lower wear rates than metals in many applications. The exceptional wear resistance of plastic bearings is due to their compatibility with steel shafts and embeddability.

Design flexibility: Bearing parts can be molded into a wide variety of shapes and can be colored, painted, or hot-stamped where appearance is important, such as in toys and baby strollers.

Electrical insulation: Plastics have lower electrical conductivity in comparison to metals. In certain applications, such as electric motors, sparks of electrical discharge can damage the bearing surfaces, and an electrical insulator, such as a plastic bearing, will prevent this problem.

Wide temperature range: Plastics can operate without lubricants, at low and high temperatures that prohibit the use of oils or greases. Some plastics have coefficients of friction that are significantly lower at very low temperatures than at room temperature. Advanced engineering plastic compounds have been developed with PV ratings as high as 1230 Pa-m/s (43,000 psi-fpm), and they can resist operating temperatures as high as 260°C. But these compounds are not as low cost as most other plastics.

11.5.1.3 Disadvantages of Plastic Bearings

A major disadvantage is low thermal conductivity, which can result in high temperatures at the bearing surface. Most low-cost plastic materials cannot operate at high temperatures because they have low melting temperatures or because they deteriorate when exposed continuously to elevated temperatures. The combination of low thermal conductivity (in comparison to metals) and low melting temperatures restricts plastic bearings to light-load applications and low-speed (low *PV* rating in comparison to metals). The adverse effect of low thermal conductivity can be reduced by using a thin plastic layer inside a metal sleeve, but this is of higher cost. The following are additional disadvantages of plastic materials in bearing applications.

Plastics have a relatively high thermal coefficient of expansion. The difference in the thermal coefficient of expansion can be 5–10 times greater for plastics than for metals. Innovative bearing designs are required to overcome this problem. Several design techniques are available, such as an expansion slot in sleeve bearings. The effect of thermal expansion can be minimized by using a thin plastic layer inside a metal sleeve so that expansion will be limited in overall size. If thermal expansion must be completely restrained, structural materials can be added, such as glass fibers.

Another general disadvantage of plastics is creep under heavy loads, due to their relatively low yield point. Although plastics are compatible with steel shafts, they are not recommended to support nonferrous shafts, such as aluminum, due to the adhesion between the two surfaces.

11.5.1.4 PTFE (Teflon)

PTFE (Teflon) is a thermoplastic polymer material whose unique characteristics make it ideal for bearing applications (Tables 11-2 and 11-3). The chemical composition of PTFE is polytetrafluoroethylene. The molecular structure is similar to that of ethylene, but with all the hydrogen atoms replaced by fluorine (see Fig. 11-2). The characteristics of this structure include high chemical inertness due to the strong carbon-fluorine bonding and stability at low and high temperatures. It has very low surface energy and friction coefficient. At high loads and low sliding velocity, the friction coefficient against steel is as low as 0.04.

PTFE is relatively soft and has low resistance to wear and creep. However, these properties can be improved by adding fibers or particulate of harder materials. Wear resistance can be improved 1000 times by these additives.

TABLE 11-2 Bearing Design Properties of PTFE

	Max pressure		Max velocity		*PV*		Max Temp.	
Material	MPa	Psi	m/s	ft/min	psi-ft/min	Pa-m/s	°C	°F
PTFE	3.4	500	0.51	100	1000	35,000	260	500
Reinforced PTFE	17.2	2500	5.1	1000	10,000	350,000	260	500

TABLE 11-3 Physical and Mechanical Properties of PTFE

Properties	
Coefficient of thermal expansion ($10^{-5} \times$ in./in.-°F)	5.5–8.4
Specific volume (in.^{3}/lb)	13
Water absorption % (24 h,1/8 in. thick)	<0.01
Tensile strength (psi)	3350
Elongation (%)	300
Thermal conductivity (BTU-in./h-ft^2-°F)	1.7
Hardness (Shore D)	50–65
Flexural modulus (10^5 psi)	0.5–0.9
Impact strength (Izod, ft-lb./in.)	3
Thermal conductivity (BTU/h-ft-°F)	0.14

PTFE has high melt viscosity. Manufacturing processes of injection molding or extrusion without lubricity additives cannot be used for PTFE, because the melt viscosity is too high for such processes. The common manufacturing process is sintering from powder (similar to powder metallurgy). Also, it can be extruded by adding lubricant to reduce the melt viscosity (lubricated extrusion).

PTFE has exceptionally low friction against all materials. However, it is relatively soft, and under load it would creep even at room temperature, and it has low wear resistance. In practice, the problem of low wear resistance, which is unacceptable for a bearing material, is solved by adding materials such as glass fiber, carbon, bronze, and metallic oxides. This reinforcement can reduce the wear rate by three orders of magnitude, while the friction coefficient is only slightly increased. Although reinforced PTFE is more expensive, it has superior properties as a bearing material relative to other plastics. PTFE has a volume expansion of about 1% at a temperature transition crossing above 65°F. This unusual property should be considered whenever precision of parts of close tolerances is required.

The friction coefficient of PTFE decreases with increasing load and sliding speed. In addition, it is not significantly affected by temperature. PTFE is also used successfully as a solid lubricant, similar to graphite powder and molybdenum disulfide (MoS_2). PTFE solid lubricants are compounded with binders and are used for bonded coatings on wear surfaces, which are effective in reducing friction and wear. PTFE is used in journal and sliding bearings as well as in components of rolling-element bearings (bearing cage). In addition, it is used in many other applications, such as gaskets, seals, packing and piston rings. It has the lowest dry coefficient of friction against any sliding material. Dry coefficients of friction of PTFE against steel have been measured in the range from 0.05 to

0.1. It has a wide operating temperature range, and can be applied at higher temperatures relative to other plastics and white metal. In addition, PTFE can be added to other materials in order to decrease friction. A thin layer, referred to as a third body layer, is formed on the surface and acts as a solid lubricant. Another important advantage of PTFE is its ability to resist corrosion, including that by strong acids.

The advantages of PTFE as a bearing material can be summarized as follows.

1. It has the lowest dry friction in comparison to any other solid material.
2. It has self-lubricating property and acts as a thin layer of a third body to lower the friction when added to other materials.
3. It retains strength at high temperature relative to white metals and other plastics.
4. There is no cold-welding, which causes seizure in metal contacts.
5. It is chemically inert and therefore resists corrosion.
6. It can elongate elastically up to 400% and then return to its original dimensions; thus PTFE bearings are useful in applications that require better resistance to impact loading.

However, PTFE also has several disadvantages in bearings. The two major disadvantages are its high cost and its relatively low load capacity. In addition, it has a tendency to creep under load. In order to overcome the last problem, PTFE resin is usually applied in modified forms, such as reinforced by glass fibers or graphite fibers. Unmodified PTFE has a *PV* rating of only 35,000 Pa-N/m^2 (1000 psi-fpm), whereas PTFE filled with glass or graphite fibers has a *PV* rating of more than 10,000 psi-fpm. It means that the *PV* as well as the maximum sliding speed of PTFE filled with glass or graphite fibers is ten times that of PTFE without reinforcement. Additional disadvantages are its low stiffness as well as its relatively high coefficient of thermal expansion.

The most important disadvantages of PTFE as bearing material can be summarized as follows.

1. It is relatively expensive because it is difficult to manufacture. In particular, it is difficult to control its molecular weight and the degree of cross-linking, which determines its rigidity.
2. It exhibits low load capacity (low resistance to deformation) and has very high rates of creep and fatigue wear relative to other plastics.
3. It has very high thermal expansion.

PTFE has many applications in machinery for sliding contacts. It finds application in journal and sliding bearings, as well as in rolling-element bearing cages. Also, it is used for gaskets, seals, and piston rings. It is modified and added to porous metals, such as in sintered bronze. It is often reinforced by various materials, such as fiberglass, metal powders, ceramics, and graphite fibers. Other fillers, such as polyester, cotton, and glass, are also used. The fillers do not eliminate the low-friction characteristic due to the formation of a third body, which demonstrates very low friction against other solid materials. These combinations improve the properties and enable the manufacture of bearings and sliding parts with improved friction properties. Examples include automotive joints, aircraft accessories, textile machines, and business machines. Its chemical inertness is an important advantage in chemical and food-processing machinery.

Reinforced PTFE has strong bonds to steel and other rigid backing material. Reinforced PTFE liners are used in high-load, low-speed bearings to eliminate oil lubrication. Woven fabrics impregnated with PTFE are used in automotive thrust washers, ball-and-socket joints, aircraft controls and accessories, bridge bearings, and electrical switches. Woven PTFE fabrics are easily applied to bearing surfaces, they resist creep and are used for relatively higher loads.

11.5.1.5 Nylon

Nylons (polyamides) are widely used thermoplastic engineering polymers. Nylon is a crystalline material that has a variety of compositions and that can be formed by various processes, including injection molding, extrusion, and sintering. The most widely used composition is nylon 6/6, which is used primarily for injection molding and extrusion (Tables 11-4 and 11-5). Nylons are used in the form of reinforced compounds, such as glass-fiber composites, to improve strength and toughness as well as other properties.

Generally, the nylons have relatively high toughness and wear resistance as well as chemical resistance, and excellent fatigue resistance. Their low friction coefficient makes them a very good choice as bearing materials. However, they absorb water and expand. This property causes them to have low dimensional stability in comparison to other engineering plastics. Moisture adversely affects their strength and rigidity while improving their impact resistance. Nylon has the widest use of all engineering plastics in bearings. Nylon bearings are used mostly in household appliances, such as mixers and blenders, and for other lightly loaded applications. Nylon resins are used extensively in the automobile industry because they are resistant to fuels and heat and can be used under the hood of

TABLE 11-4 Design Properties of Nylon

	Max pressure		Max velocity		*PV*		Max Temp.	
Material	MPa	Psi	m/s	ft/min	psi-ft/min	Pa-m/s	°C	°F
Nylon	6.9	1000	5.1	1000	3000	105,000	93	200

TABLE 11-5 Physical and Mechanical Properties of Nylon

Characteristic	Nylon 6/6	Nylon 6
Coefficient of thermal expansion (10^{-5} × in./in.-°F)	4	4.5
Specific volume (in.^{3}/lb)	24.2	24.5
Water absorption % (24 h, 1/8 in. thick)	1.2	1.6
Tensile strength (psi)	12,000	11,800
Elongation (%)	60	200
Tensile modulus (10^5 psi)	4.2	3.8
Hardness (Rockwell *R*)	120	119
Flexural modulus (10^5 psi)	4.1	3.9
Impact strength (Izod, ft-lb./in.)	1.0	0.8
Thermal conductivity (BTU/h-ft^2-°F)	0.14	0.14

motor vehicles. Examples are cooling fans, speedometer gears, and a variety of wiring connectors. They are implemented widely for wear applications, such as plastic gears, cams, and liners, for wear protection. A major advantage is that nylon can operate without lubrication. Where the bearings must run dry, such as in the food industry, nylon is widely used. In farm equipment, greases or oils can cause dust to stick to the bearing, and nylon brushings are applied without lubricants. Also, they can be used with a wide variety of lubricants.

Nylon is commonly used in bearing materials as an injection-molded sleeve or sintered as a layer inside a metal sleeve. The molded form is stronger than the sintered one. But the sintered form can operate at higher loads and speeds (relatively higher *PV* value). A commonly used engineering material for molding and extrusion is nylon 6/6, which has a *PV* rating of 3000 psi-fpm. Characteristics of nylon include an operaring temperature of 200°F, low coefficients of friction, no requirement for lubrication, good abrasion resistance, low wear rate, and good embeddability, although it is harder than PTFE, which has lower coefficient of friction when operating against steel. Like most plastics, nylon has low thermal conductivity, and failure is usually the result of overheating. Fiber fillers, such as graphite, can improve wear resistance and strength. Glass can reduce the amount of cold flow, a major problem, which occurs in all nylons, and PTFE fibers can improve frictional properties.

In many bearing applications, molded nylon is mixed with powder fillers such as graphite and molybdenum disulfide (MoS_2). These fillers increase load capacity, stiffness, and wear resistance of the bearing as well as its durability at elevated temperatures. Reinforced nylon can withstand a maximum operating temperature of 300°F (in comparison, the operating temperature of unreinforced nylon is only 200°F).

Nylon is not adversely affected by petroleum oils and greases, food acids, milk, or other types of lubricants. The process fluid can act as the lubricant as well as the coolant. This design frequently avoids the necessity for fluid sealing and prevents contamination. Like most plastics, nylon has good antiseizure properties and softens or chars, rather than seizing.

Like most plastics, nylon has a low thermal conductivity (0.24 W/m °C), which is only about 0.5% of the conductivity of low carbon steel (54 W/m °C). The heat generated in the bearing by friction is not transferred rapidly through the nylon sleeve, resulting in high operating temperatures of the bearing surface. Therefore, these bearings usually fail under conditions of high *PV* value. In hydrodynamic bearings, the heat transfer can be enhanced by a large flow rate of oil for cooling.

The main disadvantage of the nylon bearing is creep, although the creep is not as large as in other, less rigid thermoplastics, such as PTFE. Creep is the plastic deformation of materials under steady loads at high temperatures for long periods of time. To minimize this problem, nylon bearings are supported in metal sleeves or filled with graphite. The added graphite improves wear resistance and strength. A second important disadvantage is nylon's tendency to absorb water.

11.5.1.6 Phenolics

Phenolic plastics are the most widely used thermosetting materials. They are used primarily in reinforced form, usually containing organic or inorganic fibers. Compression molding, injection molding, and extrusion can process phenolics. They are low-cost plastics and have good water and chemical resistance as well as heat resistance (Tables 11-6 and 11-7). As bearing materials, phenolics exhibit very good resistance to seizure.

Phenolics have excellent resistance to water, acids, and alkali solutions. Phenolic bearings can be lubricated by a variety of fluids, including process fluids, due to their chemical resistance. However, these bearings have a disadvantage in their thermal conductivity. The thermal conductivity of phenolics is low (0.35 W/m °C). The heat generated in the bearing by friction cannot be easily transferred through the phenolic sleeve, resulting in slow heat transfer and a high temperature of the rubbing surfaces. Phenolic bearings usually fail under conditions of high *PV* value. This problem can he solved by proper designs. Large, heavily loaded bearings must have a large feed of lubricating oil for cooling.

TABLE 11-6 Bearing Design Properties of Phenolics

Max pressure		Max velocity		*PV*		Max Temp.	
MPa	Psi	m/s	ft/min	psi-ft/min	Pa-m/s	°C	°F
41.4	6000	12.7	2500	15,000	525,000	93	200

TABLE 11-7 Physical and Mechanical Properties of Phenolics

Characteristic	General purpose	Special purpose
Coefficient of thermal expansion (10^{-5} × in./in.-°F)	3.95	
Specific gravity (in.^{3}/lb)	1.35–1.46	1.37–1.75
Water absorption % (24 h, 1/8 in. thick)	0.6–0.7	0.20–0.40
Tensile strength (psi)	6500–7000	7000–9000
Elongation (%)	×60	×200
Tensile modulus (10^5 psi)	11–13	10
Hardness (Rockwell *E*)	70–95	76
Flexural modulus (10^5 psi)	11–14	10–19
Impact strength (Izod, ft-lb./in.)	0.30–0.35	0.50
Thermal conductivity (BTU/h-ft-°F	0.2	0.25

An additional disadvantage of phenolic bearings is that they tend to swell or expand. The reason is that phenolics contain fillers, which can absorb liquids. Large bearings require large radial clearances due to swelling and warping. To correct this problem, designers must allow greater clearances or add elastic support to the bearings, such as springs. The springs allow for clearance during operation and additional space for absorption.

In bearings, this material is usually found in the form of laminated phenolics. Mixing filler sheets of fabric with phenolic resin produces the bearing. Finally, the bearing goes through a curing process of high temperature and pressure. Laminated phenolics work well with steel or bronze bearings, with oil, water, or other liquid as a lubricant. They also have good conformability, by having a low modulus of elasticity (3.45–66.9 × 10^3 MPa). These plastics also have a high degree of embeddability. This property is advantageous in ship stern tube bearings, which are lubricated by water containing sand and other sediments.

Phenolics are used as composite materials that consist of cotton fabric, asbestos, or other fillers bonded with phenolic resin. Phenolics have relatively high strength and shock resistance. They have a *PV* rating of 525,000 Pa-m/s (15,000 psi-fpm), and the limiting maximum temperature is 93°C (200°F). Phenolic bearings have been replacing metal bearings in many applications, such as propeller shaft bearings in ships, electrical switch gears, and water turbine bearings. Laminated phenolics can serve as a bearing material in small instruments.

11.5.1.7 Polyamide (Polyphenelen Sulfide)

Polyamide is used whenever the bearing operates at higher temperature. Polyamide performs well under relatively high temperatures, up to 500°F for continuous operation and up to 900°F for intermittent operation. However, exposure to 500°F reduces its tensile strength from 9600 to 7500 psi, but continuous exposure (up to 4000 hours) would not cause further deterioration (Table 11-8). It has very low thermal expansion for plastic, about twice the expansion rate of aluminum. When operating dry, it has a relatively low coefficient of friction.

Polyamide is expensive relative to other engineering plastics; and, similar to nylon, it has a tendency to absorb water. Dry bearings, bushings, thrust washers, piston rings, gears, and ball bearing cages are often manufactured using polyamide, most of them designed for high-temperature operation.

This group of engineering plastics has varying properties. They have excellent resistance to chemical attack and to burning. Polyamide is noted for its high surface toughness and its long service life. A disadvantage of this group is that they tend to absorb moisture. Polyamides are usually used with fillers. The use of fillers creates useful materials for bearings with *PV* factors of 20,000–30,000 psi-fpm. Polyamide with 15% graphite filler by weight is widely used. The filler improves the characteristics, raises the limiting temperature of 550°F, and increases its *PV* rating 10-folds to 300,000 psi-fpm. The reinforced polyamide is also more resistant to wear and creep in comparison to unfilled polyamide. Polyamides are often filled with glass fibers to improve their compressive strength and resistance to creep.

TABLE 11-8 Bearing Design Properties of Polyamide

Characteristic	General purpose	Bearing grade
Coefficient of thermal expansion ($10^{-5} \times$ in./in.-°F)	2	1.3–1.5
Specific gravity	1.40	1.45
Water absorption % (24 h, 1/8 in. thick)	0.28	0.20
Tensile strength (psi) @ 300°F	15,200	9,600
Elongation (%) @ 300°F	17	7
Tensile modulus (10^5 psi)		
Hardness (Rockwell *R*)		
Flexural modulus (10^5 psi) @ 300°F	5.2	7.3
Impact strength (Izod, ft-lb./in.)	2.5	1.1

11.5.1.8 Acetal

Acetal is a rigid plastic that is used as a bearing material due to its low cost, particularly for light-duty (low-load) applications. Acetal has a low density that is important in certain applications, such as aviation. It is tough over a wide range of temperatures; however, it has a maximum useful temperature of only 185°F.

TABLE 11-9 Bearing Design Properties of Acetal

	Max pressure		Max velocity		*PV*		Max Temp.	
Material	MPa	Psi	m/s	ft/min	psi-ft/min	Pa-m/s	°C	°F
Acetal	6.9	1000	5.1	1000	3000	105,000	82	180

Although strong acids and bases can attack it, acetal is inert to many chemicals, particularly organic solvents (Tables 11-9 and 11-10).

Acetal has higher compressive strength in comparison to unfilled polyamides. It has a *PV* rating of 3000 psi-fpm, similar to that of nylon 6/6. Acetal on steel does not demonstrate stick-slip friction, which is important for quiet bearing operation. This property is associated with the friction coefficient of acetal on steel, which increases with the sliding speed. However, its major disadvantages are that it has a higher wear rate and is not as abrasion resistant, e.g., in comparison to polyamides.

Acetal demonstrates stability in a wet environment, due to its low moisture absorption as well as its resistance to wet abrasion. Acetal is commonly used in a wide variety of automotive, appliance, and various other industrial applications. When acetal is filled with internal lubricants, it yields lower friction than nonlubricated acetal; however, this significantly raises its cost and lowers its compressive strength. Applications include ball bearing cages and a number of aviation applications that take advantage of its very low density.

TABLE 11-10 Physical and Mechanical Properties of Acetal

Characteristic	Copolymer	Homopolymer
Coefficient of thermal expansion (10^{-5} × in./in.-°C)	8.5	10
Specific volume (in.^{3}/lb)	19.7	19.5
Water absorption % (24 h, 1/8 in. thick)	0.22	0.25
Tensile strength (psi) @ 73°F	8000	10,000
Elongation (%)	60	40
Thermal conductivity (BTU-in./h-ft^2-°F)	1.6	2.6
Tensile modulus (10^5 PSI)	4.1	5.2
Hardness (Rockwell *M*)	80	94
Flexural modulus (10^5 psi) @ 73°F	3.75	4.1
Impact strength (Izod, ft-lb./in.) (notched)	1.3	1.4
Thermal conductivity (BTU/h-ft-°F)	0.46	0.74

11.5.1.9 High-Density Polyethylene (UHMWPE)

Ultra high-molecular-weight polyethylene (UHMWPE) (Table 11-11) has good abrasion resistance as well as a smooth, low-friction surface. The friction coefficient of UHMWPE on steel increases with the sliding speed. This characteristic is important, because it does not result in stick-slip friction and produces quiet bearing operation. It is often used in place of acetal, nylon, or PTFE materials. One important application of UHMWPE is in artificial joint implants. For example, in hip joint replacement, the socket (bearing) is made of UHMWPE, while the matching femur head replacement is fabricated from cobalt-chromium steel. More information on artificial joint implants is in Chap. 20.

TABLE 11-11 Physical and Mechanical Properties of UHMWPE

Characteristic	Low-density	UHMWPE
Specific volume (in.^{3}/lb)	30.4–29.9	29.4
Water absorption % (24 h, 1/8 in. thick)	<0.01	<0.02
Tensile strength (psi)	600–2300	4000–6000
Elongation (%)	90–800	200–500
Tensile modulus (10^5 psi)	0.14–0.38	0.2–1.1
Hardness (Rockwell *R*)	10	55
Flexural modulus (10^5 psi)	0.08–0.60	1.0–1.7
Impact strength (Izod, ft-lb./in.)	No break	No break
Thermal conductivity (BTU/h-ft-°F)	0.67	0.92

11.5.1.10 Polycarbonate

Polycarbonate plastics are used as a bearing material in applications where the bearing is subjected to high-impact loads. Polycarbonates have exceptionally high impact strength over a wide temperature range. They are high-molecular-weight, low-crystalline thermoplastic polymers unaffected by greases, oils, and acids (Tables 11-12 and 11-13).

Polycarbonates have been tested for balls in plastic ball bearings for use in army tank machine gun turrets, where its toughness is a prime consideration.

11.5.2 Ceramic Materials

There is an increasing interest in bearing materials that can operate at elevated temperatures, much higher than the temperature limit of metal bearings. Ceramic materials are already used in many applications in machinery, and there is continuous work to develop new ceramic materials. The most important qualities

TABLE 11-12 Bearing Design Properties of Polycarbonate

	Max pressure		Max velocity		PV		Max Temp.	
Material	MPa	Psi	m/s	ft/min	psi-ft/min	Pa-m/s	°C	°F
Polycarbonate (Lexan)	6.9	1000	5.1	1000	3000	105,000	104	220

of ceramics are their high strength and hardness, which are maintained at high temperature.

During the last two decades, there was significant progress in the development of advanced engineering ceramics. New manufacturing processes, such as sintered hot-pressing and hot isostatically pressing (HIP), have improved the characteristics of engineering ceramics. Silicon nitride manufactured by a hot isostatically pressed sintering process has a much better resistance to wear and fatigue in comparison to silicon nitride from previous manufacturing processes. It has been demonstrated that rolling elements made of silicon nitride have considerable benefits, and they are already applied in many critical applications. More detailed discussion of the merits of silicon nitride for rolling bearings is included in Chapter 13. The following is a discussion of research and development in ceramics for potential future applications in sliding bearings.

The general characteristics of engineering ceramics established them as potential candidates for materials in bearing design. They are much lighter than steel, and at the same time they have twice the hardness of steel. Their operating temperature limit is several times higher than that of steel. In addition, they require minimal lubrication. They are chemically inert and electrically nonconductive. In addition, ceramics are nonmagnetic and have lower thermal conduc-

TABLE 11-13 Physical and Mechanical Properties of Polycarbonate

Characteristic	General purpose	20% glass reinforced
Specific volume (in.^{3}/lb)	23	20.5
Water absorption % (24 h, 1/8 in. thick)	0.15	0.16
Tensile strength (psi)	9000–10,500	16,000
Elongation (%)	110–125	4–6
Tensile modulus (10^5 psi)	3.4	8.6
Hardness (Rockwell *M*)	62–70	91
Flexural modulus (10^5 psi)	3.0–3.4	8.0
Impact strength (Izod, ft-lb./in.)	12–16	2
Thermal conductivity (BTU/h-ft-°F)	0.11	0.12

tivity in comparison to metals. In rolling-contact applications, rolling elements made of silicone nitride exhibit much better fatigue resistance in comparison to steel.

However, ceramics have also several disadvantages. Ceramic bearings are much more expensive to manufacture than their steel counterparts. The manufacturing cost of ceramic parts is two to five times the cost of similar steel parts. Ceramic parts are very hard, and expensive diamond-coated tools are required for machining and cutting ceramic parts. Ceramic materials are brittle and do not have plastic deformation; therefore, ceramic parts cannot be shaped by plastic deformation like metals. In addition, ceramics are not wear-resistant materials, because they are very sensitive to the traction force during rubbing (although silicone nitride has demonstrated wear resistance in rolling contact).

Ceramics have additional disadvantages. A major characteristic is their brittleness and low tensile stress. They exhibit no plastic behavior; therefore, stress concentration will cause failure. Ceramics deform elastically up to their fracture point. Their high modulus of elasticity results in fracture at relatively small strains. When used in conjunction with steel shafts and housings, they create fitting problems due to their low linear thermal expansion in comparison to steel. In very high- and very low-temperature environments, these problems are exacerbated.

11.5.2.1 Hot Isostatic Pressing (HIP)

Ceramic parts are manufactured from powder by hot-press shaping or by the improved manufacturing technique of hot isostatic pressing (HIP). This last technique offers many advantages of improved mechanical characteristics of the parts. The early manufacturing process of ceramics was by hot-pressing. The parts did not have a uniform structure, and many surface defects remained, to be removed later by various expensive processes. The parts were not accurate and required finishing machining by diamond-coated cutting tools. Moreover, the finished parts did not have the characteristics required for use in rolling-element bearings.

The recently introduced hot isostsatic pressing (HIP) process offered many advantages over the previous hot-pressing process. The HIP process uses very high pressure of inert gas at elevated temperatures to eliminate defects of internal voids. High-pressure argon, nitrogen, helium, or air is applied to all grain surfaces under uniform temperature. Temperatures up to 2000°C (3630°F) and pressures up to 207 MPa (30,000 psi) are used. The temperature and pressure are accurately controlled. This process bonds similar and dissimilar materials, to form parts very close to the final shape from metals, ceramic, and graphite powders. The term *isostatlc* means that the static pressure of the hot gas is equal in all directions throughout the part.

This process is already widely used for shaping parts from ceramic powders as well as other mixtures of metals and nonmetal powders. This process minimizes surface defects and internal voids in the parts. The most important feature of this process is that it results in strong bonds between the powder boundaries of similar or dissimilar materials. In turn, it improves significantly the characteristics of the parts for many engineering applications.

In addition, the HIP process reduces the cost of manufacturing, because it forms net or near-net shapes (close to the required dimensions of the part). The cost is reduced because the parts are near final and very little machining is required.

11.5.2.2 Engineering Ceramics

Ceramics have been used in bearings for many years in low-load applications where lubrication is limited or not possible. An example is the small sphere of alumina used in accurate instruments or mechanical watches. These bearings are commonly referred to as "watch jewels." However, these are not engineering ceramics. There is a need for bearings for high-temperature applications at high speed and load, such as aircraft engines. For this purpose, a lot of research and development has been conducted in engineering ceramics.

Engineering ceramics are dissimilar to regular ceramics, such as porcelain or clay bricks, because of their much higher strength and toughness at high operating temperatures. Engineering ceramics are compounds of metallic and nonmetallic elements, such as Si_3N_4, SiC, and Al_2O_3, that are sintered at high pressure and temperature by HIP process in order to form a strong bond between the particles. In conventional ceramics (porcelain), the particles are linked by weak bonds (mechanical linkage of particles). On the other hand, sintered engineering ceramics that are made by a hot-pressed or hot-isostatically-pressed process have particles that are bonded together by a stronger bond. The physical explanation for the improved properties of engineering ceramics is that they have strong bonds at the grain boundaries. The energy of equilibrium of these bonds is similar to that of metal grains.

During the last two decades, there were a lot of expectations that ceramics materials would be used in engines, including piston and sleeve materials. This would allow operation of the engines at higher temperatures, resulting in improved efficiency. We have to keep in mind that according to the basic principles of thermodynamics (Carnot cycle), efficiency increases with engine temperature. These expectations did not materialize at this time, because ceramic surfaces require liquid lubricants (in a similar way to metals). However, liquid lubricants that can operate at elevated temperatures in conjunction with ceramics are still not available. Research was conducted in an attempt to operate ceramic bearings at high temperatures with various types of solid powders as lubricants.

However, this work did not reach the stage of successful implementation in bearings.

The ceramic materials proved to have inferior tribological properties, such as relatively high sliding wear in comparison to compatible metal bearing materials. A large volume of research was conducted, and is still taking place, in an attempt to improve the tribological properties of engineering ceramics by various coatings or surface treatment. During the sliding of ceramics against steel, adhesive wear has been identified as the major wear mechanism. Let us recall that adhesive wear is affected by the lowest shear strength of the sliding bodies and by the compatibility of the two materials. Although some improvements have been reported, ceramic bearing materials are still not widely used for replacement of metals in plain bearings.

The main engineering ceramics in use, or in the development stage, are silicon nitride, silicon carbide, zirconia, alumina oxide, and ruby sapphire. Silicon carbide and silicon nitride are already used in various applications as high-temperature and high-strength engineering ceramics. The purpose of the following discussion is to provide the bearing designer a summary of the unique characteristics of engineering ceramics that can make them useful candidates as bearing materials. Bearings made of ceramics are of significant interest for the following reasons.

1. *High-temperature performance*: Ceramics retain high strength at elevated temperature and have a relatively high yield-point stress at temperatures above 1000°C (1832°F). This is a major advantage for potential high-temperature bearing materials. However, the available lubricants do not resist a similarly high temperature, and it is necessary to operate the bearing without lubricant. Air is considered a potential fluid film that can operate at elevated temperatures and can be used for hydrostatic or hydrodynamic lubrication. In some unique applications, such as high-speed small turbines, ceramic air bearings are already in use. In the future, it is expected that ceramics will be increasingly used for high-temperature applications, such as in aircraft engines.
2. *Low density*: An additional advantage of ceramic materials is that their density is relatively low. The density of ceramics is about a third that of steel. For example, low density reduces the centrifugal forces of rotating rolling elements. It reduces the contact stresses between the rolling elements and the outer race. This fact allows the operating of rolling bearings at higher speed.
3. *Low coefficient of expansion*: In general, a low coefficient of expansion is desirable, in particular in rolling-element bearings. Thermal stresses, due to thermal expansion, cause seizure in sleeve bearings as well as in rolling-element bearings. However, other parts in the

machine are usually made of steel, and there is a compatibility problem due to the difference in thermal expansion. A major problem is the difficulty of mounting ceramic bearings on metallic shafts and inside metallic housings, because of the large differences in thermal expansion coefficient between metals and ceramics.

4. *Corrosion resistance*: Ceramics are chemically inert and have good corrosion resistance. Ceramics can be used in a corrosive environment, such as strong acids, where steel alloys fail due to fast corrosion.

Although ceramics are not ideal tribological materials, they are considered by many to be the future technology. Ceramics enable operation at temperature ranges not previously accessible. They allow grease lubrication or unlubricated applications where steel requires more expensive and complex oil lubrication equipment.

In general, ceramics are expected in the future to replace metals in order to reduce weight and allow higher temperatures to increase engine efficiency and speed of operation. By using ceramics in gas turbines, the operating temperature will increase from the metal limited range of 1800–2100°F to 2500°F. This would result in considerable fuel savings. The characteristics of ceramic bearings, when fully used, not only would reduce fuel consumption, but also will reduce maintenance cost and increase power density (engine weight), which is particularly important in aircraft. The ability of ceramic rolling elements to operate with very little or no lubrication offers great potential for improving aircraft safety.

11.5.2.3 Ceramics for Plain Bearings

In ceramics, adhesive wear has been identified as the most significant mechanism of sliding wear against bearing steel. As was discussed in Sec. 11.2.1, adhesive wear is influenced by two factors—the shear strength of the sliding pair and the compatibility of the two materials.

Attempts were made to modify ceramic material surfaces to minimize the adhesive wear so that they would be competitive with metal plain bearings. For example, experiments have been conducted to improve the wear resistance of silicon nitride by adding whiskers of silicon carbide in various amounts (Ueki, 1993). The results showed that silicon carbide improves fracture toughness, which is correlated with reduced wear.

It is already known that water is a good lubricant for ceramic sliding friction. Tests of the effect of water on ceramic slideways were conducted*. For all ceramics it was found that the coefficient of friction decreases with increasing humidity. Zirconia demonstrated the highest friction coefficient and wear rates, relative to other ceramics. Zirconia's wear rate increased dramatically with increasing sliding velocity when submersed in water, while the wear rates of

---

*Fischer and Tomizawa, 1985; Tomizawa and Fischer, 1987; and Ogawa and Aoyama, 1991.

alumina oxide, silicon carbide, and silicon nitride decreased with increasing sliding velocities. The minimum friction coefficient of silicon carbide and silicon nitride is as low as 0.01 in the presence of water.

Experiments were conducted in oil-lubricated ceramic journal bearings. The experiments showed lower friction coefficient for silicon nitride journals (in comparison to steel journals) for bearings made of tin-coated Al-Si alloy, forged steel, and cast aluminium matrix composite with silicon carbide reinforcement (cast MMC). All bearings were lubricated with SAE 10W-30 oil (Wang et al., 1994).

11.5.2.3.1 Silicon Carbide and Silicon Nitride

As discussed earlier, these are the best-performing high-temperature, high-strength ceramics. They have high-temperature-oxidation resistance and the highest strength in structures. Silicon nitride has higher strength than silicon carbide up to 2600°F, but above this temperature silicon carbide is stronger. Silicon carbide and silicon nitride are inert to most chemicals, and for most applications they exhibit similar corrosion resistance. An important design consideration is that they have the lowest thermal expansion coefficient in comparison to other ceramics. In addition, they have the lowest density. They also have the highest compressive strengths. Silicon nitride is the only ceramic material used as a roller bearing material (see Chap. 13).

Ceramic journal bearings are widely used in very corrosive environments where metals cannot be used. For example, sealed pumps driven by magnetic induction are used for pumping corrosive chemicals. Most sealed pumps operate with ceramic sleeve bearings of silicon carbide. The ceramic sleeves are used because of their corrosion resistance and for their nonmagnetic properties.

However, the use of a silicon carbide sleeve in a sliding bearing was not successful in all cases. These bearings operate with the process fluid as lubricant. These fluids, such as gas and water often have low viscosity. These bearings perform well as hydrodynamic bearings with a full fluid film only at the high rated speeds. During starting and stopping there is direct contact of the journal with the ceramic sleeve. The silicon carbide sleeve is brittle and suffers severe wear from a direct contact; it does not have long life in pumps that operate with frequent start-ups. In such cases, all-ceramic rolling bearings made of silicon nitride proved to be a better selection. The silicone nitride rolling bearings are not so sensitive to frequent start-ups and show good corrosion resistance to chemicals.

11.5.2.3.2 Alumina Oxide

Alumina oxide has a high maximum useful temperature and good compressive strength. It was the first ceramic to be investigated as an advanced bearing material. Currently, it is being researched and developed as a candidate for plain bearings. It is also currently used in certain plain bearings. It was reported that in the presence of lubricant it has a low coefficient of friction similar to PTFE. However, for the sliding of dry ceramics, the coefficient of friction is higher than that of compatible metals (Ogawa and Aoyama, 1991).

11.5.2.3.3 Zirconia

Zirconia has the highest friction coefficient and wear rates relative to other ceramics. Zirconia's wear rate increased dramatically with increasing sliding velocity. Zirconia has the lowest operating temperature and the lowest modulus of elasticity and a very high hardness.

Researchers have been trying to modify the Zirconia manufacturing process in order to increase its compressive strength and temperature range by reducing its grain size. Zirconia was considered a good candidate for roller-element bearings because it has relatively low modulus of elasticity. Low-modulus elasticity allows ceramics to flake like metals when failing as rolling elements. Silicon nitride is the only other ceramic that flakes. The others fail catastrophically.

11.5.2.3.4 Ruby Sapphire

Ruby sapphire has the highest hardness and maximum useful temperature. It is being investigated for use in plain and rolling bearings. It is the engineering ceramic with the least reported data.

11.5.3 Other Nonmetallic Bearing Materials

11.5.3.1 Cemented Carbide

This generally consists of tungsten carbide (97%) and Co (3.0%). It can withstand extreme loading and high speeds. It must have good alignment and good lubrication. This material is used in high-speed precision grinders.

11.5.3.2 Rubber

Rubber bearings are used mostly on propeller shafts and rudders of ships, in hydraulic turbines, and in other industrial equipment that processes water or slurries. The compliance of the rubber helps to isolate vibration, provide quiet operation, and compensate for misalignment (see Chapt. 9).

11.5.3.3 Wood

Wood bearings have been replaced by plastic and rubber bearings. The main advantage of wood bearings are their clean operation, low cost, and self-lubrication properties. Common wood materials are rock maple and oak.

11.5.3.4 Carbon Graphite

Carbon graphite has good self-lubricating properties. Carbon graphite bearings are stable over a wide range of temperatures and are resilient to chemical attack. In some cases, metal or metal alloys are added to the carbon graphite composition to improve such properties as compressive strength and density. Carbon graphite has poor embeddability; therefore, filtered and clean lubricants should be used. Usually, carbon graphite does not require lubrication. In most cases, it is used in textile and food-handling machinery.

11.5.3.5 Molybdenum Disulfide (MoS_2)

Molybdenum disulfide is similar to graphite in appearance, and it has very low friction coefficient. In many applications it is mixed with a binder, such as a thermosetting plastic, in order to ensure retention of the lubricant on the surface. It has a satisfactory wear life.

11.5.3.6 Polymer–Metal Combination

Other types of bearings in the plastic family are the polymer–metal combination. These are very well known and considered quite valuable as far as bearing materials are concerned. One variety is made of a porous bronze film layer coated with a Teflon-lead mixture, plus an all-steel backing. This configuration results in favorable conductive heat transfer as well as low-friction properties. Industry reports application of temperatures up to 530°F, which indicate this bearing is desirable. This bearing material is used in bushings, thrust washers, and flat strips for handling rotating, oscillating, sliding, radial, and thrust loads. Acetal copolymer is being applied in small gears that need structural strength, while still providing low friction and wear.

Problems

11-1 List the plastic bearing materials according to the following:

a. Increasing *PV* value
b. Increasing allowed temperature

11-2 White metal (babbitt) is currently used as very thin layer. Give an example of two applications where it would be beneficial to have a thicker white metal.

11-3 What materials are used in car engine bearings? Explain the reasons for the current selection.

11-4 Select a bearing material for a low-cost mass-produced food mixer. Explain your selection.

11-5 Summarize the characteristics of nylon 6 that are significant for the selection of a bearing material. Give three examples of machines where bearings of nylon 6 can be used and three examples where this material would not be appropriate.

12

Rolling-Element Bearings

12.1 INTRODUCTION

Rolling-element bearings, or, in short, rolling bearings, are commonly used in machinery for a wide range of applications. In the past, rolling bearings were referred to as *antifriction bearings*, since they have much lower friction in comparison to sliding bearings. Many types of rolling-element bearings are available in a variety of designs that can be applied for most arrangements in machinery for supporting radial and thrust loads. The rolling elements can be balls, cylindrical rollers, spherical rollers, and conical rollers.

12.1.1 Advantages of Rolling-Element Bearings

One important advantage of rolling-element bearings is their low friction. It is well known that the rolling motion has lower friction in comparison to that of sliding. In addition to friction, the rolling action causes much less wear in comparison to sliding. For most applications, rolling-element bearings require less maintenance than hydrodynamic bearings. To minimize maintenance cost in certain cases, prepacked rolling-element bearings are available with grease that is permanently sealed inside the bearing. Ultrahigh-precision rolling bearings are available for precision machinery, such as precision machine tools and measuring equipment. It is possible to completely eliminate the clearance and even to prestress the bearings. This results in a higher stiffness of the bearings, and the shaft centerline is held tightly in its concentric position. Prestressing the bearings would minimize vibrations as well as reduce any undesired radial displacement of the shaft.

12.1.2 Fatigue Life

A major limitation of rolling-element bearings is that they are subjected to very high alternating stresses at the rolling contacts. High-speed rotation involves a large number of stress cycles per unit of time, which leads to a limited fatigue life. In fact, prestressing and centrifugal forces at high-speed operation significantly increase the contact stresses and further reduce the fatigue life.

12.1.3 Terminology

A standard rolling-element bearing has two rings, an *outer ring* and *inner ring*, which enclose the rolling elements, such as *balls*, *cylindrical rollers*, and *tapered rollers*. The rolling areas on the rings are referred to as *raceways*. An example is a *deep-groove ball bearing*, which has concave raceways (an *outer ring raceway* and an *inner ring raceway*) that form the rolling areas. A *cage* holds the rolling elements at equal distance from one another and prevents undesired contact and rubbing friction among them. The terminology of bearing parts is shown in Fig. 12-1 for various bearing types. This terminology has been adopted by the Anti-Friction Bearing Manufacturers Association (AFBMA).

12.1.4 Rolling Contact Stresses

There is a theoretical point or line contact between the rolling elements and races. But due to elastic deformation, the contact areas are actually of elliptical or rectangular shape. In machinery that involves severe shocks and vibrations, the contact stresses can be very high.

In the United States, the standard bearing material is SAE 52100 steel hardened to 60 RC. This steel has a high content of carbon and chromium. It is manufactured by an induction vacuum melting process, which minimizes porosity due to gas released during the casting process.

Stainless steel AISI 440C hardened to 58 RC is the standard rolling bearing material for corrosive environments. The allowed limit of rolling contact stress for SAE 52100 is 4.2 GPa (609,000 psi). For rolling bearings made of AISI 440C stainless steel, the allowed limit of compression stress is only 3.5 GPa (508,000 psi). Discussion of other rolling bearing materials and manufacturing processes is included at the end of Chapter 13.

The theory of elasticity indicates that the maximum shear stress of rolling contact is below the surface. Due to repeated cyclic stresses, scaly particles eventually separate from the rolling surfaces. Fatigue failure is evident in the form of metal removal, often referred to as *flaking* (or *spalling*), at the rolling contact surfaces of raceways and rolling elements.

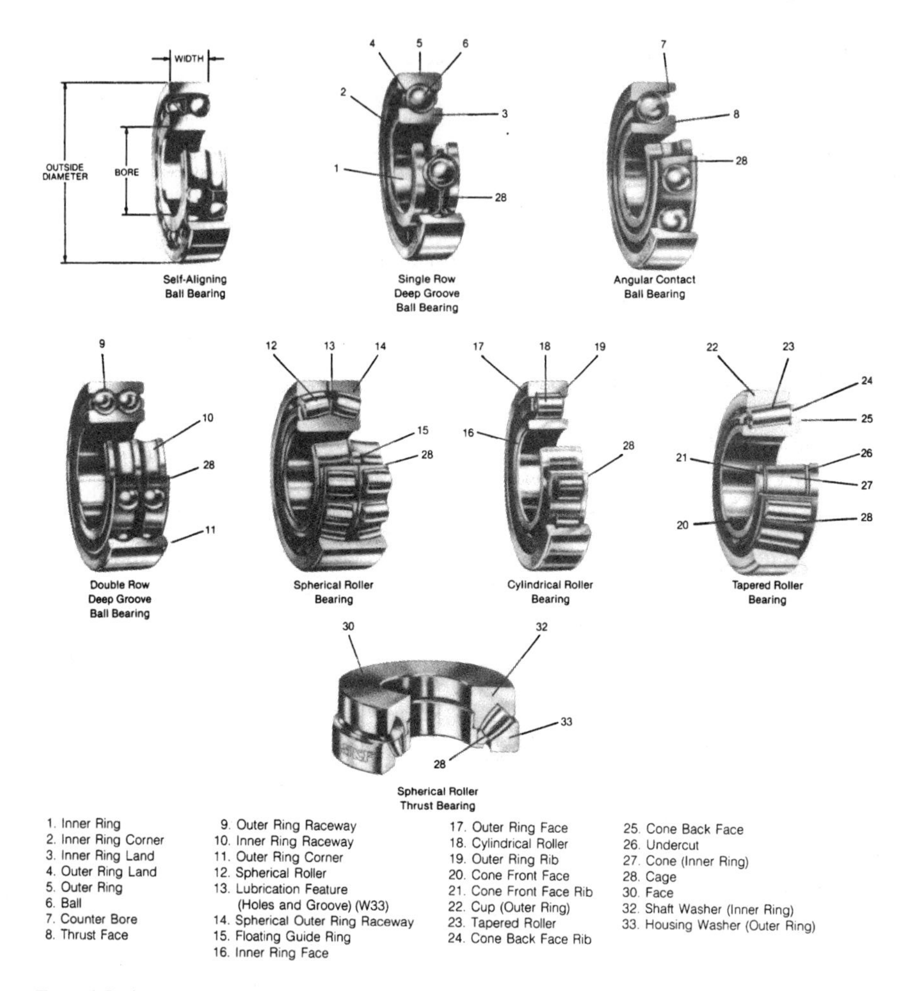

FIG. 12-1 Bearing types and terminology (with permission from SKF).

In order to make the bearing durable to the high stress levels, the contact surfaces and subsurfaces of the rolling elements and raceways are hardened by heat treatment to a minimum hardness of Rockwell C58. During bearing operation, the friction elevates the temperature of the contact surfaces. Therefore, the yield stress and hardness must be retained to elevated temperatures up to 120°C (250°F). For applications at higher temperature, rolling bearings are available that are made of better materials that can retain the desired properties at higher temperatures, The need for fatigue-resistant and hard materials at elevated temperature for rolling bearings has been the motivation for continual research and development, which has resulted in improved materials for high-speed and heavy-duty operation.

Due to the rolling action, the rolling elements and races are subjected to periodic stress cycles that can result in material failure due to fatigue. The fatigue life of rolling bearings is statistically distributed. The data for these statistics must be obtained only by many experiments for each bearing type over a long period of time. Fatigue life depends on the material and its processing methods, such as heat treatment. Fatigue life is also a function of the magnitude of the maximum stress and temperature at the contacts between the raceways and the rolling elements during operation. If stresses are low, fatigue life can be practically unlimited.

The stresses in dry contacts can be calculated via the theory of elasticity (Hertz equations). In addition to fatigue, the high stresses result in considerable wear. However, the surfaces are usually lubricated, and under favorable conditions there is a very thin lubrication film at very high pressure that separates the rolling surfaces. Whenever this film is thicker than the surface asperities, it would prevent any direct contact. In this way, this fluid film plays an important role in reducing wear. The analysis of this film is based on elastohydrodynamic (EHD) theory. The analysis of the fluid flow and pressure wave inside this thin film is performed in a similar way to that for hydrodynamic bearings. But in addition, EHD analysis considers the elastic deformation near the contact area and the increasing function of viscosity versus pressure.

Recent developments include investigation of the thermal effects in the EHD film. This analysis is referred to as *thermoelastohydrodynamic* (TEHD) analysis. This thermal analysis is quite complex because it solves for the temperature distribution by considering the dissipation of heat due to viscous friction, heat transfer, and finally the viscosity dependence on temperature and pressure distribution. Dedicated computer programs have been developed that assist in better understanding the phenomena involved in rolling contact.

Although there has been considerable progress in the analysis of stresses and fluid films, for design purposes the life of the rolling-element bearings must be estimated by means of empirical equations based on experiment. Due to the statistical nature of bearing life, bearings are selected to have a very high probability of operation without failure for a certain reasonable period of the life of the machine (such as 5 or 10 years). The life-period requirement is usually determined before the design of a machine is initiated.

Failure due to fatigue is only one possible failure mode among other, more frequent failure modes, which have a variety of causes. Proper lubrication, mounting, and maintenance of the bearing can prevent most of them. It is interesting to note that although most rolling bearings are selected by considering their fatigue life, only 5–10% of the bearings actually fail by fatigue. The causes for most bearing failures are misalignment, improper mounting, corrosion, penetration of dust or other hard particles into the bearing and lack of proper lubrication (oil starvation or not using an appropriate lubricant).

In addition to fatigue and the other reasons just mentioned, overheating can be a frequent cause of rolling bearing failure. Bearing overheating can be caused by heat sources outside the bearing, such as in the case of a steam turbine or aircraft engine. Also, friction-energy losses are converted to heat, which is dissipated in the bearing. In most cases overheating is due to heavy load and high speed. Higher bearing temperatures have an adverse effect on the lubricant. As the temperature increases, the oil oxidation process is accelerated. A rule of thumb is that the lubrication life is halved for every 10°C increase in temperature.

Thermal analyses as well as measurements have indicated that during operation there is a temperature gradient inside the bearing. Heat is transferred better from the outer ring through the housing than from the inner ring. In most practical cases of moderate load and speed, the outer ring temperature is lower than that of the inner ring by 5–10°C. This difference in temperature results in uneven thermal expansion. If the bearing has a small internal clearance or no clearance, it would result in extra thermal stresses. The thermal stresses are in the form of a tight fit and higher contact stress between the rolling elements and the races. During high-speed operation, the additional stresses further increase the temperature. There is a risk that this sequence of events can result in an unstable closed-loop process of rising temperature and stress that can lead to failure in the form of thermal seizure. For this reason, standard rolling-element bearings are manufactured with sufficient internal clearance to reduce the risk of thermal seizure.

At high speeds, the centrifugal forces of the rolling elements combine with the external load and thermal stresses to increase the maximum total contact stress between a rolling element and the outer ring race. Therefore, a combination of heavy load and high speed reduces the bearing fatigue life. In extreme cases, this combination can cause a catastrophic failure in the form of bearing seizure. The risk of failure is high whenever the product of bearing load and speed is high, because the amount of heat generated by friction in the bearing is proportional to this product. In conclusion, the load and speed are two important factors that must be considered in the selection of the proper bearing type in order to achieve reliable operation during the expected bearing life.

Developments in aircraft turbine engines resulted in a requirement for increasing power output at higher shaft speed. As discussed earlier (see Chap. 1), rolling bearings are used for aircraft engines because of the high risk of an interruption in the oil supply of hydrodynamic or hydrostatic bearings. At the very high speed required for gas turbines, the centrifugal force of the rolling elements is a major factor in limiting the fatigue life of the bearings. The centrifugal forces of the rotating rolling elements increase the contact stresses at the outer race and shorten the bearing fatigue life.

The contact force on the outer race increases due to the centrifugal force of a rolling element. The centrifugal force F_c [N] is

$$F_c = m_r \omega_c^2 R_c \tag{12-1}$$

Here, m_r is the mass of the rolling element [kg] and ω_c is the angular speed [rad/s] of the center of a rolling element in its circular orbit (equivalent to the cage angular speed, which is lower than the shaft speed). The radius R_c [m] is of the circular orbit of the rolling-element center. The units indicated are SI, but other unit systems, such as the Imperial unit system, can be applied. In a deep-groove bearing, centrifugal force directly increases the contact force on the outer race. But in an angular contact ball bearing, which is often used in high-speed turbines, the contact angle results in a higher resultant reaction force on the outer raceway.

Equation (12-1) indicates that centrifugal force, which is proportional to the second power of the angular speed, will become more significant in the future in view of the ever-increasing speeds of gas turbines in aircraft and other applications. Similarly, bearing size increases the centrifugal force, because rolling elements of larger bearings have more mass as well as larger-orbit radius. Therefore, the centrifugal forces are approximately proportional to the second power of the DN value (rolling bearing bore in millimeters times shaft speed in RPM). The centrifugal force of the rolling elements is one important consideration for limiting aircraft turbine engines to 2 million DN. A future challenge will be the development of the technology in order to break through the DN limit of 2 million.

12.1.5 Misalignment

Bearings in machines are subjected to a certain degree of angular misalignment between the shaft and bearing centerlines. A bearing misalignment can result from inaccuracy of assembly and machining (within the tolerance limits). Even if the machining and assembly are very precise, there is a certain misalignment due to the shaft's bending under load. Certain bearing types are more sensitive to angular misalignment than others. For example, cylindrical and tapered roller bearings are very sensitive to excessive misalignment, which results in uneven pressure over the roller length.

In applications where there is a relatively large degree of misalignment, the designer can select a self-aligning roller-element bearing. In most cases, it is more economical to use a self-aligning bearing than to specify close tolerances that involve the high cost of precision manufacturing. Self-aligning bearings allow angular errors in machining and assembly and reduce the requirement for very close tolerances. Self-aligning bearings include self-aligning ball bearings and spherical roller bearings. The design of a self-aligning bearing is such that the shape of the cross section of the outer raceway is circular, which allows the rolling elements to have an angular degree of freedom and self-alignment between the inner and outer rings.

12.2 CLASSIFICATION OF ROLLING-ELEMENT BEARINGS

Ball bearings can operate at higher speed in comparison to roller bearings because they have lower friction. In particular, the balls have less viscous resistance when rolling through oil or grease. However, ball bearings have lower load capacity compared with roller bearings because of the high contact pressure of point contact. There are about 50 types of ball bearings listed in manufacturer catalogues. Each one has been designed for specific applications and has its unique characteristics. The following is a description of the most common types.

12.2.1 Ball Bearings

12.2.1.1 Deep-Groove Ball Bearing

The deep-groove ball bearing (Fig. 12-2) is the most common type, since it can be used for relatively high radial loads. Deep-groove radial ball bearings are the most widely used bearings in industry, and their market share is about 80% of industrial rolling-element bearings. Owing to the deep groove in the raceways, they can support considerable thrust loads (in the axial direction of the shaft) in

FIG. 12-2 Deep-groove ball bearing.

addition to radial loads. A deep-groove bearing can support a thrust load of about 70% of its radial load. The radial and axial load capacity increases with the bearing size and number of balls.

For maximum load capacity, a filling-notch type of bearing can be used that has a larger number of balls than the standard bearing. In this design, there is a notch on one shoulder of the race. The circular notch makes it possible to insert more balls into the deep groove between the two races. The maximum number of balls can be inserted if the outer ring is split. However, in that case, external means must be provided to hold and tighten the two ring halves together.

12.2.1.2 Self-Aligning Ball Bearings

It is very important to compensate for angular machining and assembly errors between the centerlines of the bearing and the shaft. The elastic deflection of the shaft is an additional cause of misalignment. In the case of a regular deep-groove ball bearing, the misalignment causes a bending moment in the bearing and additional severe contact stresses between the balls and races. However, in the self-aligning bearing (Fig. 12-3), the spherical shape of the outer race allows an additional angular degree of freedom (similar to that of a universal joint) that prevents the transfer of any bending moment to the bearing and prevents any additional contact stresses.

Self-aligning ball bearings have two rows of balls, and the outer ring has a common spherical raceway that allows for the self-aligning characteristic. The

FIG. 12-3 Self-aligning ball bearing.

inner ring is designed with two *restraining ribs* (also known as *lips*), one at each side of the roller element, for accurately locating the rolling elements' path on the inner raceway. But the outside ring has no ribs, in order to allow for self-alignment. A wide spherical outer race allows for a higher degree of self-alignment.

Self-aligning ball bearings are widely used in applications where misalignment is expected due to the bending of the shaft, errors in the manufacture of the shaft, or mounting errors. The design engineer must keep in mind that there are always tolerances due to manufacturing errors. Self-aligning bearings can be applied for radial loads combined with moderate thrust loads. The feature that self-aligning bearings do not exert any bending moment on the shaft is particularly important in applications that require high precision (low radial run-out) at high speeds, because shaft bending causes imbalance and vibrations. The concept of self-alignment is useful in all types of bearings, including sleeve bearings.

12.2.1.3 Double-Row Deep-Groove Ball Bearing

This bearing type (Fig. 12-4) is used for relatively high radial loads. It is more sensitive to misalignment errors than the single row and should be used only for applications where minimal misalignment is expected. Otherwise, a self-alignment bearing should be selected.

The design of double-row ball bearings is similar to that of single-row ball bearings. Since double-row ball bearings are wider and have two rows, they can

FIG. 12-4 Double-row deep-groove ball bearing.

carry higher radial loads. Unlike the deep-groove bearing, designs of split rings (for the maximum number of balls) are not used, and each ring is made from one piece. However, double-row bearings include groups with larger diameters and a larger number of balls to further improve the load capacity.

12.2.1.4 Angular Contact Ball Bearing

This bearing type (Fig. 12-5) is used to support radial and thrust loads. Contact angles of up to 40° (from the radial direction) are available from some bearing manufacturers, but 15° and 25° are the more standard contact angles. The contact angle determines the ratio of the thrust to radial load.

Angular contact bearings are widely used for adjustable arrangements, where they are mounted in pairs against each other and preloaded. In this way, clearances in the bearings are eliminated or even preload is introduced in the rolling contacts. This is often done to stiffen the bearings for a rigid support of the shaft. This is important for reducing the amplitude of shaft vibrations under oscillating forces. This type of design has significant advantages whenever precision is required (e.g., in machine tools), and it reduces vibrations due to imbalance. This is particularly important in high-speed applications. An adjustable arrangement is also possible in tapered bearings; however, angular contact ball bearings have lower friction than do tapered bearings. However, the friction of angular contact ball bearings is somewhat higher than that of radial ball bearings. Angular contact ball bearings are the preferred choice in many important applications, such as high-speed turbines, including jet engines.

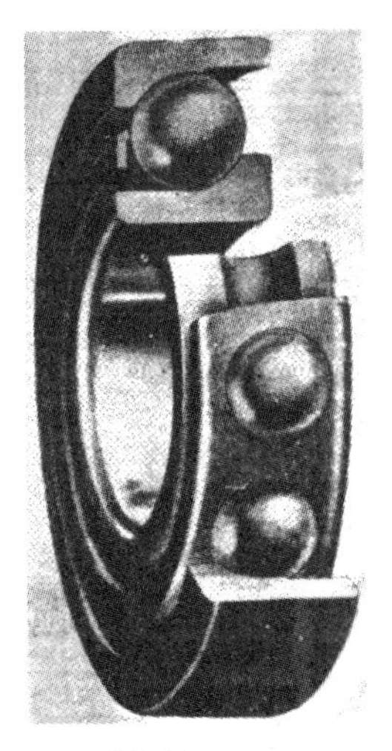

FIG. 12-5 Angular contact ball bearing.

Single-row angular contact ball bearings can carry considerable radial loads combined with thrust loads in one direction. Prefabricated mountings of two or more single-row angular contact ball bearings are widely used for two-directional thrust loads. Two bearings in series can be used for heavy unidirectional thrust loads, where two single-row angular contact ball bearings share the thrust load. Precise axial internal clearance and high-quality surface finish are required to secure load sharing of the two bearings in series. The bearing arrangement of two or more angular contact bearings facing the same direction is referred to as *tandem arrangement*. The bearings are mounted adjacent to each other to increase the thrust load carrying capacity.

12.2.2 Roller Bearings

Roller bearings have a theoretical line contact between the unloaded cylindrical rollers and races. This is in comparison to ball bearings, which have only a theoretical point contact with the raceways. Under load, there is elastic deformation, and line contact results in a larger contact area than that of a point contact in ball bearings. Therefore, roller bearings can support higher radial loads. At the same time, the friction force and friction-energy losses are higher for a line contact; therefore, roller bearings are usually not used for high-speed applications.

Roller bearings can be classified into four categories: cylindrical roller bearings, tapered roller bearings, needle roller bearings and spherical roller bearings.

12.2.2.1 Cylindrical Roller Bearings

The cylindrical roller bearing (Fig. 12-6) is used in applications where high radial load is present without any thrust load. Various types of cylindrical roller bearings are manufactured and applied in machinery. In certain applications where diameter space is limited, these bearings are mounted directly on the shaft, which serves as the inner race. For direct mounting, the shaft must be hardened to high Rockwell hardness, similar to that of the bearing race. For direct mounting, the radial load must be high in order to prevent slipping between the rollers and the shaft during the start-up. It is important to keep in mind that cylindrical roller bearings cannot support considerable thrust loads. Thus, for applications where both radial and thrust loading are present, it is preferable to use ball bearings.

12.2.2.2 Tapered Roller Bearing

The tapered roller bearing is used in applications where a high thrust load is present that can be combined with a radial load. The bearing is shown in Fig. 12-7. The races of inner and outer rings have a conical shape, and the rolling elements between them have a conical shape as well. In order to have a rolling motion, the contact lines formed by each of the various tapered roller elements

FIG. 12-6 Cylindrical roller bearing.

and the two races must intersect at a common point on the bearing axis. This intersection point is referred to as an *apex point*. The apex point is closer to the bearing when the cone angle is steeper. A steeper cone angle can support a higher thrust load relative to a radial load.

The inner ring is referred to as *cone*, while the outer ring is referred to as *cup*. The cone is designed with two retaining *ribs* (also known as *lips*) to confine

FIG. 12-7 Tapered roller bearing.

the tapered rollers as shown in Fig. 12-7. The ribs also align the rollers between the races. In addition, the larger rib has an important role in supporting the axial load. A cage holds the cone and rollers together as one unit, but the cup (outer ring) can be pulled apart.

A single-row tapered roller bearing can support a thrust load in only one direction. Two tapered roller bearings are usually mounted in opposition, to allow for thrust support in both directions (in a similar way to opposing angular contact ball bearings). Moreover, double or four-row tapered roller bearings are applied in certain applications to support a high bidirectional thrust load as well as radial load.

The reaction force on the cup acts in the direction normal to the line of contact of the rolling elements with the cup race (normal to the cup surface). This force can be divided into axial and radial load components. The intersection of the resultant reaction force (which is normal to the cup angle) with the bearing centerline is referred to as the effective center. The location of the effective center is useful in bearing load calculations.

For example, when a radial load is applied on the bearing, this produces both radial and thrust reactions. The thrust force component, which acts in the direction of the shaft centerline, can separate the cone from the cup by sliding the shaft in the axial direction through the cone or by the cup's sliding axially in its seat. To prevent such undesired axial motion, a single-row tapered bearing should be mounted with another tapered bearing in the opposite direction. This arrangement is also very important for adjusting the clearance.

One major advantage of the tapered roller bearing is that it can be applied in adjustable arrangement where two tapered roller bearings are mounted in opposite directions (in a similar way to the adjustable arrangement of the angular contact ball bearing that was discussed earlier). This arrangement allows one to eliminate undesired clearance and to provide a preload (interference or negative clearance). Bearing preload increases the bearing stiffness, resulting in reduced vibrations as well as a lower level of run-out errors in precision machining. However, the disadvantage of bearing preloading is additional contact stresses and higher friction. Preload results in lowering the speed limit because the higher friction causes overheating at high speeds.

The adjustment of bearing clearance can be done during assembly and even during steady operation of the machine. The advantage of adjustment during operation is the precise elimination of the clearance after the thermal expansion of the shaft.

12.2.2.3 Multirow Tapered Roller Bearings

The multirow tapered roller bearing (Fig. 12-8) is manufactured with a predetermined adjustment that enables assembly into a machine without any further adjustment. The multirow arrangement includes spacers and is referred to as a

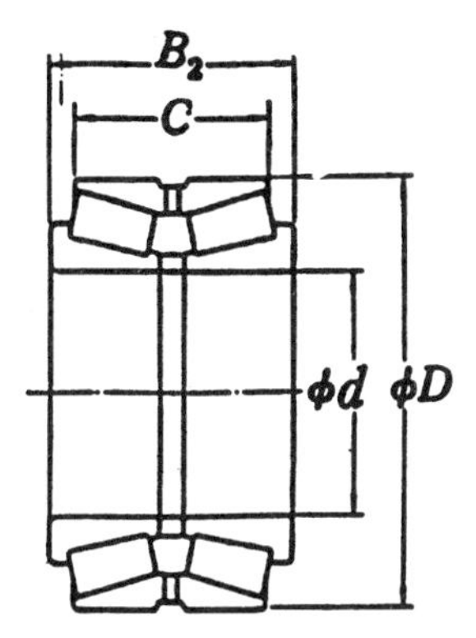

FIG. 12-8 Multirow tapered roller bearings.

spacer assembly. The spacer is matched with a specific bearing assembly during manufacturing. It is important to note that components of these assemblies are not interchangeable. Other types, without spacers, are manufactured with predetermined internal adjustment, and their components are also not interchangeable.

12.2.2.4 Needle Roller Bearing

These bearings (Fig. 12-9) are similar to cylindrical roller bearings, in the sense that they support high radial load. This type of bearing has a needlelike appearance because of its higher length-to-diameter ratio.

The objective of a needle roller bearing is to save space. This is advantageous in applications where bearing space is limited. Furthermore, in certain applications needle roller bearings can also be mounted directly on the

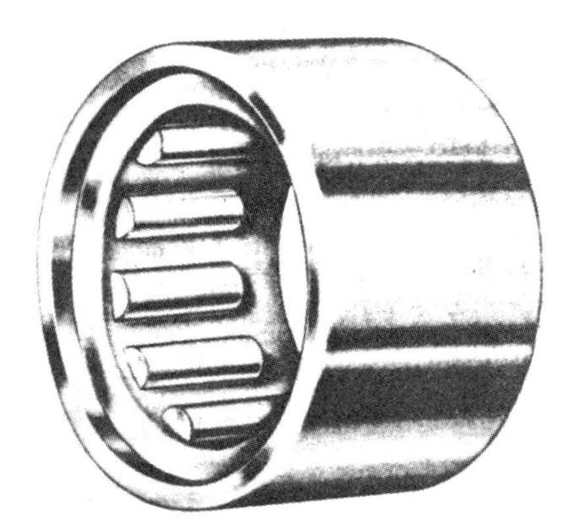

FIG. 12-9 Needle roller bearing.

shaft. For a direct mounting, the shaft must be properly hardened to a similar hardness of a bearing ring.

Two types of needle roller bearings are available. The first type, referred to as *full complement*, does not include a cage; the second type has a cage to separate the needle rollers in order to prevent them from sliding against each other. The full-complement bearing has more rollers and can support higher radial load. The second type has a lower number of rollers because it has a cage to separate the needle rollers to prevent them from rubbing against each other. The speed of a full-complement bearing is limited because it has higher friction between the rollers. A full-complement needle bearing may comprise a maximum number of needle rollers placed between a hardened shaft and a housing bore. An outer ring may not be required in certain situations, resulting in further saving of space.

12.2.2.5 Self-Aligning Spherical Roller Bearing

This bearing has barrel-shaped rollers (Fig. 12-10). It is designed for applications that involve misalignments due to shaft bending under heavy loads and due to manufacturing tolerances or assembly errors (in a similar way to the self-aligning ball bearing). The advantage of the spherical roller bearing is its higher load capacity in comparison to that of a self-aligning ball bearing, but it has higher frictional losses.

Spherical roller bearings are available as single-row, double-row, and thrust types. The single-row thrust spherical roller bearing is designed to support only thrust load, and it is not recommended where radial loads are present. Double-row

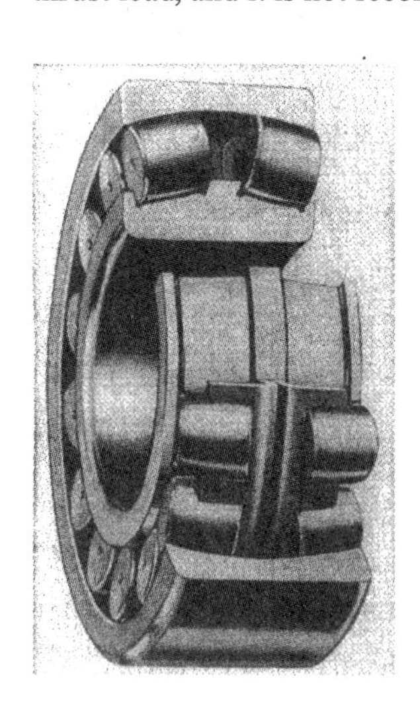

FIG. 12-10 (a) Spherical roller bearing, (b) spherical roller thrust bearing.

spherical roller bearings are commonly used when radial as well as thrust loads are present.

The double-row spherical roller bearing has the highest load capacity of all rolling bearings. This is due to the relatively large radius of contact of the rolling element. It can resist impact and other dynamic forces. It is used in heavy-duty applications such as ship shafts, rolling mills, and stone crushers.

12.3 HERTZ CONTACT STRESSES IN ROLLING BEARINGS

Hertz theory considers the elastic deformation and stress distribution near the contact of the rolling elements and races. Under load, due to an elastic deformation, the line or point contact becomes a contact area, This area is very small, resulting in a very high maximum contact pressure of the order of 1–5 GPa (1 GPa = 145,040 psi).

The calculations of the maximum contact pressure and deformation at the contact area of the rolling clement and raceways are according to Hertz's equations, which are based on the following assumptions.

1. The materials of the two bodies in contact are homogeneous and isotropic.
2. The yield point of the material is not exceeded, so plastic deformation is negligible and only elastic deformation is considered for Hertz's theory. Elastic deformation is recoverable after the load is removed.

In fact, assumption 2 is not completely accurate in practice. In heavily loaded rolling bearings there is a small plastic deformation at the contacts. However, experiments have verified Hertz's analysis. The actual stresses do not have any significant deviation from the values predicted by Hertz's theory.

3. In the contact area, only normal stresses are transmitted. Shear stresses due to friction on the surface are not considered in Hertz's theory.
4. The contact area is flat. The effect of any actual curvature can be disregarded for the analysis of stress distribution.

Concerning the last assumption, Hertz's theory is less accurate for bearings with a small radius of curvature, such as at the contact of a deep-groove ball bearing. The theory is more accurate for bearings with a large radius of curvature, e.g., at the contact of the outer ring of a self-aligning ball bearing. However, in all applications, deviations from the actual stresses are not significant for practical purposes of bearing design.

The ISO 281 standard refers to the limiting static load, C_0, of rolling-element bearings. Manufacturers' catalogues include this limit for rolling bearing static radial capacity, C_{or}. In Manufacturers' catalogues, this value is based on a limit stress of 4.2 GPa (609,000 psi) for 52100 steel and 3.5 GPa (508,000 psi) for 440C stainless steel. The static radial capacity, C_{or}, is based on the peak load, W_{max}, on one rolling element as well as additional transient and momentary overload on the same rolling element during start-up and steady operation.

At these ultimate pressure levels, the assumption of pure elastic deformation is not completely correct, because a minute plastic (irreversible) deformation occurs. For most applications, the microscopic plastic depression does not create a noticeable effect, and it does not cause a significant microcracking that can reduce the fatigue life. However, in applications that require extremely quiet or uniform rotation, a lower stress limit is usually imposed. For example, for bearings in satellite antenna tracking actuators, a static stress limit of only 2.2 GPa (320,000 psi,) is allowed on bearings made of 440C steel. This limit is because plastic deformation must be minimized for accurate functioning of the mechanism.

12.4 THEORETICAL LINE CONTACT

If a load is removed, there is only a line contact between a cylinder and a plane. However, under load, there is an elastic deformation at the contact, and the line contact becomes a rectangular contact area. The width of the contact is $2a$, as shown in Fig. 12-11. The magnitude of a (half-contact width) can be determined by the equation

$$a = R_x \left(\frac{8\bar{W}}{\pi} \right)^{1/2} \tag{12-2a}$$

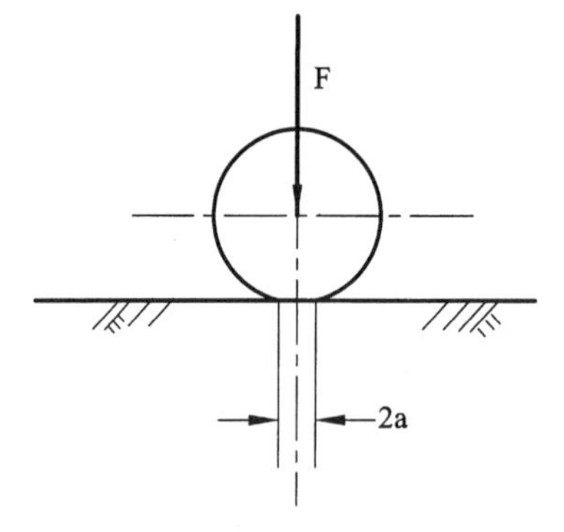

FIG. 12-11 Contact area of a cylinder and a plane.

Here, the dimensionless load, $\bar{W}$, is defined by

$$\bar{W} = \frac{W}{LE_{eq}R_x} \tag{12-2b}$$

The load W acts on the contact area. The effective length of the cylinder is L, and E_{eq} is the equivalent modulus of elasticity. In this case, R_x is an equivalent contact radius, which will be discussed in Sec. 12.4.2. For a contact of two different materials, the equivalent modulus of elasticity, E_{eq}, is determined by the following expression:

$$\frac{2}{E_{eq}} = \frac{1-\nu_1^2}{E_1} + \frac{1-\nu_2^2}{E_2} \tag{12-3a}$$

Here ν_1 and ν_2 are Poisson's ratio and E_1, E_2 are the moduli of elasticity of the two materials in contact, respectively. If the two surfaces are made of identical materials, such as in standard rolling bearings, the equation is simplified to the form

$$E_{eq} = \frac{E}{1-\nu^2} \tag{12-3b}$$

For a contact of a cylinder and plane, R_x is the cylinder radius. The subscript x defines the direction of the coordinate x along the cylinder axis. In the case of a line contact between two cylinders, R_x is an equivalent radius (defined in Sec. 12.4.2) that replaces the cylinder radius.

The bearing load is distributed unevenly on several rolling elements. The maximum load on a single cylindrical roller, W_{max}, can be approximated by the following equation, which is based on the assumption of zero radial clearance in the bearing:

$$W_{max} \approx \frac{4W_{bearing}}{n_r} \tag{12-4}$$

Here, n_r is the number of cylindrical rolling elements around the bearing and $W_{bearing}$ is the total radial load on a bearing. For design purposes, the maximum load, W_{max}, is substituted in Eq. (12-2b) for the calculation of the contact width, which is used later in the equations of the maximum deformation and maximum contact pressure.

12.4.1 Effective Length

The actual line contact is less than the length of the cylindrical rolling element because the corners are rounded. The rounded part on each side is of an approximate length equal to the cylinder radius. For determining the effective length, the cylindrical roller diameter is subtracted from the actual length of the cylindrical rolling element, $L = L_{\text{actual}} - d$.

12.4.2 Equivalent Radius

Equations (12-2) are for a contact of a cylinder and a plane, as shown in Fig. 12-12. However, in cylindrical roller bearings, there is always a contact between two cylinders of different curvature. A theoretical line contact can be between convex or concave curvatures. In all these cases, an equivalent radius, R_x, of contact curvature can be used that replaces the cylinder radius in Eq. (12-2a) and (12-2b).

Case 1: Roller on a Plane. As stated earlier, for the simple example of a contact between a plane and cylinder of radius R (roller on a plane), the equivalent contact radius is $R_x = R$.

Case 2: Convex Contact. The second case is that of a convex line contact of two cylinders, as shown in Fig. 12-13. An example of this type of contact is that between a cylindrical roller and the inner ring race. The equivalent contact radius, R_x, that is substituted in Eqs. (12.2a and b) is derived from the following expression:

$$\frac{1}{R_x} = \frac{1}{R_1} + \frac{1}{R_2} \tag{12-5}$$

Here, R_1 and R_2 are the radii of two curvatures of the convex contact, and the subscript x defines the direction of the axis of the two cylinders.

Case 3: Concave Contact. A concave contact is shown in Fig. 12-14. An example of this type of contact is that between a cylindrical rolling element and the outer ring race. For a concave contact, radius R_1 is negative, because the contact is inside this circle. The result is that the equivalent radius is derived

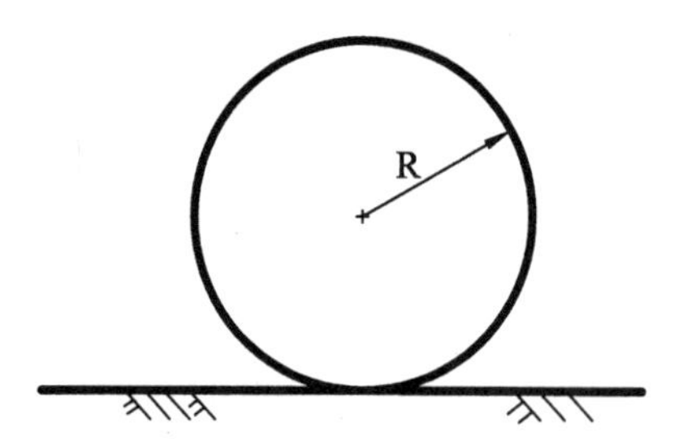

FIG. 12-12 Case 1: roller on a plane.

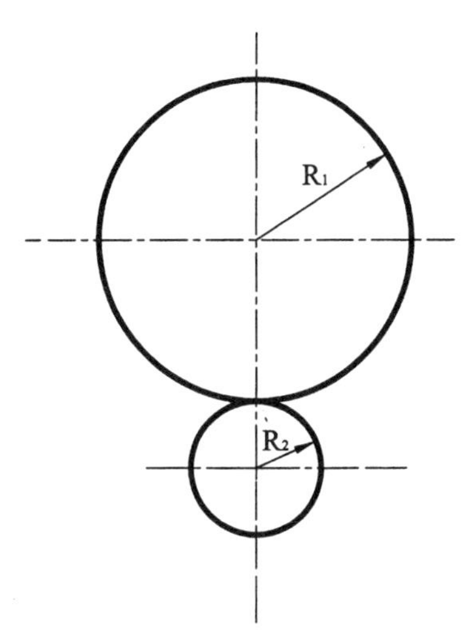

FIG. 12-13 Case 2: convex contact.

according to the following equation:

$$\frac{1}{R_x} = \frac{1}{R_2} - \frac{1}{R_1} \tag{12-6}$$

In a concave contact, the equivalent radius of contact is larger than each of the two radii in contact.

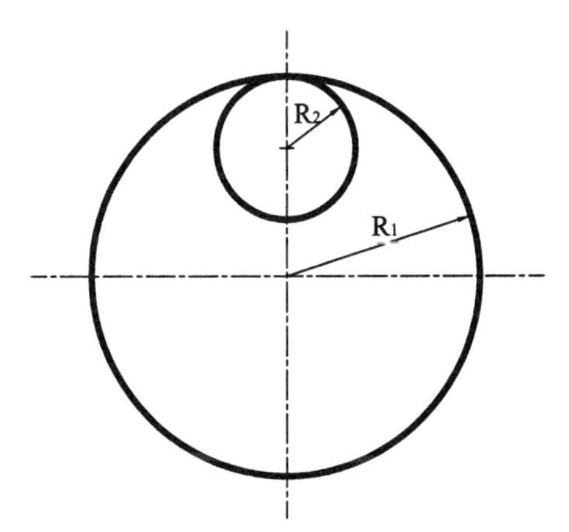

FIG. 12-14 Case 3: concave contact.

12.4.3 Deformation and Stresses in Line Contact

For a line contact, the maximum deformation of the roller in the direction normal to the contact area (vertical direction in Fig. 12-11) is

$$\delta_m = \frac{2\bar{W}R_x}{\pi}\left[\ln\left(\frac{2\pi}{\bar{W}}\right) - 1\right] \tag{12-7}$$

According to Hertz's theory, there is a parabolic pressure distribution at the contact area, as shown in Fig. 12-15. The maximum contact pressure is at the center of the contact area, and it is equal to

$$p_{\max} = E_{\text{eq}}\left(\frac{\bar{W}}{2\pi}\right)^{1/2} \tag{12-8}$$

12.4.4 Subsurface Stress Distribution

An important feature in contact stresses is that the maximum shear stress is below the surface. In many cases, this is the reason for the development of subsurface fatigue cracks and eventually fatigue failure in rolling bearings. Three curves of dimensionless stress distributions below the surface and below the center of contact are shown in Fig. 12-16. The maximum pressure at the contact area center normalizes the stresses. The maximum shear, $\tau_{\max}$, is considered an important cause of failure. The ratio of the maximum shear stress to the maximum surface pressure ($\tau_{\max}/p_{\max}$) is plotted (maximum shear, $\tau_{\max}$, is at an angle of 45° to the z axis). The maximum value of this ratio is at a depth of $z = 0.78a$, and its magnitude is $\tau_{\max} = 0.3p_{\max}$.

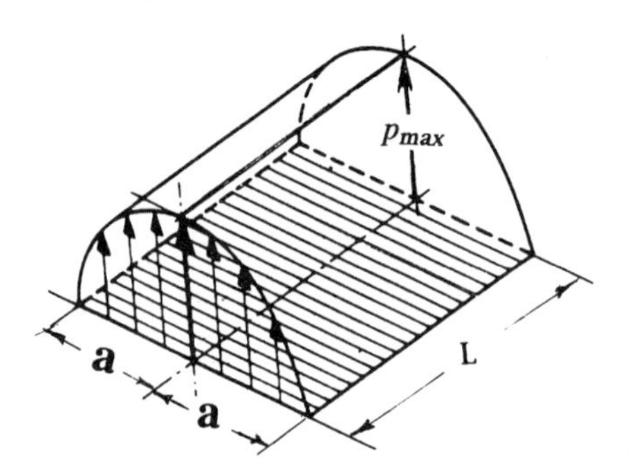

FIG. 12-15 Pressure distribution in a rectangular contact area.

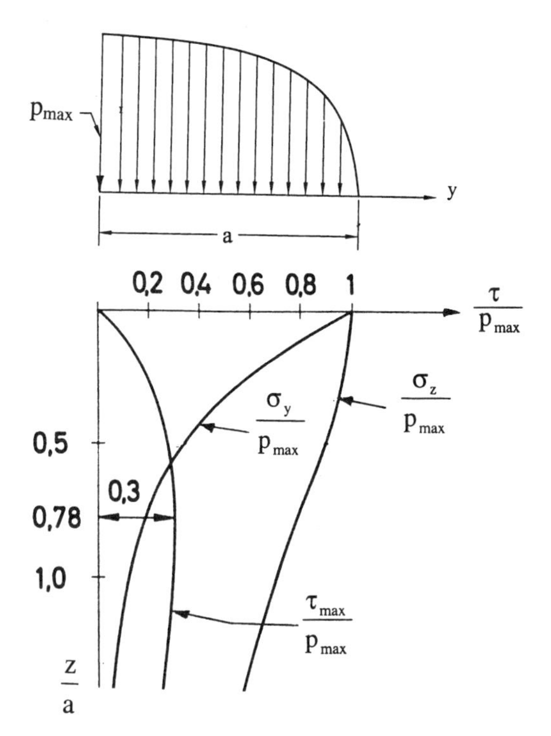

FIG. 12-16 Subsurface stresses in a rectangular contact area under p_{max}.

Example Problem 12-1

A cylindrical rolling bearing has an external load of $F = 11{,}000$ N. The bearing has the following dimensions: The diameter of the inner raceway D_{in} is 120 mm, and the diameter of the outer raceway D_{out} is 160 mm. The diameter of the cylindrical roller d_{roller} is 20 mm, and its effective length is $L = 10$ mm. There are 14 rolling elements around the bearing. The bearing (rollers and rings) is made of steel. The modulus of elasticity of the steel is $E = 2.05 \times 10^{11}$ N/m^2, and its Poisson ratio is $\nu = 0.3$.

Find the maximum pressure at the contact of the rolling elements and the raceways. Compare the maximum pressure where the rolling element contacts the inner and the outer raceways.

Solution

The radii of the contacting curvatures are

$$R_{roller} = 0.01 \text{ m}, \qquad R_{outer\ raceway} = 0.08 \text{ m}, \qquad R_{inner\ raceway} = 0.06 \text{ m}$$

The contact between the rolling elements and the inner raceway is convex, and the equivalent contact curvature, $R_{x,\mathrm{in}}$ is derived according to the equation

$$\frac{1}{R_{x,\mathrm{in}}} = \frac{1}{R_{\mathrm{roller}}} + \frac{1}{R_{\mathrm{inner\ raceway}}}$$

$$\frac{1}{R_{x,\mathrm{in}}} = \frac{1}{0.01} + \frac{1}{0.06} \qquad \Rightarrow R_{\mathrm{x,in}} = 0.0085 \mathrm{\ m}$$

However, the contact between the rolling elements and the outer raceway is concave, and the equivalent contact curvature, $R_{x,\mathrm{out}}$ is derived according to the equation

$$\frac{1}{R_{x,\mathrm{out}}} = \frac{1}{R_{\mathrm{roller}}} - \frac{1}{R_{\mathrm{outer\ raceway}}}$$

$$\frac{1}{R_{x,\mathrm{out}}} = \frac{1}{0.01} - \frac{1}{0.08}; \qquad R_{x,\mathrm{out}} = 0.0114 \mathrm{\ m}$$

If we assume that the radial clearance between rolling elements and raceways is zero, the maximum load on one cylindrical rolling element can be approximated by Eq. (12-4):

$$W_{\max} = \frac{4W_{\mathrm{bearing}}}{n} = \frac{4(11{,}000)}{14} = 3142 \mathrm{\ N}$$

Here, n is the number of cylindrical rolling elements in the bearing, W_{bearing} is the total bearing load capacity, and $W_{\max}$ is the maximum load capacity of one rolling element. The shaft and the bearing are made of identical material, so the equivalent modulus of elasticity is calculated as follows:

$$E_{\mathrm{eq}} = \frac{E}{1-\nu^2} \Rightarrow E_{\mathrm{eq}} = \frac{2.05 \times 10^{11}}{1-0.3^2} = 2.25 \times 10^{11} \mathrm{\ N/m^2}$$

The dimensionless maximum force at the contact between one rolling element and the inner ring race is

$$\text{Inner ring:} \quad \bar{W} = \frac{1}{E_{\mathrm{eq}} R_{x,\mathrm{in}} L} W_{\max} = \frac{1}{2.25 \times 10^{11} \times 0.0085 \times 0.01} \times 3142$$
$$= 1.64 \times 10^{-4}$$

In comparison, the dimensionless maximum force at the contact between one rolling element and the outer ring race is

$$\text{Outer ring:} \quad \bar{W} = \frac{1}{E_{\mathrm{eq}} R_{\mathrm{eq,out}} L} W_{\max} = \frac{1}{2.25 \times 10^{11} \times 0.0114 \times 0.01}$$
$$\times 3142 = 1.22 \times 10^{-4}$$

Comparison of the inner and outer dimensionless loads indicates a higher value for the inner contact. This results in higher contact stresses, including maximum pressure at the convex contact with the inner ring. The maximum pressure at the contact with the inner ring race is obtained via Eq. (12-8).

$$p_{\max,\mathrm{in}} = E_{\mathrm{eq}}\left(\frac{\bar{W}}{2\pi}\right)^{1/2} = 2.25 \times 10^{11} \times \left(\frac{1.64 \times 10^{-4}}{2\pi}\right)^{1/2} = 1.15 \times 10^{9}\ \mathrm{Pa}$$
$$= 1.15\ \mathrm{GPa}$$

At the same time, the maximum pressure at the contact with the outer ring race is lower:

$$p_{\max,\mathrm{in}} = E_{\mathrm{eq}}\left(\frac{\bar{W}}{2\pi}\right)^{1/2} = 2.25 \times 10^{11} \times \left(\frac{1.22 \times 10^{-4}}{2\pi}\right)^{1/2} = 0.99 \times 10^{9}\ \mathrm{Pa}$$
$$= 0.99\ \mathrm{GPa}$$

For regular-speed operation, it is sufficient to calculate the maximum pressure at the inner contact, because the stresses at the outer contact are lower (due to the concave contact). However, at high speed the centrifugal force of the rolling element increases the maximum pressure at the contact with the outer ring race relative to that of the inner ring. Therefore, at high speed, the centrifugal force is considered and the maximum pressure at the inner and outer contact should be calculated.

12.5 ELLIPSOIDAL CONTACT AREA IN BALL BEARINGS

If there is no load, there is a point contact between a sphere and a flat plane. Under load, the point contact becomes a circular contact area. However, in ball bearings the races have different curvatures in the direction of rolling and in the axial direction of the bearing. Therefore, the two bodies form an elliptical contact area. The elliptical contact area has radii a and b, as shown in Fig. 12-17.

12.5.1 Race and Ball Conformity

In a deep-groove radial ball bearing, the radius of the deep groove is always a little larger than that of the ball. A race conformity is the ratio R_r, defined as (see Hamrock and Anderson, 1973)

$$R_r = \frac{r}{d} \tag{12-9a}$$

Here, r and d are the deep-groove radius and ball diameter, respectively, as shown in Fig. 12-18a. A perfect conformity is $R_r = 0.5$. However, in order to reduce the

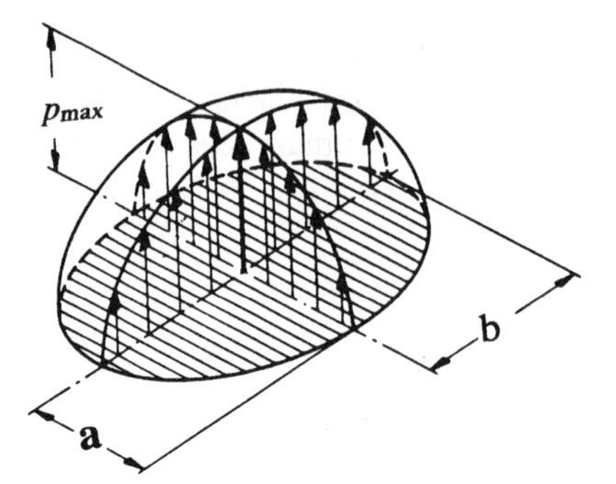

FIG. 12-17 Pressure distribution in an elliptical contact area.

friction, bearings are manufactured with a conformity ratio in the range $0.51 \leq R_r \leq 0.54$. On the other hand, if the conformity ratio is too high, it results in higher maximum contact stresses.

12.5.2 Equivalent Radius in Ball Bearing Contacts

In Fig. 12-18b, two orthogonal cross sections are shown. Each cross section shows a plane of contact of two curvatures. The left-hand cross section is the *x-z* plane, which is referred to as the *y* plane, because it is normal to the *y* coordinate. The left-hand cross section is of the inner ring in contact with a rolling ball. The contact between a ball and deep-groove curvature of the inner ring is a concave contact.

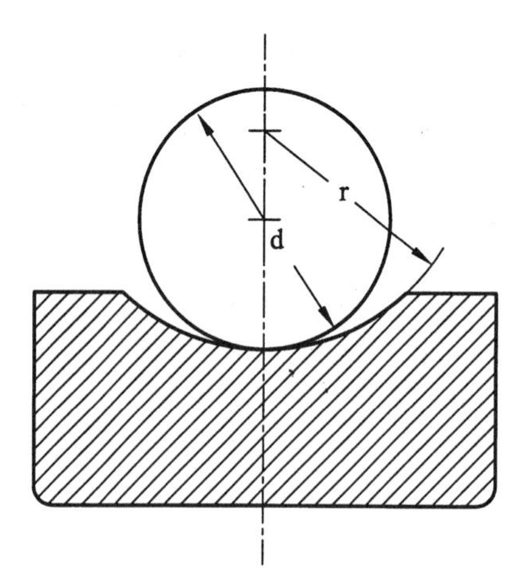

FIG. 12-18a Race and ball conformity.

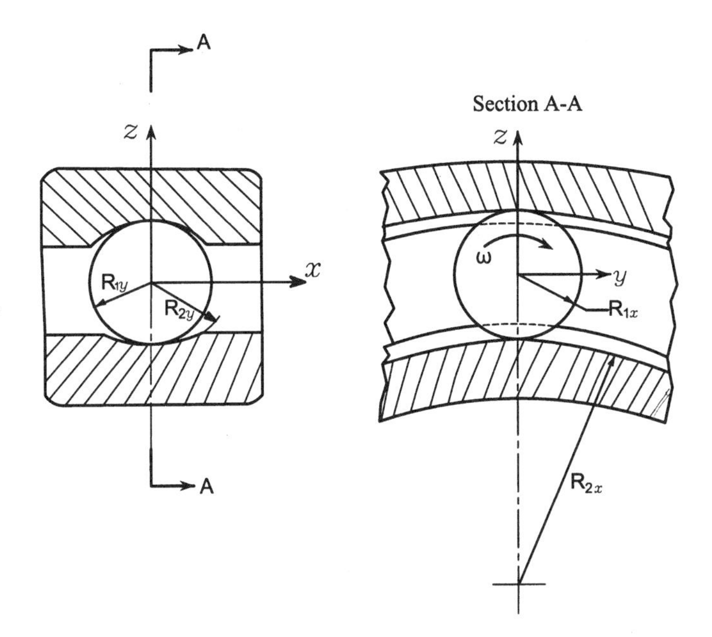

FIG. 12-18b Curvatures in contact in two orthogonal cross sections of a ball bearing (*x*-*z* and *y*-*z* planes).

In the y plane (left-hand cross section), the two radii of contact have a notation of subscript y. The radii of the curvatures in contact are R_{1y} and R_{2y}, respectively. The small radius, R_{1y}, is of the rolling ball:

$$R_{1y} = \frac{d}{2} \tag{12-9b}$$

Here, d is the rolling ball diameter. The concave curvature of the deep groove is of the somewhat larger radius R_{2y}. The equivalent contact radius in the y plane is R_y (equal to r in Fig. 12-18a). For a concave contact, the equivalent radius of contact between the ball and the deep groove of the inner ring race is calculated by the equation

$$\frac{1}{R_y} = \frac{1}{R_{1y}} - \frac{1}{R_{2y}} \tag{12-10}$$

The right-hand cross section in Fig. 12-18b is in the y-z plane. This plane is referred to as the x plane, because it is normal to the x direction. It shows a cross section of the rolling plane where a ball is rolling around the inner ring. This is a convex contact between the ball and the inner ring race. The ball has a rolling contact at the bottom diameter of the deep groove of the inner ring race.

In the x plane, the radii of the curvatures in convex contact are R_{1x} and R_{2x}, which are of the ball and of the lowest point of the inner ring deep groove, respectively. The rolling ball radius is $R_{1x} = R_{1y} = d/2$, while the inner ring race radius at the bottom of the deep groove is R_{2x}, as shown on the right-hand side of Fig. 12-18b. The equivalent contact radius, R_x, is in the x plane of the convex contact with the inner ring (the subscript x indicates that the equivalent radius is in the x plane). The equivalent contact radius, R_x, is derived from the equation

$$\frac{1}{R_x} = \frac{1}{R_{1x}} + \frac{1}{R_{2x}} \tag{12-11}$$

The radius ratio, α_r, is defined as

$$\alpha_r = \frac{R_y}{R_x} \tag{12-12}$$

This is the ratio of the larger radius to the smaller radius, $\alpha_r > 1$. The equivalent contact radius R_{eq} is obtained from combining the equivalent radius in the two orthogonal planes, as follows:

$$\frac{1}{R_{eq}} = \frac{1}{R_x} + \frac{1}{R_y} \tag{12-13}$$

The combined equivalent radius of curvature, R_{eq}, is derived from the contact radii of curvature R_x and R_y, in the two orthogonal cross sections shown in Fig. 12-18b. The equivalent radius R_{eq} is used in Sec. 12.5.4 to calculate the deformation and pressure distribution in the contact area.

12.5.3 Stresses and Deformation in an Ellipsoidal Contact

According to Hertz's theory, the equation of the pressure distribution in an ellipsoidal contact area in a ball bearing is

$$p = \left(1 - \frac{x^2}{a^2} - \frac{y^2}{b^2}\right)^{1/2} p_{max} \tag{12-14}$$

Here, a and b are the small radius and the large radius, respectively, of the ellipsoidal contact area, as shown in Fig. 12-17. The maximum pressure at the center of an ellipsoidal contact area is given by the following Hertz equation:

$$p_{max} = \frac{3}{2}\frac{W}{\pi ab} \tag{12-15}$$

Here, $W = W_{max}$ is the maximum load on one spherical rolling element. The maximum pressure is proportional to the load, and it is lower when the contact area is larger. The contact area is proportional to the product of a and b of the ellipsoidal contact area. The contact area is inversely proportional to the modulus of elasticity of the material. For example, soft materials such as rubbers have a large contact area and the maximum pressure is relatively low. In contrast, steel has high elasticity modulus, resulting in a small area and high stresses. The equations for calculating a and b are given in Sec. 12.5.4.

12.5.4 Ellipsoidal Contact Area Radii

The ellipticity parameter, k, is defined as the ratio of the large radius to the small radius:

$$k = \frac{b}{a} \tag{12-16}$$

The exact solution for the ellipsoid radii a and b is quite complex. For design purposes, Hamrock and Brewe (1983) suggested an approximate solution. The equations allow a simplified solution for the deformation and pressure distribution in the contact area. The ellipticity parameter, k, is estimated by the equation

$$k \approx \alpha_r^{2/\pi} \tag{12-17}$$

The ratio α_r is defined in Eq. 12-12. The following parameter, q_a, is used to estimate the dimensionless variable $\hat{E}$ that is used for the approximate solution of the ellipsoid radii:

$$q_a = \frac{\pi}{2} - 1 \tag{12-18}$$

$$\hat{E} \approx 1 + \frac{q_a}{\alpha_r} \qquad \text{for } \alpha_r \geq 1 \tag{12-19}$$

The ellipsoid radii a and b can now be estimated

$$a = \left(\frac{6\hat{E}WR_{\text{eq}}}{\pi k E_{\text{eq}}} \right)^{1/3} \tag{12-20a}$$

$$b = \left(\frac{6k^2\hat{E}WR_{\text{eq}}}{\pi E_{\text{eq}}} \right)^{1/3} \tag{12-20b}$$

Here, W is the load on one rolling element. The rolling elements do not share the load equally. At any time there is one rolling element that carries the maximum load, W_{max}. The maximum pressure is at the contact of the rolling element of maximum load.

The maximum deformation in the direction normal to the contact area is calculated by means of the following expression, which includes estimated terms:

$$\delta_m = \hat{T}\left[\frac{9}{2\hat{E}R_{eq}}\left(\frac{W}{\pi k E_{eq}}\right)^2\right]^{1/3} \tag{12-21}$$

Here, the following estimation for $\hat{T}$ is used:

$$\hat{T} \approx \frac{\pi}{2} + q_a \ln \alpha_r \qquad \text{for } \alpha_r \geq 1 \tag{12-22}$$

For ball bearings, the maximum load, W_{max}, on one rolling element can be estimated by the equation

$$W_{max} \approx \frac{5W_{bearing}}{n_r} \tag{12-23}$$

Here, n_r is the number of balls in the bearing. The maximum load, W_{max}, on one ball is substituted for the load W in Eqs. (12-15), (12-20a), (12-20b), and (12-21).

12.5.5 Subsurface Shear

Fatigue failure develops from subsurface cracks. These cracks propagate whenever there are alternating stresses and the maximum shear stress is high. It is important to evaluate the shear stresses below the surface that can cause fatigue failure.

The following is the maximum value of the shear, τ_{yz}, in the orthogonal direction (acting below the surface on a vertical plane y-z; see Fig. 12-18b. In fact, the maximum shear is in a plane inclined 45° to the vertical plane. However, the classical work of Lundberg and Palmgren (1947) on estimation of rolling bearing fatigue life is based on the maximum value of the orthogonal shear stress, τ_{yz}, acting on a vertical plane:

$$\tau_{xy} = p_{max}\frac{(2t^* - 1)^{1/2}}{2t^*(t^* + 1)} \tag{12-24}$$

where the following estimation can be applied:

$$t^* \approx 1 + 0.16 \,\text{csc}h\left(\frac{k}{2}\right) \tag{12-25}$$

12.5.6 Comment on Precise Solution by Elliptical Integrals

The calculation in Sec. 12.5.4 approximates a complex solution procedure that is based on the following equations [see Harris (1966)]:

$$k = \left[\frac{2\hat{T} - \hat{E}(1 + R_d)}{\hat{E}(1 - R_d)} \right]^{1/2} \tag{12-26}$$

where R_d is the curvature difference defined by the equation

$$R_d = R_{eq} \left(\frac{1}{R_x} - \frac{1}{R_y} \right) \tag{12-27}$$

The two elliptical integrals are defined as follows:

$$\hat{T} = \int_0^{\pi/2} \left[1 - \left(1 - \frac{1}{k^2} \right) \sin^2 \phi \right]^{-1/2} d\phi \tag{12-28}$$

$$\hat{E} = \int_0^{\pi/2} \left[1 - \left(1 - \frac{1}{k^2} \right) \sin^2 \phi \right]^{1/2} d\phi \tag{12-29}$$

An iteration method has developed by Hamrock and Anderson (1973) to solve for the two elliptical integrals and k. The solutions of the elliptical integrals and k are presented by graphs as a function of the radius ratio α_r. The use of the graphs or tables allows a precise solution. However, the approximate solution in Sec. 12.5.4 is sufficient for design purposes.

Example Problem 12-2

Maximum Contact Pressure in a Deep-Groove Ball Bearing

Find the maximum contact pressure and maximum deformation, δ_m, of a deep-groove ball bearing in the direction normal to the contact area. The bearing speed is low, so the centrifugal forces of the rolling elements are negligible. Therefore, calculate only the maximum values at the contact with the inner ring race.

The radial load on the bearing is $W = 10{,}500$ N. The bearing has 14 balls of diameter $d = 19.04$ mm. The radius of curvature of the inner deep groove (in cross section x-z in Fig. 12-18b) is 9.9 mm. The inner race diameter (at the bottom of the deep groove) is $d_i = 76.5$ mm (cross section y-z).

The rolling elements and rings are made of steel. The modulus of elasticity of the steel is $E = 2 \times 10^{11}$ N/m^2, and its Poisson ratio is $\nu = 0.3$.

Compare the maximum pressure to the allowed compression stress of 3.5 GPa (3.5×10^9 N/m^2) for a bearing made of 52100 steel.

Solution

Referring to the right-hand side of Fig. 12-18b, the two radii of contact curvatures at the inner ring race in the y-z plane (referred to as the x plane) are:

Inner ring radius: $R_{2x} = \frac{d_i}{2} = 38.25$ mm

Ball radius: $R_{1x} = \frac{d}{2} = 9.52$ mm

Equivalent Radius of Contact, R_x, in the x Plane at the Inner Ring

$$\frac{1}{R_x} = \frac{1}{R_{1x}} + \frac{1}{R_{2x}} \Rightarrow \frac{1}{R_x} = \frac{1}{9.52} + \frac{1}{38.25}; \qquad R_x = 7.62 \text{ mm}$$

On the left-hand side of Fig. 12.18b, the two radii of contact curvatures at the inner ring race in the x-z plane (referred to as the y plane) are:

Ball radius: $R_{1y} = 9.52$ mm
Deep-groove radius: $R_{2y} = 9.9$ mm

Equivalent Radius of Contact, R_y, in the y Plane at the Inner Ring

$$\frac{1}{R_y} = \frac{1}{R_{1y}} - \frac{1}{R_{2y}} \Rightarrow \frac{1}{R_y} = \frac{1}{9.52} - \frac{1}{9.9}; \qquad R_y = 248.0 \text{ mm}$$

The combined equivalent contact radius of curvature, R_{eq}, of the inner ring and ball contact is derived from the equivalent contact radii in the x plane and y plane according to the equation

$$\frac{1}{R_{eq}} = \frac{1}{R_x} + \frac{1}{R_y} \Rightarrow \frac{1}{R_{eq}} = \frac{1}{7.62} + \frac{1}{248}; \qquad R_{eq} = 7.4 \text{ mm}$$

and the ratio α_r becomes

$$\alpha_r = \frac{R_y}{R_x} = \frac{248}{7.62} = 32.55$$

The dimensionless coefficient k is derived directly from the ratio α_r:

$$k = \alpha_r^{2/\pi} = 32.55^{2/\pi} = 9.18$$

It is also necessary to calculate $\hat{E}$, which will be used for the calculation of the ellipsoid radii. For that purpose, q_a is required:

$$q_a = \frac{\pi}{2} - 1 = 0.57$$

$$\hat{E} \approx 1 + \frac{q_a}{\alpha_r} \Rightarrow 1 + \frac{0.57}{32.55} = 1.02$$

The shaft and the bearing are made of identical material, so the equivalent modulus of elasticity is calculated as follows:

$$E_{eq} = \frac{E}{1-\nu^2} \Rightarrow E_{eq} = \frac{2 \times 10^{11}}{1-0.3^2} = 2.2 \times 10^{11}\ \mathrm{N/m^2}$$

The maximum load at the contact of one rolling element can be estimated by the following equation:

$$W_{max} \approx \frac{5W_{shaft}}{n_r} = \frac{5(10{,}500)}{14} = 3750\ \mathrm{N}$$

Here, $n_r = 14$ is the number of balls in the bearing.

The ellipsoid radii a and b can now be determined by substitution of the values already calculated (in SI units):

$$a = \left(\frac{6\hat{E}W_{max}R_{eq}}{\pi k E_{eq}}\right)^{1/3} = \left(\frac{6 \times 1.02 \times 3750 \times 0.0074}{\pi \times 9.18 \times 2.2 \times 10^{11}}\right)^{1/3}$$
$$= 0.3 \times 10^{-3}\mathrm{m} = 0.3\ \mathrm{mm}$$

$$b = \left(\frac{6k^2\hat{E}W_{max}R_{eq}}{\pi E_{eq}}\right)^{1/3} = \left(\frac{6 \times 9.18^2 \times 1.02 \times 3750 \times 0.0074}{\pi \times 2.2 \times 10^{11}}\right)^{1/3}$$
$$= 2.75 \times 10^{-3}\mathrm{m} = 2.75\ \mathrm{mm}$$

The contact load W_{max} taken here is the maximum load on one rolling element. The maximum pressure at the contact with the inner ring race can now be determined from Eq. (12-15):

$$p_{max} = \frac{3}{2}\frac{W_{max}}{\pi ab} = \frac{3}{2}\frac{3750}{\pi \times 0.3 \times 2.87} = 2170.3\ \mathrm{N/mm^2} = 2.17 \times 10^9\ \mathrm{N/m^2}$$

or

$$p_{max} = 2.17\ \mathrm{GPa}$$

In this case, the calculated maximum pressure is below the allowed compression stress of 3.5 GPa ($3.5 \times 10^9\ \mathrm{N/m^2}$) for rolling bearings made of 52100 steel.

Deformation Normal to the Contact Area

For the purpose of calculating the maximum deformation, δ_m, the approximation for $\hat{T}$ is determined from the following equation:

$$\hat{T} \approx \frac{\pi}{2} + q_a \ln \alpha_r = \frac{\pi}{2} + 0.57 \times \ln 32.55 = 3.56$$

All the variables are now known and can be substituted in Eq. (12-21) to solve for the maximum elastic deformation at the contact (of one rolling element) in the direction normal to the contact area:

$$\delta_m = \hat{T}\left[\frac{9}{2\hat{E}R_{eq}}\left(\frac{W_{max}}{\pi k E_{eq}}\right)^2\right]^{1/3}$$

$$\delta_m = 3.56\left[\frac{9}{2\times 1.02\times 0.0074}\times\left(\frac{3750}{\pi\times 9.18\times 2.2\times 10^{11}}\right)^2\right]^{1/3}$$

$$= 2.12\times 10^{-5}\ \text{m} = 21.2\ \mu\text{m}$$

12.6 ROLLING-ELEMENT SPEED

The velocity of a rolling-element center, U_r, is important for the calculation of the centrifugal forces. In addition, the rolling angular speed is required for elastohydrodynamic fluid film computations. The rolling speed is the velocity at which the rolling-element contact is progressing relative to a fixed point on the race.

12.6.1 Velocity of the Rolling-Element Center

The velocity diagram of a rolling element is shown in Fig. 12-19. It is for a stationary outer ring and rotating inner ring. The inner ring rotates together with the shaft at an angular speed ω. The velocity of a rolling-element center is shown

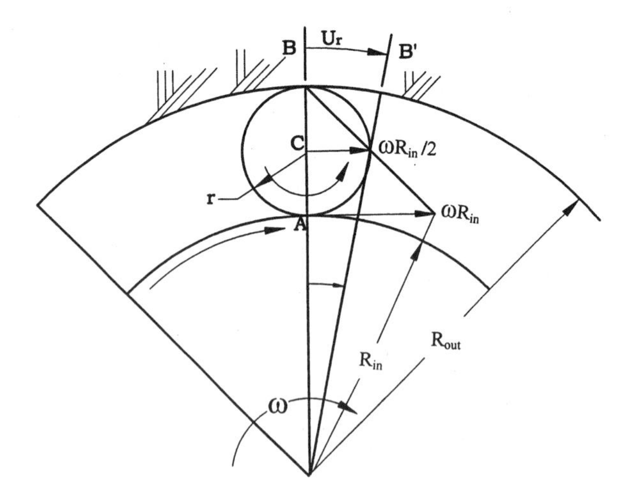

FIG. 12-19 Velocity diagram of a rolling element.

as a vector in the tangential direction; see point C. The rolling element has pure rolling (no slip over the races). The inner ring radius (at the contact) is R_{in}, and that of the outer ring is R_{out}. The velocity of the inner ring at contact point A is

$$U_A = \omega R_{in} \tag{12-30}$$

If the outer ring is stationary, point B is an instantaneous center of rotation. There is a linear velocity distribution along the line AB of the rolling element. For pure rolling, the velocity of point A on the inner ring is equal to that of point A on the rolling element U_A because there is no relative sliding between the two.

The velocity U_c of the rolling element center, point C, is half of that of point A, being at half the distance from the center of rotation B. The velocity of point C is

$$U_C = \frac{\omega R_{in}}{2} \tag{12-31}$$

12.6.2 Angular Velocity of the Rolling-Element Center

The angular velocity of the rolling-element center ω_C, together with the cage, is lower than the bearing (or shaft) speed. The angular velocity ω_C of the rolling element center, point C, is equal to the velocity U_C divided by the distance $(R_{in} + r)$ of point C from the bearing center O. The angular velocity ω_C of the rolling-element center is given by

$$\omega_C = \frac{R_{in}}{2(R_{in} + r)} \omega = \frac{R_{in}}{R_{in} + R_{out}} \omega \tag{12-32}$$

12.6.3 Rolling Velocity

Contact point B is moving due to the rolling action. Point B moves around the outer ring race at a rolling speed U_r. The rolling speed of point B can be determined by the angular motion of point C, because the contact point B is always on the line OC; therefore,

$$U_r = U_{\text{rolling of point } B} = \omega_C \overline{OB} = \omega_C R_{out} \tag{12-33}$$

Substituting the value at ω_C from Eq. (12-32) into Eq. (12-33), the expression for the rolling speed becomes

$$U_{rolling} = \frac{R_{in} R_{out}}{2(R_{in} + r)} \omega = \frac{R_{in} R_{out}}{R_{in} + R_{out}} \omega \tag{12-34}$$

Here, R_{out} is the radius of the outer ring raceway and R_{in} is the radius of the inner ring raceway; see Fig. 12-19. The angular speed ω (rad/s) is of a stationary outer ring and rotating inner ring, or vice versa.

The rolling speed U_r can also be written as a function of the inside and outside diameters, d_i and d_o respectively,

$$U_r = \frac{1}{2}\frac{d_i d_o}{d_i + d_o}\omega \tag{12-35}$$

Equations (12-32) and (12-35) also apply to the case where the inner ring is stationary and the outer ring rotates at angular speed ω. In addition, the rolling speeds at points A and B are equal, because there is no slip.

12.6.4 Centrifugal Forces of Rolling Elements

The angular velocity, ω_C, is used for solving for the centrifugal force, F_c, of an individual rolling element:

$$F_c = m_r \omega_C^2 (R_{\text{in}} + r) \tag{12-36}$$

The angular speed of the rolling-element center, ω_C is determined from Eq. (12-32). The volume of a rolling ball and its material density, ρ, determine the ball mass, m_r:

$$m_r = \frac{\pi d^3}{6}\rho \tag{12-37}$$

Here, d is the ball diameter. For a standard bearing, the density of steel is about $\rho = 7800\ \text{kg/m}^3$. In comparison, silicon nitride has a much lower density, $\rho = 3200\ \text{kg/m}^3$.

12.7 ELASTOHYDRODYNAMIC LUBRICATION IN ROLLING BEARINGS

Elastohydrodynamic (EHD) lubrication theory is concerned with the formation of a thin fluid film at the contact area of a rolling element and a raceway (see Dowson and Higginson, 1966). Under favorable conditions of speed, load, and fluid viscosity, the elastohydrodynamic fluid film can be of sufficient thickness to separate the rolling surfaces. Rolling bearings operating with a full EHD film have significant reduction of wear. However, even mixed lubrication would be beneficial in wear reduction, and much longer bearing life is expected in comparison to dry bearings.

As has been discussed in the previous sections, in cylindrical rolling bearings there is a theoretical line contact between rolling elements and raceways, whereas in ball bearings there is a point contact. However, due to elastic deformation, there is a small contact area where a thin fluid film is generated due to the rotation of the rolling elements. In a similar way to the formation of fluid film in plain bearings, the oil adheres to the surfaces, resulting in a squeeze-film effect between the rolling surfaces.

If the film thickness exceeds the size of the surface asperities, it can completely separate the rolling surfaces and thus eliminate wear due to a direct contact. Theoretically, the stresses under an EHD fluid film are similar to the Hertz stresses of dry contact, and the fluid film is not expected to improve the fatigue life. However, in practice the fluid-film acts as a damper and reduces dynamic stresses due to impact and vibrations. In this way, it improves the fatigue life as well as wear resistance. Tests indicated (Tallian, 1967) that the fatigue life is significantly improved for rolling bearings operating with a full EHD film.

In hydrodynamic journal bearings, the minimum film thickness depends only on one fluid property, the viscosity of the lubricant. In comparison, the formation of a elastohydrodynamic fluid film is more complex and depends on several physical properties of the fluid as well as of the solid material in contact. In addition to the effect of viscosity, two additional important effects are involved in the formation of a elastohydrodynamic fluid film: elastic deformation of the surfaces at the contact and a rise of viscosity with pressure. In summary, the film thickness depends on the following properties of the rolling bearing material and lubricant.

The elastohydrodynamic film thickness depends on the elastic properties of the two materials in contact, namely, their elastic modulus E and Poisson's ratio ν.

The pressure at the contact area is high. The viscosity of lubricant increases significantly with pressure. Therefore, the elastohydrodynamic film thickness depends on the pressure–viscosity coefficient $\alpha[m^2/N]$ as well as the absolute viscosity at atmospheric pressure.

The fluid film and the hydrodynamic pressure wave are shown in Fig. 12-20. The fluid film thickness increases with the rolling speed. The clearance thickness, h_0, is nearly constant along the fluid film, and it reduces only near the outlet (right side), where the minimum film thickness is h_{min}.

The left-hand side of Fig. 12-20 is a comparison between the EHD pressure wave and the Hertz pressure of a dry contact. The fluid film pressure wave increases from the inlet and reaches a peak equal to the maximum Hertz pressure, p_{max}, at the contact area center. After that, the pressure decreases, but rises again with a sharp spike near the outlet side, where the gap narrows to h_{min}. Under high loads and low speeds, the EHD pressure distribution is similar to that of Hertz theory, because the influence of the elastic deformation dominates the pressure distribution. However, at high speed the hydrodynamic effect prevails, and the EHD pressure spike is relatively high.

Generally, the minimum thickness of the lubricant film is of the order of a few tenths to 1 micrometer; however optimal conditions of high speed and adequate viscosity, a film thickness of several micrometers can be generated. For critical applications, such as high-speed turbines, it is very important to minimize

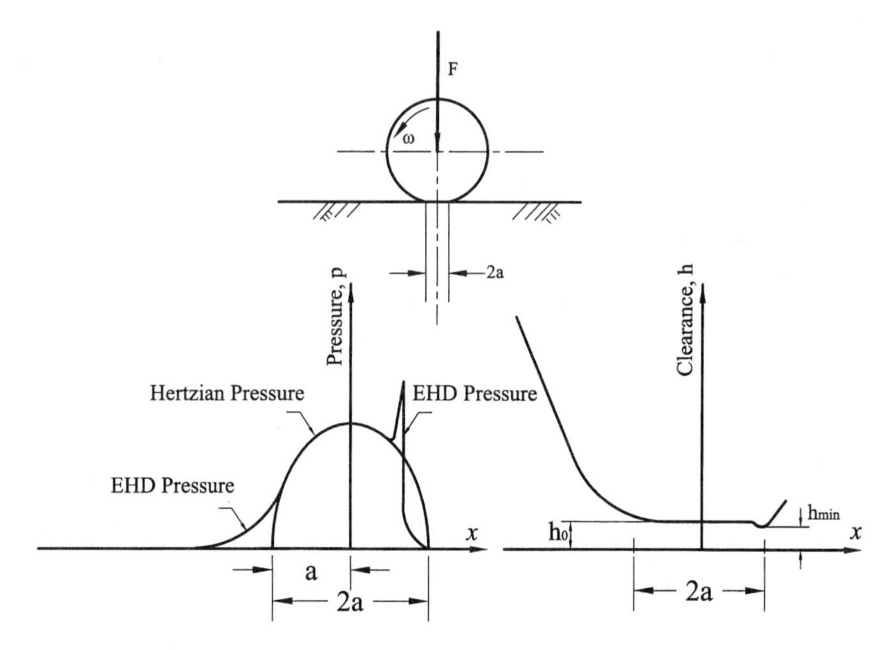

FIG. 12-20 Pressure wave and film thickness distribution in elastohydrodynamic lubrication.

wear and optimize the bearing life. In such cases, it is possible to have an optimal design that operates with a full EHD fluid film. For this purpose, the designer can calculate the minimum film thickness, h_{min}, according to the equations discussed in the following sections.

For optimal conditions, the minimum film thickness, h_{min}, must be greater than the surface roughness. The equivalent surface roughness at the contact, R_s (RMS), is obtained from the roughness of the two individual surfaces in contact, R_{s1} and R_{s2}, from the equation (see Hamroch, 1994)

$$R_s = (R_{s1}^2 + R_{s2}^2)^{1/2} \tag{12-38}$$

The surface roughness is measured by a profilometer, often referred to as *stylus measurement*. Microscope devices are used when higher-precision measurements are needed. The ratio Λ of the film thickness to the size of surface asperities, R_s, is:

$$\Lambda = \frac{h_{min}}{R_s} \tag{12-39}$$

For a full EHD lubrication of rolling bearings, Λ is usually between 3 and 5. The desired ratio Λ is determined according to the expected level of vibrations and other disturbances in the machine. Although h_{min} is very important for successful bearing operation, calculation of h_0 is often required for determining the viscous shear force, referred to as *traction force*, which is the resistance to relative sliding.

12.8 ELASTOHYDRODYNAMIC LUBRICATION OF A LINE CONTACT

A theoretical line contact is formed between two cylinders, such as in a cylindrical rolling bearing. For this case, Pan and Hamrock (1989) introduced the following empirical equation for the minimum film thickness, h_{min}:

$$\frac{h_{min}}{R_x} = \frac{1.714\bar{U}_r^{0.694}(\alpha E_{eq})^{0.568}}{\bar{W}^{0.128}} \tag{12-40}$$

Here, α is the viscosity–pressure coefficient. Although theoretical equations were derived earlier, Eq. (12-40) is more accurate, because it is based on actual measurements. The equation is a function of dimensionless terms. The advantage in using dimensionless terms is that any system of units can be used, as long as the units are consistent and result in dimensionless terms.

The dimensionless terms $\bar{U}_r$ and $\bar{W}$ are the dimensionless rolling velocity and load per unit of cylinder length, respectively. The dimensionless terms are defined by:

$$\bar{U}_r = \frac{\mu_0}{E_{eq}R_x}U_r \qquad \bar{W} = \frac{1}{E_{eq}R_xL}W \tag{12-41}$$

Here, U_r is the rolling velocity, η_0 is the viscosity of the lubricant at atmospheric pressure, W is the load on one rolling element, E_{eq} is the equivalent modulus of elasticity, and R_x is an equivalent radius of contact in the x plane (direction of axis of rotation of rolling element), as defined in Sec. 12.5.2.

Jones et al. (1975) measured the viscosity–pressure coefficient α [m^2/N] for various lubricants. The data is summarized in Table 12-1.

The rolling velocity $\bar{U}_r$ in Eq. (12-41) is for pure rolling (no relative sliding) (Eq. (12-34)). However, in many other problems, such as in gears and cams, there is a combination of rolling and sliding. If the ratio of the rolling to the sliding is ξ and U_s is the sliding velocity, then the dimensionless rolling velocity $\bar{U}_r$ used in Eq. (12-41) is replaced by

$$\bar{U}_r = \frac{\mu_0}{2E_{eq}R_x}U_s(1+\xi) \tag{12-42}$$

Equation (12-40) indicates that the viscosity and rolling speed have a more significant effect on the minimum film thickness, h_{min}, in comparison to the load

TABLE 12-1 Viscosity–Pressure Coefficient (α [m^2/N]) for Various Lubricants

Fluid	Temperature t_m		
	38°C	99°C	149°C
Ester	1.28×10^{-8}	0.987×10^{-8}	0.851×10^{-8}
Formulated ester	1.37	1.00	0.874
Polyalkyl aromatic	1.58	1.25	1.01
Synthetic paraffinic oil	1.77	1.51	1.09
Synthetic paraffinic oil	1.99	1.51	1.29
Synthetic paraffinic oil plus antiwear additive	1.81	1.37	1.13
Synthetic paraffinic oil plus antiwear additive	1.96	1.55	1.25
C-ether	1.80	0.980	0.795
Superrefined naphthenic mineral oil	2.51	1.54	1.27
Synthetic hydrocarbon (traction fluid)	3.12	1.71	0.937
Fluorinated polyether	4.17	3.24	3.02

Source: Jones et al. (1975).

(the viscosity and speed are of higher power in this equation). This means that a high rolling velocity, U_r, is essential for a full EHD lubrication.

On the right-hand side of Fig. 12-20, we can see that the fluid film thickness is nearly uniform along the contact, except where the clearance narrows at the exit. The film thickness h_0 at the center of the contact is used for calculation of the traction force (resistance to relative sliding), and it can be derived from the following equation of Pan and Hamrock (1989):

$$\frac{h_o}{R_x} = \frac{2.922\bar{U}_r^{0.694}(\alpha E_{eq})^{0.470}}{\bar{W}^{0.166}} \tag{12-43}$$

Example Problem 12-3

Calculation of Oil Film Thickness in a Cylindrical Roller Bearing

The radial load on a cylindrical rolling bearing is $W = 11{,}000$ N, and the inner ring speed is $N = 5000$ RPM. The rolling bearing has the following dimensions: The diameter of the inner raceway, d_{in}, is 120 mm, the diameter of the outer raceway, d_{out}, is 160 mm, and the diameter of the cylindrical roller, d_{roller}, is 20 mm. The effective length of the cylindrical rolling element is $L = 10$ mm. There are 14 rolling elements around the bearing. The bearing (rollers and rings) is made of steel. The modulus of elasticity of the steel is $E = 2.05 \times 10^{11}$ N/m^2, and its Poisson ratio is $\nu = 0.3$. The bearing is lubricated by oil having an absolute viscosity of $\mu_0 = 0.01$ N-s/m^2 at atmospheric pressure and bearing operating temperature. The viscosity–pressure coefficient is $\alpha = 2.2 \times 10^{-8}$ m^2/N.

Find the minimum elastohydrodynamic film thickness at the following points:

a. At the contact of the rolling elements with the inner raceway
b. At the contact of the rolling elements with the outer raceway

Solution

The radii of the contacting curvatures are:

Roller radius: $R_{\text{roller}} = 0.01$ m,
Outer ring raceway: $R_{\text{outer raceway}} = 0.08$ m,
Inner ring raceway: $R_{\text{inner raceway}} = 0.06$ m

The contact between the rolling elements and the inner raceway is convex, and the equivalent contact curvature, $R_{x,\text{in}}$, is derived according to the equation

$$\frac{1}{R_{x,\text{in}}} = \frac{1}{R_{\text{roller}}} + \frac{1}{R_{\text{inner raceway}}}$$

$$\frac{1}{R_{x,\text{in}}} = \frac{1}{0.01} + \frac{1}{0.06} \Rightarrow R_{\text{eq,in}} = 0.0085 \text{ m}$$

However, the contact between the rolling elements and the outer raceway is concave, and the equivalent contact curvature, $R_{x,\text{out}}$, is derived according to the equation

$$\frac{1}{R_{x,\text{out}}} = \frac{1}{R_{\text{roller}}} - \frac{1}{R_{\text{outer raceway}}}$$

$$\frac{1}{R_{x,\text{out}}} = \frac{1}{0.01} - \frac{1}{0.08}; \qquad R_{x,\text{out}} = 0.0114 \text{ m}$$

If we assume that there is no radial clearance, then the maximum load on one cylindrical rolling element can be calculated from the following formula:

$$W_{\max} = \frac{4W_{\text{shaft}}}{n_r} = \frac{4 \times 11{,}000}{14} = 3142 \text{ N}$$

where n_r is the number of rollers in the bearing. The shaft and the bearing are made of identical material, so the equivalent modulus of elasticity is

$$E_{\text{eq}} = \frac{E}{1 - \nu^2} \Rightarrow E_{\text{eq}} = \frac{2.05 \times 10^{11}}{1 - 0.3^2} = 2.25 \times 10^{11} \text{ N/m}^2$$

The dimensionless load on the inner race is

$$\bar{W} = \frac{1}{E_{eq} R_{x,\text{in}} L} W_{\max} = \frac{1}{2.25 \times 10^{11} \times 0.0085 \times 0.01} 3142 = 1.64 \times 10^{-4}$$

In comparison, the dimensionless load on the outer race is

$$\bar{W} = \frac{1}{E_{eq} R_{x,\text{out}} L} W_{\max} = \frac{1}{2.25 \times 10^{11} \times 0.0114 \times 0.01} 3142 = 1.22 \times 10^{-4}$$

Comparison of the inner and outer dimensionless loads indicates a higher value for the inner contact. This results in higher contact stresses, including maximum pressure, at the inner contact. The rolling velocity, U_r, for a cylindrical roller is calculated via Eq. (12-34).

$$U_r = \frac{R_{\text{in}} R_{\text{out}}}{R_{\text{out}} + R_{\text{in}}} \omega$$

where the angular shaft speed, ω, is equal to

$$\omega = \frac{2\pi N}{60} = \frac{2 \times \pi \times 5000}{60} = 523 \text{ rad/s}$$

Hence, the rolling speed is

$$U_r = \frac{R_{\text{in}} R_{\text{out}}}{R_{\text{out}} + R_{\text{in}}} \omega = \frac{0.06 \times 0.08}{0.06 + 0.08} 523 = 17.93 \text{ m/s}$$

The dimensionless rolling velocity of the inner surface is

$$\bar{U}_r = \frac{\mu_0}{E_{eq} R_{x,\text{in}}} U_r = \frac{0.01}{2.25 \times 10^{11} \times 0.0085} 17.93 = 9.375 \times 10^{-11}$$

In comparison, the dimensionless rolling velocity at the outer ring race is

$$\bar{U}_r = \frac{\mu_0}{E_{eq} R_{x,\text{out}}} U_r = \frac{0.01}{2.25 \times 10^{11} \times 0.0114} 17.93 = 6.99 \times 10^{-11}$$

The elastohydrodynamic minimum film thickness is derived from Eq. (12.40) for a fully lubricated ball bearing.

$$\frac{h_{\min}}{R_x} = \frac{1.714 \bar{U}^{0.694} (\alpha E_{eq})^{0.568}}{\bar{W}^{0.128}}$$

The minimum oil film thickness at the contact with the inner raceway is

$$\frac{h_{\min}}{R_{x,\text{in}}} = \frac{1.714(9.3 \times 10^{-11})^{0.694} (2.2 \times 10^{-8} \times 2.25 \times 10^{11})^{0.568}}{(1.64 \times 10^{-4})^{0.128}}$$

$$h_{\min,\text{in}} = 0.609 \ \mu\text{m}$$

The minimum oil film thickness at the contact with the outer raceway is

$$\frac{h_{\min}}{R_{x,\mathrm{out}}} = \frac{1.714(6.99 \times 10^{-11})^{0.694}(2.2 \times 10^{-8} \times 2.25 \times 10^{11})^{0.568}}{(1.22 \times 10^{-4})^{0.128}}$$

$$h_{\min,\mathrm{out}} = 0.695\ \mu\mathrm{m}$$

The minimum film thickness at the contact with the inner ring race is lower because the equivalent radius of convex curvatures is lower. Therefore, it is sufficient to calculate the minimum film thickness at the contact with the inner ring race. However, for a rolling bearing operating at high speed, the centrifugal force of the rolling element is added to the contact force at the contact with the outer raceway. In such cases, $h_{\min,\mathrm{out}}$ may be lower than $h_{\min,\mathrm{in}}$, and the EHD fluid film should be calculated at the inner and outer ring races.

Example Problem 12-4

Elastohydrodynamic Fluid Film in a Cam and a Follower

A cam and a follower operate in a car engine as shown in Fig. 12-21. The cam radius at the tip of the cam is $R = 20$ mm, the distance of this radius center from

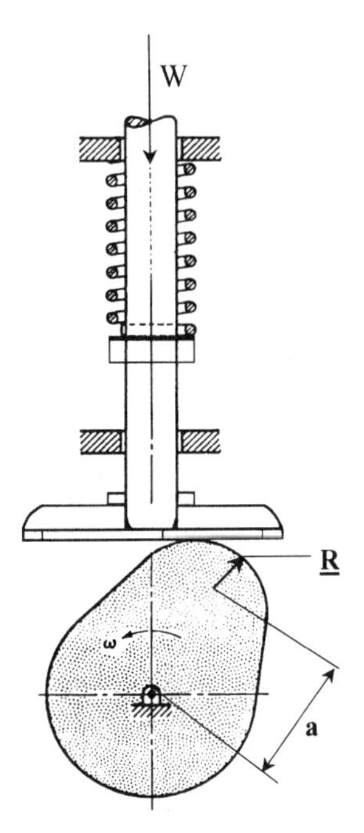

FIG. 12-21 Cam and follower.

the cam center of rotation is $a = 40$ mm, and the width of the cam (effective length of contact) is 10 mm. There is a maximum reaction force $W = 1200$ N between the cam and the follower when the follower reaches its maximum height. The rotation speed of the cam is $N = 600$ RPM. The cam and the follower are made of steel. The steel modulus of elasticity is $E = 2.05 \times 10^{11}$ N/m^2, and its Poisson ratio ν is 0.3. The contact between the cam and follower is fully lubricated. The absolute viscosity at atmospheric pressure and engine temperature is $\mu_0 = 0.01$ N-s/m^2, and the viscosity–pressure coefficient is $\alpha = 2.2 \times 10^{-8}$ m^2/N.

Find the minimum oil film thickness at the contact between the cam and the shaft.

Solution

The contact is between a plane and a curvature of radius $R = 20$ mm at the tip of the cam. In that case, the equivalent radius is R. The equation for the equivalent modulus is

$$E_{eq} = \frac{E}{1 - \nu^2} = \frac{2.05 \times 10^{11}}{1 - 0.3^2} = 2.2 \times 10^{11} \text{ N/m}^2$$

The dimensionless load becomes

$$\bar{W} = \frac{1}{E_{eq} R L} W_{max} = \frac{1}{2.2 \times 10^{11} \times 0.020 \times 0.01} 1200 = 2.727 \times 10^{-5}$$

The sliding velocity U is equal to the tangential velocity at the tip of the cam:

$$U = \omega(a + R)$$

Here, ω is equal to

$$\omega = \frac{2\pi N}{60} = \frac{2\pi 600}{60} = 62.83 \text{ rad/s}$$

This problem is one of pure sliding, $\xi = 0$, and the sliding velocity is

$$U_s = (0.04 + 0.02)62.83 = 3.76 \text{ m/s}$$

For pure sliding, $\xi = 0$, the dimensionless equivalent rolling velocity is [Eq. (12-42)]

$$\bar{U}_r = \frac{\mu_0}{2E_{eq} R} U_s(1 + \xi) = \frac{0.01}{2 \times 2.2 \times 10^{11} \times 0.02} 3.76 = 4.27 \times 10^{-12}$$

For line contact and in the presence of sufficient lubricant, the elastohydrodynamic minimum film thickness is derived according to the equation

$$\frac{h_{min}}{R} = \frac{1.714\bar{U}^{0.694}(\alpha E_{eq})^{0.568}}{\bar{W}^{0.128}}$$

The oil film thickness in the inner surface is

$$\frac{h_{min}}{0.02} = \frac{1.714(4.27 \times 10^{-12})^{0.694}(2.2 \times 10^{-8} \times 2.22 \times 10^{11})^{0.568}}{(2.727 \times 10^{-5})^{0.128}}$$

$$h_{min} = 0.21\ \mu m$$

12.9 ELASTOHYDRODYNAMIC LUBRICATION OF BALL BEARINGS

Under load, the theoretical point contact between a rolling ball and raceways becomes an elliptical contact area. The elliptical contact area has radii a and b, as shown in Fig. 12-17.

In a similar way to the theoretical line contact, there is a minimum elastohydrodynamic film thickness, h_{min}, near the exit from a uniform film thickness, h_0. For hard surfaces, such as steel in rolling bearings, and sufficient lubricant, Hamrock and Dowson (1977) obtained the following formula for the minimum film thickness, h_{min}:

$$\frac{h_{min}}{R_x} = 3.63\frac{\bar{U}_r^{0.68}(\alpha E_{eq})^{0.49}}{\bar{W}^{0.073}}(1 - e^{-0.68k}) \tag{12-44}$$

Here, α is the viscosity–pressure coefficient and $\bar{U}_r$ and $\bar{W}$ are dimensionless velocity and load, respectively, defined by:

$$\bar{U}_r = \frac{\mu_o U_r}{E_{eq}R_x} \quad \text{and} \quad \bar{W} = \frac{W}{E_{eq}R_x^2} \tag{12-45}$$

Here, U_r is the rolling velocity, μ_0 is the viscosity of the lubricant at atmospheric pressure and bearing operating temperature, W is the load on one rolling element, and E_{eq} is the equivalent modulus of elasticity.

The equations for the equivalent modulus of elasticity and equivalent contact radius are used for calculating the Hertz stresses at a point contact. These equations were discussed in Secs. 12.4 and 12.5. For convenience, these equations are repeated here.

Recall that E_{eq} is determined from equation

$$\frac{2}{E_{eq}} = \frac{1-\nu_1^2}{E_1} + \frac{1-\nu_2^2}{E_2}$$

Here, ν is poisson's ratio and E is the modulus of elasticity of the respective two materials. For identical materials, the equation becomes

$$E_{eq} = \frac{E}{1 - \nu^2}$$

The equivalent radius of curvature in the plane of rotation, R_x, for the contact with the inner ring race is

$$\frac{1}{R_x} = \frac{1}{R_{1x}} + \frac{1}{R_{2x}}$$

where R_{1x} and R_{2x} are as shown on the right-hand side of Fig. 12-18b for the contact at the inner ring race and $R_{1x} = d/2$, where d is the ball diameter.

On the left-hand side of Fig. 12-18b, the contact at the inner ring race is concave, and the equivalent contact curvature radius in this plane is

$$\frac{1}{R_y} = \frac{1}{R_{1y}} - \frac{1}{R_{2y}}$$

The radius ratio, α_r (the ratio of the larger radius to the smaller radius, $\alpha_r > 1$) is defined as

$$\alpha_r = \frac{R_y}{R_x}$$

The ellipticity parameter, k, is the ratio

$$k = \frac{b}{a}$$

The parameter k can be estimated from

$$k \approx \alpha_r^{2/\pi}$$

For hard surfaces, sufficient lubricant, and for pure rolling, Hamrock and Dowson (1981) obtained the following formula for the central film thickness, h_c (at the center of the fluid film):

$$\frac{h_c}{R_x} = 2.69 \frac{\bar{U}_r^{0.67} (\alpha E_{eq})^{0.53}}{\bar{W}^{0.067}} (1 - 0.61 e^{-0.73k}) \tag{12-46}$$

This equation is useful for the calculation of the traction force (resistance to relative sliding) where an EHD film is separating the surfaces.

For soft surfaces, such as rubber, the contact area is relatively large, resulting in a lower contact pressure. In addition, the viscosity does not increase as much as predicted by the preceding equations.

Example Problem 12-5

Find the minimum film thickness for a rolling contact of a deep-groove ball bearing having the following dimensions: The bearing has 14 balls of diameter $d = 19.04$ mm. The radius of curvature of the inner-deep groove (in cross section x-z on the left-hand side of Fig. 12-18b) is 9.9 mm. The inner race diameter, R_{2x} (at the bottom of the deep groove), is $d_i = 76.5$ mm (cross section y-z on the right-hand side of Fig. 12-18b). The radial load on the bearing is $W = 10{,}500$ N, and the bearing speed is $N = 5000$ RPM. The rolling elements and rings are made of steel. The modulus of elasticity of the steel for rollers and rings is $E = 2 \times 10^{11}$ N/m^2, and Poisson's ratio is $\nu = 0.3$. The properties of the lubricant are: The absolute viscosity at ambient pressure and bearing operating temperature is $\mu_0 = 0.04$ N-s/m^2, and the viscosity–pressure coefficient is $\alpha = 2.3 \times 10^{-8}$ m^2/N.

Solution

Referring to Fig. 12-18b, the radius of curvature in the y-z plane is

$$R_{2x} = 38.25 \text{ mm} \qquad \text{and } R_{1x} = \frac{d}{2} = 9.52 \text{ mm}$$

Equivalent Radius for Inner Raceway Convex Contact in y-z Plane (x Plane)

$$\frac{1}{R_x} = \frac{1}{R_{1x}} + \frac{1}{R_{2x}} \Rightarrow \frac{1}{R_x} = \frac{1}{9.52} + \frac{1}{38.25}; \qquad R_x = 7.62 \text{ mm}$$

Equivalent Inner Radius in y Plane (Concave Contact)

The curvatures in this plane are: $R_{1y} = 9.52$ mm and $R_{2y} = 9.9$ mm. The equivalent inner radius in the y plane is

$$\frac{1}{R_y} = \frac{1}{R_{1y}} - \frac{1}{R_{2y}} \Rightarrow \frac{1}{R_y} = \frac{1}{9.52} - \frac{1}{9.9}; \qquad R_y = 248.0 \text{ mm}$$

Equivalent Curvature of Inner Ring and Ball Contact

$$\frac{1}{R_{eq}} = \frac{1}{R_x} + \frac{1}{R_y} \Rightarrow \frac{1}{R_{eq}} = \frac{1}{7.62} + \frac{1}{248}; \qquad R_{eq} = 7.4 \text{ mm}$$

$$\alpha_r = \frac{R_y}{R_x} = 32.55 \qquad \text{and} \qquad k = \alpha_r^{2/\pi} = 32.54^{2/\pi} = 9.18$$

The shaft and the bearing are made of identical steel, so the equivalent modulus of elasticity is:

$$E_{eq} = \frac{E}{1-\nu^2} \Rightarrow E_{eq} = \frac{2 \times 10^{11}}{1-0.3^2} = 2.2 \times 10^{11}\ \mathrm{N/m^2}$$

For a ball bearing without radial clearance, the maximum load at the contact of one rolling element can be estimated with the following equation:

$$W_{max} \approx \frac{5W_{shaft}}{n_r} = \frac{5(10{,}500)}{14} = 3750\ \mathrm{N}$$

where $n_r = 14$ is the number of balls in the bearing.

The angular velocity, ω, is

$$\omega = \frac{2\pi N}{60} = \frac{2\pi 5000}{60} = 523.6\ \mathrm{rad/s}$$

The rolling speed is derived according to Eq. (12-34):

$$U_{rolling} = \frac{R_{in}R_{out}}{2(R_{in}+r)}\omega = \frac{R_{in}R_{out}}{R_{in}+R_{out}}\omega$$

or

$$U_r = \frac{R_{2x}(R_{2x}+d)}{2R_{2x}+d}\omega$$

$$U_r = \frac{38.25 \times 10^{-3}(38.25 \times 10^{-3} + 19.05 \times 10^{-3})}{(2 \times 38.25 \times 10^{-3}) + 19.04 \times 10^{-3}} 523.6 = 12\ \mathrm{m/s}$$

The dimensionless rolling velocity at the inner ring race is

$$\bar{U}_r = \frac{\mu_0}{E_{eq}R_x}U_r = \frac{0.01}{2.2 \times 10^{11} \times 0.00762} 12 = 7.17 \times 10^{-11}$$

The dimensionless load on the inner ring race is

$$\bar{W} = \frac{1}{E_{eq}R_x^2}W_{max} = \frac{1}{2.2 \times 10^{11} \times 0.00762^2} 3750 = 2.94 \times 10^{-4}$$

Finally, the oil film thickness between the inner race and the roller is

$$\frac{h_{min}}{R_x} = 3.63 \frac{\bar{U}_r^{0.68}(\alpha E_{eq})^{0.49}}{\bar{W}^{0.073}}(1 - e^{-0.68k}) \Rightarrow$$

$$\frac{h_{min}}{0.00762} = 3.63 \frac{(7.17 \times 10^{-11})^{0.68}(2.3 \times 10^{-8} \times 2.2 \times 10^{11})^{0.49}}{(2.94 \times 10^{-4})^{0.073}} \times (1 - e^{-0.68 \times 9.18})$$

Thus, at the inner ring race

$$h_{min} = 0.412\ \mu m$$

Example Problem 12-6

Centrifugal Force

The deep-groove ball bearing in Example Problem 12-5 is used in a high-speed turbine where the average shaft speed is increased to $N = 30{,}000$ RPM. The radial bearing load is equal to that in Example Problem 12-5, $W = 10,500$ N. The lubricant is also equivalent to that in Example Problem 12-5. The properties of the lubricant are: The absolute viscosity at ambient pressure and bearing operating temperature is $\mu_0 = 0.04$ N-s/m^2, and the viscosity–pressure coefficient is $\alpha = 2.3 \times 10^{-8}$ m^2/N.

Consider the centrifugal force at this high-speed operation, and

a. find the maximum contact pressure at the inner and outer ring raceways.
b. calculate the minimum film thickness at the contact with the outer ring.

Solution

a. Maximum Contact Pressure

For the contact with the inner ring race, the maximum pressure and minimum film thickness was calculated in Example Problems 12-2 and 12-5. In this problem, the maximum pressure will be compared with that at the outer ring race in the presence of a centrifugal force.

Contact with Outer Ring Race

The contact radii in the x plane (y-z plane) in Fig. 12-18b are:

$$R_{2x} = 57.29 \text{ mm} \qquad \text{and} \qquad R_{1x} = \frac{d}{2} = 9.52 \text{ mm}$$

The equivalent radius for the outer raceway concave contact in the x plane (y-z plane) is

$$\frac{1}{R_x} = \frac{1}{R_{1x}} - \frac{1}{R_{2x}} \Rightarrow \frac{1}{R_x} = \frac{1}{9.52} - \frac{1}{57.29}; \qquad R_x = 11.42 \text{ mm}$$

Equivalent Outer Radius in y Plane

The curvatures in this plane are $R_{1y} = 9.52$ mm and $R_{2y} = 9.9$ mm. The equivalent inner radius in the y (or x-z) plane is

$$\frac{1}{R_y} = \frac{1}{R_{1y}} - \frac{1}{R_{2y}} \Rightarrow \frac{1}{R_y} = \frac{1}{9.52} - \frac{1}{9.9}; \qquad R_y = 248.0 \text{ mm}$$

The equivalent radius of the outer ring and ball contact is

$$\frac{1}{R_{\text{eq}}} = \frac{1}{R_x} + \frac{1}{R_y} \Rightarrow \frac{1}{R_{\text{eq}}} = \frac{1}{11.42} + \frac{1}{248}; \qquad R_{\text{eq}} = 10.92 \text{ mm}$$

$$\alpha_r = \frac{R_y}{R_x} = 21.72$$

and

$$k = \alpha_r^{2/\pi} = 21.72^{2/\pi} = 7.1$$

It is also necessary to calculate $\hat{E}$, which will be used to calculate the ellipsoid radii. First q_a will be determined

$$q_a = \frac{\pi}{2} - 1 = 0.57$$

$$\hat{E} \approx 1 + \frac{q_a}{\alpha_r} \Rightarrow 1 + \frac{0.57}{21.72} = 1.03$$

From Example Problem 12-5, the equivalent modulus of elasticity is

$$E_{\text{eq}} = 2.2 \times 10^{11} \text{ N/m}^2$$

The load on the bearing is divided unevenly between the rolling elements. The approximate equation for zero bearing clearance is used. The maximum load, $W_{\max}$, at a contact of one rolling element can be estimated by the following equation:

$$W_{\max \text{ (one ball)}} \approx \frac{5W_{\text{shaft}}}{n_r} = \frac{5(10{,}500)}{14} = 3750 \text{ N}$$

where $n_r = 14$ is the number of balls in the bearing.

The total maximum load at the contact of one-roller and the outer raceway is equal to the transmitted shaft load plus the centrifugal force generated.

The angular velocity ω is

$$\omega = \frac{2\pi N}{60} = \frac{2 \times \pi \times 30{,}000}{60} = 3141.59 \text{ rad/s}$$

The angular velocity of the rolling-element center, point C, is given by

$$\omega_C = \frac{R_{in}}{R_{in} + R_{out}}\omega = \frac{38.25}{38.25 + 57.29}3141.59 = 1257.75 \text{ rad/s}$$

This angular velocity is used to calculate the centrifugal force, F_c:

$$F_c = m_r R_c \omega_C^2$$

The density of steel and the ball volume determine its mass:

$$m_r = \frac{\pi}{6}d^3\rho = \frac{\pi}{6} \times 0.01904^3 \times 7800 = 0.028 \text{ kg}$$

After substituting these values in the equation for the centrifugal force, we obtain

$$F_c = 0.028 \times 0.04777 \times 1257.75^2 = 2115.93 \text{ N}$$

The maximum force is at the outer raceway. It is equal to the sum of the transmitted shaft load and the centrifugal force:

$$W_{max} = W_{max,load} + F_c = 3750 + 2115.93 = 5865.93 \text{ N}$$

The ellipsoid radii a and b can now be determined by substituting the values already calculated:

$$a = \left(\frac{6\hat{E}W_{max}R_{eq}}{\pi k E_{eq}}\right)^{1/3}$$

$$= \left(\frac{6 \times 1.03 \times 5865.93 \times 0.01092)}{\pi \times 7.1 \times 2.2 \times 10^{11}}\right)^{1/3} = 0.43 \text{ mm}$$

$$b = \left(\frac{6k^2\hat{E}W_{max}R_{eq}}{\pi E_{eq}}\right)^{1/3}$$

$$= \left(\frac{6 \times 7.1^2 \times 1.03 \times 5865.93 \times 0.01092}{\pi \times 2.2 \times 10^{11}}\right)^{1/3} = 3.07 \text{ mm}$$

The maximum pressure at the outer contact is

$$p_{max} = \frac{3}{2}\frac{W_{max}}{\pi ab} = \frac{3}{2}\frac{5865.93}{\pi \times 0.43 \times 3.07} = 2121.64 \text{ N/mm}^2 = 2.12 \text{ GPa}$$

The maximum pressure is very close to that at the inner ring contact, $p_{max} = 2.17$ GPa (see Example Problem 12.2).

b. Minimum Film Thickness at the Outer Race Contact

Rolling occurs only in the x plane, so

$$U_r = \frac{R_{2x}(R_{2x} + d)}{2R_{2x} + d}\omega$$

$$U_r = \frac{57.29 \times 10^{-3}(57.29 \times 10^{-3} + 19.04 \times 10^{-3})}{(2 \times 57.29 \times 10^{-3}) + 19.04 \times 10^{-3}} 3141.59 = 102.81 \text{ m/s}$$

The dimensionless velocity for the outer surface is

$$\bar{U}_r = \frac{\mu_0}{E_{eq}R_x} U_r = \frac{0.01}{2.2 \times 10^{11} \times 0.01142} 102.81 = 4.09 \times 10^{-10}$$

where U_r is the rolling velocity of the ball on the race and ω is the angular velocity of the shaft, in rad/s. The dimensionless load on the outer race is

$$\bar{W} = \frac{1}{E_{eq}R_x^2} W_{max} = \frac{1}{2.2 \times 10^{11} \times (0.01142)^2} \times 5865.93 = 2.04 \times 10^{-4}$$

Finally, the oil film thickness between the outside race and the ball is

$$\frac{h_{min}}{R_x} = 3.63 \frac{\bar{U}_r^{0.68} (\alpha E_{eq})^{0.49}}{\bar{W}^{0.073}} (1 - e^{-0.68k})$$

$$\frac{h_{min}}{0.01142} = 3.63 \frac{(4.09 \times 10^{-10})^{0.68} (2.3 \times 10^{-8} \times 2.2 \times 10^{11})^{0.49}}{(2.04 \times 10^{-4})^{0.073}} \times (1 - e^{-0.68 \times 7.1})$$

$$h_{min} = 2.06 \ \mu\text{m}$$

Example Problem 12-7

Ceramic Rolling Elements

For the high-speed turbine given in Example Problem 12-6, the bearing is replaced by a deep-groove ball bearing with equivalent geometry. However, the bearing is hybrid and the rolling elements are made of silicone nitride. The rings are made of steel. The lubricant is also equivalent to that in Example Problems 12-5 and 12-6. The shaft speed is $N = 30{,}000$ RPM. The radial bearing load is equal to that in Example Problem 12-5, $W = 10{,}500$ N.

a. Find the maximum pressure at the outer race contact
b. Compare the maximum pressure to that of all steel bearings in Example Problem 12-6.

The properties of silicon nitride are:

$E = 3.14$ GPa (3.14×10^{11} Pa), in comparison to steel, with $E = 2.00$ GPa.

$\rho = 3200\ \mathrm{kg/m^3}$, in comparison to steel, with $\rho = 7800\ \mathrm{kg/m^3}$.

$\nu = 0.24$, in comparison to steel, with $\nu = 0.3$.

Solution

a. Maximum Pressure at the Outer Race

It is possible to decrease the centrifugal force by lowering the density of the rolling elements. One important advantage of a hybrid bearing where the rolling elements are made of silicone nitride (and the rings are made of steel) is that it lowers the density of the rolling elements. However, we have to keep in mind that the modulus of elasticity of silicone nitride is higher than that of steel, and this may result in a higher maximum pressure. In this problem, the maximum pressure is calculated and compared with those of a conventional steel bearing.

Radius of Curvature in the x Plane

$$R_{2x} = 57.29\ \mathrm{mm} \qquad R_{1x} = \frac{d}{2} = 9.52\ \mathrm{mm}$$

The equivalent radius for the outer raceway contact in the x plane is

$$\frac{1}{R_x} = \frac{1}{R_{1x}} - \frac{1}{R_{2x}} \Rightarrow \frac{1}{R_x} = \frac{1}{9.52} - \frac{1}{57.29}; \qquad R_x = 11.42\ \mathrm{mm}$$

Equivalent Outer Radius in y Plane

The curvatures in this plane are $R_{1y} = 9.52$ mm and $R_{2y} = 9.9$ mm. The equivalent inner radius in the y-z plane is

$$\frac{1}{R_y} = \frac{1}{R_{1y}} - \frac{1}{R_{2y}} \Rightarrow \frac{1}{R_y} = \frac{1}{9.52} - \frac{1}{9.9}; \qquad R_y = 248.0\ \mathrm{mm}$$

The equivalent radius of the curvature of the outer ring and ball contact is

$$\frac{1}{R_{eq}} = \frac{1}{R_x} + \frac{1}{R_y} \Rightarrow \frac{1}{R_{eq}} = \frac{1}{11.42} + \frac{1}{248}; \qquad R_{eq} = 10.92\ \mathrm{mm}$$

$$\alpha_r = \frac{R_y}{R_x} = 21.72$$

and

$$k = \alpha_r^{2/\pi} = 21.72^{2/\pi} = 7.1$$

It is also necessary to calculate $\hat{E}$, which will be used to calculate the ellipsoid radii. First, q_a will be determined:

$$q_a = \frac{\pi}{2} - 1 = 0.57$$

$$\hat{E} \approx 1 + \frac{q_a}{\alpha_r} \Rightarrow 1 + \frac{0.57}{21.72} = 1.03$$

For a hybrid bearing, we must consider the equivalent modulus of elasticity of two different materials (steel and silicon nitride), given by

$$\frac{2}{E_{eq}} = \frac{1 - \nu_{21}^2}{E_1} + \frac{1 - \nu_2^2}{E_2} \Rightarrow$$

$$\frac{2}{E_{eq}} = \frac{1 - (0.24)^2}{3.14 \times 10^{11}} + \frac{1 - (0.3)^2}{2.0 \times 10^{11}} \Rightarrow E_{eq} = 2.65 \times 10^{11} \text{ Pa}$$

The maximum load transmitted by one rolling element at the contact with the ring (assuming zero clearance) is estimated by the following equation:

$$W_{max} \approx \frac{5W_{shaft}}{n_r} = \frac{5(10{,}500)}{14} = 3750 \text{ N}$$

The number of balls in the bearing is $n_r = 14$.

The total maximum load at the contact of one roller and the outer raceway is equal to the transmitted shaft load plus the centrifugal force generated. The angular velocity ω is equal to

$$\omega = \frac{2\pi N}{60} = \frac{2\pi 30{,}000}{60} = 3141.59 \text{ rad/s}$$

The angular velocity of the rolling element center, point C, is given by

$$\omega_C = \frac{R_{in}}{R_{in} + R_{out}} \omega = \frac{38.25}{38.25 + 57.29} 3141.59 = 1257.75 \text{ rad/s}$$

This angular velocity is used to calculate the centrifugal force, F_c:

$$F_c = m_r(R_i + r)\omega_C^2$$

The density (silicon nitride) and volume determine the mass of the ball:

$$m_r = \frac{\pi}{6} d^3 \rho \Rightarrow \frac{\pi}{6} \times 0.01904^3 \times 3200 = 0.012 \text{ kg}$$

Substituting these values in the equation for centrifugal force we obtain

$$F_c = 0.012 \times 0.04777 \times 1257.75^2 = 906.76 \text{ N}$$

Therefore, the maximum force occurs at the outer raceway. It is equal to the sum of the shaft load and the centrifugal force:

$$W_{\max} = W_{\max,\text{load}} + F_c \Rightarrow 3750 + 906.76 = 4656.76 \text{ N}$$

In comparison, $w_{max} = 5866\ N$, for all steel-bearing and identical conditions (Example Problem 12-6).

The ellipsoid radii a and b can now be determined by substitution of the values already calculated:

$$a = \left(\frac{6\hat{E}W_{\max}R_{\text{eq}}}{\pi k E_{\text{eq}}}\right)^{1/3} = \left(\frac{6 \times 1.03 \times 4656.76 \times 0.01092}{\pi \times 7.1 \times 2.65 \times 10^{11}}\right)^{1/3} = 0.38 \text{ mm}$$

$$b = \left(\frac{6k^2\hat{E}W_{\max}R_{\text{eq}}}{\pi E_{\text{eq}}}\right)^{1/3}$$

$$= \left(\frac{6 \times 7.1^2 \times 1.03 \times 4656.76 \times 0.01092}{\pi \times 2.65 \times 10^{11}}\right)^{1/3} = 2.67 \text{ mm}$$

The maximum pressure at the outer race contact is

$$p_{\max} = \frac{3}{2}\frac{W_{\max}}{\pi ab} = \frac{3}{2}\frac{4656.76}{\pi \times 0.38 \times 2.67} = 2190 \text{ N/mm}^2 = 2.19 \text{ GPa}$$

b. Comparison with All-Steel Bearing

The bearing made of steel has a maximum pressure of 2.12 GPa on the outer raceway, while the hybrid bearing has a maximum pressure of 2.19 GPa. By using rollers made of silicon nitride, it is possible to reduce the centrifugal force by lowering the density of the rolling element. Although the centrifugal force is reduced, the modulus of elasticity of the ceramic rolling element is higher. The combination of these two factors results in a slight increase in pressure on the outer raceway, which was unexpected. In conclusion, for this specific application there is no reduction in the maximum pressure by replacing the steel with silicon nitride ceramic rolling elements.

12.10 FORCE COMPONENTS IN AN ANGULAR CONTACT BEARING

In an angular contact bearing, the resultant contact forces of a ball with the inner and outer ring races, W_i and W_o, are shown in Fig. 12-22. At low speed, the centrifugal forces of the rolling element are small and can be neglected in comparison to the external load. In that case, the contact angle of the inner ring, α_i is equal to the contact angle of the outer ring, α_o; see Fig. 12-22. However, at high speed the centrifugal forces are significant and should be considered. In that

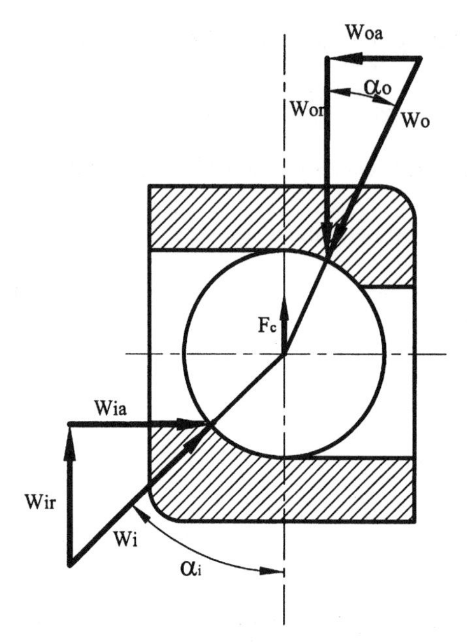

FIG. 12-22 Force components in an angular contact bearing.

case, the contact angle of the inner ring, α_i, is no longer equal to the contact angle of the outer ring, α_o.

For design purposes, the maximum contact stresses at the inner and outer ring races should be calculated. For this purpose, the first step is to solve for the resultant contact forces with the inner and outer rings, W_i and W_o, in the direction normal to the contact area.

Let us consider an example where the following bearing data is given:

a. Inner ring contact angle is α_i.
b. Thrust load on each rolling element is W_{ia}.
c. The centrifugal force of each rolling element is F_c.

We solve for the contact angle of the outer ring, α_o, and the resultant normal contact forces of the inner and outer ring races, W_i and W_o, respectively. The resultant contact force at the inner ring, W_i, can be directly solved from the thrust load and inner contact angle:

$$\frac{W_{ia}}{W_i} = \sin \alpha_i \qquad W_i = \frac{W_{ia}}{\sin \alpha_i} \tag{12-47}$$

The two unknowns, the contact angle with the outer ring, α_o, and the reaction force, W_o, are solved from the two equations of balance of forces on a rolling element. The balance of forces in the x (horizontal) direction is

$$\begin{aligned} \Sigma F_x &= 0 \\ W_{oa} &= W_{ia} \end{aligned} \tag{12-48}$$

In the y (vertical) direction, the balance of forces is

$$\begin{aligned} \Sigma F_y &= 0 \\ W_{ir} + F_c &= W_{or} \end{aligned} \tag{12-49}$$

Example Problem 12-8

An angular contact ball bearing has a contact angle with the inner ring of $a_i = 40°$. The thrust load is 2400 N divided exactly evenly on 12 balls (thrust load of 200 N on each ball). The ball has a diameter of $d_r = 14$ mm. The shaft speed is $N = 30{,}000$ RPM. The diameter of the outer race at the contact point with the balls is 78 mm, and the radius of the race groove, in an axial cross section, is 8 mm. The balls and rings are made of steel having the following properties:

$$\begin{aligned} E &= 2 \times 10^{11}\ \text{N/m}^2 \\ \nu &= 0.3 \\ \rho &= 7870\ \text{kg/m}^3 \end{aligned}$$

Find the contact angle α_o and the contact force with the inner and outer rings.

Solution

Given:

$$\begin{aligned} \alpha_i &= 40° \\ N &= 30{,}000\ \text{RPM} \\ W_a &= 200\ \text{N} \\ d_r &= 14\ \text{mm} = 0.014\ \text{m} \\ R_{\text{out}} &= 39\ \text{mm} = 0.039\ \text{m} \\ R_{\text{in}} &= 25\ \text{mm} = 0.025\ \text{m} \\ R_{c(\text{ball center})} &= 32\ \text{mm} = 0.032\ \text{m} \end{aligned}$$

Volume of a Sphere

$$V_{\text{sphere}} = \frac{4}{3}\pi r^3$$
$$V_{\text{sphere}} = \frac{4}{3}\pi(0.7)^3 = 1.43\ \text{cm}^3$$

Mass of the Ball

$$m_r = \rho V = 7.87\ \text{g/cm}^3 \times 1.43\ \text{cm}^3 = 11.25\ \text{g} = 0.011\ \text{kg}$$

The shaft and inner ring angular speed is

$$\omega = \frac{2\pi N}{60}$$
$$\omega = 30{,}000\frac{\text{rev}}{\text{min}} \times \frac{1\ \text{min}}{60\ \text{s}} \times \frac{2\pi\text{rad}}{1\ \text{rev}} = 3142\ \text{rad/s}$$

Angular Speed of Rolling Elements and Cage

$$\omega_C = \frac{R_{\text{in}}}{R_{\text{in}} + R_{\text{out}}}\omega_{\text{shaft}} = \frac{0.025\ \text{m}}{0.025\ \text{m} + 0.039\ \text{m}} \times 3142\ \text{rad/s} = 1227\ \text{rad/s}$$

Centifugal Force, F_c

$$F_c = m_r\omega_C^2 R_r = 0.011\ \text{kg} \times (1227\ \text{rad/s})^2 \times 0.032\ \text{m} = 530\ \text{N}$$

Thrust Force Component at Outer Ring Contact

$$\Sigma F_x = 0$$
$$W_{oa} = W_{ia} = 200\ \text{N}$$

Radial Component of Inner Ring Contact

$$\frac{W_{ir}}{W_{ia}} = c\tan\alpha_i$$
$$W_{ir} = 200 \times c\tan 40^\circ = 238.4\ \text{N}$$

Radial Component of Outer Ring Contact

$$\Sigma F_y = 0$$
$$W_{or} = W_{ir} + F_c$$
$$W_{or} = 238.4 + 530 = 768.4 \text{ N}$$

Outer Ring Contact Angle

$$\tan \alpha_o = \frac{W_{oa}}{W_{or}}$$
$$\tan \alpha_o = \frac{200 \text{ N}}{768.4 \text{ N}} = 0.26$$
$$\alpha_o = 14.5^\circ$$

Resultant Force on Outer Ring Contact

The resultant force on the outer ring race, W_o, is in the direction normal to surface, as shown in Fig. 12-22. It has two components in the axial and radial directions:

$$\frac{W_{oa}}{W_o} = \sin 14.5^\circ$$
$$W_o = \frac{200 \text{ N}}{\sin 14.5^\circ} = 798.8 \text{ N}$$

Normal Contact Force on Inner Ring Race

The resultant force on the inner ring race, W_i, is in the direction normal to surface, as shown in Fig. 12-22. It has two components in the axial and radial directions:

$$\frac{W_{ia}}{W_o} = \sin 40^\circ$$
$$W_i - \frac{200 \text{ N}}{\sin 40^\circ} = 311.2 \text{ N}$$

Example Problem 12-9

An angular contact ball bearing has a contact angle with the inner ring of $\alpha_i = 30^\circ$. The thrust load of $W = 11{,}500$ N is divided evenly on 14 balls. The ball has a diameter of $d_r = 18$ mm. The shaft speed is $N = 33{,}000$ RPM. The conformity ratio $R_r = 0.52$. The diameter of the outer race at the contact point with the balls is 118 mm. The balls and rings are made of steel having the following properties: modulus of elasticity $E = 2 \times 10^{11}$ N/m^2, Poisson's ratio $\nu = 0.3$, density $\rho = 7870$ kg/m^3.

a. Find contact angle, α_o, with the outer ring.
b. Find the contact force with the inner and outer rings.
c. Find the centrifugal force of each rolling element.
d. Find the maximum pressure at the contact with the outer ring.
e. Find the maximum pressure at the contact with the outer ring, given rings made of steel and balls made of silicone nitride (hybrid bearing).

The properties of silicone nitride are: modulus of elasticity $E = 3.14 \times 10^{11}$ N/m^2, Poisson's ration $\nu = 0.24$, density $\rho = 3200$ kg/m^3.

Solution

a. Contact Angle, α_o, with Outer Ring

The first step is to find the centrifugal force, $F_c = m_r \omega_C^2 R_c$, where m_r is the mass of a ball, ω_C is the angular speed of the ball center (or cage), and R_c is the radius of the ball center circular orbit. The volume and mass of a ball in the bearing are:

$$V_{\text{sphere}} = \frac{4}{3}\pi r^3 = \frac{4}{3} \times \pi \times 0.009^3 \text{ m}^3 = 3.05 \times 10^{-6} \text{ m}^3$$
$$m_r = \rho V = 7870 \text{ kg/m}^3 \times 3.05 \times 10^{-6} \text{ m}^3 = 0.024 \text{ kg}$$

The shaft and inner ring angular speed is

$$\omega = \frac{2\pi \times 33{,}000}{60} = 3456 \text{ rad/s}$$

The outer diameter is

$$R_{\text{out}} = \frac{118}{2} \text{ mm} = 59 \times 10^{-3} \text{ m}$$

In order to find the inner diameter, we assume that $\alpha_o \sim \alpha_i$ and that the difference between the outer and inner radius is $d_r \cos \alpha_i$. In that case,

$$R_{\text{in}} = (R_{\text{out}} - d_r) \cos \alpha_i = (59 \text{ mm} - 18 \text{ mm}) \cos 30^\circ = 43.4 \text{ mm}$$
$$= 43.4 \times 10^{-3} \text{ m}$$

Now it is possible to find the angular speed of a rolling-element center (or cage), ω_c:

$$\omega_c = \frac{R_{\text{in}}}{R_{\text{in}} + R_{\text{out}}} \omega_{\text{shaft}} = \frac{43.4}{43.4 + 59} \times 3456 \text{ rad/s} = 1465 \text{ rad/s}$$

The distance between the ball center and the bearing center as R_c, is required for the calculation of the centrifugal force.

$$R_c = \frac{R_{\text{in}} + R_{\text{out}}}{2} = 51.2 \text{ mm}$$

The centrifugal force, F_c, of each rolling element is in the radial direction:

$$F_c = m_r \omega_c^2 R_c = 0.024 \text{ kg} \times 1465^2 \times 1/s^2 \times 0.0512 \text{ m} = 2637 \text{ N}$$

The thrust force component of each ball at the outer ring contact is equal to that of the inner ring:

$$W_{oa} = W_{ia} = \frac{11{,}500 \text{ N}}{14} = 821 \text{ N}$$

The radial component of inner ring contact is

$$\frac{W_{ir}}{W_{ia}} = c\tan \alpha_i$$

$$W_{ir} = 821 \times c\tan 30^\circ = 1422 \text{ N}$$

The radial component of outer ring contact force is

$$W_{or} = W_{ir} + F_c = 1421 \text{ N} + 2637 \text{ N} = 4059 \text{ N}$$

The outer ring contact angle is solved as follows:

$$\tan \alpha_o = \frac{W_{oa}}{W_{or}} = \frac{821 \text{ N}}{4058 \text{ N}} = 0.202, \qquad \alpha_o = 11.43^\circ$$

b. Contact Force with Inner and Outer Rings

The resultant (normal) component of the outer ring contact is

$$\frac{W_{oa}}{W_o} = \sin 11.43^\circ$$

$$W_o = \frac{821 \text{ N}}{\sin 11.43^\circ} = 4143 \text{ N}$$

The resultant (normal) contact force at the inner ring race is

$$\frac{W_{ia}}{W_i} = \sin 30^\circ$$

$$W_i = \frac{W_{ia}}{\sin 30^\circ} = 1642 \text{ N}$$

c. Rolling-Element Centrifugal Force

From part (a) the centrifugal force is 2637 N, in the radial direction.

d. Maximum Pressure at Contact of Outer Ring

The maximum force is at the race of the outer ring, $W_o = 4143\,\text{N}$. The radius of the contact curvatures is

$$R_{1x} = R_{1y} = 9\ \text{mm}$$
$$R_{2y} = R_r d_r = 18\ \text{mm} \times 0.52 = 9.36\,\text{mm}$$
$$R_{2x} = 59\ \text{mm}$$

The equivalent radius of the outer raceway contact in the y-z plane (referred to as the x plane) is

$$\frac{1}{R_x} = \frac{1}{R_{1x}} - \frac{1}{R_{2x}} = \frac{1}{9\ \text{mm}} - \frac{1}{59\ \text{mm}}$$
$$R_x = \frac{9 \times 59\ \text{mm}}{43.4 - 9} = 10.62\ \text{mm}$$

The equivalent inner radius in the y plane is

$$\frac{1}{R_y} = \frac{1}{R_{1y}} - \frac{1}{R_{2y}} = \frac{1}{9\ \text{mm}} - \frac{1}{9.36\ \text{mm}}$$
$$R_y = \frac{R_{2y} \cdot R_{1y}}{R_{2y} - R_{1y}} = \frac{9 \times 9.36\ \text{mm}}{9.36 - 9} = 234\ \text{mm}$$

The equivalent curvature of the inner ring and ball contact is

$$\frac{1}{R_{eq}} = \frac{1}{R_x} + \frac{1}{R_y} = \frac{1}{10.62\ \text{mm}} + \frac{1}{234\ \text{mm}}$$
$$R_{eq} = \frac{R_x \cdot R_y}{R_x + R_y} = \frac{10.62 \times 234\ \text{mm}}{234 + 10.62} = 10.16\ \text{mm}$$

The ratio α_r becomes

$$\alpha_r = \frac{R_y}{R_x} = \frac{234\ \text{mm}}{10.62\ \text{mm}} = 22.03$$

The dimensionless coefficient k is estimated from the ratio α_r:

$$k = \alpha_r^{2/\pi} = 7.16$$

For the calculation of the ellipsoid radii, the following values are required:

$$q_a = \frac{\pi}{2} - 1 = 0.57$$

$$\hat{E} \approx 1 + \frac{q_a}{\alpha_r}$$

$$\hat{E} \approx 1 + \frac{0.57}{22.03} = 1.025$$

The shaft and the bearing are made of identical material, so the equivalent modulus of elasticity is

$$E_{eq} = \frac{E}{1 - \nu^2} = \frac{2 \times 10^{11} \text{N/m}^2}{1 - 0.3^2} = 2.2 \times 10^{11} \text{ N/m}^2$$

Now the ellipsoid radii a and b can be determined:

$$a = \left(\frac{6 \cdot \hat{E} \cdot W_{max} \cdot R_{eq}}{\pi \cdot k \cdot E_{eq}} \right)^{1/3}$$

$$= \left(\frac{6 \times 1.025 \times 4143 \text{ N} \times 10.82 \times 10^{-3} \text{ m}}{\pi \times 7.16 \times 2.2 \times 10^{11} \text{ N/m}^2} \right)^{1/3} = 0.37 \text{ mm}$$

$$b = \left(\frac{6 \cdot k^2 \cdot \hat{E} \cdot W_{max} \cdot R_{eq}}{\pi \cdot E_{eq}} \right)^{1/3}$$

$$= \left(\frac{6 \times 7.16^2 \times 1.025 \times 4143 \text{ N} \times 10.16 \times 10^{-3} \text{ m}}{\pi \times 2.2 \times 10^{11} \text{ N/m}^2} \right)^{1/3} = 2.68 \text{ mm}$$

The maximum pressure at the contact with outer ring is

$$p_{max} = \frac{3}{2} \cdot \frac{W_{max}}{\pi \cdot ab} = \frac{3}{2} \cdot \frac{4134 \text{ N}}{\pi \times 0.37 \times 2.68 \times 10^{-6} \text{ m}^2} = 1.99 \text{ GPa}$$

e. Maximum stress of hybrid bearing

The rings are made of steel and the balls are made of silicone nitride, which has the following properties:

Modulus of elasticity $E = 3.14 \times 10^{11}$ N/m^2
Poisson's ratio $\nu = 0.24$
Density $\rho = 3200$ kg/m^3

The major advantage of a hybrid bearing is a lower centrifugal force due to lower density of the rolling element. However, the modulus of elasticity of silicone nitride is higher than that of steel, and this can result in a higher maximum pressure.

For a hybrid bearing, it is necessary to consider the equivalent modulus of elasticity of silicone nitride on steel:

$$\frac{2}{E_{eq}} = \frac{1-\nu_{21}^2}{E_1} + \frac{1-\nu_2^2}{E_2} = \frac{1-0.24^2}{3.14\times10^{11}\ \mathrm{N/m^2}} + \frac{1-0.3^2}{2.0\times10^{11}\ \mathrm{N/m^2}}$$
$$= 0.755\times10^{-11}\ \mathrm{m^2}/N$$
$$E_{eq} = 2.65\times10^{11}\ \mathrm{N/m^2}$$

Due to the low density of the balls, the centrifugal force is lower:

$$m_r = \rho V = 3200\frac{kg}{m^3}\times 3.05\times10^{-6}\ \mathrm{m^3} = 0.0098\ \mathrm{kg}$$

The centrifugal force is

$$F_c = m_r\omega_c^2 R_c = 0.0098\ \mathrm{kg}\times 1465^2\ 1/s^2 \times 0.0512\ \mathrm{m} = 1077\ \mathrm{N}$$

The radial component of the inner ring contact is the same as for a steel bearing:

$$W_{ir} = 1422\ \mathrm{N}$$

The radial component on the outer ring contact becomes

$$W_{or} = W_{\mathrm{ir}} + F_c = 1422\ \mathrm{N} + 1077\ \mathrm{N} = 2499\ \mathrm{N}$$

The outer ring contact angle is

$$\tan\alpha_o = \frac{W_{oa}}{W_{or}} = \frac{821\ \mathrm{N}}{2499\ \mathrm{N}} = 0.329, \qquad \alpha_o = 18.19^\circ$$

The resultant (normal) component on the outer ring contact becomes

$$\frac{W_{oa}}{W_o} = \sin 18.19^\circ \qquad \text{and} \qquad W_o = \frac{821\ \mathrm{N}}{\sin 18.19^\circ} = 2630\ \mathrm{N}$$

The ellipsoid radii a and b can now be determined by substituting the values already calculated:

$$a = \left(\frac{6\, E\, W_{max}\, R_{eq}}{\pi \times k \times E_{eq}} \right)^{1/3}$$

$$= \left(\frac{6 \times 1.025 \times 2630\ \text{N} \times 10.16 \times 10^{-3}\ \text{m}}{\pi \times 6.86 \times 2.65 \times 10^{11}\ \text{N/m}^2} \right)^{1/3} = 0.30\ \text{mm}$$

$$b = \left(\frac{6\, k^2\, E\, W_{max}\, R_{eq}}{\pi \times E_{eq}} \right)^{1/3}$$

$$= \left(\frac{6 \times 6.86^2 \times 1.028 \times 2630\ \text{N} \times 10.82 \times 10^{-3}\ \text{m}}{\pi \times 2.65 \times 10^{11}\ \text{N/m}^2} \right)^{1/3}$$

$$= 2.16\ \text{mm}$$

The maximum pressure at the contact is

$$p_{max} = \frac{3}{2} \frac{W_{max}}{\pi ab} = \frac{3}{2} \frac{2630\ \text{N}}{\pi \times 0.3 \times 2.16 \times 10^{-6}\ \text{m}^2} = 1.94\ \text{GPa}$$

Conclusion

In this example of a high-speed bearing, the maximum stress is only marginally lower for the hybrid bearing. The maximum pressure at the contact with outer ring is

For an all-steel bearing: $p_{max} = 1.99$ GPa
For a hybrid bearing: $p_{max} = 1.94$ GPa

Example Problem 12-10

For the bearing in Example Problem 12-9, find h_{min} at the contact with the outer race and the inner race for both (a) an all-steel bearing and (b) a hybrid bearing. Use oil SAE 10 at 70°C and a viscosity–pressure coefficient $\alpha = 2.2 \times 10^{-8}\ \text{m}^2/\text{N}$.

Solution

The analysis of the forces is identical to that of Example Problem 12-9. In this problem, the EHD minimum film thickness is calculated.

Bearing Data from Example Problem 12-9:

Inner contact diameter, $R_{in} = 43.4$ mm
Outer contact diameter, $R_{out} = 59$ mm

Outer ring equivalent radius in the x plane, $R_x = 10.62$ mm (contact at outer race)
Shaft speed, $\omega = 3456$ rad/s
$R_x = 9$ mm, $R_{2x} = 43.4$ mm, $R_y = 234$ mm

Data for Steel Bearing from Example Problem 12.9:

Equivalent modulus of elasticity: $E_{eq} = 2.2 \times 10^{11}$ N/m^2
Resultant component of outer ring contact: $W_o = 4143$ N
Normal contact force at inner ring race: $W_i = 1642$ N

Data for Hybrid Bearing:

Equivalent modulus of elasticity: $E_{eq} = 2.65 \times 10^{11}$ N/m^2
Resultant component of outer ring contact: $W_o = 2630$ N
Normal contact force at inner ring race: $W_i = 1642$ N

For hard surfaces, such as steel in rolling bearings, the equation for calculating the minimum film thickness, h_{min}, is presented in dimensionless form, as follows:

$$\frac{h_{min}}{R_x} = 3.63 \frac{\bar{U}_r^{0.68}(\alpha \cdot E_{eq})^{0.49}}{\bar{W}^{0.073}}(1 - e^{-0.68k})$$

Here, α is the viscosity–pressure coefficient and $\bar{U}_r$ and $\bar{W}$ are dimensionless velocity and load, respectively, defined by the following equations:

$$\bar{U}_r = \frac{\mu_o U_r}{E_{eq} R_x} \qquad \text{and} \qquad \bar{W} = \frac{W}{E_{eq} R_x^2}$$

Here, U_r is the rolling velocity, μ_o is the viscosity of the lubricant at atmospheric pressure and bearing operating temperature, W is the reaction force of one rolling element, and E_{eq} is the equivalent modulus of elasticity.

All steel and hybrid bearings have the same rolling velocity, which is calculated from the shaft speed via Eq. (12-34):

$$U_{rolling} = \frac{R_{in} R_{out}}{2(R_{in} + r)} \omega = \frac{R_{in} R_{out}}{R_{in} + R_{out}} \omega$$

$$U_r = \frac{0.0434 \times 0.059 \text{ m}^2}{0.0434 \text{ m} + 0.059 \text{ m}} \times 3456 \text{ rad/s} = 86.42 \text{ m/s}$$

The first step is to calculate R_x and k for the inner and outer contacts. The radius of curvature at the inner contact is

$$\frac{1}{R_x} = \frac{1}{R_{1x}} + \frac{1}{R_{2x}} = \frac{1}{9\ \text{mm}} + \frac{1}{43.4\ \text{mm}}$$

$$R_x = \frac{9 \times 43.4\ \text{mm}}{43.4 + 9} = 7.45\ \text{mm}$$

The ratio α_r is

$$\alpha_r = \frac{R_y}{R_x} = \frac{234\ \text{mm}}{7.45\ \text{mm}} = 31.2$$

The dimensionless coefficient k is derived directly from the ratio α_r:

$$k = \alpha_r^{2/\pi} = 8.94$$

For the outer contact, R_x and k can be taken from Example Problem 12-9:

$$R_x = 10.62\ \text{mm} \qquad \text{and} \qquad k = 7.16$$

a. All-Steel Bearing

Equivalent modulus of elasticity: $E_{eq} = 2.2 \times 10^{11}\ \text{N/m}^2$
Resultant component of outer ring contact is: $W_o = 4143\ \text{N}$
Normal contact force at inner ring race: $W_i = 1642\ \text{N}$

Inner Race Contact: The dimensionless rolling velocity is

$$\bar{U}_r = \frac{0.01\ \text{N-s/m}^2 \times 86.42\ \text{m/s}}{2.2 \times 10^{11}\ \text{N/m}^2 \times 7.45 \times 10^{-3}\ \text{m}^2} = 52.7 \times 10^{-11}$$

The dimensionless load at the inner race is

$$\bar{W} = \frac{1642\ \text{N}}{2.2 \times 10^{11}\ \text{N/m}^2 \times 7.45^2 \times 10^{-6}\ \text{m}^2} = 13.44 \times 10^{-5}$$

Substituting these values in the formula for the minimum thickness, we get

$$\frac{h_{min}}{7.45 \times 10^{-3}} = 3.63 \times \frac{(52.7 \times 10^{-11})^{0.68}(2.2 \times 10^{-8} \times 2.2 \times 10^{11})^{0.49}}{(13.44 \times 10^{-5})^{0.073}}$$
$$\times (1 - e^{-0.68 \times 8.94}) = 217.8 \times 10^{-6}$$

The minimum thickness for the inner race is

$$h_{min} = 217.8 \cdot 10^{-6} \times 7.45 \cdot 10^{-3}\ \text{m} = 1.62 \cdot 10^{-6}\ \text{m}$$
$$h_{min} = 1.62\ \mu\text{m}$$

Outer Race Contact: The dimensionless rolling velocity is

$$\bar{U}_r = \frac{0.01\ \text{N-s/m}^2 \times 86.42\ \text{m/s}}{2.2 \times 10^{11}\ \text{N/m}^2 \times 10.62 \times 10^{-3}\ \text{m}^2} = 37 \times 10^{-11}$$

The dimensionless load at the outer race is

$$\bar{W} = \frac{4134\ \text{N}}{2.2 \times 10^{11}\ \text{N/m}^2 \times 10.62^2 \times 10^{-6}\ \text{m}^2} = 16.7 \times 10^{-5}$$

Substituting these values in the formula for the minimum thickness, we get

$$\frac{h_{\min}}{10.62 \times 10^{-3}} = 3.63 \times \frac{(37 \times 10^{-11})^{0.68}(2.2 \times 10^{-8} \times 2.2 \times 10^{11})^{0.49}}{16.7 \times 10^{-5})^{0.073}} \times (1 - e^{-0.68 \times 7.16}) = 167.1 \times 10^{-6}$$

The minimum thickness at the outer race contact is

$$h_{\min} = 167.1 \times 10^{-6} \times 10.62 \times 10^{-3}\ \text{m} = 1.78 \times 10^{-6}\ \text{m}$$
$$h_{\min} = 1.78\ \mu\text{m}$$

b. Hybrid Bearing

Equivalent modulus of elasticity: $E_{eq} = 2.65 \times 10^{11}\ \text{N/m}^2$
Resultant force component on outer race contact: $W_o = 2630$ N
Resultant force component on inner ring contact: $W_i = 1642$ N

Inner Race: The value of dimensionless velocity is

$$\bar{U}_r = \frac{0.01\ \text{N-s/m}^2 \times 86.42\ \text{m/s}}{2.65 \times 10^{11}\ \text{N/m}^2 \times 7.45 \times 10^{-3}\ \text{m}^2} = 43.8 \times 10^{-11}$$

The dimensionless load on the inner race is

$$\bar{W} = \frac{1642\ \text{N}}{2.65 \times 10^{11}\ \text{N/m}^2 \times 7.45^2 \times 10^{-6}\ \text{m}^2} = 11.16 \times 10^{-5}$$

Substituting these values in the equation for the minimum film thickness, we get

$$\frac{h_{\min}}{7.45 \times 10^{-3}} = 3.63 \cdot \frac{(43.8 \times 10^{-11})^{0.68}(2.2 \times 10^{-8} \times 2.65 \times 10^{11})^{0.49}}{(11.16 \times 10^{-5})^{0.073}} \times (1 - e^{-0.68 \times 8.94}) = 213.24 \times 10^{-6}$$

The minimum fluid film thickness at the inner race contact is

$$h_{\min} = 213.24 \times 10^{-6} \times 7.45 \times 10^{-3}\ \text{m} = 1.59 \times 10^{-6}\ \text{m}$$
$$h_{\min} = 1.59\ \mu\text{m}$$

Outer Race: The value of dimensionless velocity is

$$\bar{U}_r = \frac{0.01 \text{ N-s/m}^2 \times 86.42 \text{ m/s}}{2.65 \times 10^{11} \text{ N/m}^2 \times 10.62 \times 10^{-3}} \text{m}^2 = 30.7 \times 10^{-11}$$

The dimensionless load on the outer race is

$$\bar{W} = \frac{2630 \text{ N}}{2.65 \times 10^{11} \text{ N/m}^2 \times 10.62^2 \times 10^{-6} \text{ m}^2} = 8.8 \times 10^{-5}$$

Substituting these values in the equation for the minimum thickness, we get

$$\frac{h_{\min}}{10.62 \times 10^{-3}} = 3.63 \times \frac{(30.7 \times 10^{-11})^{0.68}(2.2 \times 10^{-8} \times 2.65 \times 10^{11})^{0.49}}{(8.8 \times 10^{-5})^{0.073}} \times (1 - e^{-0.68 \times 7.16}) = 169.4 \times 10^{-6}$$

The minimum fluid-film thickness at the outer race is

$$h_{\min} = 169.4 \times 10^{-6} \times 10.62 \times 10^{-3} \text{ m} = 1.80 \times 10^{-6} \text{ m}$$

Conclusion

The following is a summary of the results.

Steel bearing

Minimum thickness for inner race: $h_{\min} = 1.62\,\mu\text{m}$
Minimum thickness for outer race: $h_{\min} = 1.78\,\mu\text{m}$

Hybrid Bearing

Minimum thickness for inner race: $h_{\min} = 1.59\ \mu\text{m}$
Minimum thickness for outer race: $h_{\min} = 1.80\ \mu\text{m}$

In this case, the results show only a marginal difference. The minimum thickness for the inner race is a little thinner. It means that the hybrid bearing has only a marginal adverse effect on the EHD lubrication. In this problem, the shaft speed is $N = 30{,}000$ RPM. A significant effect of the hybrid bearing is apparent only at much higher speed.

Problems

12-1 A deep-groove ball bearing has the following dimensions: The bearing has 12 balls of diameter $d = 16$ mm. The radius of curvature of the inner groove (in cross section x-z) is 9 mm. The inner race diameter (at the bottom of the deep groove) is $d_i = 62$ mm (in cross section y-z). The radial load on the bearing is $W = 15{,}000$ N, and the bearing speed is $N = 8000$ RPM. The bearing, rolling elements, and rings are made of steel. The modulus of elasticity of the steel for rollers and rings is $E = 2 \times 10^{11}$ N/m^2, and Poisson's ratio for the steel is $\nu = 0.3$. The properties of the lubricant are: The absolute viscosity at ambient pressure and bearing operating temperature is $\mu_o = 0.01$ N-s/m^2, and the viscosity–pressure coefficient is $\alpha = 2.31 \times 10^{-8}$ m^2/N.

Find the minimum film thickness at the contact with the inner ring.

12-2 A cam and follower are shown in Fig. 12-21. For the same cam and follower, find the maximum stress when there is no rotation. The cam and follower are made of steel. The steel modulus of elasticity is $E = 2.05 \times 10^{11}$ N/m^2 and its Poisson ratio ν is 0.3. The follower is in contact with the tip radius, under the load of $W = 1200$ N.

12-3 For the cam and follower shown in Fig. 12-21, find the viscous torque when the follower is in contact with the tip radius under the same load and fluid viscosity as in Example Problem 12-4. Hint: Find h_o and the contact area, and consider it as a simple shear problem.

12-4 A deep-groove ball radial bearing has 10 balls of diameter $d = 20$ mm. The radius of curvature of the deep grooves (in cross section y-z) is determined by the conformity ratio of $R_r = 0.54$. The inner race diameter (at the bottom of the deep groove) is $d_i = 80$ mm (cross section x-z). The radial load on the bearing is $W = 20{,}000$ N, and the bearing speed is $N = 3000$ RPM. The bearing, rolling elements, and rings are made of steel. The modulus of elasticity of the steel for rollers and rings is $E = 2 \times 10^{11}$ N/m^2, and Poisson ratio for the steel is $\nu = 0.3$.

a. Find the maximum rolling contact pressure at the deep-groove contact.
b. Suppose this bearing is to be used in a high-speed turbine where the average shaft speed is increased to $N = 40{,}000$ RPM. Find the maximum contact pressure.
c. For the preceding two cases, given rolling elements made of silicone nitride, find the maximum pressure in each case of low and high speed. Find the maximum pressure at the inner and outer raceways.

Note: Centrifugal force must be considered at this high rotational speed.

12-5 A deep-groove ball radial bearing has 10 balls of diameter

$d = 20$ mm. The radius of curvature of the deep grooves (in cross section *y-z*) is determined by the conformity ratio of $R_r = 0.54$. The inner race diameter (at the bottom of the deep groove) is $d_i = 80$ mm (cross section *x-z*). The radial load on the bearing is $W = 20{,}000$ N, and the bearing speed is $N = 3000$ RPM. The bearing, rolling elements, and rings are made of steel. The modulus of elasticity of the steel for rollers and rings is $E = 2 \times 10^{11}$ N/m^2, and Poisson ratio for the steel is $\nu = 0.3$. The absolute viscosity of the lubricant at ambient pressure and bearing operating temperature is $\mu_o = 0.015$ N-s/m^2. The viscosity–pressure coefficient is $\alpha = 2.2 \times 10^{-8}$ m^2/N.

a. Find the minimum fluid film thickness of a rolling contact at the deep-groove contact.
b. Suppose this bearing is to be used in a high-speed turbine where the average shaft speed is increased to $N = 40{,}000$ RPM. Find the minimum film thickness (consider centrifugal force).
c. For the preceding two cases, given rolling elements made of silicone nitride, find the maximum pressure in each case of low and high speed. Find the minimum film thickness at the inner and outer raceways.

13

Selection and Design of Rolling Bearings

13.1 INTRODUCTION

Several factors must be considered for an appropriate selection of a rolling bearing for a particular application. The most important factors are load characteristics, speed, lubrication, and environmental conditions. Load characteristics include steady or oscillating load and the magnitude of the radial and axial load components. In selecting a rolling-element bearing, the first two steps are (a) to check whether the bearing can resist the static load and (b) the level of fatigue under oscillating stresses. The fatigue life is estimated in order to ensure that the bearing will not fail prematurely.

In Chapter 12, bearing selection based on basic principles was discussed. The selection was based on stress calculations using Hertz equations for calculating the maximum normal compression stress (maximum pressure) of a rolling contact. The purpose of the calculations is to make sure that the actual maximum contact pressure does not exceed a certain allowed stress limit, which depends on the bearing material. In this chapter a simplified approach for bearing selection is presented for bearings made of standard materials. This approach is based on empirical and analytical data that is provided in manufacturers' catalogues.

In addition to stress calculations, Chapter 12 discussed the EHD fluid film equations. The EHD equations are used for optimum selection of a rolling bearing in combination with an appropriate lubricant. However, in this chapter, the selection of a rolling bearing and lubricant are simplified for standard bearings. Empirical data in the form of charts is used for the selection of bearing and lubricant.

13.1.1 Static Load

Selection of bearings by means of static load calculations is necessary only for slow speeds, because at higher speeds, the requirement for fatigue resistance is much more demanding. This means that bearings that are selected via fatigue calculations are usually loaded much below the static load limits.

For standard rolling bearings that are made of hardened steel, the ISO standard has been set to limit the maximum stress in order to prevent excessive permanent (plastic) deformation. In fact, the ISO standard is limiting the total plastic deformation of the rolling element and raceway. The total plastic deformation limits to ($10^{-4} D_c$), where D_c is the circular orbit diameter of a rolling-element center.

A modified calculation method, which limits the maximum contact stresses, was adopted recently. The international standard ISO 76 was revised, and the Antifriction Bearing Manufacturers Association (AFBMA) in 1986 adopted this method of calculation of static loads. In order to satisfy the plastic deformation limit, the compression stress limits applied to various rolling element bearings made of standard hardened steel are:

Ball bearings:	4200 MPa
Self-aligning ball bearings:	4600 MPa
Roller bearings:	4000 MPa

The basic static load rating C_0 is defined as the static load that results in the calculated stress limit at the center of the rolling contact area where there is maximal compression stress. For radial bearings, the radial static load, C_{0r}, is limited; for thrust bearings, the axial load C_{0a} is limited.

For standard steel bearings, the values of the static load rating C_0 are given in manufacturers' catalogues. Examples are in Tables 13-1 through 13-4. These values are helpful because the designer can rely on the maximum load without resorting to calculations based on Hertz equations. However, better materials are often used; in such cases the designer should use Hertz equations for determining the maximum allowed bearing load from basic principles.

TABLE 13-1 Dimensions and Load Ratings for Deep Ball Bearing Series 6300. (From FAG Bearing Catalogue, with permission)

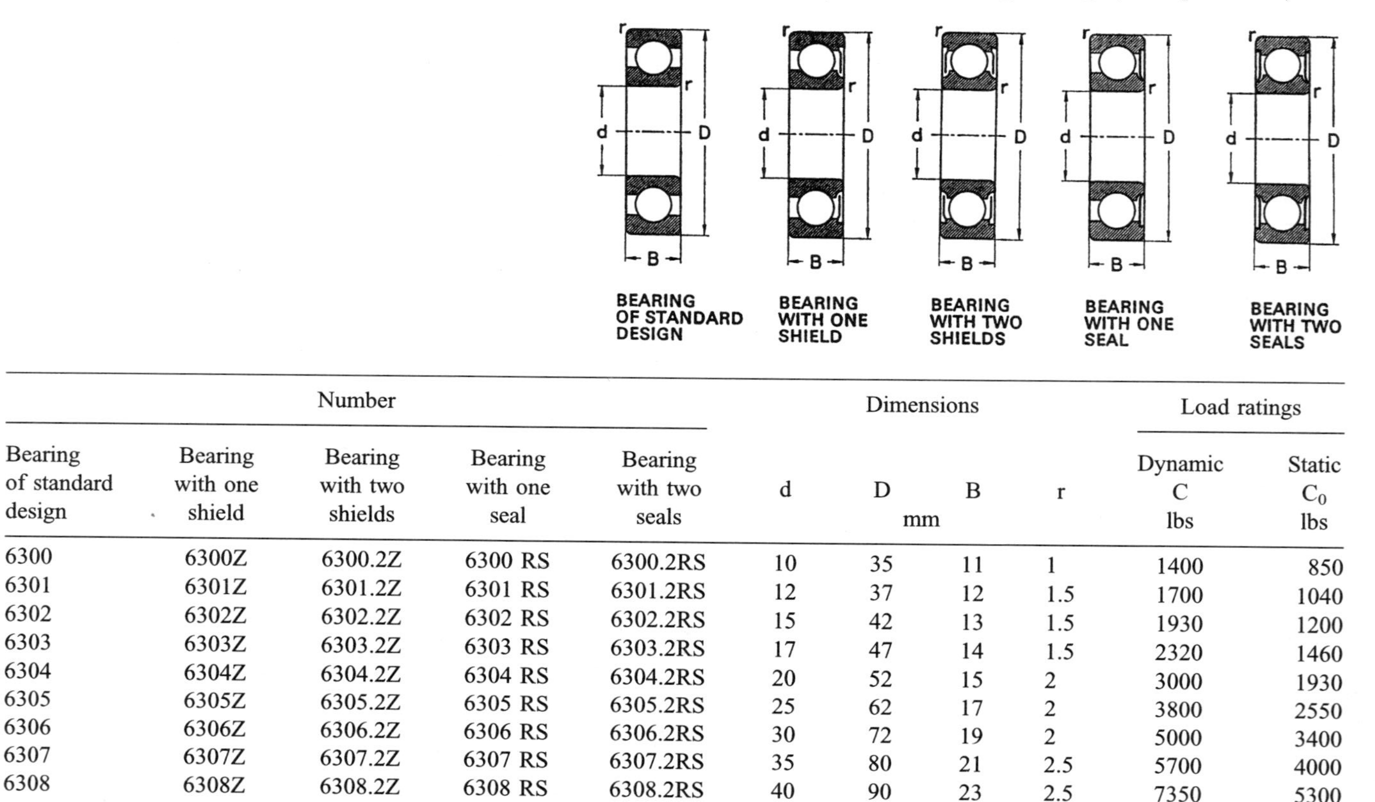

Number					Dimensions				Load ratings	
Bearing of standard design	Bearing with one shield	Bearing with two shields	Bearing with one seal	Bearing with two seals	d	D (mm)	B	r	Dynamic C lbs	Static C_0 lbs
6300	6300Z	6300.2Z	6300 RS	6300.2RS	10	35	11	1	1400	850
6301	6301Z	6301.2Z	6301 RS	6301.2RS	12	37	12	1.5	1700	1040
6302	6302Z	6302.2Z	6302 RS	6302.2RS	15	42	13	1.5	1930	1200
6303	6303Z	6303.2Z	6303 RS	6303.2RS	17	47	14	1.5	2320	1460
6304	6304Z	6304.2Z	6304 RS	6304.2RS	20	52	15	2	3000	1930
6305	6305Z	6305.2Z	6305 RS	6305.2RS	25	62	17	2	3800	2550
6306	6306Z	6306.2Z	6306 RS	6306.2RS	30	72	19	2	5000	3400
6307	6307Z	6307.2Z	6307 RS	6307.2RS	35	80	21	2.5	5700	4000
6308	6308Z	6308.2Z	6308 RS	6308.2RS	40	90	23	2.5	7350	5300

6309	6309Z	6309.2Z	6309 RS	6309.2RS	45	100	25	2.5	9150	6700
6310	6310Z	6310.2Z	6310 RS	6310.2RS	50	110	27	3	10600	8150
6311	6311Z	6311.2Z			55	120	29	3	12900	10000
6312	6312Z	6312.2Z			60	130	31	3.5	14000	10800
6313	6313Z	6313.2Z			65	140	33	3.5	16000	12500
6314	6314Z	6314.2Z			70	150	35	3.5	18000	14000
6315					75	160	37	3.5	19300	16300
6316					80	170	39	3.5	19600	16300
6317					85	180	41	4	21600	18600
6318					90	190	43	4	23200	20000
6319					95	200	45	4	24500	22400
6320					100	215	47	4	28500	27000
6321					105	225	49	4	30500	30000
6322					110	240	50	4	32500	32500
6324					120	260	55	4	36000	38000
6326					130	280	58	5	39000	43000
6328					140	300	62	5	44000	50000
6330					150	320	65	5	49000	60000

TABLE 13-2 Angular Contact Bearing of Series 909 $\alpha = 25°$, separable. (From FAG Bearing Catalogue, with permission)

EQUIVALENT DYNAMIC LOAD

$P = F_r$ when $\frac{F_a}{F_r} \le 0.68$

$P = 0.41\ F_r + 0.87\ F_a$ when $\frac{F_a}{F_r} > 0.68$

EQUIVALENT STATIC LOAD

$Po = F_r$ when $\frac{F_a}{F_r} \le 1.3$

$Po = 0.5\ F_r + 0.38\ F_a$ when $\frac{F_a}{F_r} > 1.3$

Number	d	D	B	C	T	a	Max. fillet radius inch	Load ratings dynamic C lbs	static C_0 lbs
	Dimensions inch								
909001	.7503	2.0800	.5950	.6080	.7080	.65	.060	3250	2240
909002	1.1904	2.9630	.8700	.7700	1.1450	.91	.060	6100	4400
909003	.8128	2.4370	.6880	.7290	.8290	.73	.100	4550	3200
909004	1.2815	3.3750	.9640	.9330	1.3080	1.06	.010	8300	6300
909007	.9379	3.0300	.8440	.9310	1.0310	.96	.060	7500	5600
909008	1.4384	3.9300	1.0580	1.0950	1.4700	1.20	.100	11400	9150
909021	.6875	1.8750	.5630	.5630	.6880	.63	.060	3050	2040
909022	1.1250	2.5000	.8440	.6250	.9840	.75	.060	4150	2900
909023	.7503	2.2500	.6590	.6900	.7900	.73	.100	4500	3200

909024	1.3128	3.1496	.9170	.8510	1.2260	.98	.100	7350	5600
909025	.8440	2.2500	.6590	.6900	.7900	.75	.060	4300	3100
909026	1.4065	3.1496	.9170	.8510	1.2260	.98	.060	6550	5500
909027	.9379	2.8125	.8000	.8500	.9100	.91	.100	6300	4550
909028	1.5000	3.7500	1.0700	1.0150	1.4500	1.18	.100	9800	7650
909029	1.1250	3.1875	.8750	.9730	1.0730	1.02	.100	8300	6300
909030	1.6250	4.0625	1.1875	1.0950	1.5620	1.20	.100	11400	9150
909052	1.2815	2.9630	.8700	.7700	1.1450	.91	.060	6100	4400
909062	1.3750	2.9630	.8700	.7700	1.1450	.91	.060	6000	4550
909067	.7502	2.0800	.4690	.4690	.7080	.69	.040	3900	2750
909070	1.2500	2.6500	.7000	.5150	.8000	.81	.040	5300	3800

TABLE 13-3 Angular Contact Ball Bearings Series 73B, $\alpha = 40°C$, Non-separable. (From FAG Bearing Catalogue, with permission)

EQUIVALENT DYNAMIC LOAD	
$P = F_r$	when $\frac{F_a}{F_r} \leq 1.14$
$P = 0.35\ F_r + 0.57\ F_a$	when $\frac{F_a}{F_r} > 1.14$

EQUIVALENT STATIC LOAD	
$P_o = F_r$	when $\frac{F_a}{F_r} \leq 1.9$
$P_o = 0.5\ F_r + 0.26\ F_a$	when $\frac{F_a}{F_r} > 1.9$

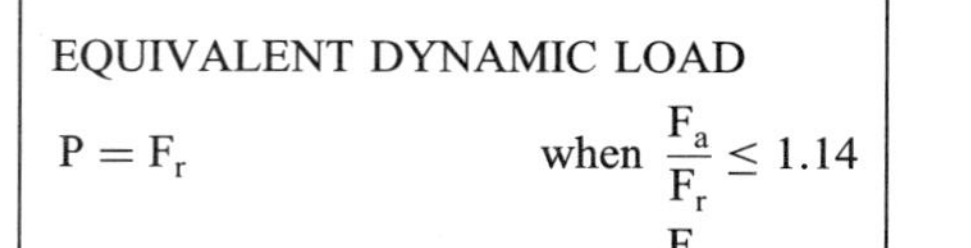

Number	Dimensions d (mm)	D (mm)	B (mm)	r (mm)	r_1 (mm)	a (mm)	d (inch)	D (inch)	B (inch)	a (inch)	Max. fillet radius for r (inch)	Max. fillet radius for r_1 (inch)	Load ratings dynamic C lbs	Load ratings static C_0 lbs
7300B	10	35	11	1	.5	15	.3937	1.3780	.4331	.59	.025	.012	1460	830
7301B	12	37	12	1.5	.8	16	.4724	1.4567	.4724	.63	.040	.020	1830	1080
7302B	15	42	13	1.5	.8	18	.5906	1.6535	.5118	.71	.040	.020	2240	1340
7303B	17	47	14	1.5	.8	20	.6693	1.8504	.5512	.79	.040	.020	2750	1730
7304B	20	52	15	2	1	23	.7874	2.0472	.5906	.91	.040	.025	3250	2120
7305B	25	62	17	2	1	27	.9842	2.4409	.6693	1.06	.040	.025	4500	3050
7306B	30	72	19	2	1	31	1.1811	2.8346	.7480	1.22	.040	.025	5600	3900
7307B	35	80	21	2.5	1.2	35	1.3780	3.1496	.8268	1.38	.060	.030	6800	4800
7308B	40	90	23	2.5	1.2	39	1.5748	3.5433	.9055	1.54	.060	.030	8650	6300
7309B	45	100	25	2.5	1.2	43	1.7716	3.9370	.9842	1.69	.060	.030	10200	7800
7310B	50	110	27	3	1.5	47	1.9685	4.3307	1.0630	1.85	.080	.040	12000	9300

7311B	55	120	29	3	1.5	51	2.1654	4.7244	1.1417	2.01	.080	.040	13400	10800
7312B	60	130	31	3.5	2	55	2.3622	5.1181	1.2205	2.17	.080	.040	15600	12500
7313B	65	140	33	3.5	2	60	2.5590	5.5118	1.2992	2.36	.080	.040	17600	14300
7314B	70	150	35	3.5	2	64	2.7559	5.9055	1.3780	2.52	.080	.040	19600	16300
7315B	75	160	37	3.5	2	68	2.9528	6.2992	1.4567	2.68	.080	.040	22000	19000
7316B	80	170	39	3.5	2	72	3.1496	6.6929	1.5354	2.83	.080	.040	24000	22000
7317B	85	180	41	4	2	76	3.3464	7.0866	1.6142	2.99	.10	.040	26000	24000
7318B	90	190	43	4	2	80	3.5433	7.4803	1.6929	3.15	.10	.040	28000	27000
7319B	95	200	45	4	2	84	3.7402	7.8740	1.7716	3.31	.10	.040	30000	29000
7320B	100	215	47	4	2	90	3.9370	8.4646	1.8504	3.54	.10	.040	33500	34000
7321B	105	225	49	4	2	94	4.1338	8.8582	1.9291	3.70	.10	.040	35500	38000
7322B	110	240	50	4	2	98	4.3307	9.4488	1.9685	3.86	.10	.040	38000	43000

TABLE 13-4 Angular Contact Ball Bearings. (From FAG Bearing Catalogue, with permission)

TANDEM ARRANGEMENT   BACK-TO-BACK (O ARRANGEMENT)   FACE-TO-FACE (X ARRANGEMENT)

EQUIVALENT DYNAMIC LOAD

Tandem arrangement $P = F_r$ when $\frac{F_a}{F_r} \le 1.14$

$P = 0.35\ F_r + 0.57\ F_a$ when $\frac{F_a}{F_r} > 1.14$

O and X arrangements $P = F_r + 0.55\ F_a$ when $\frac{F_a}{F_r} \le 1.14$

$P = 0.57\ F_r + 0.93\ F_a$ when $\frac{F_a}{F_r} > 1.14$

EQUIVALENT STATIC LOAD

Tandem arrangement $P_o = F_r$ when $\frac{F_a}{F_r} \le 1.9$

$P = 0.5\ F_r + 0.26\ F_a$ when $\frac{F_a}{F_r} > 1.9$

O and X arrangements $P_o = F_r + 0.52\ F_a$

Bearing pair number			Dimensions (mm)						Dimensions (inch)				Max. fillet radius for (inch)		Load ratings for bearing pair	
			d	D	2B	r	r_1	2a	d	D	2B	2a	r	r_1	dynamic C^1 lbs	static C_0 lbs
2 × 7300 B.UA	2 × 7300 B.UO	2 × 7300 B.UL	10	35	22	1	.5	30	.3937	1.3780	.8661	1.18	.025	.012	2360	1660
2 × 7301 B.UA	2 × 7301 B.UO	2 × 7301 B.UL	12	37	24	1.5	.8	33	.4724	1.4567	.9449	1.26	.040	.020	3050	2160
2 × 7302 B.UA	2 × 7302 B.UO	2 × 7302 B.UL	15	42	26	1.5	.8	37	.5906	1.6535	1.0236	1.42	.040	.020	3600	2750
2 × 7303 B.UA	2 × 7303 B.UO	2 × 7303 B.UL	17	47	28	1.5	.8	41	.6693	1.8504	1.1024	1.57	.040	.020	4500	3400
2 × 7304 B.UA	2 × 7304 B.UO	2 × 7304 B.UL	20	52	30	2	1	45	.7874	2.0472	1.1811	1.81	.040	.025	5300	4250
2 × 7305 B.UA	2 × 7305 B.UO	2 × 7305 B.UL	25	62	34	2	1	53	.9842	2.4409	1.3386	2.13	.040	.025	7350	6100
2 × 7306 B.UA	2 × 7306 B.UO	2 × 7306 B.UL	30	72	38	2	1	62	1.1811	2.8346	1.4961	2.44	.040	.025	9150	7800

2 × 7307 B.UA	2 × 7307 B.UO	2 × 7307 B.UL	35	80	42	2.5	1.2	69	1.3780	3.1496	1.6535	2.76	.060	.030	11000	9650
2 × 7308 B.UA	2 × 7308 B.UO	2 × 7308 B.UL	40	90	46	2.5	1.2	78	1.5748	3.5433	1.8110	3.07	.060	.030	13700	12500
2 × 7309 B.UA	2 × 7309 B.UO	2 × 7309 B.UL	45	100	50	2.5	1.2	86	1.7716	3.9370	1.9685	3.39	.060	.030	17000	15300
2 × 7310 B.UA	2 × 7310 B.UO	2 × 7310 B.UL	50	110	54	3	1.5	94	1.9685	4.3307	2.1260	3.70	.080	.040	19600	18600
2 × 7311 B.UA	2 × 7311 B UO	2 × 7311 B.UL	55	120	58	3	1.5	102	2.1654	4.7244	2.2835	4.02	.080	.040	22000	21600
2 × 7312 B.UA	2 × 7312 B UO	2 × 7312 B.UL	60	130	62	3.5	2	111	2.3622	5.1181	2.4409	4.33	.080	.040	25000	25000
2 × 7313 B.UA	2 × 7313 B.UO	2 × 7313 B.UL	65	140	66	3.5	2	119	2.5590	5.5118	2.5984	4.72	.080	.040	28500	28500
2 × 7314 B.UA	2 × 7314 B.UO	2 × 7314 B.UL	70	150	70	3.5	2	127	2.7559	5.9055	2.7559	5.04	.080	.040	32000	32500
2 × 7315 B.UA	2 × 7315 B.UO	2 × 7315 B.UL	75	160	74	3.5	2	136	2.9528	6.2992	2.9134	5.35	.080	.040	35500	37500
2 × 7316 B.UA	2 × 7316 B.UO	2 × 7316 B.UL	80	170	78	3.5	2	144	3.1496	6.6929	3.0709	5.67	.080	.040	39000	44000
2 × 7317 B.UA	2 × 7317 B.UO	2 × 7317 B.UL	85	180	82	4	2	152	3.3464	7.0866	3.2283	5.98	.10	.040	42500	48000
2 × 7318 B.UA	2 × 7318 B.UO	2 × 7318 B.UL	90	190	86	4	2	160	3.5433	7.4803	3.38583	6.30	.10	.040	45000	54000
2 × 7319 B.UA	2 × 7319 B.UO	2 × 7319 B.UL	95	200	90	4	2	169	3.7402	7.8740	3.5433	6.61	.10	.040	48000	58500
2 × 7320 B.UA	2 × 7320 B.UO	2 × 7320 B.UL	100	215	94	4	2	179	3.9370	8.4646	3.7008	7.09	.10	.040	54000	69500
2 × 7321 B.UA	2 × 7321 B.UO	2 × 7321 B.UL	105	225	98	4	2	187	4.1338	8.8582	3.8583	7.40	.10	.040	58500	76500
2 × 7322 B.UA	2 × 7322 B.UO	2 × 7322 B.UL	110	240	100	4	2	197	4.3307	9.4488	3.9370	7.72	.10	.040	64000	86500

13.1.2 Permissible Static Load and Safety Coefficients

The operation of most machines is associated with vibrations and disturbances. The vibrations result in dynamic forces: in turn, the actual maximum stress can be much higher than that calculated by the static load. Therefore, engineers always use a safety coefficient, f_s. In addition, whenever there is a requirement for low noise, the maximum permissible load is reduced to much lower value than C_0. Low loads would result in a significant reduction of permanent deformation of the races and rolling-element surfaces. Plastic deformation distorts the bearing geometry and causes noise during bearing operation.

The permissible static load on a bearing, P_0, is usually less than the basic static load rating, C_0, according to the equation

$$P_0 = \frac{C_0}{f_s} \tag{13-1}$$

The safety coefficient, f_s, depends on the operating conditions and bearing type. Common guidelines for selecting a safety coefficient, f_s are in Table 13-5.

13.1.3 Static Equivalent Load

Most bearings in machinery are subjected to combined radial and thrust loads. It is necessary to establish the combination of radial and thrust loads that would result in the limit stress of a particular bearing. Static equivalent load is introduced to allow bearing selection under combined radial and thrust forces. It is defined as a hypothetical load (radial or axial) that results in a maximum contact stress equivalent to that under combined radial and thrust forces. In radial bearings, the static equivalent load is taken as a radial equivalent load, while in thrust bearings the static equivalent load is taken as a thrust equivalent load.

TABLE 13-5 Safety Coefficient, f_s for Rolling Element Bearings (From FAG 1998)

	For ball bearings	For roller bearings
Standard operating conditions	$f_s = 1$	$f_s = 1.5$
Bearings subjected to vibrations	$f_s = 1.5$	$f_s = 2$
Low-noise applications	$f_s = 2$	$f_s = 3$

13.1.4 Static Radial Equivalent Load

For radial bearings, the higher of the two values calculated by the following two equations is taken as the static radial equivalent load:

$$P_0 = X_0 F_r + Y_0 F_a \tag{13-2}$$

$$P_0 = F_r \tag{13-3}$$

Here,

P_0 = static equivalent load
F_r = static radial load
F_a = static thrust (axial) load
X_0 = static radial load factor
Y_0 = static thrust load factor

Values of X_0 and Y_0 for several bearing types are listed in Table 13-6.

13.1.5 Static Thrust Equivalent Load

For thrust bearings, the static thrust equivalent load is obtained via the following equation:

$$P_0 = X_0 F_r + F_a \tag{13-4}$$

This equation can be applied to thrust bearings for contact angles lower than 90°. The value of X_0 is available in bearing tables in catalogues provided by bearing

TABLE 13-6 Values of Coefficients X_0 and Y_0 (From SKF, 1992, with permission)

Bearing type	Single row bearings X_0	Single row bearings Y_0	Double row bearings X_0	Double row bearings Y_0
Deep groove ball bearings*	0.6	0.5	0.6	0.5
Angular contact ball bearings				
$\alpha = 15°$	0.5	0.46	1	0.92
$\alpha = 20°$	0.5	0.42	1	0.84
$\alpha = 25°$	0.5	0.38	1	0.76
$\alpha = 30°$	0.5	0.33	1	0.66
$\alpha = 35°$	0.5	0.29	1	0.58
$\alpha = 40°$	0.5	0.26	1	0.52
$\alpha = 45°$	0.5	0.22	1	0.44
Self-aligning ball bearings	0.5	0.22ctgα	1	0.44ctgα

*Permissible maximum value of F_a/C_0 depends on bearing design (internal clearance and raceway groove depth).

manufacturers. For a contact angle of 90°, the static thrust equivalent load is $P_0 = F_a$.

13.2 FATIGUE LIFE CALCULATIONS

The rolling elements and raceways are subjected to dynamic stresses. During operation, there are cycles of high contact stresses oscillating at high frequency that cause metal fatigue. The fatigue life—that is, the number of cycles (or the time in hours) to the initiation of fatigue damage in identical bearings under identical load and speed—has a statistical distribution. Therefore, the fatigue life must be determined by considering the statistics of the measured fatigue life of a large number of dimensionally identical bearings.

The method of estimation of fatigue life of rolling-element bearings is based on the work of Lundberg and Palmgren (1947). They used the fundamental theory of the maximum contact stress, and developed a statistical method for estimation of the fatigue life of a rolling-element bearing. This method became a standard method that was adopted by the American Bearing Manufacturers' Association (ABMA). For ball bearings, this method is described in standard ANSI/ABMA-9, 1990; for roller bearings it is described in standard ANSI/ABMA-11, 1990.

13.2.1 Fatigue Life, L_{10}

The *fatigue life*, L_{10}, (often referred to as *rating life*) is the number of revolutions (or the time in hours) that 90% of an identical group of rolling-element bearings will complete or surpass its life before any fatigue damage is evident. The tests are conducted at a given constant speed and load.

Extensive experiments have been conducted to understand the statistical nature of the fatigue life of rolling-element bearings. The experimental results indicated that when fatigue life is plotted against load on a logarithmic scale, a negative-slope straight line could approximate the curve. This means that fatigue life decreases with load according to power-law function. These results allowed the formulation of a simple equation with empirical parameters for predicting the fatigue life of each bearing type.

The following fundamental equation considers only bearing load. Life adjustment factors for operating conditions, such as lubrication, will be discussed later. The fatigue life of a rolling-element bearing is determined via the equation

$$L_{10} = \left(\frac{C}{P}\right)^k \quad \text{[in millions of revolutions]} \tag{13-5}$$

Here, C is the *dynamic load rating* of the bearing (also referred to as the *basic load rating*), P is the equivalent radial load, and k is an empirical exponential parameter ($k = 3$ for ball bearings and 10/3 for roller bearings). The units of C and P can be pounds or newtons (SI units) as long as the units for the two are consistent, since the ratio C/P is dimensionless.

Engineers are interested in the life of a machine in hours. In industry, machines are designed for a minimum life of five years. The number of years depends on the number of hours the machine will operate per day. Equation (13-5) can be written in terms of hours:

$$L_{10} = \frac{10^6}{60N}\left(\frac{C}{P}\right)^k \qquad \text{[in hours]} \qquad (13\text{-}6)$$

13.2.2 Dynamic Load Rating, C

The dynamic load rating, C, is defined as the radial load on a rolling bearing that will result in a fatigue life of 1 million revolutions of the inner ring. Due to the statistical distribution of fatigue life, at least 90% of the bearings will operate under load C without showing any fatigue damage after 1 million revolutions. The value of C is determined empirically, and it depends on bearing type, geometry, precision, and material. The dynamic load rating C is available in bearing catalogues for each bearing type and size. The actual load on a bearing is always much lower than C, because bearings are designed for much longer life than 1 million revolutions.

The dynamic load rating C has load units, and it depends on the design and material of a specific bearing. For a radial ball bearing, it represents the experimental steady radial load under which the radial bearing endured a fatigue life, L_{10}, of 10^6 revolutions.

To determine the dynamic load rating, C, a large number of identical bearings are subjected to fatigue life tests. In these tests, a steady load is applied, and the inner ring is rotating while the outer ring is stationary. The fatigue life of a large number of bearings of the same type is tested under various radial loads.

13.2.3 Combined Radial and Thrust Loads

The *equivalent radial load* P is the radial load, which is equivalent to combined radial and thrust loads. This is the constant radial load that, if applied to a bearing with rotating inner ring and stationary outer ring, would result in the same fatigue life the bearing would attain under combined radial and thrust loads, and different rotation conditions.

In Eq. (13-5), P is the equivalent dynamic radial load, similar to the static radial load. If the load is purely radial, P is equal to the bearing load. However, when the bearing is subjected to combined radial and axial loading, the equivalent load, P, is determined by:

$$P = XVF_r + YF_a \tag{13-7}$$

Here,

P = equivalent radial load
F_r = bearing radial load
F_a = bearing thrust (axial) load
V = rotation factor: 1.0 for inner ring rotation, 1.2 for outer ring rotation and for a self-aligning ball bearing use 1 for inner or outer rotation
X = radial load factor
Y = thrust load factor

The factors X and Y differ for various bearings (Table 13-7).

The equivalent load (P), is defined by the Anti-Friction Bearings Manufacturers Association (AFBMA). It is the constant stationary radial load that, if applied to a bearing with rotating inner and stationary outer ring, would give the same life as what the bearing would attain under the actual conditions of load and rotation.

13.2.4 Life Adjustment Factors

Recent high-speed tests of modern ball and roller bearings, which combine improved materials and proper lubrication, show that fatigue life is, in fact, longer than that predicted previously from Eq. (12-5). It is now commonly accepted that an improvement in fatigue life can be expected from proper lubrication, where the rolling surfaces are completely separated by an elastohydrodynamic lubrication film. In Sec. 13.4 the principles of rolling-element bearing lubrication are discussed. For a rolling bearing with adequate EHD lubrication, adjustments to the fatigue life should be applied. The adjustment factor is dependent on the operating speed, bearing temperature, lubricant viscosity, size and type of bearing, and bearing material.

In many applications, higher reliability is required, and 10% probability of failure is not acceptable. Higher reliability, such as L_5 (5% failure probability) or L_1 (failure probability of 1%), is applied. As defined in the AFBMA Standards, fatigue life is calculated according to the equation

$$L_{na} = a_1 a_2 a_3 \left(\frac{C}{P}\right)^P \times 10^6 \text{ (revolutions)} \tag{13-8}$$

TABLE 13-7 Factors X and Y for Radial Bearings. (From FAG Bearing Catalogue, with permission)

Bearing type			Single row bearings[1]		Double row bearings[2]				e
			$\frac{F_a}{F_r} > e$[1]		$\frac{F_a}{F_r} \le e$		$\frac{F_a}{F_r} > e$		
	3 $\frac{F_a}{C_0}$	3 $\frac{F_a}{iZD_w^2}$	X	Y	X	Y_1	X	Y_2	
Radial									
Contact	0.014	25		2.30				2.30	0.19
Groove	0.028	50		1.99				1.99	0.22
Ball	0.056	100		1.71				1.71	0.26
Bearings									
	0.084	150		1.55				1.55	0.28
	0.11	200	0.56	1.45	1	0	0.56	1.45	0.30
	0.17	300		1.31				1.31	0.34
	0.28	500		1.15				1.15	0.38
	0.42	750		1.04				1.04	0.42
	0.56	1000		1.00				1.00	0.44
20°			0.43	1.00		1.09	0.70	1.63	0.57
25°			0.41	0.87		0.92	0.67	1.44	0.68

(continued)

TABLE 13-7 Continued.

	Single row bearings[1]		Double row bearings[2]				
	$\frac{F_a}{F_r} > e^1$		$\frac{F_a}{F_r} \le e$		$\frac{F_a}{F_r} > e$		e
Bearing type	X	Y	X	Y_1	X	Y_2	
30°	0.39	0.76	1	0.78	0.63	1.24	0.80
35°	0.37	0.66		0.66	0.60	1.07	0.95
40°	0.35	0.57		0.55	0.57	0.93	1.14
Self-Aligning[6] Ball Bearings	0.40	$0.4 \cot\alpha$	1	$0.42 \cot\alpha$	$0.65 \cot\alpha$	$1.5 \tan\alpha$	
Spherical[6] and Tapered[4,5] Roller Bearings	0.40	$0.4 \cot\alpha$	1	$0.45 \cot\alpha$	0.67	$0.67 \cot\alpha$	$1.5 \tan\alpha$

[1] For single row bearings, when $\frac{F_a}{F_r} \le e$ use $X = 1$ and $Y = 0$.

For two single row angular contact ball or roller bearings mounted "face-to-face" or "back-to-back" use the values of X and Y which apply to double row bearings. For two or more single row bearings mounted "in tandem" use the values of X and Y which apply to single row bearings.

[2] Double row bearings are presumed to be symmetrical.

[3] C_0 = static load rating, i = number of rows of rolling elements. Z = number of rolling elements/row, D_w = ball diameter.

[4] Y values for tapered roller bearings are shown in the bearing tables.

[5] $e = \frac{0.6}{Y}$ for single row tapers, and $e = \frac{1}{Y_2}$ for double tow tapers.

TABLE 13-8 Life Adjustment Factor a_1 for Different Failure Probabilities

Failure probability, n					
10	5	4	3	2	1
1	0.62	0.53	0.44	0.33	0.1

where

L_{na} = adjusted fatigue life for a reliability of $(100 - n)\%$, where n is a failure probability (usually, $n = 10$)
a_1 = life adjustment factor for reliability ($a_1 = 1.0$ for $L_n = L_{10}$) (Table 13-8)
a_2 = life adjustment factor for bearing materials made from steel having a higher impurity level
a_3 = life adjustment factor for operating conditions, particularly lubrication (see Sec. 13.4)

Example Problem 13-2 demonstrates the calculation of adjusted rating life; see Sec. 13.4 on bearing lubrication. Experience indicated that the value of the two parameters a_2 and a_3 ultimately depends on proper lubrication conditions. Without proper lubrication, better materials will have no significant benefit in improvement of bearing life. However, better materials have merit only when combined with adequate lubrication. Therefore, the life adjustment factors a_2 and a_3 are often combined, $a_{23} = a_2 a_3$.

13.3 BEARING OPERATING TEMPERATURE

Advanced knowledge of rolling bearing operating temperature is important for bearing design, lubrication, and sealing. Attempts have been made to solve for the bearing temperature at steady-state conditions. The heat balance equation was used, equating the heat generated by friction (proportional to speed and load) to the heat transferred (proportional to temperature rise). It is already recognized that analytical solutions do not yield results equal to the actual operating temperature, because the bearing friction coefficient and particularly the heat transfer coefficients are not known with an adequate degree of precision. For these reasons, we can use only approximations of average bearing operating temperature for design purposes. The temperature of the operating bearing is not uniform. The point of maximum temperature is at the contact of the races with the rolling elements. At the contact with the inner race, the temperature is higher than that of the contact with the outer race. However, for design purposes, an average (approximate) bearing temperature is considered. The average oil temperature is lower than that of the race surface. It is the average of inlet and outlet oil temperatures.

Several attempts to present precise computer solutions are available in the literature. Harris (1984) presented a description of the available numerical methods for solving the temperature distribution in a rolling bearing. Numerical calculation of the bearing temperature is quite complex, because it depends on a large number of heat transfer parameters.

For simplified calculations, it is possible to estimate an average bearing temperature by considering the bearing friction power losses and heat transfer. Friction power losses are dissipated in the bearing as heat and are proportional to the product of friction torque and speed. The heat is continually transferred away by convection, radiation, and conduction. This heat balance can be solved for the temperature rise, bearing temperature minus ambient (atmospheric) temperature $(T_b - T_a)$.

More careful consideration of the friction losses and heat transfer characteristics through the shaft and the housing can only help to estimate the bearing temperature rise. This data can be compared to bearings from previous experience where the oil temperature has been measured. It is relatively easy to measure the oil temperature at the exit from the bearing. (The oil temperature at the contact with the races during operation is higher and requires elaborate experiments to be determined).

It is possible to control the bearing operating temperature. In an elevated-temperature environment, the oil circulation assists in transferring the heat away from the bearing. The final bearing temperature rise, above the ambient temperature, is affected by many factors. It is proportional to the bearing speed and load, but it is difficult to predict accurately by calculation. However, for predicting the operating temperature, engineers rely mostly on experience with similar machinery. A comparative method to estimate the bearing temperature is described in Sec. 13.3.1.

A lot of data has been derived by means of field measurements. The bearing temperature for common moderate-speed applications has been measured, and it is in the range of 40°–90°C. The relatively low bearing temperature of 40°C is for light-duty machines such as the bench drill spindle, the circular saw shaft, and the milling machine. A bearing temperature of 50°C is typical of a regular lathe spindle and wood-cutting machine spindle. The higher bearing temperature of 60°C is found in heavier-duty machinery, such as an axle box of train locomotives. A higher temperature range is typical of machines subjected to load combined with severe vibrations. The bearing temperature of motors, of vibratory screens, or impact mills is 70°C; and in vibratory road roller bearings, the higher temperature of 80°C has been measured.

Much higher bearing temperatures are found in machines where there is an external heat source that is conducted into the bearing. Examples are rolls for paper drying, turbocompressors, injection molding machines for plastics, and bearings of large electric motors, where considerable heat is conducted from the motor armature. In such cases, air cooling or water cooling is used in the bearing housing for reducing the bearing temperature. Also, fast oil circulation can help to remove the heat from the bearing.

13.3.1 Estimation of Bearing Temperature

The following derivation is useful where there is already previous experience with a similar machine. In such cases, the temperature rise can be predicted whenever there are modifications in the machine operation, such as an increase in speed or load.

The friction power loss, q, of a bearing is calculated from the frictional torque T_f $[N\text{-}m]$ and the shaft angular speed ω [rad/s]:

$$q = T_f\omega \qquad [W] \tag{13-9}$$

The angular speed can be written as a function of the speed $N[RPM]$:

$$\omega = \frac{2\pi N}{60} \tag{13-10}$$

Under steady-state conditions there is heat balance, and the same amount of heat that is generated by friction, q, must be transferred to the environment. The heat transferred from the bearing is calculated from the difference between the bearing temperature, T_b, and the ambient temperature, T_a, from the size of the heat-transmitting areas A_B $[m^2]$ and the total heat transfer coefficient U_t $[W/m^2\text{-}C]$:

$$q = U_t A_B (T_b - T_a) \qquad [W] \tag{13-11}$$

In the case of no oil circulation, all the heat is transferred through the bearing surfaces (in contact with the shaft and housing). Equating the two equations gives

$$T_b - T_a = \frac{\pi N T_f}{30 U_t A_B} \tag{13-12}$$

According to Eq. (13-12), the temperature rise, $T_b - T_a$, is proportional to the speed N and the friction torque, T_f, while all the other terms can form one constant k, which is a function of the heat transfer coefficients and the geometry and material of the bearing and housing:

$$\Delta T = T_b - T_a = k\,N\,T_f \tag{13-13}$$

The friction torque T_f is

$$T_f = f\,R\,F \tag{13-14}$$

where f is the friction coefficient, R is the rolling contact radius, and F is the bearing load. The temperature rise, in Eq. (13–13), can be expressed as

$$\Delta T = (T_b - T_a) = K f N F \tag{13-15}$$

where $K = kR$ is a constant. The result is that the temperature rise, $\Delta T = T_b - T_a$, is proportional to the friction coefficient, speed, and bearing load.

Prediction of the bearing temperature can be obtained by determining the steady-state temperature in a test run and calculating the coefficient K. If the friction coefficient is assumed to be constant, then Eq. (13-15) will allow estimation with sufficient accuracy of the steady-state temperature rise of this bearing for other operating conditions, under various speeds and loads. A better temperature estimation can be obtained if additional data is used concerning the function of the friction coefficient, f, versus speed and load.

In the case of oil circulation lubrication, the oil also carries away heat. This can be considered in the calculation if the lubricant flow rate and inlet and outlet temperatures of the bearing oil are measured.

The bearing temperature can then be calculated by equating

$$q = q_1 + q_2 \qquad [W] \tag{13-16}$$

where q_1 is the heat transferred by conduction according to Eq. (13–11) and q_2 is the heat transferred by convection via the oil circulation.

13.3.2 Operating Temperature of the Oil

For selecting an appropriate lubricant, it is important to estimate the operating temperature of the oil in the bearing. It is possible to estimate the operating oil temperature by measuring the temperature of the bearing housing. If the machine is only in design stages, it is possible to estimate the housing temperature by comparing it to the housing temperature of similar machines. During the operation of standard bearings that are properly designed, the operating temperature of the oil is usually in the range of 3°–11°C above that of the bearing housing. It is relatively simple to measure the housing temperature in an operating machine and to estimate the oil temperature. Knowledge of the oil temperature is important for optimal selection of lubricant, oil replacement, and fatigue life calculations.

Tapered and spherical roller bearings result in higher operating temperatures than do ball bearings or cylindrical roller bearings under similar operating conditions. The reason is the higher friction coefficient in tapered and spherical roller bearings.

13.3.3 Temperature Difference Between Rings

During operation, the shaft temperature is generally higher than the housing temperature. The heat is removed from the outer ring through the housing much faster than from the inner ring through the shaft. There is no good heat transfer through the small contact area between the rolling elements and rings (theoretical point or line contact). Therefore, heat from the inner ring is conducted through the shaft, and heat from the outer ring is conducted through the housing. In general, heat conduction through the shaft is not as effective as through the housing. The outer ring and housing have good heat transfer, because they are in direct contact with the larger body of the machine. In comparison, the inner ring and shaft have more resistance to heat transfer, because the cross-sectional area of the shaft is small in comparison to that of the housing as well as to its smaller surface area, which has lower heat convection relative to the whole machine.

If there is no external source of heat outside the bearing, the operating temperature of the shaft is always higher than that of the housing. For medium-speed operation of standard bearings, if the housing is not cooled, the temperatures of the inner ring are in the range of 5°–10°C higher than that of the outer ring. If the housing is cooled by air flow, the temperature of the inner ring can increase to 15°–20°C higher than that of the outer ring. An example of air cooling of the housing is in motor vehicles, where there is air cooling whenever the car is in motion. It is possible to reduce the temperature difference by means of adequate oil circulation, which assists in the convection heat transfer between the rings.

A higher temperature difference can develop in very high-speed bearings. The temperature difference depends on several factors, such as speed, load, and type of bearing and shape of the housing. This temperature difference can result in additional thermal stresses in the bearing.

13.4 ROLLING BEARING LUBRICATION

13.4.1 Objectives of Lubrication

Various types of grease, oils, and, in certain cases, solid lubricants are used for the lubrication of rolling bearings. Most bearings are lubricated with grease because it provides effective lubrication and does not require expensive supply systems (grease can operate with very simple sealing). In most applications, rolling-element bearings operate successfully with a very thin layer of oil or grease. However, for high-speed applications, such as turbines, oil lubrication is important for removing the heat from the bearing or for formation of an EHD fluid film.

The first objective of liquid lubrication is the formation of a thin elastohydrodynamic lubrication film at the rolling contacts between the rolling elements and the raceways. Under appropriate conditions of load, viscosity, and bearing speed, this film can completely separate the surfaces of rolling elements and raceways, resulting in considerable improvement in bearing life.

The second objective of lubrication is to minimize friction and wear in applications where there is no full EHD film. Experience has indicated that if proper lubrication is provided, rolling bearings operate successfully for a long time under mixed lubrication conditions. In practice, ideal conditions of complete separation are not always maintained. If the height of the surface asperities is larger than the elastohydrodynamic lubrication film, contact of surface asperities will take place, and there is a mixed friction (hydrodynamic combined with direct contact friction).

In addition to pure rolling, there is also a certain amount of sliding contact between the rolling elements and the raceways as well as between the rolling elements and the cage. At the sliding surfaces of a rolling bearing, such as the roller and lip in a roller bearing and at the guiding surface of the cage, a very thin lubricant film can be formed, resulting in mixed friction under favorable conditions. Any sliding contact in the bearing requires lubrication to reduce friction and wear.

The third objective of lubrication (applies to fluid lubricants) is to cool the bearing and reduce the maximum temperature at the contact of the rolling elements and the raceways. For effective cooling, sufficient lubricant circulation should be provided to remove the heat from the bearing. The most effective cooling is achieved by circulating the oil through an external heat exchanger. But even without elaborate circulation, a simple oil sump system can enhance the heat transfer from the bearing by convection. Solid lubricants or greases are not effective in cooling; therefore, they are restricted to relatively low-speed applications.

Additional objectives of lubrication are damping of vibrations, corrosion protection, and removal of dust and wear debris from the raceways via liquid lubricant. A full EHD fluid film plays an important role as a damper. A full EHD fluid film acts as noncontact support of the shaft that effectively isolates vibrations. The fluid film can be helpful in reducing noise and vibrations in a machine.

Lubricants for rolling bearings include liquid lubricants (mineral and synthetic oils), greases, and solid lubricants. The most common liquid lubricants are petroleum-based mineral oils with a long list of additives to improve the lubrication performance. Also, synthetic lubricants are widely used, such as ester, polyglycol, and silicone fluoride. Greases are commonly applied in relatively low-speed applications, where continuous flow for cooling is not essential for successful operation. The most important advantages of grease are that it seals the bearing from dust and provides effective protection from corrosion. To minimize maintenance, sealed bearings are widely used, where the bearing is filled with grease and sealed for the life of the bearing. The grease serves as a matrix that retains the oil. The oil is slowly released from the grease during operation.

In addition to grease, oil-saturated solids, such as oil-saturated polymer, are used successfully for similar applications of sealed bearings. The saturated solid fills the entire bearing cavity and effectively seals the bearing from contaminants. The advantage of oil-saturated polymers over grease is that grease can be filled only into half the bearing internal space in order to avoid churning. In comparison, oil-saturated solid lubricants are available that can fill the complete cavity without causing churning. The oil is released from oil-saturated solid lubricants in a similar way to grease.

Rolling bearings successfully operate in a wide range of environmental conditions. In certain high-temperature applications, liquid oils or greases cannot be applied (they oxidize and deteriorate from the heat) and only solid lubricants can be used. Examples of solid lubricants are PTFE, graphite and molybdenum disulfide (MoS_2). Solid lubricants are effective in reducing friction and wear, but obviously they cannot assist in heat removal as liquid lubricants.

In summary: Lubrication of rolling bearings has several important functions: to form a fluid film, to reduce sliding friction and wear, to transfer heat away from the bearing, to damp vibrations, and to protect the finished surfaces from corrosion. Greases and oils are mostly used. Grease packed sealing is commonly used to protect against the penetration of abrasive particles into the bearing. Reduction of friction and wear by lubrication is obtained in several ways. First, a thin fluid film at high pressure can separate the rolling contacts by forming elastohydrodynamic lubrication. Second, lubrication reduces friction of the sliding contacts that do not involve rolling, such as between the cage and the rolling elements or between the rolling elements and the guiding surfaces. Also, the contacts between the rolling elements and the raceways are not pure rolling, and there is always a certain amount of sliding. Solid lubricants are also effective in reducing sliding friction.

13.4.2 Elastohydrodynamic Lubrication

In Chapter 12, the elastohydrodynamic (EHD) lubrication equations were discussed. EHD theory is concerned with the formation of a thin fluid film at high pressure at the contact area of a rolling element and a raceway under rolling conditions. Both the roller and the raceway surfaces are deformed under the load. In a similar way to fluid film in plain bearings, the oil that is adhering to the surfaces is drawn into a thin clearance formed between the rolling surfaces. An important effect is that the viscosity of the oil rises under high pressure; in turn, a load-carrying fluid film is formed at high rolling speed. The clearance thickness, h_0, is nearly constant along the fluid film, and it is reducing only near the outlet side (Fig. 12-20).

Under high loads, the EHD pressure distribution is similar to the pressure distribution according to the Hertz equations, because the influence of the elastic deformations dominates the pressure distribution. But at high speeds, the hydrodynamic effect prevails.

In Chapter 12, the calculation of the film thickness was quite complex. For many standard applications, engineers often resort to a simplified method based on charts. The simplified approach also considers the effect of the elastohydrodynamic lubrication in improving the fatigue life of the bearing. Even if the EHD fluid film does not separate completely the rolling surfaces (mixed EHD lubrication), the lubrication improves the performance, and longer fatigue life will be obtained. In this chapter, the use of charts is demonstrated for finding the effect of lubrication in improving the fatigue life of a bearing.

13.4.3 Selection of Liquid Lubricants

The best performance of a rolling bearing is under operating conditions where the elastohydrodynamic minimum film thickness, h_{min}, is thicker than the surface asperities, R_s. The required viscosity of the lubricant, μ, for this purpose can be solved for from the EHD equations (see Chapter 12). However, for many standard applications, designers determine the viscosity by a simpler practical method. It is based on an empirical chart, where the required viscosity is determined according to the bearing speed and diameter.

For rolling bearings, the decision concerning the oil viscosity is a compromise between the requirement of low viscous friction (low viscosity) and the requirement for adequate EHD film thickness (high viscosity). The friction of a rolling bearing consists of two components. The first component is the rolling friction, which results from deformation at the contacts between a rolling element and a raceway. The second friction component is viscous resistance of the lubricant to the motion of the rolling elements. The first component of rolling friction is a function of the elastic modulus, geometry, and bearing load. The second component of viscous friction increases with lubricant viscosity, quantity of oil in the bearing, and bearing speed. The viscous component increases with speed, so it becomes a dominant factor in high–speed machinery.

It is possible to minimize the viscous resistance by applying a very small quantity of oil, just sufficient to form a thin layer over the contact surface. In addition, using low-viscosity oil can reduce the viscous resistance. However, minimum lubricant viscosity must be maintained to ensure elastohydrodynamic lubrication with adequate fluid film thickness.

For lubricant selection, a knowledge of the operating bearing temperature is required. One must keep in mind that the lubricant viscosity decreases with temperature. In applications where the bearing temperature is expected to rise significantly, lubricant of higher initial viscosity should be selected. It is possible to reduce the bearing operating temperature via oil circulation for removing the heat and cooling the bearing. The final bearing temperature rise, above the ambient temperature, is affected by many factors, such as speed and load. A simplified method for estimating the bearing temperature was discussed earlier. For predicting the operating temperature, this method relies mostly on experience with similar machinery for determining the heat transfer coefficients.

For bearings that do not dissipate heat from outside the bearing and that operate at moderate speeds and under average loads, it is possible to estimate the oil temperature by measuring the housing temperature. During operation, the temperature of the oil is usually in the range of 3°–11°C above that of the bearing housing. This simple temperature estimation is widely used for lubricant selection.

In order to simplify the selection of oil viscosity, charts based on bearing speed and bearing average diameter are used. Figure 13-1 is used for determining the minimum oil viscosity for lubrication of rolling-element bearings as a function of bearing size and speed.

The ordinate on the left side shows the kinematic viscosity in metric units, mm^2/s (cSt). The ordinate on the right side shows the viscosity in Saybolt universal seconds (SUS). The abscissa is the pitch diameter, d_m, in mm, which is the average of internal bore, d, and outside bearing diameter, D.

$$d_m = \frac{d + D}{2} \tag{13-17}$$

The diagonal straight lines in Fig. 13-1 are for the various bearing speed N in *RPM* (revolutions per minute). The dotted lines show examples of determining the required lubricant viscosity.

Example Problem 13-1

Calculation of Minimum Viscosity

A rolling bearing has a bore diameter $d = 45$ mm and an outside diameter $D = 85$ mm. The bearing rotates at 2000 RPM. Find the required minimum viscosity of the lubricant.

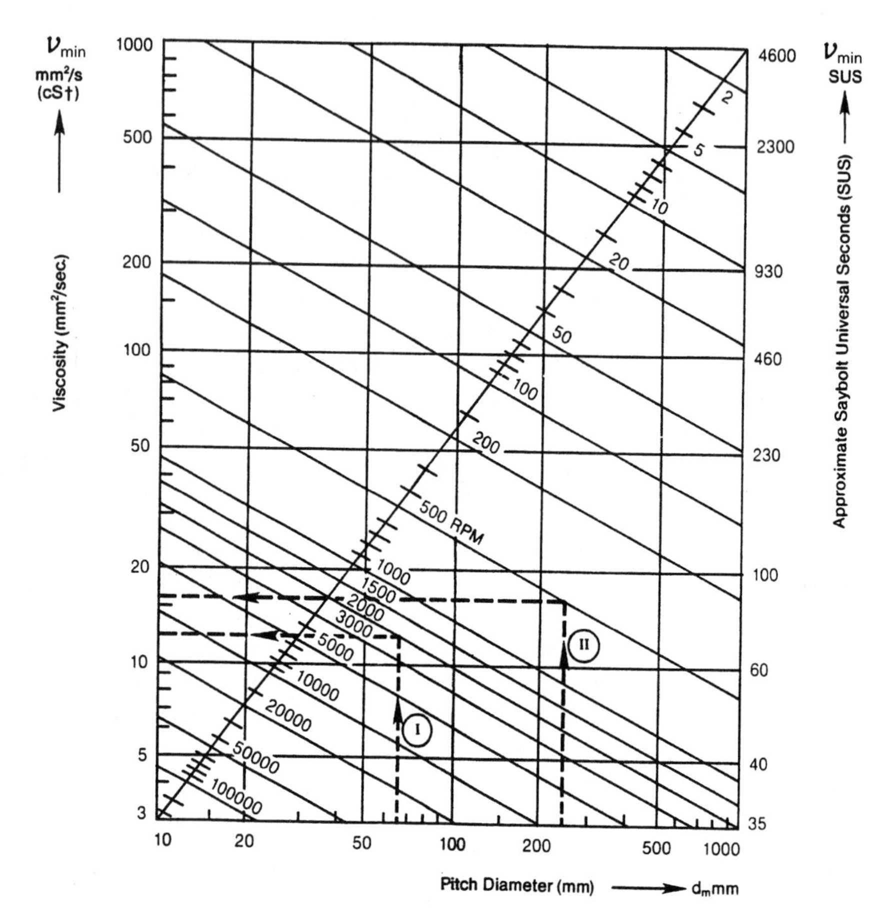

FIG. 13-1 Requirement for minimum lubricant viscosity in rolling bearings (from SKF, 1992, with permission).

Solution

The pitch diameter according to Eq. (13-17) is

$$d_m = \frac{45 + 85}{2} = 65 \text{ mm}$$

Line I in Fig. 13-1 shows the intersection of $d_m = 65$ with the diagonal straight line of 2000 RPM. The horizontal dotted line indicates a minimum viscosity required of 13 cSt (mm^2/s).

Based on the required viscosity, the oil grade should be selected. The oil viscosity decreases with temperature, and the relation between the oil grade and

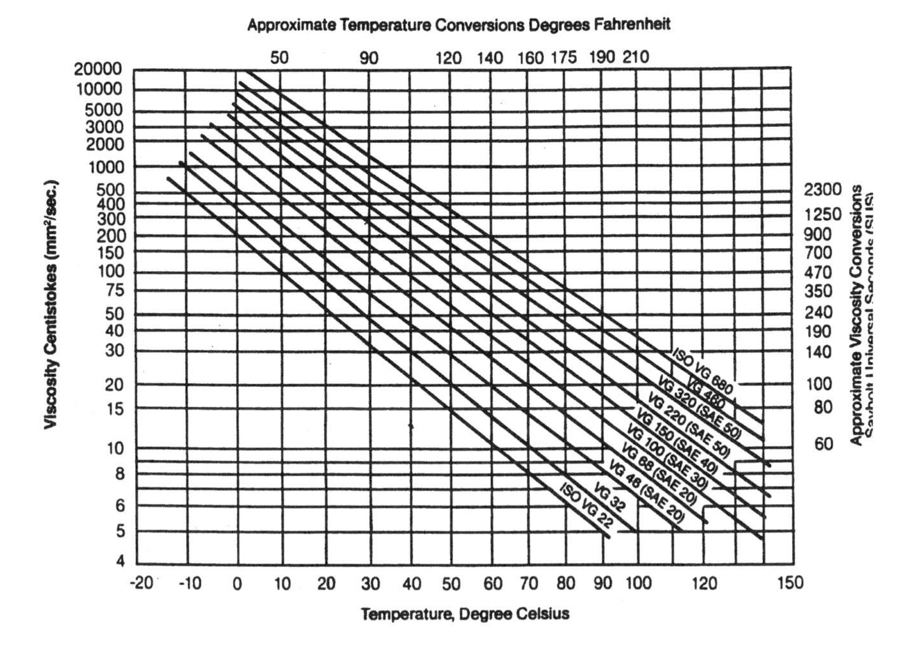

FIG. 13-2 Viscosity–temperature charts (from SKF, 1992, with permission).

its viscosity depends on the oil temperature. In Fig. 13-2, viscosity–temperature charts for several rolling bearing oil grades are presented. Estimation of the oil temperature inside the operating bearing is required before one can select the oil grade according to Fig. 13-2.

It is preferable to estimate the temperature with an error on the high side. This would result in higher viscosity, which can ensure a full EHD fluid film at the rolling contact, although the friction resistance can be slightly higher. If a lubricant with higher-than-required viscosity is selected, an improvement in bearing life can be expected. However, since a higher viscosity raises the bearing operating temperature, there is a limit to the improvement that can be obtained in this manner.

The improvement in the bearing fatigue life due to higher lubricant viscosity (above the minimum required viscosity) is shown in Fig. 13-3. The life adjustment factor a_3 (sec. 13.2.4) is a function of the viscosity ratio, κ, defined as

$$\kappa = \frac{\nu}{\nu_{min}} \tag{13-18}$$

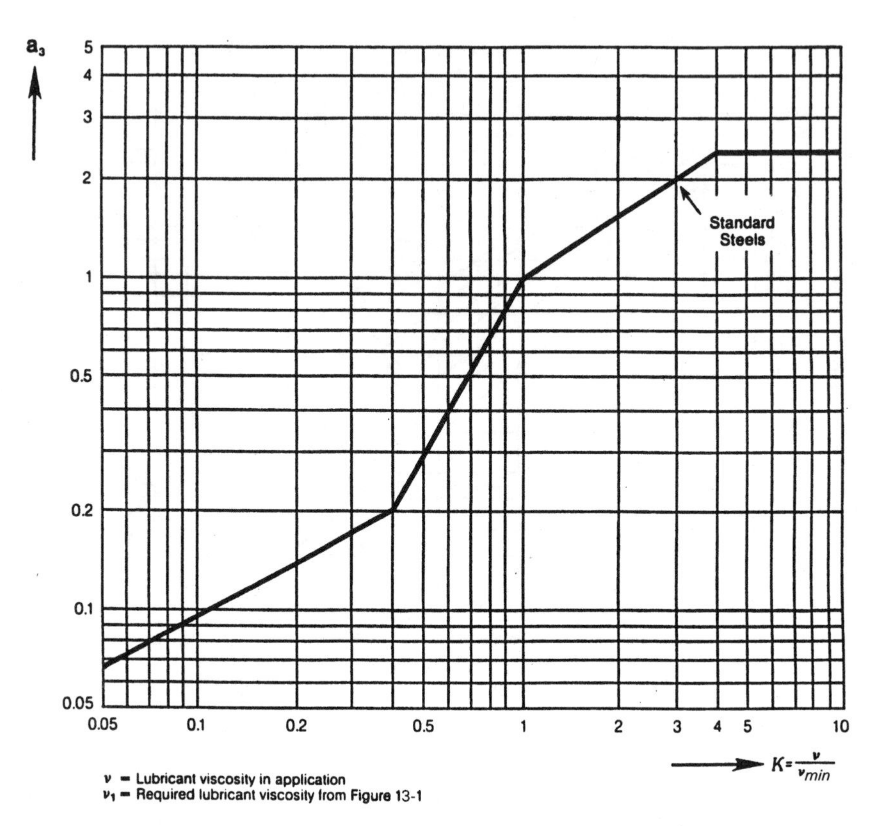

FIG. 13-3 Fatigue life adjustment factor for lubrication (from SKF, 1992, with permission).

Here, ν is the actual viscosity of the lubricant (at the operating temperature) and ν_{min} is the minimum required lubricant viscosity from Fig. 13-1.

According to Fig. 13-3, the life adjustment factor a_3 is an increasing function of the viscosity ratio κ. This means that there is an improvement in fatigue-life due to improvement in EHD lubrication at higher viscosity. However, there is a limit to this improvement. For ν higher than 4, Fig. 13-3 indicates that there is no additional improvement in fatigue life from using higher-viscosity oil. This is because higher viscosity has the adverse effect of higher viscous friction, which in turn results in higher bearing operating temperature.

In conclusion, there is a limit on the benefits obtained from increasing oil viscosity. Moreover, oils with excessively high viscosity introduce a higher operating temperature and in turn a higher thermal expansion of the inner ring.

This results in extra rolling contact stresses, which counteract any other benefits obtained from using high-viscosity oil.

The fatigue life adjustment factor a_3 in Fig. 13-3 is often used as $a_{23} = a_2 a_3$. This is because experience indicated that there is no significant improvement in fatigue life due to better bearing steel if there is inadequate lubrication.

Example Problem 13-2

Calculation of Adjusted Fatigue Life

Find the life adjustment factor and adjusted fatigue life of a deep-groove ball bearing. The bearing operates in a gearbox supporting a 25-mm shaft. The bearing is designed for 90% reliability. The shaft speed is 3600 RPM, and the gearbox is designed to transmit a maximum power of 10 kW. The lubricant is SAE 20 oil, and the maximum expected surrounding (ambient) temperature is 30°C. One helical gear is mounted on the shaft at equal distance from both bearings. The rolling bearing data is from the manufacturer's catalog:

Designation bearing:	No. 61805
Bore diameter:	$d = 25$ mm
Outside diameter:	$D = 37$ mm
Dynamic load rating:	$C = 4360$ N
Static load rating:	$C_0 = 2600$ N

The gear data is

Helix angle $\psi = 30°$
Pressure angle (in a cross section normal to the gear) $\phi = 20°$
Diameter of pitch circle = 5 in.

Solution

Calculation of Radial and Thrust Forces Acting on Bearing: Given:

Power transmitted by gear:	$\dot{E} = 10\ \text{kW} = 10^4$ N-m/s
Rotational speed of shaft:	$N = 3600$ RPM
Helix angle:	$\psi = 30°$
Pressure angle:	$\phi = 20°$
Pitch circle diameter of gear:	$d_p = 5$ in. $= 0.127$ m

The angular velocity of the shaft, ω, is

$$\omega = \frac{2\pi N}{60} = \frac{2\pi 3600}{60} = 377\ \text{rad/s}$$

Torque produced by the gear is

$$T = \frac{F_t d_p}{2}$$

Substituting this into the power equation, $\dot{E} = T\omega$, yields

$$\dot{E} = \frac{F_t d_p}{2} \omega$$

Solving for the tangential force, F_t, results in

$$F_t = \frac{2\dot{E}}{d_p \omega} = \frac{2 \times 10{,}000 \text{ N-m/s}}{0.127 \text{ m} \times 377 \text{ rad/s}} = 418 \text{ N}$$

Once the tangential component of the force is solved, the radial force F_r, and the thrust load (axial force), F_a, can be calculated, as follows:

$$F_a = F_t \tan \psi$$

$$F_a = 417 \text{ N} \times \tan 30^\circ = 241 \text{ N}$$

$$F_r = F_t \tan \phi$$

$$F_r = 418 \text{ N} \times \tan 20^\circ$$

$$F_r = 152 \text{ N}$$

The force components F_t and F_r are both in the direction normal to the shaft centerline. The bearing force reacting to these two gear force components, W_r, is the radial force component of the bearing. The gear is in the center, and the bearing radial force is divided between the two bearings. The resultant, W_r, for each bearing is calculated by the equation

$$2W_r = \sqrt{F_t^2 + F_r^2} = \sqrt{418^2 + 152^2} = 445 \text{ N}$$

The resultant force of the gear is supported by the two bearings. It is a radial bearing reaction force, because it is acting in the direction normal to the shaft centerline. Since the helical gear is mounted on the shaft at equal distance from each bearing, each bearing will support half of the radial load:

$$W_r = \frac{445 \text{ N}}{2} = 222.25 \text{ N}$$

However, the thrust load will act on one bearing only. The direction of the thrust load depends on the gear configuration and the direction of rotation. Therefore, each bearing should be designed to support the entire thrust load:

$$F_a = 241 \text{ N}$$

Calculation of Adjusted Fatigue Life of Rolling Bearing. In this example, combined radial and thrust loads are acting on a bearing. In all cases of combined load, it is necessary to determine the equivalent radial load, P, from Eq. (13-7). The radial and thrust load factors X and Y in the following table are available in manufacturers manuals. The values of X and Y differ for different bearings. Table 13-7 includes the factors X and Y of a deep-groove ball bearing.

For $F_a/F_r > e$, the values are

F_a/C_o	e	X	Y
0.025	0.22	0.56	2
0.04	0.24	0.56	1.8
0.07	0.27	0.56	1.6
0.13	0.31	0.56	1.4
0.25	0.37	0.56	1.2
0.5	0.44	0.56	1

The ratio of the axial load, F_a, and the basic static load rating C_0 must be calculated:

$$\frac{F_a}{C_0} = \frac{241.17}{2600} = 0.093$$

Then, by interpolation, the values for e, X, and Y can be determined:

$$e = 0.29 \qquad X = 0.56 \qquad Y = 1.5$$

Also, the ratio of F_a to F_r is

$$\frac{F_a}{F_r} = \frac{241}{222} = 1.09 \quad 1.09 > e$$

Therefore

$$P = XF_r + YF_a$$

$$P = (0.56)(222.2) + (1.5)(241.17)$$

$$P = 486 \text{ N}$$

Using the bearing life equation, the bearing life is determined from the equation

$$L_{10} = \left(\frac{C}{P}\right)^{p} \times 10^{6}$$

$$L_{10} = \left(\frac{4360}{486}\right)^{3} \times 10^{6}$$

$$L_{10} = 722 \times 10^{6} \text{ (revolutions)} = \frac{722 \times 10^{6} \text{ rev}}{3600 \text{ rev/min} \times 60 \text{ min/hr}} = 3343 \text{ hr}$$

This is the fatigue life without adjustment for lubrication. Following is the selection of the minimum required viscosity and the adjustment for the bearing fatigue life when operating with lubricant SAE 20. There is improvement in the fatigue life when the lubricant is of higher viscosity than the minimum required viscosity.

Selection of Oil. The selection of an appropriate oil is an important part of bearing design. The most important property is the oil's viscosity, which is inversely related to temperature. The minimum required viscosity is determined according to the size and rotational speed of the bearing. The bearing size is determined by taking the average of the inner (bearing bore) and outer diameters of the bearing.

The pitch diameter of the bearing is

$$d_m = \frac{d + D}{2}$$

$$d_m = \frac{25 + 37}{2}$$

$$d_m = 31 \text{ mm}$$

From Fig 13-1, at a speed of 3600 RPM and a pitch diameter of 31 mm, the minimum required viscosity is 14 mm^2/s.

As discussed earlier, the temperature of the oil of an operating bearing is usually 3°–11°C above the housing temperature. In this problem, the maximum expected surrounding (ambient) temperature is 30°C, and it is assumed that 5°C should be added for the maximum operating oil temperature. From Fig. 13-2 (viscosity–temperature charts), the viscosity of SAE 20 (VG 46) oil at 35°C is approximately 52 mm^2/s. The viscosity ratio, κ, is

$$\kappa = \frac{\nu}{\nu_{min}} = \frac{52}{14} = 3.7$$

For 90% reliability, the life adjustment factor a_1 is given a value of 1. The life adjustment factor a_2 for material is also given a value of 1 (standard material).

Based on the viscosity ratio, the operating conditions factor, a_3, can be obtained using Fig. 13-3: $a_3 = 2.2$.

The adjusted rating life is:

$$L_{10a} = a_1 a_2 a_3 (L_{10}), \qquad \text{where } a_1 a_2 = 1$$

$$L_{10a} = 2.2 \times 3343 \text{ hr} = 7354 \text{ hr}$$

Discussion. The fatigue life of an industrial gearbox must be at least five years. If we assume operation of eight hours per day, the minimum fatigue life must be for 14,400 hours. In this case, the tested bearing has an adjusted life of only 7354 hr. The conclusion is that the bearing tested in this example is not a suitable selection for use in an industrial gearbox. The adjusted life is much shorter than required. A more appropriate bearing, therefore, should be selected, of higher dynamic load rating C, and the foregoing procedure should be repeated to verify that the selection is adequate.

13.5 BEARING PRECISION

Manufacturing tolerances specify that the actual dimensions of a bearing be within specified limits. For precision applications, such as precise machine tools and precision instruments, ultrahigh-precision rolling bearings are available with very narrow tolerances. In high-speed machinery, it is important to reduce vibrations, and high-precision bearings are often used. In precision applications and high-speed machinery, it is essential that the center of a rotating shaft remain at the same place, with minimal radial displacement during rotation. During rotation under steady load, any variable eccentricity between the shaft center and the center of rotation is referred to as *radial run-out*. At the same time, any axial displacement of the shaft during its rotation is referred to as *axial run-out*.

If the shaft is precise and centered, the radial and axial run-outs depend on manufacturing tolerances of the rolling-element bearing. Radial run-out depends on errors such as eccentricity between the inside and outside diameters of the rings, deviation from roundness of the races, and deviations in the actual diameters of the rolling elements. For running precision, it is necessary to distinguish between run-out of the inner ring and that of the outer ring, which are not necessarily equal.

There are many applications where different levels of precision of run-out and dimensions are required, such as in machine tools of various precision levels. Of course, higher precision involves higher cost, and engineers must not specify higher precision than really required. The Annular Bearing Engineering Committee (ABEC) introduced five precision grades (ABEC 1, 3, 5, 7, and 9). Each precision grade has an increasing grade of smaller tolerance range of all bearing dimensions. ABEC 1 is the standard bearing and has the lowest cost; it has about

80% of the bearing market share. Bearings of ABEC 3 and 5 precision have very low market share. Bearings of ABEC 7 and 9 precision are for ultraprecision applications. The American Bearing Manufactures Association (ABMA) has adopted this standard for bearing tolerances, ANSI/ABMA-20, 1996, which is accepted as the international standard.

The most important characteristics of precise bearings are the inner ring and outer ring run-outs. However, tolerances of all dimensions are more precise, such as inside and outside diameters, and width. All bearing manufacturers produce standard bearings that conform to these standard dimensions and tolerances.

13.5.1 Inner Ring Run-Out

The inner ring run-out of a rolling bearing is measured by holding the outer ring stationary by means of a fixture and turning the inner ring under steady load. The radial inner ring run-out is measured via an indicator normal to the inner ring surface (inner ring bore). The axial inner ring run-out is measured by an indicator in contact with the face of the inner ring, in a direction normal to the face of the inner ring (parallel to the bearing centerline). In both cases, the run-out is the difference between the maximum and minimum indicator readings.

In machine tools where the shaft is turning, such as in a lathe, the inner ring radial run-out is measured by turning a very precise shaft between two centers, under steady load. A precise dial indicator is fixed normal to the shaft surface. The shaft rotates slowly, and the radial run-out is the difference between the maximum and minimum indicator readings. The axial run-out is measured by an indicator normal to the face of the shaft.

13.5.2 Outer Ring Run-Out

In a similar way, an indicator measures the outer ring run-out. But in that case, the inner ring is constrained by a fixture and the run-out is measured when the outer ring is rotating. In the two cases, a small load, or gravity, is applied to cancel the internal clearance. In this way, the clearance does not affect the run-out measurement, because the run-out depends only on the precision of the bearing parts. For example, an eccentricity between the bore of the inner ring and its raceway will result in a constant inner-ring radial run-out but will not contribute to any outer ring radial run-out.

In precision applications, such as machine tools, there is a requirement for bearings with very low levels of run-out. In addition, there is a requirement for low run-out for high-speed rotors, where radial run-out would result in imbalance and excessive vibrations.

For machine tools, it is important to understand the effect of various types of run-out on the precision of the workpiece. Also, it is necessary to distinguish between a bearing where the inner ring is rotating, such as an electric motor, and a bearing where the outer ring is rotating, such as in car wheels. In the case of an electric motor, the inner ring radial run-out will cause radial run-out of the rotor centerline. If, instead, there is only outer ring radial run-out, there would be no influence on the rotor, because the rotor continues to operate with a new, steady center of rotation (although not concentric with the bearing outer ring).

The opposite applies to a car wheel, where only the outer ring radial run-out is causing run-out of the wheel, while the radial inner ring run-out does not affect the running of the car wheel. In machine tools the precision is measured by the axial and radial run-out of a spindle. However, it is necessary to distinguish between machinery where the workpiece is rotating and where the cutting tool is rotating.

It is necessary to distinguish between steady and time-variable run-out. Steady radial run-out is where the spindle axis has a constant run-out (resulting from eccentricity between the inner ring bore and inner ring raceway). If the workpiece is turning, a steady radial run-out does not result in machining errors, because the workpiece forms its own, new center of rotation, and it will not result in a deviation from roundness. But the workpiece must not be reset during machining, because the center would be relocated. An example is a lathe where the bearing inner ring is rotating together with the spindle and workpiece while the outer ring is stationary. In this configuration, if the spindle has a constant radial displacement, the cutting tool will form a round shape with a new center of rotation without any deviation from roundness. However, any deviation from roundness of the two races, in the form of waviness or elliptical shape, will result in a similar deviation from roundness in the workpiece.

In contrast to a lathe, in a milling machine the cutting tool is rotating while the workpiece is stationary. In this case, operating the rotating tool with a constant eccentricity can result in manufacturing errors. This is evident when trying to machine a planar surface: A wavy surface would be produced, with the wave level depending on the cutting tool run-out (running eccentricity). Such manufacturing error can be minimized by a very slow feed rate of the workpiece.

The load affects the concentric running of a shaft, due to elastic deformations in the contact between rolling elements and raceways. During operation, radial and axial run-outs in rolling bearings are caused not only by deviations from the ideal dimensions, but also by elastic deformation in the bearing—whenever there are rotating forces on the bearing. Most of the elastic deformation is at the contact between the rolling elements and the raceways. Roller-element bearings, such as cylindrical roller bearings, have less elastic deformation than ball bearings.

By designing an adjustable arrangement of two opposing angular ball bearings or tapered bearings, it is possible to eliminate clearance and introduce preload in the bearings (negative clearance). In this way, the bearings are stiffened and the run-out due to elastic deformation or clearance is significantly reduced.

13.6 INTERNAL CLEARANCE OF ROLLING BEARINGS

Rolling bearings are manufactured with internal clearance. The internal clearance is between the rolling elements and the inner and outer raceways. In the absence of clearance, the rolling elements will fit precisely into the space between the raceways of the outer and inner rings. However, in practice this space is always a little larger than the diameter of the rolling elements, resulting in a small clearance. The radial and axial clearances are measured by the displacements in the radial and axial directions that one ring can have relative to the other ring.

The purpose of the clearance is to prevent excessive rolling contact stresses due to uneven thermal expansion of the inner and outer rings. In addition, the clearance prevents excessive rolling contact stresses due to tight-fit assembly of the rings into their seats. During operation, the temperature of the inner ring is usually higher than that of the outer ring, resulting in uneven thermal expansion. In addition, for most bearings the inner and outer rings are tightly fitted into their seats. Tight fit involves elastic deformation of the rings that can cause *negative clearance* (bearing preload), which results in undesired extra contact stresses between the raceways and the rolling elements. The extra stresses can be prevented if the bearing is manufactured with sufficient internal clearance.

In the case of negative clearance, the uneven thermal expansion and elastic deformation due to tight-fit mounting are combined with the bearing load to cause excessive rolling contact stresses. It can produce a chain reaction where the high stresses result in higher friction and additional thermal expansion, which in turn can eventually lead to bearing seizure. Therefore, in most cases bearing manufacturers provide internal clearance to prevent bearing seizure due to excessive contact stresses.

Catalogues of rolling-element bearings specify several standard classes of bearing clearance. This specification of internal bearing clearance is based on the ABMA standard. The specification is from tight, ABMA Class 2, to extra-loose clearance, ABMA Class 5.

Standard bearings have five classes of precision. Table 13-9 shows the classes for increasing levels of internal clearance: C2, Normal, C3, C4, and C5. C2 has the lowest clearance, while C5 has the highest clearance. It should be noted that the normal class is between C2 and C3. The clearance in each class increases with bearing size. In addition, there is a tolerance range for the clearance in each class and bearing size.

After the selection of bearing type and size has been completed, the selection of appropriate internal clearance is the most important design decision. Appropriate internal clearance is important for successful bearing operation. Internal clearance can be measured by displacement of the inner ring relative to the outer ring. This displacement can be divided into radial and axial components. The clearance is selected according to the bearing type and diameter as well as the level of precision required in operation.

TABLE 13-9 Classes of Radial Clearance in Deep Groove Ball Bearing

Normal bore diameter d (mm)		Radial clearance μm									
		C2		Normal		C3		C4		C5	
Over	incl.	min.	max.	min.	max.	min.	max.	min.	max.	min.	max.
d < 10		0	7	2	13	8	23	14	29	20	37
10	18	0	9	3	18	11	25	18	33	25	45
18	24	0	10	5	20	13	28	20	36	28	48
24	30	1	11	5	20	13	28	23	41	30	53
30	40	1	11	6	20	15	33	28	46	40	64
40	50	1	11	6	23	18	36	30	51	45	73
50	65	1	15	8	28	23	43	38	61	55	90
65	80	1	15	10	30	25	51	46	71	65	105
80	100	1	18	12	36	30	58	53	84	75	120
100	120	2	20	15	41	36	66	61	97	90	140
120	140	2	23	18	48	41	81	71	114	105	160
140	160	2	23	18	53	46	91	81	130	120	180
160	180	2	25	20	61	53	102	91	147	135	200
180	200	2	30	25	71	63	117	107	163	150	230
200	225	2	35	25	85	75	140	125	195	175	265
225	250	2	40	30	95	85	160	145	225	205	300
250	280	2	45	35	105	90	170	155	245	225	340
280	315	2	55	40	115	100	190	175	270	245	370
315	335	3	60	45	125	110	210	195	300	275	410
355	400	3	70	55	145	130	240	225	340	315	460
400	450	3	80	60	170	150	270	250	380	350	510
450	500	3	90	70	190	170	300	280	420	390	570
500	550	10	100	80	210	190	330	310	470	440	630
560	630	10	110	90	230	210	360	340	520	490	690
630	710	20	130	110	260	240	400	380	570	540	760
710	800	20	140	120	290	270	450	430	630	600	840

The operating clearance is usually lower than the bearing clearance before its assembly. The clearance is reduced by tight fit and thermal expansion during operation. Much care should be taken during design to secure an appropriate operating clearance. Example Problem 13-3 presents a calculation for predicting the operating clearance.

Low operating clearance is very important in precision machines and high-speed machines. Operating clearance affects the concentric running of a shaft, it causes radial and axial run-out, and reduction of precision. Therefore, elimination of clearance is particularly important for precision machinery, such as machine tools and measuring machines, and high-speed machines, such as turbines, where radial clearance can cause imbalance and vibrations.

Moreover, a small amount of preload (small negative bearing clearance) is desirable in order to stiffen the support of the shaft. This is done in order to reduce the level of vibrations, particularly in high-speed machines or high-precision machines, such as machine tools. In such cases, much care is required in the design, to secure that the preload is not excessive during operation. In certain applications, it is desirable to run the bearing as close as possible to clearance-free operation. Experiments have indicated that clearance-free operation reduces the bearing noise. Clearance-free operation requires accurate design that includes calculations of the housing and shaft tolerances.

In the following sections, it is shown that a widely used method for eliminating the clearance and providing an accurate preload is an adjustable bearing arrangement. In this arrangement, two angular contact ball bearings or tapered roller bearings are mounted in opposite directions on one shaft. This arrangement is designed to allow, during mounting, for one ring to be forced to slide in its seat, in the axial direction, to adjust the bearing clearance or even provide preload inside the bearing.

An adjustable arrangement is not appropriate in all applications. In order to reduce the operating clearance, the designer can select high-precision bearings manufactured with relatively low clearances. Another method often used in precision machines is to incorporate a tapered bore or tapered housing. It is possible to design a tapered-bore bearing, which reduces radial clearance by either expanding elastically the inner ring with a tapered shaft or press-fitting the outer ring with a tapered housing bore.

In many cases, a floating bearing is required to prevent thermal stresses due to thermal elongation of the shaft (thermal expansion in the axial direction). If, at the same time, precise concentric shaft running is required, the designer should not specify a sliding fit (radial clearance) between the bore of the inner ring and the shaft or outer ring and housing. Instead, it is possible to use a cylindrical roller bearing, which can be installed with interference fit to both the shaft and housing and still allow small axial displacements.

13.7 VIBRATIONS AND NOISE IN ROLLING BEARINGS

Theoretically, vibrations can be generated in rolling bearings even if the bearing is manufactured with great precision and its geometry does not significantly deviate from the ideal dimensions. Under external load, the rotation of a rolling element induces periodic cycles of variable elastic deformation, which result in audible noise. In practice, however, vibrations resulting from inaccuracy in dimensions generate most of the noise. Deviation from roundness of the races or rolling elements results in vibrations and noise. Dimensional inaccuracy results in imbalances in the bearing that induce noise.

The noise level of rolling bearings increases with the inaccuracy of the dimensions of the rolling elements and races. It is well known that there are always small deviations from the ideal geometry. Even the most precise manufacturing processes can only reduce dimensional errors, not eliminate them completely. The level of noise increases with the magnitude of the deviation from the nominal dimensions, particularly deviation from roundness of rolling elements or races or uneven diameters of the rolling elements, which can vary within a certain tolerance, depending on the specified tolerances. In addition, the waviness of the surface finish of the raceways and rolling-element surfaces can result in audible noise. In order to reduce the noise, it is important to minimize dimensional errors as well as the magnitude of the surface waviness, which is specified as deviation from roundness. Additional parameters that play a role in the noise level and vibrations include: elastic deformations at the contacts, design and material of the cage, and clearances between the rolling elements and the cage.

A steady radial run-out (resulting from eccentricity of the inner and outer diameters of the rotating ring) generates a vibrational frequency equal to that of the shaft frequency. This frequency is usually too low to cause audible noise, but deviations from roundness, in the form of a few oscillations over the circumference of the race, can generate audible noise. One must keep in mind that a machine is a dynamic system, and vibrations, which originate in the rolling bearing, can excite vibrations and audible noise in other parts of the machine. In the same way, vibrations originating in other parts of the machine can excite audible noise in the rolling bearing. In particular, a high level of vibrations is expected whenever the exciting frequency is close to one of the natural frequencies in the machine.

High-frequency vibrations generate audible noise as well as ultrasonic sound waves. Ultrasonic sound waves can be measured and used for predictive maintenance. The ultrasonic measurements are analyzed and the results used to predict the condition of rolling bearings while the machine is running. A significant change in the level or frequency of ultrasonic sound is an indication of possible damage on the surface of the races or rolling elements. Experiments have indicated that when the bearing runs as close as possible to clearance free, the operating conditions are optimal for minimum noise and vibrations. This can be achieved by proper tight fit mounting to eliminate bearing clearance; however, thermal expansion must be considered, to avoid excessive preload.

Noise generated by roller bearings is a concern in many applications, particularly in office or hospital machines. Moreover, there is an increasing interest in improving the manufacturing environment and reducing the noise level in all machinery. Bearing manufacturers supply low-noise bearings. These bearings are of high precision and pass quality control tests. The tests include a plot of the actual roundness of rings in polar coordinates, where deviations from theoretical roundness are magnified. In addition, samples are tested in operation, the level of noise is measured and recorded, and the correlation between certain dimensional deviations and noise level can be established. In this way, better low-noise bearings are developed.

In additional to manufacturing precision, low-noise bearings must be mounted properly. Manufacturing errors in the form of deviations from roundness can be transferred by elastic deformation of the thin rings from the housing and shaft seats to the races. Therefore, for low-noise applications, high precision is required for the housing and shaft seats. Also, minimal misalignment is required.

There are many parameters that influence vibrations in bearings, including bearing clearance. To minimize noise, it is desirable to have a clearance-free operating bearing. On the one hand, large clearance generates a characteristic hollow noise. On the other hand, a bearing that is excessively press-fitted (negative clearance) induces another characteristic high-pitch noise. For optimum results, an effort must be made to operate the bearing as closely as possible to a clearance-free condition. But in practice this is difficult to achieve, because the operating clearance is affected by several factors and there are variable conditions during operation. The design engineer must take into account the reduction of clearance after installation due to the interference fits at the bearing seats, and thermal expansion must be considered (usually the operating temperature of the inner ring is higher than that of the outer ring). Example Problem 13-3 covers a radial clearance calculation during operation.

Another method to achieve clearance-free running is to include adjustable arrangements of two angular ball bearings or tapered rolling bearings in opposition. By axial movement of the inner or outer ring, either through the use of a nut, which is tightened once the machine has reached thermal equilibrium, or with the use of spring washers, it is possible to adjust the clearance. The operating temperatures of the inner and outer rings are not always known, and an adjustable arrangement that is adjusted during machine assembly does not ensure clearance-free operation. Therefore, the best results can be achieved by designs that allow for clearance adjustment while the bearing is running and after thermal steady state has been reached.

In addition to precise geometry, noise reduction is often achieved by using special greases with better damping characteristics for noise and vibrations. Grease manufacturers recommend certain greases that have been tested and proved to reduce noise more effectively than other types.

13.8 SHAFT AND HOUSING FITS

For successful bearing operation, care must be taken to specify tolerances of the appropriate fits between the bearing bore and the shaft seat as well as the outer diameter and the housing seat. It is important that the rotating shaft always be tightly fitted into the bearing bore, because a loose fit will damage the bearing bore as well as the shaft seat. The type of fit, such as tight fit or loose fit, depends on the application. Since the tolerances for the bore and outside bearing diameter are standardized, the required fits are obtained by selecting the proper tolerances for the shaft diameter and the housing bore. The fits and tolerances are selected based on the ISO standard, which contains a very large selection of shaft and housing tolerances for any desirable fit, from a tight fit to a loose fit.

For the housing and shaft, the ISO standard provides for various degrees of tightness: very tight, moderately tight, sliding, and loose fit. According to the ISO standard, a letter and a number are used for specifying tolerances; capital letters are used for housing bores, and small letters for shaft diameters. The letter specifies the location of the *tolerance zone* (range between minimum and maximum dimensions) relative to the *nominal dimension*. In fact, the letter determines the degree of clearance or tightness of the housing or shaft in relation to the outside diameter or bore of the bearing. At the same time, the number specifies the size of the *tolerance zone*.

A demonstration of the tolerance zone of the housing and shaft is shown in Fig. 13-4. The tolerance zone of the housing and shaft is relative to the bearing tolerance, for various tolerance grades, from the loosest, G7, to the tightest, P7. Table 13-10 lists the tolerances of the shaft for various nominal diameters; Table 13-11 lists the tolerances of the housing for various nominal diameters. Table 13-12 provides recommendations for shaft tolerances for various applications; Table 13-13 gives similar recommendations for housing tolerances.

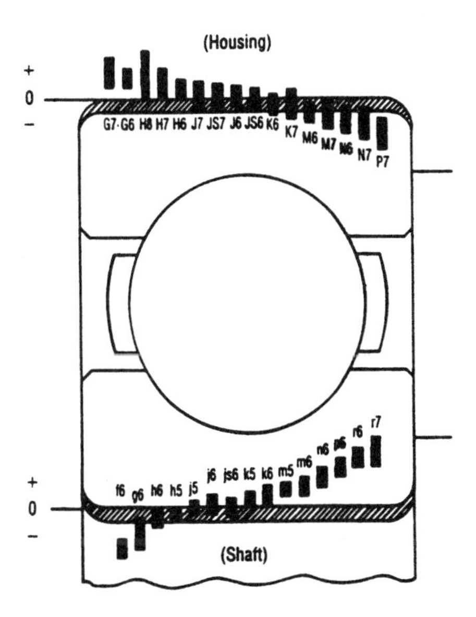

FIG. 13-4 Illustration of tolerance grades (from SKF, 1992, with permission).

TABLE 13-10 Tolerances of the Shaft (from FAG (1999) with permission of FAG and Handel AG)

		Dimensions in mm																								
Nominal shaft dimension	over	3		6		10		18		30		50		65		80		100		120		140		160		180
	to	6		10		18		30		50		65		80		100		120		140		160		180		200
		Tolerance in microns (0.001 mm) (normal tolerance)																								
Bearing bore diameter deviation		0		0		0		0		0		0		0		0		0		0		0		0		0
	Δ_{dmp}	−8		−8		−8		−10		−12		−15		−15		−20		−20		−25		−25		−25		−30
Diagram of fit Shaft		Shaft tolerance, interference or clearance in microns (0.001 mm)																								
f6 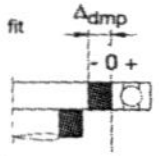			2		5		8		10		13		15		15		16		16		18		18		18	20
		−10	8	−13	11	−16	15	−20	17	−25	22	−30	26	−30	26	−36	30	−36	30	−43	34	−43	34	−43	34	−50 40
		−18	18	−22	22	−27	27	−33	33	−41	41	−49	49	−49	49	−58	58	−58	58	−68	68	−68	68	−68	68	−79 79
g5 			4		3		2		3		3		5		5		8		8		11		11		11	15
		−4	0	−5	2	−6	3	−7	3	−9	5	−10	4	−10	4	−12	4	−12	4	−14	3	−14	3	−14	3	−15 2
		−9	9	−11	11	−14	14	−16	16	−20	20	−23	23	−23	23	−27	27	−27	27	−32	32	−32	32	−32	32	−35 35
g6 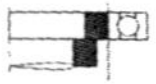			4		3		2		3		3		5		5		8		8		11		11		11	15
		−4	1	−5	3	−6	4	−7	5	−9	6	−10	6	−10	6	−12	6	−12	6	−14	6	−14	6	−14	6	−15 5
		−12	12	−14	14	−17	17	−20	20	−25	25	−29	29	−29	29	−34	34	−34	34	−39	39	−39	39	−39	39	−44 44
h5 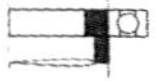			8		8		8		10		12		15		15		20		20		25		25		25	30
		0	4	0	3	0	3	0	4	0	4	0	6	0	6	0	8	0	8	0	11	0	11	0	11	0 13
		−5	5	−6	6	−8	8	−9	9	−11	11	−13	13	−13	13	−15	15	−15	15	−18	18	−18	18	−18	18	−20 20
h6 			8		8		8		10		12		15		15		20		20		25		25		25	30
		0	3	0	2	0	2	0	2	0	3	0	4	0	4	0	6	0	6	0	8	0	8	0	8	0 10
		−8	8	−9	9	−11	11	−13	13	−16	16	−19	19	−19	19	−22	22	−22	22	−25	25	−25	25	−25	25	−29 29
j5 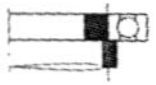			11		12		13		15		18		21		21		26		26		32		32		32	37
		+3	7	+4	7	+5	8	+5	9	+6	10	+6	12	+6	12	+6	14	+6	14	+7	18	+7	18	+7	18	+7 20
		−2	2	−2	2	−3	3	−4	4	−5	5	−7	7	−7	7	−9	9	−9	9	−11	11	−11	11	−11	11	−13 13
j6 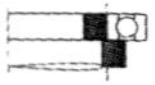			14		15		16		19		23		27		27		33		33		39		39		39	46
		+6	8	+7	9	+8	10	+9	11	+11	14	+12	16	+12	16	+13	19	+13	19	+14	22	+14	22	+14	22	+16 26
		−2	2	−2	2	−3	3	−4	4	−5	5	−7	7	−7	7	−9	9	−9	9	−11	11	−11	11	−11	11	−13 13
js5 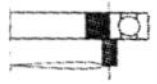			11		11		12		15		18		22		22		28		28		34		34		34	40
		+2.5	6	+3	6	+4	6	+4.5	9	+5.5	10	+6.5	13	+6.5	13	+7.5	16	+7.5	16	+9	20	+9	20	+9	20	+10 23
		−2.5	3	−3	3	−4	4	−4.5	5	−5.5	6	−6.5	7	−6.5	7	−7.5	8	−7.5	8	−9	9	−9	9	−9	9	−10 10
js6 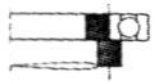			12		13		14		17		20		25		25		31		31		38		38		38	45
		+4	7	+4.5	7	+5.5	8	+6.5	9	+8	11	+9.5	13	+9.5	13	+11	17	+11	17	+12.5	21	+12.5	21	+12.5	21	+14.5 25
		−4	4	−4.5	5	−5.5	6	−6.5	7	−8	8	−9.5	10	−9.5	10	−11	11	−11	11	−12.5	13	−12.5	13	−12.5	13	−14.5 15
k5 			14		15		17		21		25		30		30		38		38		46		46		46	54
		+6	9	+7	10	+9	12	+11	15	+13	17	+15	21	+15	21	+18	26	+18	26	+21	32	+21	32	+21	32	+24 37
		+1	1	+1	1	+1	1	+2	2	+2	2	+2	2	+2	2	+3	3	+3	3	+3	3	+3	3	+3	3	+4 4

Fit																										
k6		17		18		20		25		30		36		36		45		45		53		53		53		63
	+9	11	+10	12	+12	14	+15	17	+18	21	+21	25	+21	25	+25	31	+25	31	+28	36	+28	36	+28	36	+33	43
	+1	1	+1	1	+1	1	+2	2	+2	2	+2	2	+2	2	+3	3	+3	3	+3	3	+3	3	+3	3	+4	4
m5		17		20		23		27		32		39		39		48		48		58		58		58		67
	+9	13	+12	15	+15	18	+17	21	+20	24	+24	30	+24	30	+28	36	+28	36	+33	44	+33	44	+33	44	+37	50
	+4	4	+6	6	+7	7	+8	8	+9	9	+11	11	+11	11	+13	13	+13	13	+15	15	+15	15	+15	15	+17	17
m6		20		23		26		31		37		45		45		55		55		65		65		65		76
	+12	15	+15	17	+18	20	+21	23	+25	27	+30	34	+30	34	+35	42	+35	42	+40	48	+40	48	+40	48	+46	56
	+4	4	+6	6	+7	7	+8	8	+9	9	+11	11	+11	11	+13	13	+13	13	+15	15	+15	15	+15	15	+17	17
n5 (Δ_{dmp} - 0 +)		21		24		28		34		40		48		48		58		58		70		70		70		81
	+13	17	+16	19	+20	23	+24	28	+28	32	+33	39	+33	39	+38	46	+38	46	+45	56	+45	56	+45	56	+51	64
	+8	8	+10	10	+12	12	+15	15	+17	17	+20	20	+20	20	+23	23	+23	23	+27	27	+27	27	+27	27	+31	31
n6		24		27		31		38		45		54		54		65	65	65		77		77		77		90
	+16	19	+19	21	+23	25	+28	30	+33	36	+39	43	+39	43	+45	51	+45	51	+52	60	+52	60	+52	60	+60	70
	+8	8	+10	10	+12	12	+15	15	+17	17	+20	20	+20	20	+23	23	+23	23	+27	27	+27	27	+27	27	+31	31
p6		28		32		37		45		54		66		66		79		79		93		93		93		109
	+20	23	+24	26	+29	31	+35	37	+42	45	+51	55	+51	55	+59	65	+59	65	+68	76	+68	76	+68	76	+79	89
	+12	12	+15	15	+18	18	+22	22	+26	26	+32	32	+32	32	+37	37	+37	37	+43	43	+43	43	+43	43	+50	50
p7		32		38		44		53		63		77		77		92		92		108		108		108		126
	+24	25	+30	30	+36	35	+43	43	+51	51	+62	62	+62	62	+72	73	+72	73	+83	87	+83	87	+83	87	+96	101
	+12	12	+15	15	+18	18	+22	22	+26	26	+32	32	+32	32	+37	37	+37	37	+43	43	+43	43	+43	43	+50	50
r6		31		36		42		51		62		75		77		93		96		113		115		118		136
	+23	25	+28	30	+34	35	+41	44	+50	53	+60	64	+62	66	+73	79	+76	82	+88	97	+90	99	+93	102	+106	116
	+15	15	+19	19	+23	23	+28	28	+34	34	+41	41	+43	43	+51	51	+54	54	+63	63	+65	65	+68	68	+77	77
r7		35		42		49		59		71		86		88		106		109		128		130		133		153
	+27	28	+34	34	+41	40	+49	49	+59	59	+71	71	+73	73	+86	87	+89	90	+103	107	+105	109	+108	112	+123	128
	+15	15	+19	19	+23	23	+28	28	+34	34	+41	41	+43	43	+51	51	+54	54	+63	63	+65	65	+68	68	+77	77

Example: Shaft dia 40 j5			
Maximum material	+6	18	Interference or clearance when upper shaft deviations coincide with lower bore deviations
		10	Probable interference or clearance
Minimum material	−5	5	Interference or clearance when lower shaft deviations coincide with upper bore deviations

Numbers in boldface print identify interference.
Standard-type numbers in right column identify clearance.

TABLE 13-11 Tolerances of the Housing (from FAG (1999) with permission of FAG OEM and Handel AG)

		Dimensions in mm								Dimensions in mm																		
Nominal housing bore	over	6		10		18		30		50		80		120		150		180		250		315		400		500		
	to	10		18		30		50		80		120		150		180		250		315		400		500		630		
		Tolerance in microns (0.001mm) (normal tolerance)												Tolerance in microns (0.001 mm) (normal tolerance)														
Bearing outside diameter deviation		0		0		0		0		0		0		0		0		0		0		0		0		0		
	Δ_{dmp}	−8		−8		−9		−11		−13		−15		−18		−25		−30		−35		−40		−45		−50		
Diagram of fit Housing	Δ_{Dmp} +0−	Housing tolerance, interference or clearance in microns (0.001 mm)												Housing tolerance, interference or clearance in microns (0.001 mm)														
E8	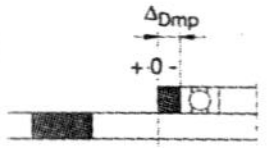		25		32		40		50		60		72		85		85		100		110		125		135		145	
		+47	35	+59	44	+73	54	+89	67	+106	79	+126	85	+148	112	+148	114	+172	134	+191	149	+214	168	+232	182	+255	199	+
		+25	55	+32	67	+40	82	+50	100	+60	119	+72	141	+85	166	+85	173	+100	202	+110	226	+125	254	+135	277	+145	305	+
F7			13		16		20		25		30		36		43		43		50		56		62		68		76	
		+28	21	+34	25	+41	30	+50	37	+60	44	+71	53	+83	62	+83	64	+96	75	+108	85	+119	94	+131	104	+146	116	+
		+13	36	+16	42	+20	50	+25	61	+30	73	+36	86	+43	101	+43	108	+50	126	+56	143	+62	159	+68	176	+76	196	+
G6	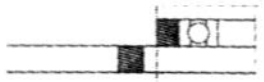		5		6		7		9		10		12		14		14		15		17		18		20		22	
		+14	11	+17	12	+20	14	+25	18	+29	21	+34	24	+39	28	+39	31	+44	35	+49	39	+54	43	+60	48	+66	54	+
		+5	22	+6	25	+7	29	+9	36	+10	42	+12	49	+14	57	+14	64	+15	74	+17	84	+18	94	+20	105	+22	116	+
G7			5		6		7		9		10		12		14		14		15		17		18		20		22	
		+20	13	+24	15	+28	17	+34	21	+40	24	+47	29	+54	33	+54	36	+61	40	+69	46	+75	50	+83	56	+92	62	+
		+5	28	+6	32	+7	37	+9	45	+10	53	+12	62	+14	72	+14	79	+15	91	+17	104	+18	115	+20	128	+22	142	+
H6			0		0		0		0		0		0		0		0		0		0		0		0		0	
		+9	6	+11	6	+13	7	+16	9	+19	11	+22	12	+25	14	+25	17	+29	20	+32	22	+36	25	+40	28	+44	32	+
		0	17	0	19	0	22	0	27	0	32	0	37	0	43	0	50	0	59	0	67	0	76	0	85	0	94	
H7			0		0		0		0		0		0		0		0		0		0		0		0		0	
		+15	8	+18	9	+21	10	+25	12	+30	14	+35	17	+40	19	+40	22	+46	25	+52	29	+57	32	+63	36	+70	40	
		0	23	0	26	0	30	0	36	0	43	0	50	0	58	0	65	0	76	0	87	0	97	0	108	0	120	0
H8			0		0		0		0		0		0		0		0		0		0		0		0		0	
		+22	10	+27	12	+33	14	+39	17	+46	20	+54	23	+63	27	+63	29	+72	34	+81	39	+89	43	+97	47	+110	54	+
		0	30	0	35	0	42	0	50	0	59	0	69	0	81	0	88	0	102	0	116	0	129	0	142	0	160	0
J6			4		5		5		6		6		6		7		7		7		7		7		7			
		+5	2	+6	1	+8	2	+10	3	+13	5	+16	6	+18	7	+18	10	+22	13	+25	15	+29	18	+33	21			
		−4	13	−5	14	−5	17	−6	21	−6	26	−6	31	−7	36	−7	43	−7	52	−7	60	−7	69	−7	78			
J7			7		8		9		11		12		13		14		14		16		16		18		20			
		+8	1	+10	1	+12	1	+14	1	+18	2	+22	4	+26	5	+26	8	+30	9	+36	13	+39	14	+43	16			
		−7	16	−8	18	−9	21	−11	25	−12	31	−13	37	−14	44	−14	51	−16	60	−16	71	−18	79	−20	88			
JS6			4.5		5.5		6.5		8		9.5		11		12.5		12.5		14.5		16		18		20		22	
		+4.5	2	+5.5	1	+6.5	0	+8	1	+9.5	0	+11	1	+12.5	1	+12.5	3	+14.5	5	+16	7	+18	6	+20	8	+22	10	

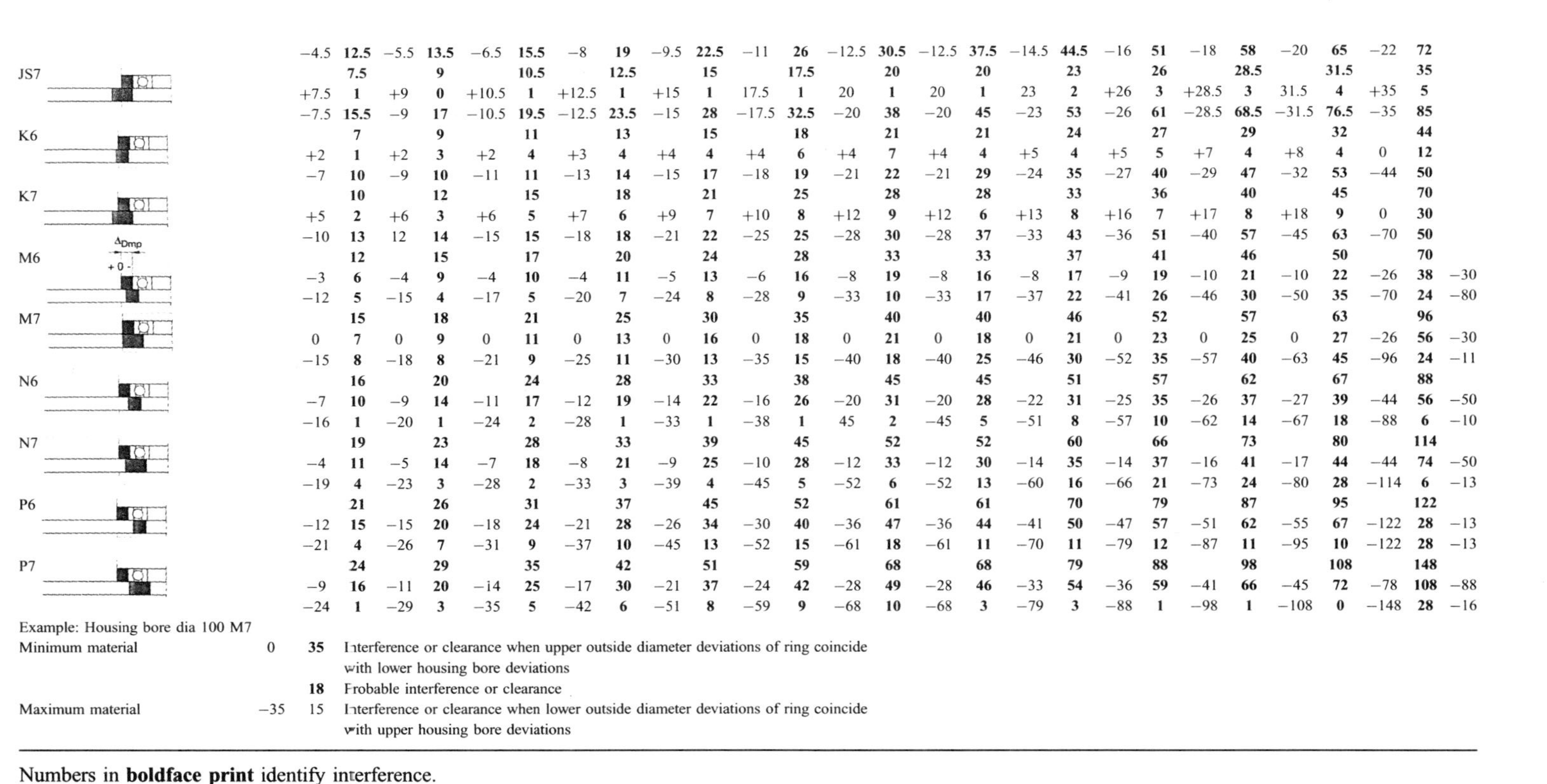

	−4.5	**12.5**	−5.5	**13.5**	−6.5	**15.5**	−8	**19**	−9.5	**22.5**	−11	**26**	−12.5	**30.5**	−12.5	**37.5**	−14.5	**44.5**	−16	**51**	−18	**58**	−20	**65**	−22	**72**	
JS7		**7.5**		**9**		**10.5**		**12.5**		**15**		**17.5**		**20**		**20**		**23**		**26**		**28.5**		**31.5**		**35**	
	+7.5	**1**	+9	**0**	+10.5	**1**	+12.5	**1**	+15	**1**	17.5	**1**	20	**1**	20	**1**	23	**2**	+26	**3**	+28.5	**3**	31.5	**4**	+35	**5**	
	−7.5	**15.5**	−9	**17**	−10.5	**19.5**	−12.5	**23.5**	−15	**28**	−17.5	**32.5**	−20	**38**	−20	**45**	−23	**53**	−26	**61**	−28.5	**68.5**	−31.5	**76.5**	−35	**85**	
K6		**7**		**9**		**11**		**13**		**15**		**18**		**21**		**21**		**24**		**27**		**29**		**32**		**44**	
	+2	**1**	+2	**3**	+2	**4**	+3	**4**	+4	**4**	+4	**6**	+4	**7**	+4	**4**	+5	**4**	+5	**5**	+7	**4**	+8	**4**	0	**12**	
	−7	**10**	−9	**10**	−11	**11**	−13	**14**	−15	**17**	−18	**19**	−21	**22**	−21	**29**	−24	**35**	−27	**40**	−29	**47**	−32	**53**	−44	**50**	
K7		**10**		**12**		**15**		**18**		**21**		**25**		**28**		**28**		**33**		**36**		**40**		**45**		**70**	
	+5	**2**	+6	**3**	+6	**5**	+7	**6**	+9	**7**	+10	**8**	+12	**9**	+12	**6**	+13	**8**	+16	**7**	+17	**8**	+18	**9**	0	**30**	
	−10	**13**	12	**14**	−15	**15**	−18	**18**	−21	**22**	−25	**25**	−28	**30**	−28	**37**	−33	**43**	−36	**51**	−40	**57**	−45	**63**	−70	**50**	
M6		**12**		**15**		**17**		**20**		**24**		**28**		**33**		**33**		**37**		**41**		**46**		**50**		**70**	
	−3	**6**	−4	**9**	−4	**10**	−4	**11**	−5	**13**	−6	**16**	−8	**19**	−8	**16**	−8	**17**	−9	**19**	−10	**21**	−10	**22**	−26	**38**	−30
	−12	**5**	−15	**4**	−17	**5**	−20	**7**	−24	**8**	−28	**9**	−33	**10**	−33	**17**	−37	**22**	−41	**26**	−46	**30**	−50	**35**	−70	**24**	−80
M7		**15**		**18**		**21**		**25**		**30**		**35**		**40**		**40**		**46**		**52**		**57**		**63**		**96**	
	0	**7**	0	**9**	0	**11**	0	**13**	0	**16**	0	**18**	0	**21**	0	**18**	0	**21**	0	**23**	0	**25**	0	**27**	−26	**56**	−30
	−15	**8**	−18	**8**	−21	**9**	−25	**11**	−30	**13**	−35	**15**	−40	**18**	−40	**25**	−46	**30**	−52	**35**	−57	**40**	−63	**45**	−96	**24**	−11
N6		**16**		**20**		**24**		**28**		**33**		**38**		**45**		**45**		**51**		**57**		**62**		**67**		**88**	
	−7	**10**	−9	**14**	−11	**17**	−12	**19**	−14	**22**	−16	**26**	−20	**31**	−20	**28**	−22	**31**	−25	**35**	−26	**37**	−27	**39**	−44	**56**	−50
	−16	**1**	−20	**1**	−24	**2**	−28	**1**	−33	**1**	−38	**1**	45	**2**	−45	**5**	−51	**8**	−57	**10**	−62	**14**	−67	**18**	−88	**6**	−10
N7		**19**		**23**		**28**		**33**		**39**		**45**		**52**		**52**		**60**		**66**		**73**		**80**		**114**	
	−4	**11**	−5	**14**	−7	**18**	−8	**21**	−9	**25**	−10	**28**	−12	**33**	−12	**30**	−14	**35**	−14	**37**	−16	**41**	−17	**44**	−44	**74**	−50
	−19	**4**	−23	**3**	−28	**2**	−33	**3**	−39	**4**	−45	**5**	−52	**6**	−52	**13**	−60	**16**	−66	**21**	−73	**24**	−80	**28**	−114	**6**	−13
P6		**21**		**26**		**31**		**37**		**45**		**52**		**61**		**61**		**70**		**79**		**87**		**95**		**122**	
	−12	**15**	−15	**20**	−18	**24**	−21	**28**	−26	**34**	−30	**40**	−36	**47**	−36	**44**	−41	**50**	−47	**57**	−51	**62**	−55	**67**	−122	**28**	−13
	−21	**4**	−26	**7**	−31	**9**	−37	**10**	−45	**13**	−52	**15**	−61	**18**	−61	**11**	−70	**11**	−79	**12**	−87	**11**	−95	**10**	−122	**28**	−13
P7		**24**		**29**		**35**		**42**		**51**		**59**		**68**		**68**		**79**		**88**		**98**		**108**		**148**	
	−9	**16**	−11	**20**	−14	**25**	−17	**30**	−21	**37**	−24	**42**	−28	**49**	−28	**46**	−33	**54**	−36	**59**	−41	**66**	−45	**72**	−78	**108**	−88
	−24	**1**	−29	**3**	−35	**5**	−42	**6**	−51	**8**	−59	**9**	−68	**10**	−68	**3**	−79	**3**	−88	**1**	−98	**1**	−108	**0**	−148	**28**	−16

Example: Housing bore dia 100 M7

Minimum material	0	**35**	Interference or clearance when upper outside diameter deviations of ring coincide with lower housing bore deviations
		18	Probable interference or clearance
Maximum material	−35	15	Interference or clearance when lower outside diameter deviations of ring coincide with upper housing bore deviations

Numbers in **boldface print** identify interference.
Standard-type numbers in right column identify clearance.

TABLE 13-12 Recommendations for Shaft Tolerances Selection of Solid Steel Shaft Tolerance Classification for Metric Radial Ball and Roller Bearings of Tolerance Classes ABEC-1, RBEC-1 (Except Tapered Roller Bearings) (From SKF, 1992, with permission)

Conditions	Examples	Shaft diameter, mm			
		ball bearings[1]	Cylindrical roller bearing	Spherical roller bearings	Tolerance symbol
Rotating inner ring load or direction of loading indeterminate					
Light loads	Conveyors, lightly	(18) to 100	≤ 40	—	j6
	loaded gearbox	(100) to 140	(40) to 100	—	k6
	bearings				
Normal loads	Bearing applications	≤ 18	—	—	j5
	generally	(18) to 100	≤ 40	≤ 40	k5 (k6)[2]
	electric motors	(100) to 140	(40) to 100	(40) to 65	m5 (m6)[2]
	turbines, pumps	(140) to 200	(100) to 140	(65) to 100	m6
	internal combustion	(200) to 280	(140) to 200	(100) to 140	n6
	engines, gearing	—	(200) to 400	(140) to 280	p6
	woodworking machines	—	—	(280) to 500	r6
		—	—	> 500	r7
Heavy loads	Axleboxes for heavy	—	(50) to 140	(50) to 100	n6[3]
	railway vehicles,	—	(140) to 200	(100) to 140	p6[3]
	traction motors,	—	> 200	> 140	r6[3]
	rolling mills				

High demands on running accuracy with light loads	Machine tools	≤ 18	—	—	$h5^4$
		(18) to 100	≤ 40	—	$j5^4$
		(100) to 200	(40) to 140	—	$k5^4$
		—	(140) to 200	—	$m5^4$
Stationary inner ring load					
Easy axial displacement of inner ring on shaft desirable	Wheels on non-rotating axles	all	all	all	g6
Easy axial displacement of inner ring on shaft unnecessary	Tension pulleys, rope sheaves	all	all	all	h6
Axial loads only					
	Bearing applications of all kinds	all	all	all	j6

TABLE 13-13 Recommendations for Housing Tolerances (From SKF, 1992, with permission)

Conditions	Examples	Tolerence symbol	Displacement of outer ring
SOLID HOUSINGS			
Rotating outer ring load			
Heavy loads on bearings in thin-walled housings, heavy shock loads	Roller bearing wheel hubs, big-end bearings	P7	Cannot be displaced
Normal loads and heavy loads	Ball bearing wheel hubs, big-end bearings, crane travelling wheels	N7	Cannot be displaced
Light and variable loads	Conveyor rollers, rope sheaves, belt tension pulleys	M7	Cannot be displaced
Direction of load indeterminate			
Heavy shock loads	Electric traction motors	M7	Cannot be displaced
Normal loads and heavy loads axial displacement of outer ring unnecessary	Electric motors, pumps, crankshaft bearings	K7	Cannot be displaced as a rule
Accurate or silent running			
	Roller bearings for machine tool work spindles	K6[1]	Cannot be displaced as a rule
	Ball bearings for grinding spindles, small electric motors	J6[2]	Can be displaced
	Small electric motors	H6	Can easily be displaced

SPLIT OR SOLID HOUSING			
Direction of load indeterminate			
Light loads and normal loads axial displacement of outer ring desirable	Medium-sized electrical machines, pumps, crankshaft bearings	J7	Can be normally displaced
Stationary outer ring load			
Loads of all kinds	Railway axleboxes	H7[3]	Can easily be displaced
Light loads and normal loads with simple working conditions	General engineering	H8	Can easily be displaced
Heat condition through shaft	Drying cylinders, large electrical machines with spherical roller bearings	G7[4]	Can easily be displaced

For a standard rolling bearing, the tolerance zones of the outside and bore diameters are below the nominal diameter. The tolerance zone has two boundaries. One boundary is the nominal dimension, and the second boundary is of lower diameter. The lower boundary, which determines the tolerance zone, depends on the bearing precision and size. For example, for a normal bearing of outside diameter $D = 100$ mm, the tolerance zone is $+0$ to $-18\,\mu$m. This means that the actual outside bearing diameter can be within the tolerance zone of the nominal 100 mm and 18 μm lower than the nominal dimension. In drawings, dimensions with a tolerance are specified in several ways, for example, $100^{+0,-18}$. For a bearing bore diameter $d = 60$ mm of a normal bearing, the tolerance is $+0$, $-15\,\mu$m. In this case, the actual bore diameter can be between the nominal 60 mm and 15 μm lower than the nominal dimension, $60^{+0,-15}$. In addition, there are various precision classes, from class 2 to class 6, where class 2 is of the highest precision. For comparison with the previous example of a normal precision class, the dimension of class 2 of the outside diameter is $D = 100^{+0,-5}$; for the bore diameter it is $d = 60^{+0,-2.5}$.

As a rule, the rotating ring of a rolling-element bearing is always tightly fitted in its seat. In most machines, the rotating ring is the inner ring, such as in a centrifugal pump, where the bearing bore is mounted by a tight fit on the shaft. In that case, the outer ring can be mounted in the housing with tight fit, or it can be mounted with a loose fit to allow for free thermal expansion of the shaft. However, if the outer ring is rotating, such as in a grinding wheel, the outer ring should be mounted with a tight fit, while the inner ring can be mounted on a stationary shaft with tight or loose fit. Tight fit of the rotating ring is essential for preventing sliding between the ring and its seat during start-up and stopping, when the rotating ring is subjected to high angular acceleration and tends to slide. Sliding of the ring will result in severe wear of the seat, and eventually the ring will be completely loose in its seat.

In the case of a rotating force, such as centrifugal forces in an unbalanced spindle of a lathe, it is important that the two rings be tightly fitted. Otherwise, the bearing will freely swing inside the free clearance, resulting in an excessive level of vibrations. Usually two or more bearings are used to support a shaft, and only the bearings at one end of the shaft can have a completely tight fit of the two rings in their seats. The radial bearing on the other end of the shaft must have one ring with a loose fit. This is essential to allow the ring to *float* on the shaft or inside the housing seat in order to prevent thermal stresses during operation due to thermal expansion of the shaft length relative to the machine.

In many designs, the bearing is located between a shoulder on the shaft and a standard locknut and lock washer, for preventing any axial bearing displacement (Fig. 13-5). Precision machining of the housing and shaft seats is required in order to prevent the bending of the bearing relative to the shaft. The shoulders on the shaft must form a plane normal to the shaft centerline, the threads on the

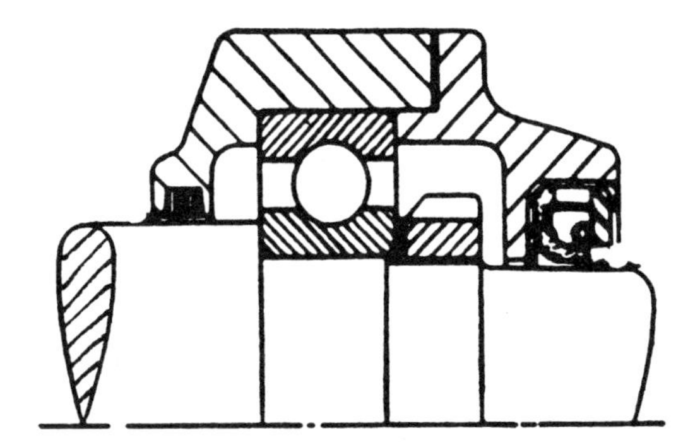

FIG. 13-5 Locating a bearing by a locknut.

locknut and shaft must be precisely cut, with very small run-out tolerance, and the washers must have parallel surfaces in order to secure uniform parallel contact with the inner ring face. Many mass-produced small machines and appliances that operate under light conditions rely only on a tight fit for locating the bearings, in order to reduce the cost of production.

13.9 STRESS AND DEFORMATION DUE TO TIGHT FITS

Tight-fit mounting is where there is an interference (negative clearance) between the housing seat and the outer ring or between the bearing bore and the shaft. A hydraulic or mechanical press is used for bearing mounting. For larger bearings, a temperature difference is used for mounting. The bearing is heated and fitted on the shaft, or the housing is heated for fitting in the bearing.

Tight-fitting involves elastic deformation. Tight-fitting shrinks the outer ring and expands the inner ring. After tight-fit mounting of the inner ring on a shaft, the bore of the inner rings slightly increases in diameter, while the shaft diameter slightly decreases. This results in tangential tensile stresses around the ring, while tight-fit mounting of the outer ring into the housing seat results in compression stresses. When the bearing housing and shaft are made of the same material (steel), the stress equation is simplified. In addition, in order to simplify the equation, it is assumed that the bearing rings have a rectangular cross section.

If the diameter interference of the inner ring on the shaft is Δd, the equation for the tensile stress in the inner ring is

$$\sigma_t = \frac{1}{2} E \left(1 + \frac{d_i^2}{d_o^2}\right) \frac{\Delta d}{d_i} \tag{13-19}$$

where

d_i = ID (inside diameter) of inner ring
d_o = OD (outside diameter) of inner ring
E = modulus of elasticity
Δd = diameter interference (negative clearance)

In a similar way, if the diameter interference of the outer ring inside the housing seat is ΔD, the equation for the compression stresses in the outer ring is

$$\sigma_t = \frac{1}{2} E \left(1 + \frac{D_i^2}{D_o^2}\right) \frac{\Delta D}{D_o} \tag{13-20}$$

where

D_i = ID of outer ring
D_o = the OD of outer ring
ΔD = diameter interference of outer ring

For the two rings, there are compression stresses in the radial direction. At the interference boundary, the compression stress is in the form of pressure between the rings and the seats. The equation for the pressure between the inner ring and the shaft (for a full shaft) is

$$p_{(\text{shaft})} = \frac{1}{2} E \left(1 + \frac{d_i^2}{d_o^2}\right) \frac{\Delta d}{d_i} \tag{13-21}$$

In a similar way, the equation for the pressure between the outer ring and the housing is

$$p_{(\text{housing})} = \frac{1}{2} E \left(1 + \frac{D_i^2}{D_o^2}\right) \frac{\Delta D}{D_o} \tag{13-22}$$

The pressure keeps the rings tight in place, and the friction prevents any sliding in the axial direction or due to the rotation of the ring. The axial load required to pull out the fitted ring or to displace it in the axial direction is

$$F_a = f \pi d L p \tag{13-23a}$$

where f is the static friction coefficient. In steel-on-steel bearings, the range of the static friction coefficient is 0.1–0.25. In a similar way, the equation for the maximum torque that can be transmitted through the tight fit by friction (without key) is

$$T_{\max} = f \pi L p \frac{d^2}{2} \tag{13-23b}$$

13.9.1 Radial Clearance Reduction Due to Interference Fit

Interference-fit mounting of the inner or outer ring results in elastic deformation and, in turn, in a reduction of the radial clearance of the bearing. The reduction of radial clearance, Δ_s, due to tight-fit mounting of interference Δd with the shaft is

$$\Delta_s = \frac{d_i}{d_o} \Delta d \tag{13-24a}$$

In a similar way, the reduction in radial clearance due to interference with the housing seat is

$$\Delta_h = \frac{D_i}{D_o} \Delta d \tag{13-24b}$$

13.9.2 Reduction of Surface Roughness by Tight Fit

The actual interference is reduced by a reduction of roughness (surface smoothing) of tight-fit mating surfaces. Roughness reduction is equivalent to interference loss. For the calculation of the stresses and radial clearance reduction by interference fit, the surface smoothing should be considered.

The greater the surface roughness of the mating parts, the greater the resulting smoothing effect, which will result in interference loss. According to DIN 7190 standard, about 60% of the roughness depth, R_s, is expected to be smoothed (reduction of the outside diameter and increase of the inside diameter) when parts are mated by a tight-fit assembly.

In rolling bearing mounting, the smoothing of the hardened fine-finish surfaces of the rolling bearing rings can be neglected in comparison to the smoothing of the softer surfaces of the shaft and housing. Table 13-14 can be

TABLE 13-14 Surface Roughness for Various Machining Qualities

	Roughnes of surfaces, R_s	
	μm	μin
Ultrafine grinding	0.8	32
Fine grinding	2	79
Ultrafine turning	4	158
Fine turning	6	236

used as a guide for determining the roughness, R_s, according to the quality of machining (Eschmann et al., 1985).

The smoothing effect is neglected for precision-ground and hardened bearing rings, because the roughness R_s is very small. However, there is interference loss to the part fitted to the bearing, such as a shaft or housing. Since 60% of the roughness depth, R_s, is smoothed, the reduction in diameter, ΔD_s, by smoothing is estimated to be

$$\Delta D_s = 1.2 R_s \tag{13-25}$$

Here,

ΔD_s = reduction in diameter due to smoothing (interference loss)
R_s = surface roughness (maximum peak to valley height)

In addition to interference loss due to smoothing, losses due to uneven thermal expansion occur. When the outer ring and housing or inner ring and shaft are made from different materials, operating temperatures will alter the original interference. Usually, the bearing housing is made of a lighter material than the bearing outer ring (higher thermal expansion coefficient), resulting in interference loss at operating temperatures higher than the ambient temperature. Interference loss due to thermal expansion can be calculated as follows:

$$\Delta D_t = D(\alpha_o - \alpha_i)(T_o - T_a) \tag{13-26}$$

Here,

ΔD_t = interference loss due to thermal expansion
D = bearing OD
α_o = coefficient of expansion of outside metal
α_i = coefficient of expansion of inside metal
T_o = operating temperature
T_a = ambient temperature

On the housing side, the effective interference after interference reduction due to surface smoothing and thermal expansion of dissimilar materials is

$$u = \Delta D_{(\text{machining interference})} - \Delta D_s - \Delta D_t \tag{13-27}$$

where

u = effective interference
ΔD = machining interference
ΔD_s = diameter reduction due to smoothing
ΔD_t = interference loss due to thermal expansion

13.9.3 Bearing Radial Clearance During Operation

Bearings are manufactured with a larger radial clearance than required for operation. The original manufactured radial clearance is reduced by tight-fit mounting and later by uneven thermal expansion of the rings during operation. The design engineer should estimate the radial clearance during operation. In many cases, the radical clearance becomes interference, and the design engineer should conduct calculations to ensure that the interference is not excessive. The interference results in extra rolling contact pressure, which can reduce the fatigue life of the bearing. However, small interference is desirable for many applications, because it increases the bearing stiffness.

The purpose of the following section is to demonstrate the calculation of the final bearing clearance (or interference). This calculation is not completely accurate, because it involves estimation of the temperature difference between the inner and outer rings.

13.9.4 Effects of Temperature Difference Between Rings

During operation, there is uneven temperature distribution in the bearing. In Sec. 13.3.3, it was mentioned that for average operation speed the temperature of the inner ring is 5°–10°C higher than that of the outer ring (if the housing is cooled by air flow, the difference increases to 15°–20°C). The temperature difference causes the inner ring to expand more than the outer ring, resulting in a reduction of the bearing radial clearance. The radial clearance reduction can be estimated by the equation

$$\Delta D_{td} = \frac{\Delta T \, \alpha(d + D)}{2} \tag{13-28}$$

Here,

ΔD_{td} = diameter clearance reduction due to temperature difference between inner and outer rings
ΔT = temperature difference between inner and outer rings
α = coefficient of linear thermal expansion
d = bearing bore diameter
D = bearing OD

Example Problem 13-3

Calculation of Operating Clearance

Find the operating clearance (or interference) for a standard deep-groove ball bearing No. 6306 that is fitted on a shaft and inside housing as shown in Fig. 13-6. During operation, the temperature of the inner ring as well as of the shaft is 10°C higher than that of the outer ring and housing. The dimensions and tolerances of the inner ring and shaft are:

Bore diameter: $d = 30\,\text{mm}\ (-10,\ +0)\,\mu\text{m}$
Shaft diameter: $d_s = 30\,\text{mm}\ (+15,\ +2)\,\mu\text{m}$ k6
OD of inner ring: $d_1 = 38.2\,\text{mm}$

The dimensions and tolerances of outer ring and housing seat are:

OD of outer ring: $D = 72\,\text{mm}\ (+0,\ -11)\,\mu\text{m}$
ID of outer ring: $D_1 = 59.9\,\text{mm}$
ID of housing seat: $D_H = 72\,\text{mm}\ (-15,\ +4)\,\mu\text{m}$ K6
Shaft finish: fine grinding
Housing finish: fine grinding
Radial clearance before mounting: C5 Group, 40–50 μm
Coefficient of linear expansion of steel: $\alpha = 0.000011$ [1/K]

Consider surface smoothing, elastic deformation, and thermal expansion while calculating the operating radial clearance.

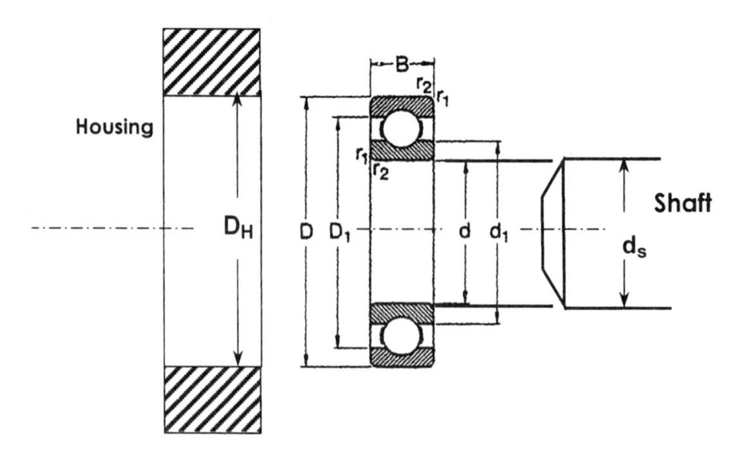

FIG. 13-6 Dimensions and tolerances of rolling bearing, shaft, and housing.

Solution

In most cases, the machining process of the rings, shaft, and housing seat will stop not too far after reaching the desired tolerance. There is high probability that the actual dimension will be near one-third of the tolerance zone, measured from the tolerance boundary close to the surface where the machining started. Common engineering practice is to take two-thirds of the tolerance range and then add to that the lowest tolerance (Eschmann et al., 1985). The result should be a value close to the side on which the machining is started. Therefore,

Shaft interference :

Bearing bore: $(-10 + 0) \times 2/3 + 0 = -7\,\mu\text{m}$

Shaft: $(15 - 2) \times 2/3 + 2 = +11\,\mu\text{m}$

Total theoretical interference fit is: $11 + 7 = +18\,\mu\text{m}$

Housing interference :

OD of outer ring: $(0 + 11) \times 2/3 - 11 = -4\,\mu\text{m}$

ID of housing seat: $(-15 - 4) \times 2/3 + 4 = -9\,\mu\text{m}$

Total theoretical interference fit: $9 - 4 = +5\,\mu\text{m}$

The bearing is made of hardened steel and is precision ground, so smoothing of the bearing inner and outer rings can be neglected. The R_s value for a finely ground surface is obtained from Table 13-5:

Smoothing to finely ground shaft: $\Delta D_s = 1.2R_s = 1.2(2) = 2.4\,\mu\text{m}$
Smoothing to finely ground housing: $\Delta D_s = 1.2R_s = 1.2(2) = 2.4\,\mu\text{m}$

In this example, the shaft and housing are both made of steel, so there is no change in interference due to different thermal expansion of two materials.
The effective interference becomes:

$$u = \text{theoretical interference} - \Delta D_s$$

Inner ring: $u_i = 18 - 2.4 = 15.6\,\mu\text{m}$

Outer ring: $u_o = 5 - 2.4 = 2.6\,\mu\text{m}$

The radial clearance reduction due to tight-fit installation of the rolling bearing is also considered. The reduction in clearance due to interference with the shaft is (Eq. 13-24)

$$\Delta_s = \frac{d}{d_1} u_i = \frac{30\,\text{mm}}{38.2\,\text{mm}} \times 15.6\,\mu\text{m} = 12.25\,\mu\text{m}$$

The reduction in clearance due to interference with the housing is

$$\Delta_H = \frac{D_1}{D} u_0 = \frac{59.9\,\text{mm}}{72\,\text{mm}} \times 2.6\,\mu\text{m} = 2.16\,\mu\text{m}$$

The total radial clearance reduction due to installation is therefore

$$\Delta_s + \Delta_H = 12.25\,\mu\text{m} + 2.16\,\mu\text{m} = 14.4\,\mu\text{m}$$

Finally, as stated in the problem, there is a temperature difference of $\Delta T = 10°\text{C}$ between the inner and outer rings. This is due to the more rapid heat transfer away from the housing than from the shaft. In turn the shaft and inner ring will have a higher operating temperature than the outer ring. This will result in higher thermal expansion of the inner ring, which will further reduce the radial clearance. The thermal clearance reduction is

$$\Delta D_{\text{th}} = \frac{\Delta T\ \alpha(d + D)}{2}$$

$$\Delta D_{\text{th}} = 10(\text{K}) \times 0.000011\ (1/\text{K}) \times \frac{(30 + 72)\,\text{mm}}{2} \times \frac{1000\,\text{m}}{\text{mm}}$$
$$= 5.6 \times 10^{-3}\,\text{m} = 5.6\,\mu\text{m}$$

In summary, the expected radial running clearance of this bearing will be:

Radial clearance before mounting:	40–50 μm
Radial clearance reduction due to mounting:	−14.4 μm
Radial clearance reduction by thermal expansion:	−5.6 μm
Expected radial clearance during operation	20–30 μm

13.10 BEARING MOUNTING ARRANGEMENTS

An important part of bearing design is the mounting arrangement, which requires careful consideration. For an appropriate design, the following aspects should be considered.

The shaft should be able to have free thermal expansion in the axial direction, due to its temperature rise during operation. This is essential for preventing extra thermal stresses.

The mounting arrangement should allow easy mounting and dismounting of the bearings. The designer must keep in mind that rolling bearings need maintenance and replacement.

The shaft and bearings are part of a dynamic system that should be designed to have sufficient rigidity to minimize vibrations and for improvement of running precision. For improved rigidity, the mounting arrangement is often designed for elimination of any clearance by preloading the bearings.

Bearing arrangements should ensure that the bearings are located in their place while supporting the radial and axial forces.

It was discussed earlier that during operation, if the housing has no cooling arrangement, the temperatures of the shaft and inner ring could be 5°–10°C higher than that of the outer ring. If the housing is cooled by air flow, the temperatures of the inner ring can increase to 15°–20°C higher than that of the outer ring. During operation, the temperature difference between the shaft and the machine frame is higher than between the rings. This results in a thermal elongation of the shaft relative to the machine frame that can cause extra stresses at the rolling contact of the bearings. In addition, due to manufacturing tolerances, the distances between the shaft seats and the housing seats are not equal. The extra stresses caused by thermal elongation and manufacturing tolerances can be very high if the shaft is long and there is a large distance between the supporting bearings.

This problem can be prevented by appropriate design of the bearing arrangement. The design must provide one bearing with a loose fit so that it will have the freedom to float in the axial direction (*floating bearing*). In most cases, the loose fit of the floating bearing is at the outer ring, which is fitted in the housing seat.

A floating bearing allows free axial elongation of the shaft. The common design is referred to as a *locating/floating* or *fixed-end/free-end* bearing arrangement. In this design, one bearing is the *locating* bearing, which is fixed in the axial direction to the housing and shaft and can support thrust (axial) as well as radial loads. On the other side of the shaft, the second bearing is *floating*, in the sense that it can slide freely, relative to its seat, in the axial direction. The floating bearing can support only radial loads, and only the locating bearing supports the entire thrust load on the shaft. In shafts supported by two or more bearings, only one bearing is designed as a locating bearing, while all the rest are floating bearings. This is essential in order to prevent extremely high thermal stresses in the bearings.

An example of a *locating/floating* bearing arrangement is shown in Fig. 13-7. Additional practical examples are presented in Sec. 13-12. The bearing on the left side of the shaft is fixed in the axial direction and can support thrust forces in the two directions as well as radial force. The bearing on the right end of the shaft can float in the axial direction and can support only radial force. Axial floating of the bearing is achieved by providing the housing with a loose fit (a clearance between the housing seat and the bearing outer ring). In certain applications, two angular contact ball bearings or tapered roller bearings that are symmetrically arranged and preloaded are used as locating bearings (see Sec. 13.11). This design provides for an accurate rigid location of the shaft.

In principle, axial floating of the shaft is also possible by means of a loose fit between the shaft and the bearing bore. However, for a rotating shaft and stationary housing, the clearance must be on the housing side, to prevent wear of the shaft surface during starting and stopping of the machine.

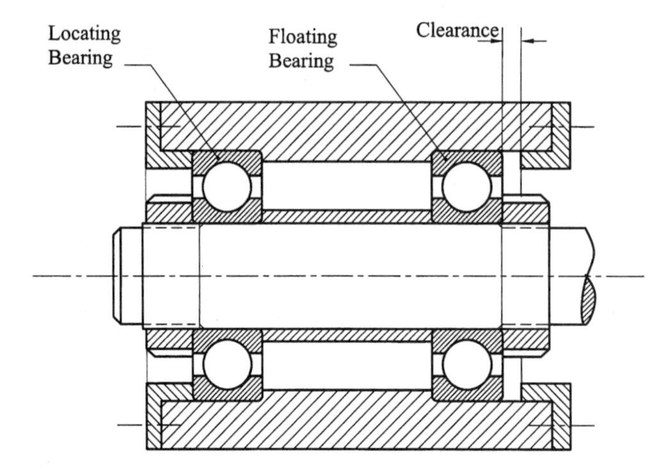

FIG. 13-7 Locating/floating bearing arrangement.

During starting and stopping, the shaft has a high angular acceleration. In turn, the moment of inertia (of the inner ring and balls of the bearing) causes inertial torque resistance (in the direction opposite to that of the shaft angular acceleration). In many cases, the inertial torque is higher than the friction and the shaft would slide during the start-up inside the bearing bore. The relative sliding can cause severe wear of the shaft surface. This undesired effect can be prevented by means of a tight fit (interference fit) of the shaft inside the bearing inner ring bore. Shafts seats are often rebuilt due to the wear during starting and stopping. Therefore, for a rotating shaft, the clearance (loose fit) of the floating arrangement is always on the housing seat, while a tight fit is on the shaft side.

In certain machines, the shaft is stationary and the outer ring rotates, e.g., a grinding machine or a centrifuge. In such applications, a tight fit must be provided on the rotating side and a sliding fit at the seat of the stationary shaft.

Compensation for shaft elongation can also be achieved by using a cylindrical roller bearing. Certain cylindrical roller bearings are designed to operate as floating bearings by allowing the roller-and-cage assembly to shift in the axial direction on the raceway of the outer ring. For this purpose, the bearing rings are designed without ribs (often named *lipless bearing rings*). All other bearing types, such as a deep-groove ball bearing or a spherical roller bearing, can operate as floating bearings only if one bearing ring has a loose fit in its seat.

13.10.1 Tandem Arrangement

Two angular contact ball bearings can be used in series for heavy unidirectional thrust loads. Precise dimensions and high quality surface finish are required to secure load sharing of the two bearings. The arrangement of two or more angular contact bearings, adjacent to each other in the same direction, is referred to as *tandem arrangement.* This arrangement is used to increase the thrust load carrying capacity as well as the radial load capacity. Tandem arrangement is often used in spindles of machine tools, where high axial stiffness and high thrust load capacity are required; examples are shown in Sec. 13.12. Bearing manufacturers provide a combination of two angular contact ball bearings that are designed and made for tandem arrangement.

13.10.2 Bearing Seat Precision

For a locating bearing, the inner and outer rings are tightly fitted into their seats. But a floating bearing has one ring that is fitted tightly, while the other ring has a loose fit to allow free axial sliding. For a floating bearing, if the shaft is rotating, only the inner ring must be mounted by interference fit. If the outer ring is rotating, only the outer ring is mounted by interference fit. The reason for a tight fit of the rotating ring is to avoid sliding and wear during start-up and stopping.

In interference fit (tight fit) there is elastic deformation of the ring that reduces the internal clearance of the bearing. Therefore, it is important to select the recommended standard fit for a proper internal radial clearance after the bearing mounting. The bearings are manufactured with internal clearance to provide for this elastic deformation and for thermal expansion of the shaft and inner ring during operation.

In the case of a tight fit, the bearing can be mounted by application of heat or cold-mounted by pressing the face of the ring that is tightly fitted (in order to prevent bearing damage, never apply force through the rolling elements). In many cases, such as a bore diameter over 70 mm, it is easier to mount via temperature difference. This can be obtained by heating one part, or heating and cooling, respectively, the two parts. An additional simple method for tight-fit mounting is the use of tapered-bore bearings combined with tapered seats. The bearing is tightened in the axial direction by a locknut. A tapered adapter sleeve is another convenient method for tight-fit mounting.

For the shaft and housing seats, precision and good surface finish are required. In fact, the precision and surface finish of the seats should be similar to those of the rolling bearing in contact with the seat. Whenever possible, a ground finish of the bearing seats on the shaft and housing is preferred. Only in exceptional cases of low speed and load—if cost saving is critical—are rougher shaft and housing seats used. In such cases, rougher ball bearings can be used as well, in order to reduce the cost in low-cost machines.

A common locating arrangement is where the ring is tightly fixed between a shaft shoulder and a locknut, as shown in Fig. 13-5. Precision of the shaft shoulder seat is required because many rolling bearings are so narrow that they are not aligned accurately by the length of the seat on the shaft. The final accurate alignment is by the shaft shoulder and nut. Precision of the seats and the nuts is particularly important for medium-and high-speed applications. The shoulder plane should be perpendicular to the shaft centerline (squareness). In the same way, locknut precision is required. A standard locknut should have precise thread having maximum face run-out within 0.05 mm (0.002 in.). Precision nuts with much lower face run-out are used for precision or high-speed applications.

Quality inspection of shaft seats and shoulders for axial and radial run-out is required for medium and high speeds. Rotating the shaft between centers, with a dial indicator placed against the seat or shoulder, is the standard inspection. Proper manufacturing practice is to grind the seats for the inner ring and shaft shoulder together, in one clamping of the shaft, and the same applies to the housing. One clamping ensures that the two surfaces are perpendicular.

The recommended height of a shaft shoulder is one-half of the inner ring face. Were the shoulder too low, it would result in a plastic deformation of the shoulder due to excessive pressure, particularly under high thrust load. On the other hand, the shaft shoulder should not be too high (more than half of the inner ring face), to allow disassembly and removal of the bearing from the shaft. A puller placed against the inner ring surface is usually used for removing the bearing.

Careful design of the corners of the shaft shoulder and bearing seat is necessary. The corner radius of the seat must be less than that of the ring. In many designs, the corner has an undercut or a shaft fillet to secure a proper fit to the bearing ring. However, an undercut weakens the shaft and causes stress concentration at the corner. Whenever weakening of the shaft is not desired, a fillet can be used. Standard fillet sizes for each particular bearing are available and are listed in bearing catalogues. In many cases, a small taper is provided on the bearing seat edge to provide a guide to assist in mounting the bearing.

13.11 ADJUSTABLE BEARING ARRANGEMENT

The bearing clearance allows a free radial or axial displacement of the inner ring relative to the outer ring. The objective of an adjustable arrangement of angular contact ball bearings or tapered roller bearings is to eliminate this undesired clearance. In addition, by using an adjustable arrangement it is possible to preload the bearing (negative clearance). Preload means that there are compression stresses and elastic deformation at the contacts of the rolling elements and the raceways before the bearing is in operation.

Bearing preload is important for many applications requiring high system rigidity. By preloading the bearing, the stiffness of the machine increases; namely, there is a reduction in the elastic deformation under external load. Bearing preload causes extra stresses at the contacts between the rolling elements and the raceways, which can reduce the fatigue life of the bearing. Therefore, the preload must be precisely adjusted, because excessive contact stresses will have an adverse effect on bearing life.

In an adjustable arrangement, angular ball bearings or tapered bearings are mounted in pairs against each other on one shaft and are preloaded. Deep-groove ball bearings are used as well for adjustable arrangements, because they act like angular contact ball bearings with a small contact angle. The arrangement is designed to allow, during mounting, for one ring to slide in its seat, in the axial direction, for adjusting the bearing clearance or even provide preload inside the bearing. This is done by tightening the inner ring by means of a nut on the shaft or via an alternative design for tightening the outer ring of the bearing in the axial direction. Examples of adjustable arrangements are shown in Figs. 13-8 and 13-9.

It was discussed earlier that by a tight fit of the bearing rings in their seats, the radial clearance can be eliminated and the bearing can be preloaded. However, better control and precision of the preload can be achieved via an adjustable arrangement using angular contact ball bearings or tapered roller bearings. Preload by tight fit of the bearings in their seats is not always precise. This is due to machining tolerances of the seats and bearing rings. However, in an

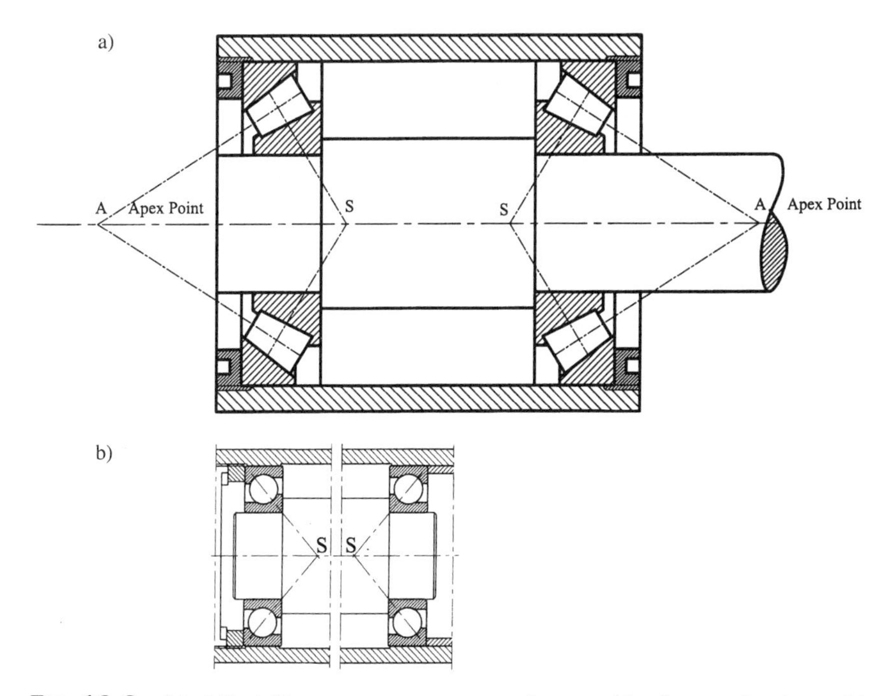

FIG. 13-8 (a) Adjustable arrangement, apex points outside the two bearings. (b) Similar adjustable arrangement for angular contact bearings.

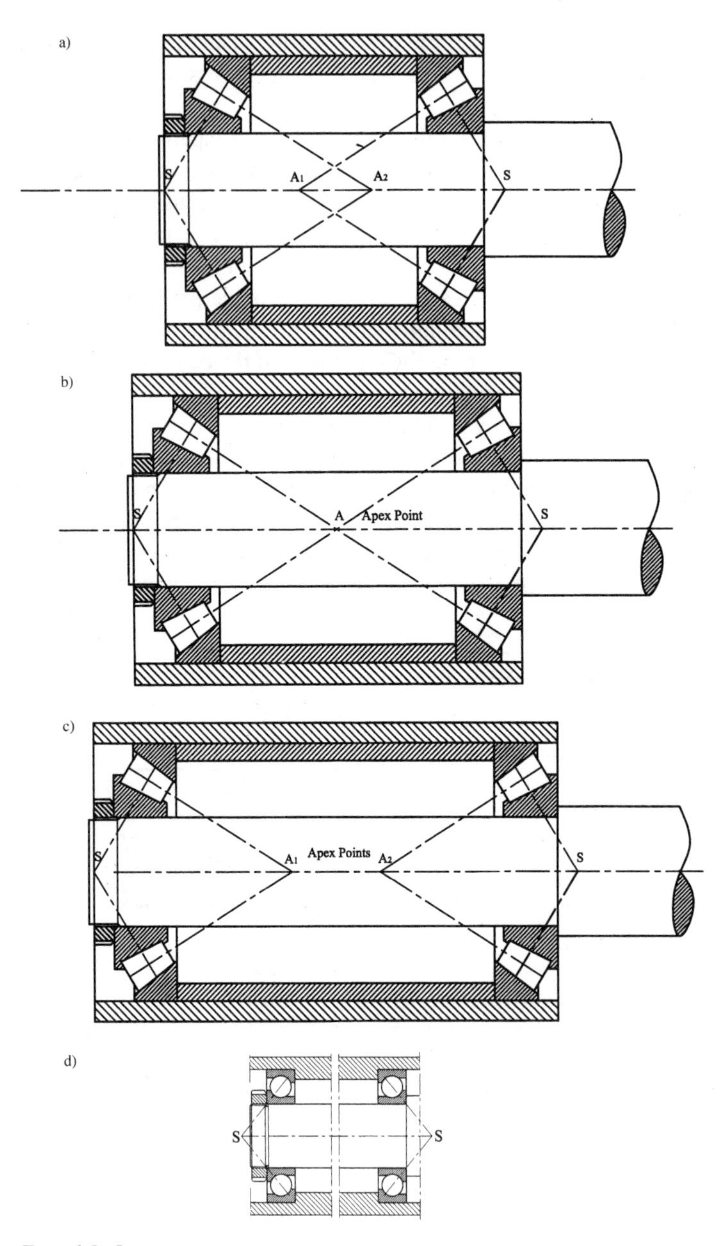

FIG. 13-9 (a) Adjustable arrangement, apex points between the bearings overlap. (b) Adjustable arrangement, apex points coincide between the two bearings. (c) Adjustable arrangement, apex points apart between the bearings. (d) Similar adjustable arrangements for angular contact bearings.

adjustable arrangement, the preload is independent of machining tolerances. Nevertheless, thermal expansion of the shaft during operation must be taken into consideration when the adjustment is performed during assembly, when the machine is cold. If the operation temperatures of the shaft and machine are known from previous experience, the thermal expansion can be calculated, and precise adjustment to the desired tightness during bearing operation is possible.

13.11.1 Thermal Effects

Whenever the operation temperatures are unknown, it is possible to reduce the thermal stresses by having the adjustable pair of bearings close to each other (a short shaft length between the two bearings). A better alternative is to design an adjustable bearing arrangement that can be adjusted after the machine is assembled and run. In such cases, the adjustment is performed after the machine has been operating for some time and thermal equilibrium has been reached.

In a tapered bearing, the rolling elements and races have a conical shape, with a line contact between them. In order to have a rolling motion, all the contact lines of the tapered rollers and raceways must meet at a common point on the axis, referred to as the *apex point*. Similarly for angular contact ball bearings, the lines of contact angle meet at the apex point on the bearing axis. There are two types of adjustable bearing arrangement, depending on the location of the apex points. The first type is where the two apex points, A, are outside the space between the two bearings (see Fig 13-8a and 8b). The second type is where the apex points are between the two bearings (see Fig. 13-9a, 9b, 9c, 9d). The designer should consider the level of thermal expansion in order to choose between these arrangements.

It was discussed in Sec. 13.3.3 that the temperature of the inner ring is higher than that of the outer ring. For the same reason, the shaft temperature is higher than the housing temperature. In turn, the shaft is thermally expanding in the axial direction more than the distance between the two outer bearing seats in the housing. The thermal expansion of the shaft relative to the housing seats is proportional to the distance between the two bearings. The diameters of the shaft and inner ring will also expand thermally more than the outer ring and housing diameters.

13.11.1.1 Apex Points Outside the Two Bearings

This bearing arrangement is often referred to as *X arrangement*, because the lines in the direction normal to the contact lines, intersecting at point S, form an X shape. These lines are the directions of the forces acting on the rolling element. In angular contact ball bearings (Fig. 13-8b), these lines form the contact angle.

The temperature rise of the shaft relative to that of the housing increases the length and diameter of the shaft as well as the diameter of the cone (inner ring) of the bearings. The first type of an adjustable bearing arrangement is shown in Fig 13-8. In this arrangement, the apex points, *A*, are outside the two bearings. As shown in Fig. 13-8, tightening a threaded ring on the housing side does the adjustment. In this way, the bearing cup (outer ring) is shifted in the axial direction and, thus, the clearance in the two bearings can be adjusted. In this arrangement, a temperature rise will always result in a tighter bearing clearance.

In the bearing arrangement of Figs. 13-8a and 13-8b, if the bearings are preloaded when the machine is cold, the temperature rise results in a higher bearing preload and rolling contact stresses. If some bearing clearance is left after the adjustment, the clearance will be reduced due to the thermal expansion. As discussed earlier, the advantage of this arrangement type is that it can be designed to allow a final adjustment during operation, after the machine has reached a steady thermal equilibrium.

13.11.1.2 Apex points between the two bearings

This arrangement is often referred to as *O arrangement*, because the lines in the direction normal to the contact lines, intersecting at point S, form an O shape. These lines are the directions of the forces acting on the rolling element. In angular contact ball bearings (see Fig. 13-9d), these lines form the contact angle.

In general, arrangement of apex points between the two bearings (*O arrangement*) is preferred whenever a strong axial guidance is required. This means that the shaft is supported more rigidly than the adjustable arrangement with apex points outside the bearings. The direction of the rolling-elements reaction force resists better any rotational vibrations of the shaft around an axis perpendicular to the shaft centerline.

The effect of a temperature rise may be different in the second arrangement type, which is shown in Figs. 13-9a, 13-9b, and 13-9c, where the apex points are between the bearings. As shown in these figures, tightening a nut on the shaft side does the adjustment. The bearing cone (inner ring) is shifted in the axial direction, and the clearance in the two bearings is adjusted.

During operation, the temperature rise increases the shaft length and at the same time increases the diameter of the inner ring (cone). In the second arrangement type (Figs. 13-9), a thermal expansion of the shaft length has a loosening effect on the two bearings; however, at the same time, the thermal expansion of the cone diameter has a tightening effect. The combined effect depends on the ratio of the shaft length to the cone diameter. The combined thermal effects can be determined by the location of the apex points. This arrangement can be divided into three cases:

1. For a short distance between the two bearings, the roller cone apex points overlap, as shown in Fig. 13-9a. In this case, the thermal expansion of the cone (inner ring) diameter has a larger effect than

the axial expansion of the shaft. The combined effect is that the thermal expansion increases the preload. This combined effect should be considered and the bearings should be adjusted with a reduced preload in comparison to the desired preload during operation.

2. The two apex points coincide, as shown in Fig. 13-9b. In this case, the axial and the radial thermal expansions compensate each other without any significant thermal effect on the clearance.
3. The distance between the bearings is large and the roller cone apices do not overlap, as shown in Fig. 13-9c. In this case, the cone (inner ring) thermal expansion is less than that of the shaft. In turn, the combined thermal effect is to increase the clearances of the two bearings (or reduce the preload). This should be considered, and the bearings are usually adjusted tighter with more preload in comparison to the desired preload during operation.

The selection of the adjustment arrangement type depends on several factors. The second type, where the roller cone apices are between the bearings, has more rigidity to keep the shaft centerline in place. In addition, it can be designed so that the thermal expansion is compensated. The first type, where the roller cone apices are outside the bearings, is often selected in order to allow a fine adjustment during operation. This is possible to do only if the threaded ring (or other adjustment design) is accessible for adjustment during the operation of the machine.

13.11.2 Inner and Outer Ring Fits

The inner or outer ring that is adjusted should move freely by means of a slightly loose fit, while the other ring is mounted with a tight fit. As with other rolling bearings, the inner ring (cone) should always be mounted with a tight fit when the cone rotates. Similarly, the outer ring (cup) should be mounted with a tight fit when it rotates. For a rotating shaft, this requirement usually favors the first type of adjustable bearing arrangement, where the apex points are outside the two bearings. However, the second type is often used for rotating shafts as well.

If the housing rotates, as in a nondriven car wheel, the cup is tightly fitted. If the bearing is subjected to severe loads, shocks, or frequent direction reversals, such as in construction equipment, both cup and cone must be tightly fitted.

For high-speed applications, such as turbines and high-speed machine tools, an adjustable arrangement of angular contact ball bearings is preferred, because tapered rolling bearings have higher friction. In a similar way to the intersection of cone apices, in angular contact bearings the arrangement type is determined by the intersection of the lines normal to the angular bearing contact lines.

Manufactured pairs of angular contact ball bearings or tapered bearings are available. The bearings are paired in a first-or second-type arrangement. Angular contact ball bearings of these designs are accurately finished and can be selected with a low axial clearance, a zero clearance, or a light preload.

13.11.3 Designs for Reduction of Thermal Effects on Bearings Preload

It is important that the bearings in an adjustable arrangement will operate with the desired precise preload force. However, the operating temperature can fluctuate resulting in a variable preload force. Excessive preload can reduce the fatigue life of the bearing, and if the preload is reduced, the bearings' stiffness may be too low. Engineers are always looking for new designs for mitigating the thermal effect, so that a precise preload will be sustained in the bearings.

It is possible to design a preloaded bearing arrangement where springs provide the thrust force. The spring force is not as sensitive to the thermal elongation in comparison to the rigid shaft in the common adjustable bearing arrangement. The advantage is that the spring force is constant, and the preload force does not increase by the temperature rise during bearing operation. Examples of designs where springs provide the preload force are shown in Sec. 13.12.

Additional example for reducing the effect of the temperature on the level of the preload force is by using two spacer sleeves between the two bearings for the outer and inner rings of the adjustable arrangement. The two spacer sleeves have only a small contact area with the rings and housing. The spacer for the inner rings has an air clearance with the shaft, and the spacer for the outer rings has an air clearance with the housing. It results that the two long spacer sleeves are nearly thermally isolated, and have approximately equal temperature during operation. In this way, the axial elongation of the two long spacer sleeves is equal without any significant effect on the preload. An example of a design where two long spacer sleeves are used for a NC Lathe spindle bearing arrangement is shown in Sec. 13.12.

13.11.4 Machine Tool Spindles

The two most important requirements for machine tool spindle bearings are

(a) high precision (very low bearing run-out)
(b) high rigidity (very low elastic deformation under load).

High precision spindle bearings are manufactured with very low tolerances, which are tested for very small radial and axial run-out. In addition, the bearing seats must have similar precision, and very good surface finish. The requirement of high stiffness can be achieved by using relatively large bearings that are precisely preloaded. For precision machining, the cutting forces should result in very small elastic deformation. Therefore, the system of spindle and bearings must be rigid. For this purpose, machine tool designs entail large diameter spindle and large bearings, in comparison to other machines with similar forces. Moreover, to ensure rigidity of the system, the bearings must be preloaded, in order to minimize the elastic deformation at the contacts of rolling elements and races.

The requirement for high stiffness results in large bearings relative to other machines with similar forces; therefore, the fatigue life is usually not a problem in machine tool spindle bearings. Spindle bearings usually do not fail by fatigue, but can wear out, and it is important to have clean lubricant to reduce wear.

In most cases, machine tools are fitted with angular contact ball bearings to support the high thrust load. The requirement for high axial stiffness under heavy thrust cutting forces is achieved in many cases by arrangement of two or more angular contact ball bearings in *tandem arrangement* (see Sec. 13.10.1). The bearings are preloaded by adjustable arrangement, and care must be taken to ensure that the preload will remain constant and will not vary due to variable bearing temperature. Examples of bearing arrangements for machine tool spindles are presented in Sec. 13.12.

13.12 EXAMPLES OF BEARING ARRANGEMENTS IN MACHINERY

13.12.1 Vertical-Pump Motor (Fig. 13-10a)

Design data
Power: 160 kW
Speed: 3000 RPM
Thrust force: 14 kN (Total of weight of rotor and impeller, pump thrust force and spring force).

Tolerances
cylindrical roller bearing shaft m5; housing M6
deep grove ball bearing: shaft k5; housing H6
angular contact bearing: shaft k5; housing E8

Lubrication: Grease lubrication with time period of 1000 hours between lubrications.

Design: This is a locating/floating bearing arrangement. The two bearings at the top form the locating side, whereas the lower cylindrical roller bearing is a floating bearing. In the locating top bearings, preload is

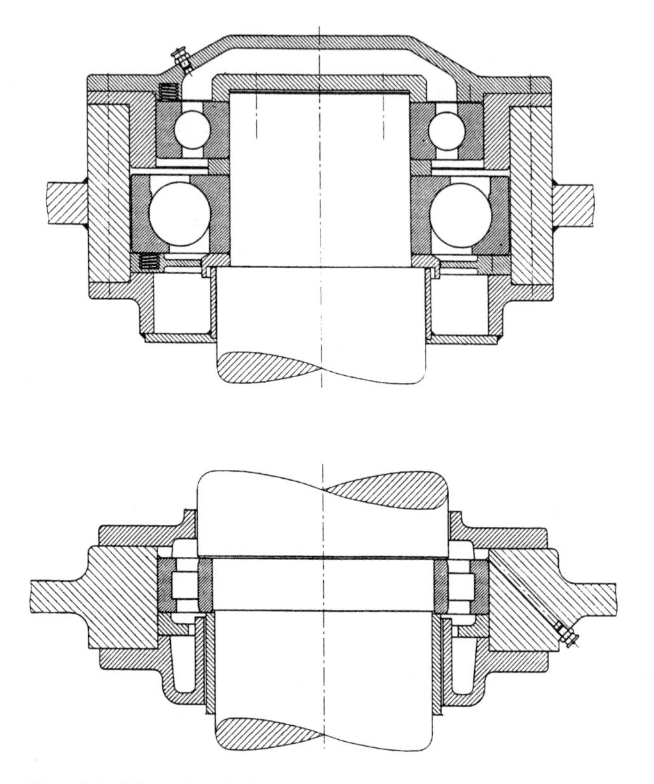

FIG. 13-10a Vertical pump motor. (From FAG, 1998, with permission of FAG and Handel AG.)

done via springs. The springs ensure a constant predetermined load (see Sec. 13.11.3). The angular contact bearing carries the thrust load, and the deep grove bearing carries any possible radial load (and the small spring axial preload). A clearance fit relieves the angular contact bearing from any radial load, which can reduce its fatigue life. The lower cylindrical roller bearing carries only radial load.

13.12.2 NC-Lathe Spindle

Figure 13-10b shows a bearing arrangement for a spindle of *numerically controlled* (NC) lathe. The bearings have adjustable arrangement and are lightly preloaded. The adjustable arrangement is of the type of apex points between the two bearings (often referred to as an *O arrangement*). As the speed is relatively high, the spindle is fitted with angular contact ball bearings (lower friction than tapered bearings). However, in order to support the high thrust load and provide

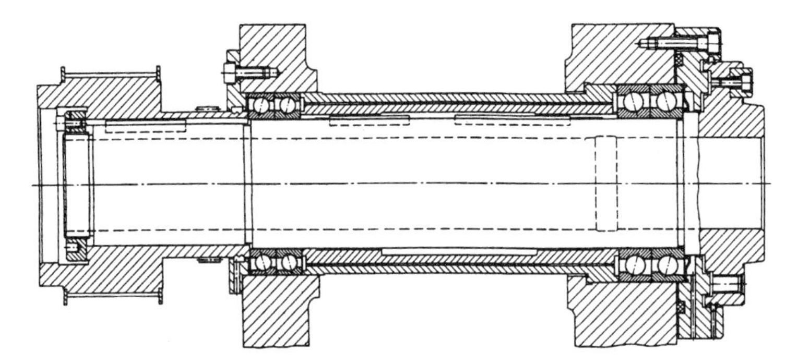

FIG. 13-10b NC-lathe main spindle. (From FAG, 1998, with permission of FAG OEM and Handel AG).

the required rigidity, two angular contact ball bearings in tandem arrangement are fitted at each side.

For mitigating the effect of the temperature rise on the preload level, the design includes two spacer sleeves of approximately equal temperature between the two bearings for axial support of the outer and inner rings of the adjustable arrangement (see Sec. 13.11.3).

Design data
Power: 27 kW
speed: 9000 RPM

Lubrication: The bearings are greased and sealed for the bearing life, and 35% of cavity is filled. Sealing is via labyrinth seals.

Tolerances: High precision spindle bearings are used. The bearings have tight fit on the shaft seat (shaft seat tolerance $+5/-5$ μm), and sliding fit on the housing seats, (housing seat tolerance $+2/+10$ μm).

13.12.3 Bore Grinding Spindle (Fig. 13-10c)

Design data
Power: 1.3 kW
Speed: 16,000 RPM

Lubrication: The bearings are greased and sealed for the bearing life. Sealing is by labyrinth seals.

Design: High rigidity is required. Angular contact ball bearings are used of 15° contact angle for high radial stiffness, and each side has a tandem arrangement for axial stiffness. Bearings have adjustable arrangement and are lightly preloaded by a coil spring.

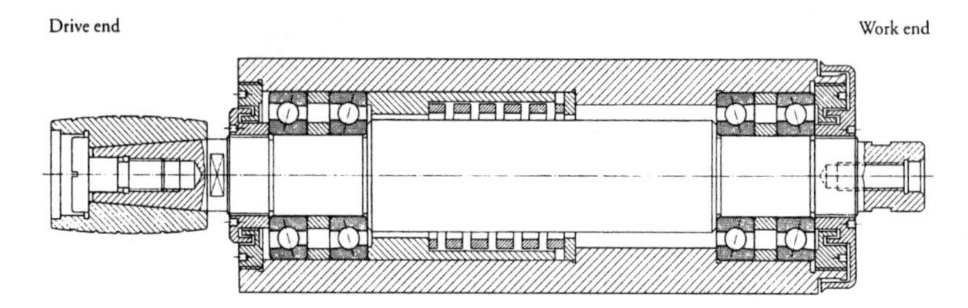

FIG. 13-10c Bore-grinding spindle. (From FAG, 1998, with permission of FAG OEM and Handel AG).

13.12.4 Rough-turning lathe (Fig. 13-10d)

Design data
Power: 75 kW
Speed: 300–3600 RPM

Machining tolerances
seats for the outer ring G6
seats for the inner ring js5

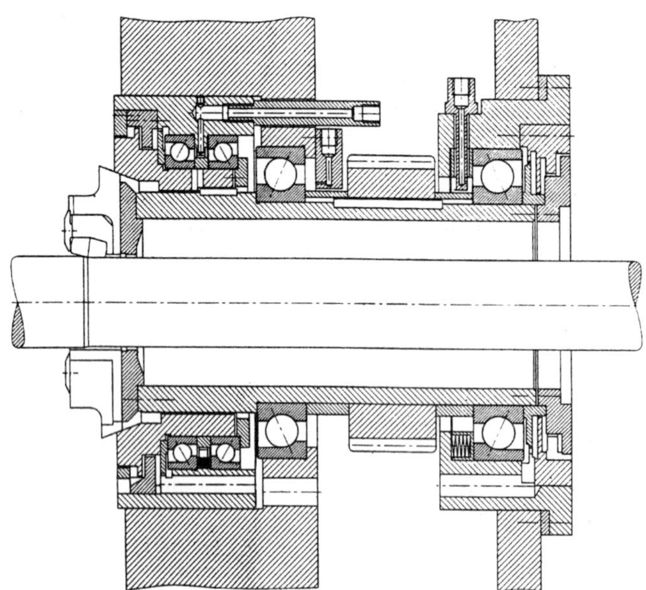

FIG. 13-10d Rough-turning lathe (from FAG, 1998, with permission of FAG OEM and Handel AG).

Lubrication: Oil injection lubrication. A well designed, non-contact labyrinth seal prevents oil leaks and protects the bearings from any penetration of cutting fluid and metal chips.

Design: The bearings have adjustable arrangement and lightly preloaded by springs.

13.12.5 Gearbox (Fig 13-10e)

Design data
Power: 135 kW
Speed: 1000 RPM

Tolerances: shaft m5, housing H6

Lubrication: Splash oil from the gears. Shaft seals are fitted at the shaft openings.

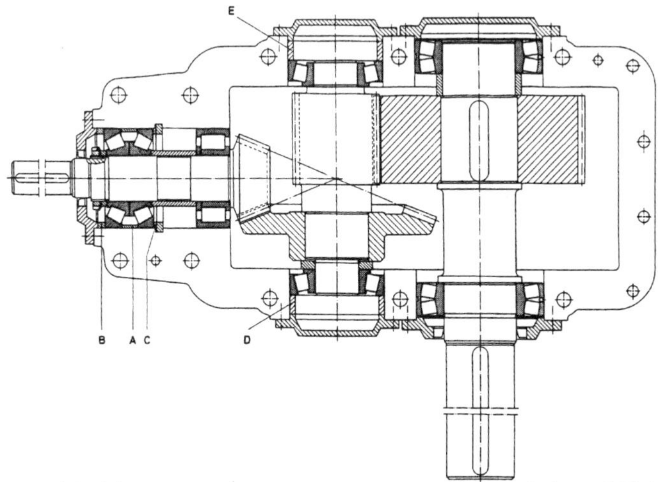

FIG. 13-10e Gearbox (from FAG, 1998, with permission of FAG OEM and Handel AG).

13.12.6 Worm Gear Transmission (Fig. 13-10f)

Design data
Power: 3.7 kW
Speed: 1500 RPM

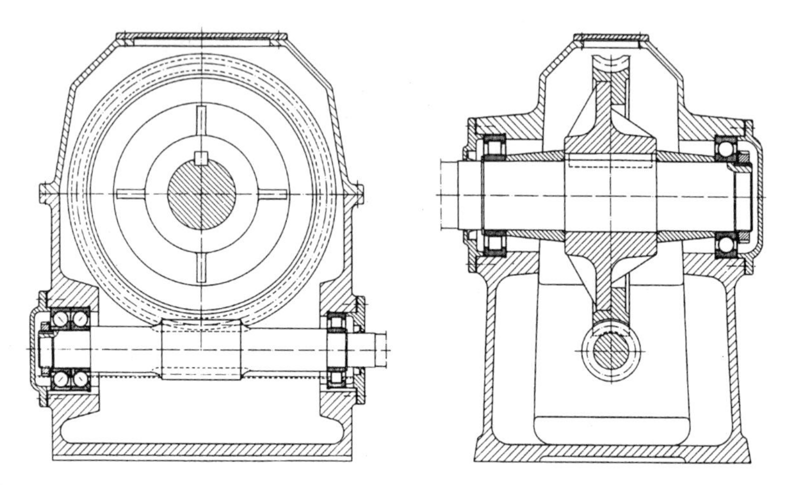

FIG. 13-10f Worm gear (From FAG 1998, with permission of FAG OEM and Handel AG).

Tolerances
Angular contact ball bearing: shaft j5; housing J6
Cylindrical roller bearing: shaft k5; housing J6
Deep groove ball bearing: shaft k5; housing K6

Lubrication: Oil. Contact sealing rings at the shaft opening prevent oil from escaping and protect from contamination.

Design: The two shafts have a locating/floating bearing arrangement.

13.12.7 Passenger Car Differential Gear (Fig. 13-10g)

Design Data
Torque: 160 N-m
Speed: 3000 RPM

Tolerances
Pinion shaft: m6 (larger size bearing)
h6 (smaller size bearing)
housing P7
Crown wheel: hollow shaft r6
housing H6

Lubrication: Gear oil

Design: The turn shafts have adjustable bearing arrangement.

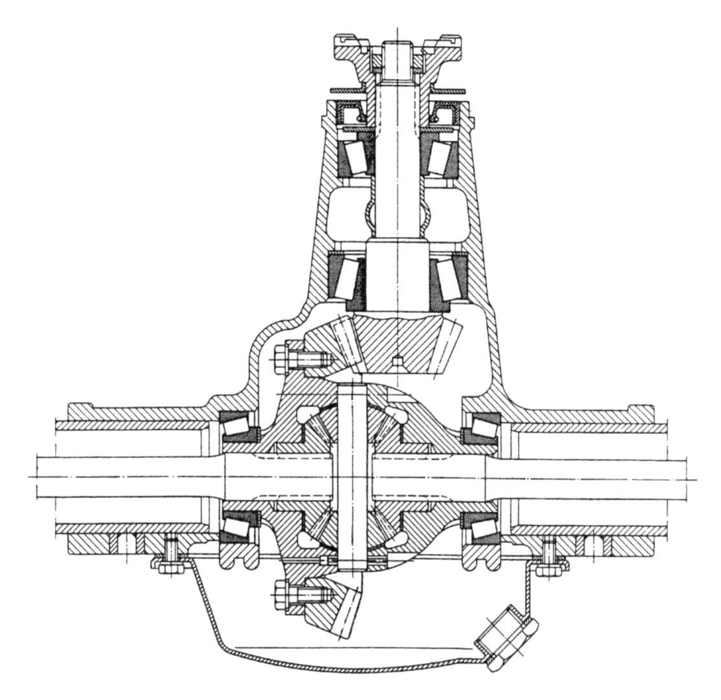

FIG. 13-10g Passenger car differential gear (with permission of FAG OEM and Handel AG).

13.12.8 Guide Roll for Paper Mill (Fig. 13-10h)

Design Data
Speed: 750 RPM
Roll weight: 80 kN
Paper pull force: 9 kN
Bearing load: 44.5 N
Bearing temperature: 105°C

Tolerances: housing G7, inner ring fitted to a tapered shaft

Lubrication: oil circulation

Sealing: double noncontact seal, as shown in Fig 13-10h. The double noncontact seals prevent oil from leaking out.

Design: Special bearings durable to the high operation temperature of the dryer are required. Bearing manufacturers offer high-temperature bear-

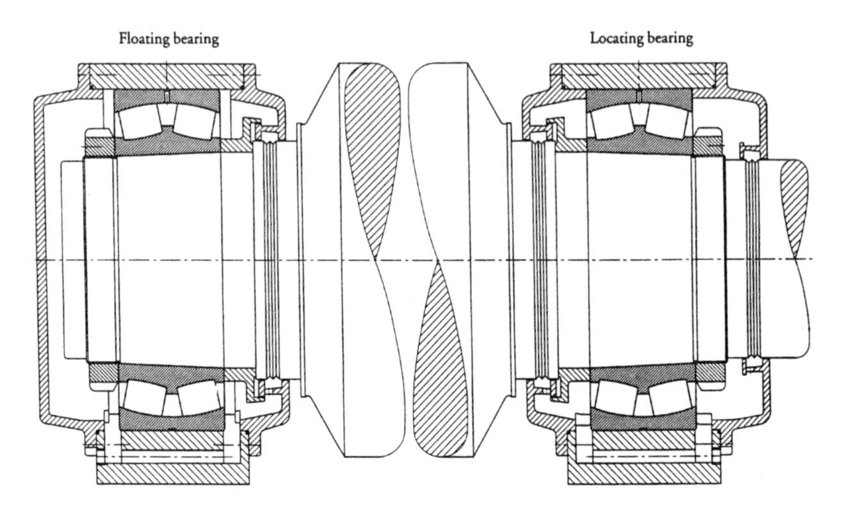

FIG. 13-10h Guide roll for paper mill (from FAG, 1998, with permission of FAG OEM and Handel AG).

ings, which passed special heat treatment, and are dimensionally stable up to 200°C.

Operating clearance is required for preventing thermal stresses, due to the large temperature rise during operation. Also, locating/floating arrangement must be included in this design of relatively high operating temperature. Self-aligning bearings are used to compensate for any misaligning due to thermal distortion.

13.12.9 Centrifugal pump (Fig. 13-10i)

Design Data
Power: 44 kW
Speed: 1450 RPM
Radial load: 6 kN
Thrust force: 7.7 kN

Lubrication: Oil bath lubrication, the oil level should be no higher than the center of the lowest rolling element.

Sealing: Contact seals are used on the two sides. At the impeller side, a noncontact labyrinth seal provides extra sealing protection.

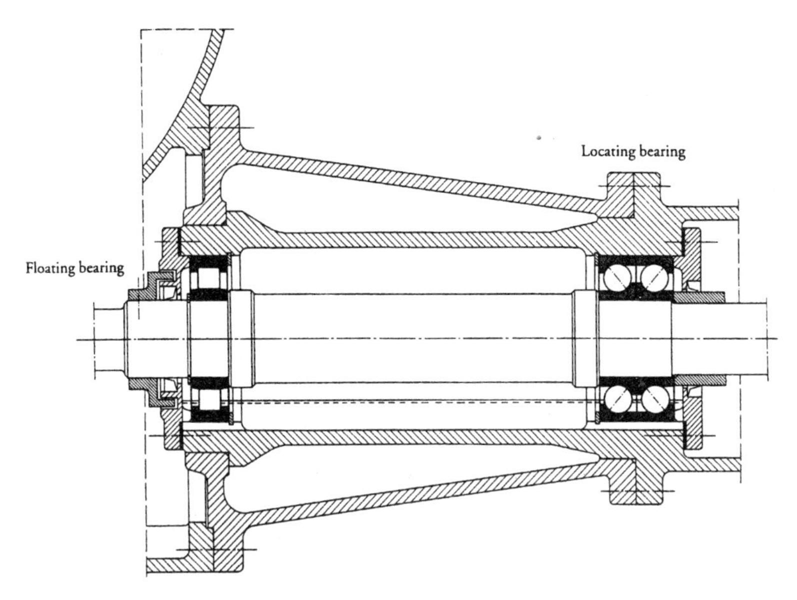

FIG. 13-10i Centrifugal pump (from FAG, 1998, with permission of FAG OEM and Handel AG).

13.12.10 Support Roller of a Rotary Kiln (Fig. 13-10j)

Design Data
Radial load: 1200 kN
Thrust load: 700 kN
Speed: 5 RPM

Tolerances
Shaft n6
Housing H7

Lubrication and Sealing: Grease lubrication with lithium soap base grease. At the roller side, the bearings are sealed with felt strips and grease packed labyrinths.

Design: The bearings are under very high load, and are exposed to a severe dusty environment. Lithium soap base grease is used for bearing lubrication and for sealing. These rollers support a large rotary kiln, which is used in cement manufacturing. Self-aligning spherical roller bearings are used. The two bearings are mounted in a floating arrange-

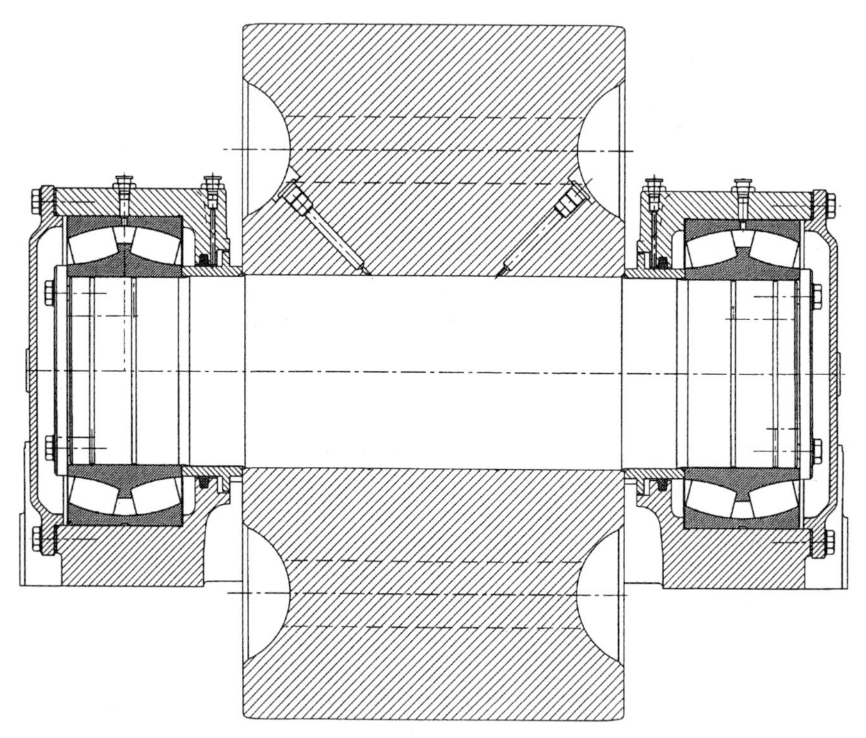

FIG. 13-10j Support roller of a rotary kiln (from FAG, 1998, with permission).

ment (to allow axial adjustment to the kiln). The bearings are mounted into split plummer block housings with a common base.

The grease is fed directly into the bearing through a grease valve and a hole in the outer rings. The grease valve restricts the grease flow and protects the bearing from overfilling. The bearings have double seal of felt strip and grease packed labyrinth. A second grease valve feeds grease directly into the labyrinth seal and prevents penetration of any contamination into the bearings.

The support roller shown in this figure has diameter of 1.6 m and width of 0.8 m. The speed is low, N = 5 RPM and the load on one bearing is high, Fr = 1200 kN. These rollers support the rotary kiln for cement production. The kiln dimensions are 150 m long and 4.4 m diameter. The supports are spaced at 30 m intervals.

13.12.11 Crane Pillar Mounting (Fig. 13-10k)

Design Data
Thrust load: 6200 kN
Radial force: 2800 kN
Speed: 1 RPM

Tolerances: Shaft j6, housing K7

Lubrication and Sealing: Oil bath lubrication with rollers fully immersed in oil.

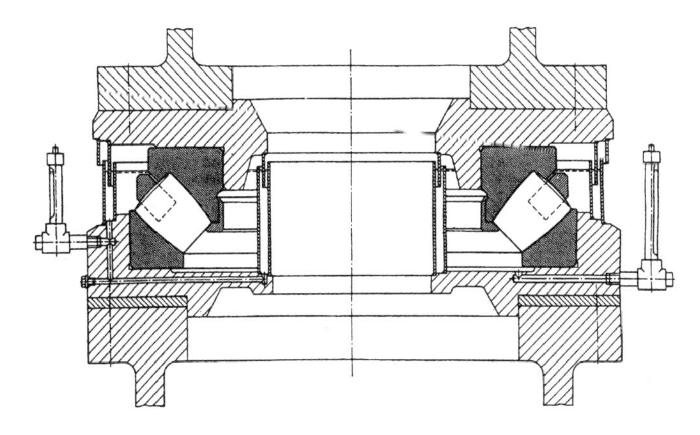

FIG. 13-10k Crane pillar mounting (from FAG, 1998, with permission).

Sealing: Noncontact labyrinth seal as shown in Fig 13-10k. The crane usually operates in severe dust environment. The labyrinth is full of oil, to prevent any penetration of dust from the environment into the bearing.

13.13 SELECTION OF OIL VERSUS GREASE

Greases and oils are widely used for lubrication of rolling-element bearings. In this section, the considerations for selection of oil versus grease are discussed. In addition to considerations directly related to bearing performance, selection depends on economic considerations as well as the ease of maintenance and effective sealing of the bearing.

Whenever possible, greases are preferred by engineers because they are easier to use and involve lower cost. For example, grease lubrication is widely selected for light-and-medium-duty industrial applications, in order to reduce the cost of maintenance. However, at high speeds, considerable amount of heat is generated in the bearings, and greases usually deteriorate at elevated temperatures. In addition, liquid oils improve the heat transfer from the bearing.

Empirical criterion that is widely used by engineers for the selection of oil versus grease is the DN value, which is the product of rolling bearing bore (equal to shaft diameter) in mm and shaft speed in RPM. Rolling bearings operating at DN value above 0.2 million usually require liquid oil, although there are special high-temperature greases that can operate above this limit. Below this limit, both greases and oils can be used. This is an approximate criterion, which considers only the bearing speed for medium loads. In fact, the load, friction coefficient, and heat sources outside the bearing also affect the bearing temperature.

In addition to the DN value, the product of speed and load is used to determine whether the bearing operates under light or heavy-duty conditions. This product is proportional to the bearing temperature rise (see estimation of the temperature rise in Sec. 13.3). The bearing operation temperature must be much lower than the temperature limit specified for the grease.

For low-speed rolling bearings, grease is the most widely used lubricant, because it has several advantages and the maintenance cost is lower. In comparison to oil, grease does not leak out easily through the seals. Prevention of leakage is essential in certain industries such as food, pharmaceuticals and textiles. Tight contact seals on the shafts are undesirable because they introduce additional friction and wear. The advantage of grease is that it can be used in bearing housings with noncontact labyrinth seals. The grease does not leak out, as oil would, and it seals the bearing from abrasive dust particles and a corrosive environment. Rolling bearings are sensitive to penetration of dust, which causes severe erosion, and the bearings must be properly sealed. Section 13.23 presents various types of contact and noncontact seals.

Contact seals are often referred to as *tight seals* or *rubbing seals*. They are tightly fitted on the shaft and are used mostly for oil lubrication. They introduce additional energy losses due to high friction between the seal and the rotating shaft, which raises the bearing temperature. Tightly fitted seals are also undesirable because they wear out and require frequent replacements; they should be avoided wherever possible. Moreover, the shaft wears out due to friction with the seal. In high-speed machines, expensive mechanical seals are often used to replace the regular contact seals. An important advantage of grease lubrication is that noncontact labyrinth seals of low friction and wear can be used effectively. In certain applications, unique designs of noncontact seals are used successfully for oil lubrication (see Sec. 13.12.11).

Grease is particularly effective where the shaft is not horizontal and oils leak easily through the seals. For grease, a relatively simple noncontact labyrinth seal with a small clearance is adequate in most applications.

A very thin layer of grease can be applied on the races to reduce the friction resistance. In such cases, the friction is lower than for oil sump lubrication. Another important advantage of grease is the low cost of maintenance in comparison to oil. Oil requires extra expense to refill and maintain oil levels. In addition, oil can be lost due to leakage, and expensive frequent inspections of oil levels must be conducted in order to prevent machine failure. In comparison, in grease lubrication, there is no need to maintain oil levels, and the addition of lubricant is less frequent. In most cases, grease lubrication results in a lower cost of maintenance.

Economic considerations favor grease lubrication. Oil lubrication systems involve higher initial cost and the long-term bearing maintenance is also more expensive. Therefore, oil is selected only where the selection can be justified based on performance. Oil has several important performance advantages over grease.

1. Unlike grease, oil flows through the bearing and assists in heat transfer from the bearing. This advantage is particularly important in applications of high speed and high temperature.
2. Continuous supply of oil is essential for the formation of an EHD fluid-film. This is very important in high-speed machinery, such as gas turbines.
3. Oil circulation through the bearing has an important function in removing wear debris.
4. Liquid oil is much easier to handle via pumps and tubes, in comparison to grease. In addition, oil is relatively simple to fill and drain; therefore it should be selected particularly when frequent replacements of lubricant are required.
5. In most applications, only a very thin lubrication layer is required. This can be obtained by introducing an accurate slow flow rate of lubricant

(measured in drops per minute) into the bearing. Flow dividers (described in Chapter 10) can be used for feeding at the desired flow rate to each bearing. A precise amount of lubricant at a steady flow rate can be supplied to the bearing and controlled only if the lubricant is oil; this is not feasible with grease.

6. Oil can provide lubrication to all the parts of a machine. An example is a gearbox, where the same oil lubricates the gears as well as the bearings.

As this discussion indicates, grease can be selected for light- and medium-duty applications, whereas oil should be selected for heavy-duty applications in which sufficient flow rates of liquid oil are essential for removing the heat from the bearing and for the formation of a fluid film.

13.14 GREASE LUBRICATION

The compositions and properties of various greases are discussed in Chapter 3. Greases are suspensions of mineral or synthetic oil in soaps, such as sodium, calcium, aluminum, lithium, and barium soaps, as well as other thickeners, such as silica and treated clays. The thickener acts as a sponge that contains and slowly releases small quantities of oil. When the rolling elements roll over the grease, the thickener structure breaks down gradually. Minute quantities of oil release and form a thin lubrication layer on the races and rolling-element surfaces. The lubrication layer is very thin and cannot generate a proper elastohydrodynamic film for separation of the rolling contacts, but it is effective in reducing friction and wear. In addition, the oil layer is too thin to play a role in cooling the bearing or in removing wear debris.

13.14.1 Design of Bearing Housings for Grease Lubrication

The design of the housing and grease supply depends mostly on the temperature, bearing size, load, and speed as well as the environment. The following is a survey of the most common designs.

13.14.1.1 Bearings Packed and Sealed for Life

If the bearing operating temperature is low and its speed and load are not high, the life of the grease can equal or exceed the bearing life. In such cases, using a bearing packed with grease and sealed for life would reduce significantly the maintenance cost. Sealed-for-life small bearings are commonly used under light-duty conditions. Sealed-for-life bearings are also used for occasional operation (not for 24 hours a day), such as in cars, domestic appliances, and pumps for occasional use. Examples are small electric motors for domestic appliances, bearings supporting the drum of a washing machine, and many bearings in passenger cars, such as water pump bearings.

The grease life is sensitive to a temperature rise, and sealed-for-life bearings are not used in machines having a heat source that can raise the bearing temperature. In some applications that involve a moderate temperature rise, such as small electric motors, sealed-for-life bearings with high-temperature grease are used successfully. We have to keep in mind that the life of sealed ball bearings is limited to the lower of bearing life and lubricant life. A method for estimating grease life is presented in Sec. 13.15.

Fig. 13-11 presents an example of the front wheel of a front-wheel-drive car. A double-row angular contact ball bearing is used. Certain cars use angular contact ball bearings or tapered roller bearings that are adjusted. The bearing is packed with grease and sealed on both sides for the life of the bearing.

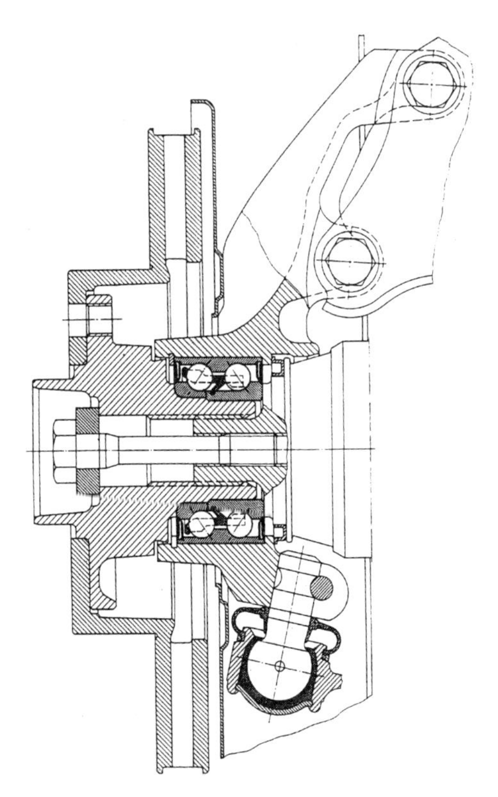

FIG. 13-11 Sealed-for-life bearing in the front wheel of a front-wheel-drive car (from FAG, 1988, with permission).

13.14.1.2 Housing Without Feeding Fittings

For industrial machines that operate for many hours, if the bearings operate at low temperature under light-to-medium duty, the original grease in the housing can last one to two years or even longer. In such cases, the grease can be replaced by new grease only during machine overhauls and the housing is not provided with grease-feeding fittings.

Housing without feeding fittings are used only if there is no heat source from any process outside the bearing and if the bearing operating temperature due to friction is low. This can be applied to light- and medium-duty bearings, namely, where the load and speed are not very high.

The advantage of elimination of any grease fittings is that it prevents overfilling of grease in the housing. The old grease is replaced by new grease only during overhauls, and this can be done manually without using grease guns. Only one-third to one-half the volume of the housing is filled with grease for regular applications. However, to minimize friction in small machines, only a very thin layer of grease is applied on the bearing surfaces, particularly if the drive motor is small and has low power.

Overfilling of grease in the bearing housing results in a high resistance to the motion of the rolling elements and grease overheating, as well as early breakdown of the grease (the grease is overworked). Therefore, the use of high-pressure guns for feeding grease into the housing of rolling bearings is undesirable, particularly for large bearings, because it packs too much grease into the housing and causes bearing overheating. Moreover, feeding under high pressure always results in grease loss.

During the assembly and periodic relubrication, it is very important to keep the bearing and lubricant completely clean from dust or even from old grease. Although less than half the volume of the housing is filled for regular applications, if the bearing is exposed to a severe environment of dust or moisture, the bearing should be fully packed to seal the bearing and prevent its contamination. Grease-feeding fittings are provided for frequent topping-up of grease. In many cases, additional grease fittings feed grease directly to the labyrinth seals (see Sec. 13.12). Fully packed bearings are used only for low- and medium-speed bearings, where the extra friction power loss is not significant.

13.14.1.3 Housings with Feeding Fittings

The common bearing design includes fittings for grease topping-up (adding grease between replacements by grease gun). Although it is desirable not to overfill the housing with grease, this is difficult to avoid. Low-cost maintenance is an important consideration, and in most cases the new grease replaces the old by pushing it out with grease guns. Experience indicates that small, light-duty bearings can operate successfully even when overfilled with grease. Overfilling initially generates extra resistance, but the extra grease is lost over time through the labyrinth seals. The housing is often designed with an outlet hole at the lower side of the housing and noncontact labyrinth seals. The temperature of the overfilled grease rises; after running a few hours (depending on bearing size and grease consistency), the surplus grease escapes through the hole and labyrinth seals. Low-consistency grease is used for this purpose.

If the bearing is exposed to an environment of dust, overfilling prevents contamination of the bearing. Frequent topping-up of grease ensures overfilling, particularly near the seals. The grease fittings must be completely clean before adding grease.

For small and medium-size bearings, it is possible to avoid overfilling and at the same time simplify the grease replacement. This is done via a simple housing design that allows one to force the old grease out completely with the new grease. The design includes a large-diameter drain outlet with a plug, in the side opposite the inlet grease fitting and at the lower side of the housing. This way the grease must pass through the bearing. In order to avoid overfilling, the replacement procedure is as follows: The outlet plug is removed; the shaft is rotating while the new grease is pumped into the housing. The old grease is worked out so that it is easier to replace. The new grease is pumped until it starts to come out of the drain. The shaft rotates for about half an hour to allow the surplus grease to drain out before locking the outlet.

This method is not applicable to large bearings because the pressure of the grease gun is not sufficient to remove all the old grease through the outlet. Also, the bearing might be overfilled, resulting in overheating during operation. Therefore, in large bearings, the grease is replaced manually during overhauls. In addition, large bearings require topping-up of grease at certain intervals, determined according to the temperature and operating conditions. It is important to avoid overfilling during relubrication. The addition of grease is done with grease guns, and it is important to design the housing and fittings to prevent overfilling. These designs involve higher cost and can be justified only for larger bearings.

13.14.2 Design Examples of Bearing Housings

It is important to ensure by appropriate design that during topping-up of grease, the new grease (fed by grease guns) will pass as much as possible through the bearings. The grease is supplied as close as possible to the bearings and discharged through the bearing into the space on the opposite side. In this way, the new grease must pass through the bearing, and the new grease will replace the old grease as much as possible.

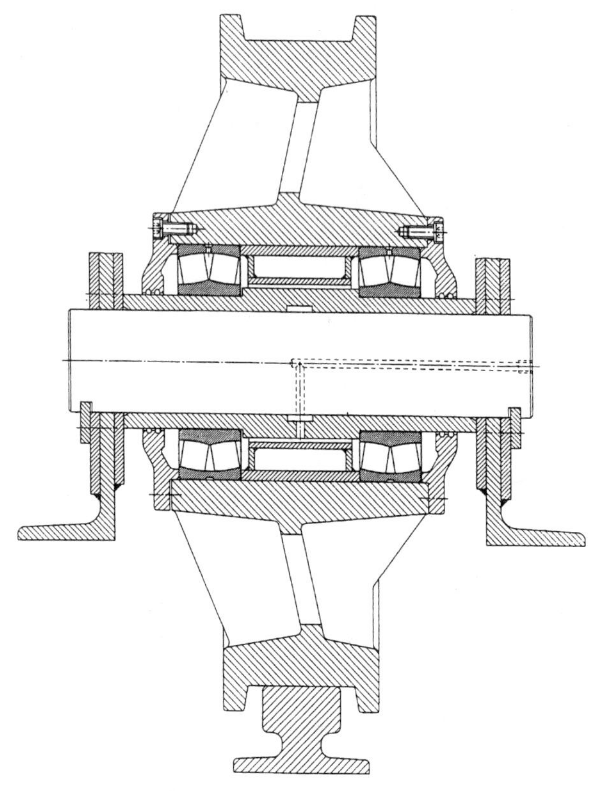

FIG. 13-12 Grease lubrication of crane wheel bearings (from FAG, 1986, with permission of FAG and Handel AG).

13.14.2.1 Crane Wheel Bearing Lubrication

An example of grease lubrication in a crane wheel is shown in Fig. 13-12. The crane wheel runs on a rail. The grease is fed through holes in the stationary shaft between two self-aligning spherical roller bearings. The design limits the grease volume between the two bearings. The grease passes through the two bearings, and the surplus grease is discharged through a double labyrinth seal clearance. Lithium soap base grease is used. The time period between grease replacements is approximately one year.

13.14.2.2 Grease-Quantity Regulators

An example of large bearing housing that is designed for avoiding overfilling during relubrication by grease guns is shown in Fig. 13-13. This design is widely used for large electric motors (SKF, 1992). The grease is fed at the bottom of the housing, near the left side of the outer ring. The design of the housing includes

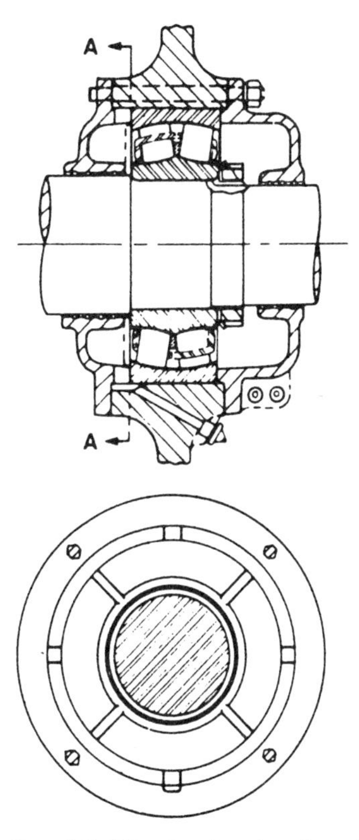

FIG. 13-13 Bearing housing for a large electric motor (from SKF, 1992, with permission).

radial ribs inside the left cover of the housing. They direct the new grease into the bearing without overfilling the space on the left side of the bearing. The old grease escapes through the bearing into the large space at the right side of the bearing. The ribs also keep the grease in place and prevent it from being worked by the rotating shaft during regular operation. In this way, the ribs prevent overheating. In this design, the cover is split to simplify the removal of the old grease during overhauls.

13.14.2.3 Grease Chamber

Another method that prevents overfilling of a bearing is shown in Fig. 13-14. It uses a double-sealed, prelubricated bearing. The concept is that only one side of the bearing housing is full of grease (fed by a grease gun). The advantage of this method is that only a small quantity of grease is gradually released from the

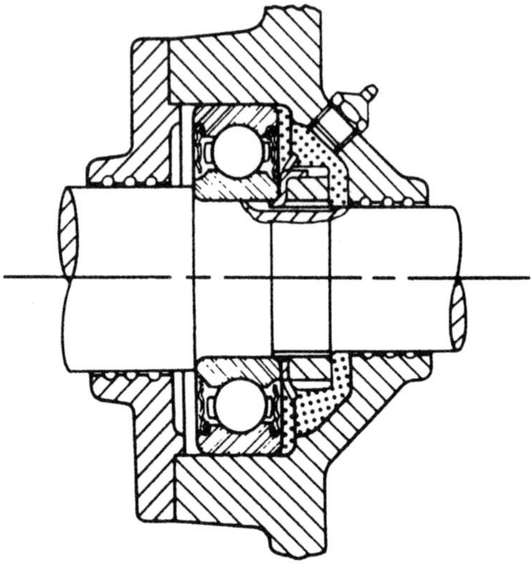

FIG. 13-14 Grease chamber for double-sealed bearings (from SKF, 1992, with permission).

grease packing and penetrates into the bearing. This design of a double-sealed bearing combined with noncontact labyrinth seals protects the bearing from dust.

13.14.2.4 Dust Environment

Small bearings in a dusty environment are fully packed with grease. However, for large bearings, it is important to prevent overfilling with grease, which results in overheating and early failure of the bearing.

An example of a double-shaft hammer mill for crushing large material (FAG, 1986) exposed to a severe dust environment is shown in Fig. 13-15. This example combines a design for a grease-quantity regulating disk that prevents overfilling and a separate arrangement for packing the grease between the labyrinth and felt seals.

13.14.2.5 Regulating Disk

The bearing housing design consists of a regulating disk that rotates together with the shaft. It is mounted at the side opposite the grease inlet side. If the grease quantity in the bearing cavity is too high, the rotating disc shears and softens part of the grease. By centrifugal action, the grease drains through the radial clearance into the volume between the disk and seals, as shown in Fig. 13-15.

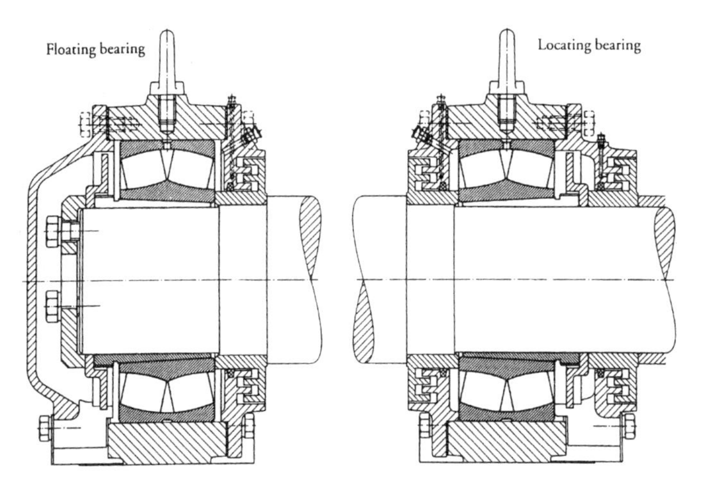

FIG. 13-15 Bearing housing of a double-shaft hammer mill (from FAG, 1986, with permission of FAG and Handel AG).

13.14.2.6 Sealing in a Severe Dust Environment

In Fig. 13-15, grease is fed directly into the bearing through a grease valve and a hole in the outer rings. The bearings have a double seal of felt strip and grease-packed labyrinth. A second grease valve feeds grease between the labyrinth and the felt seal. Frequent relubrication of the grease in the labyrinth seal prevents penetration of abrasive dust particles into the bearing.

13.15 GREASE LIFE

The life of greases and oils is limited due to oxidation. High temperature accelerates the oxidation rate, and the life of greases and oils is very sensitive to a temperature rise. An approximate rule is that grease life is divided by 2 for every 15°C (27°F) temperature rise above 70°C (160°F). In addition to oxidation, the bleeding of the oil from the grease and its evaporation limit the life of the grease at high temperature.

At low operating temperature, the life of the grease is long, and bearings packed with grease and sealed for life are widely used. Adding fresh grease to sealed-for-life bearings is not necessary, because the life of the grease is longer than the bearing life.

If the life of the grease is shorter than the life of the bearing, the grease should be replaced. Since it is difficult to precisely predict the grease life, fresh grease should be added much before the grease loses its effectiveness. The time period between lubrications (also referred to as relubrication intervals) is a function of many operating parameters, such as temperature, grease type, bearing type and size, speed, and grease contamination. The time period, Δt, between grease replacements is determined empirically. It is based on the requirement that less than 1% of the bearings not be effectively lubricated by the end of the period.

In Fig. 13-16, curves are presented of the recommended time period Δt (in hours) as a function of bearing speed N (RPM) and bearing bore diameter d (SKF, 1992). The charts are based on experiments with lithium-based greases at temperatures below 70°C (160°F). For higher temperatures, the time period Δt is divided by two for every 15°C (27°F) of temperature rise above 70°C (160°F). However, the temperature should never exceed the maximum temperature allowed for the grease. In the same way, the time period Δt can be longer at temperatures lower than 70°C (160°F), but Δt should not be more than double that obtained from the charts in Fig. 13-16. Also, one should keep in mind that at very low temperatures, the grease releases less oil.

The time period Δt between grease replacements is a function of the bearing speed N (RPM), and bearing bore diameter d (mm), and bearing type. According to the bearing type, the time period Δt is determined by one of the following scales.

Scale a: is for radial ball bearings.
Scale b: is for cylindrical and needle roller bearings.
Scale c: is for spherical roller bearings, tapered roller bearings, and thrust ball bearings.

Figure 13-16 is valid only for bearings on horizontal shafts. For vertical shafts, only half of Δt from in Fig. 13-16 is applied. The maximum time period between grease replacements, Δt should not exceed 30,000 hours. Bearings subjected to severe operating conditions, such as elevated temperature, high speed, contamination, or humidity, must have more frequent grease replacements. Under severe conditions, the best way to determine the time period between grease replacements is by periodic inspections of the grease.

The following cases require shorter periods between lubrications:

1. Full-complement cylindrical rolling bearing, 0.2 Δt (in scale c)
2. Cylindrical rolling bearing with a cage, 0.3 Δt (in scale c)
3. Cylindrical roller thrust bearing, needle roller thrust bearing, spherical roller thrust bearing. 0.5 Δt (in scale c)

Experience has indicated that large bearings, of bore diameter over $d = 300$ mm, need more frequent grease replacements than indicated in Fig. 13-16 (the large bearings are marked by dotted lines). Frequent grease replacements are required if there are high contact stresses, high speed and high temperature. Whenever the time period between grease replacements is short, a continuous grease supply can be provided via a grease pump and a grease valve. For a continuous grease supply, the grease mass per unit of time, G, fed into the

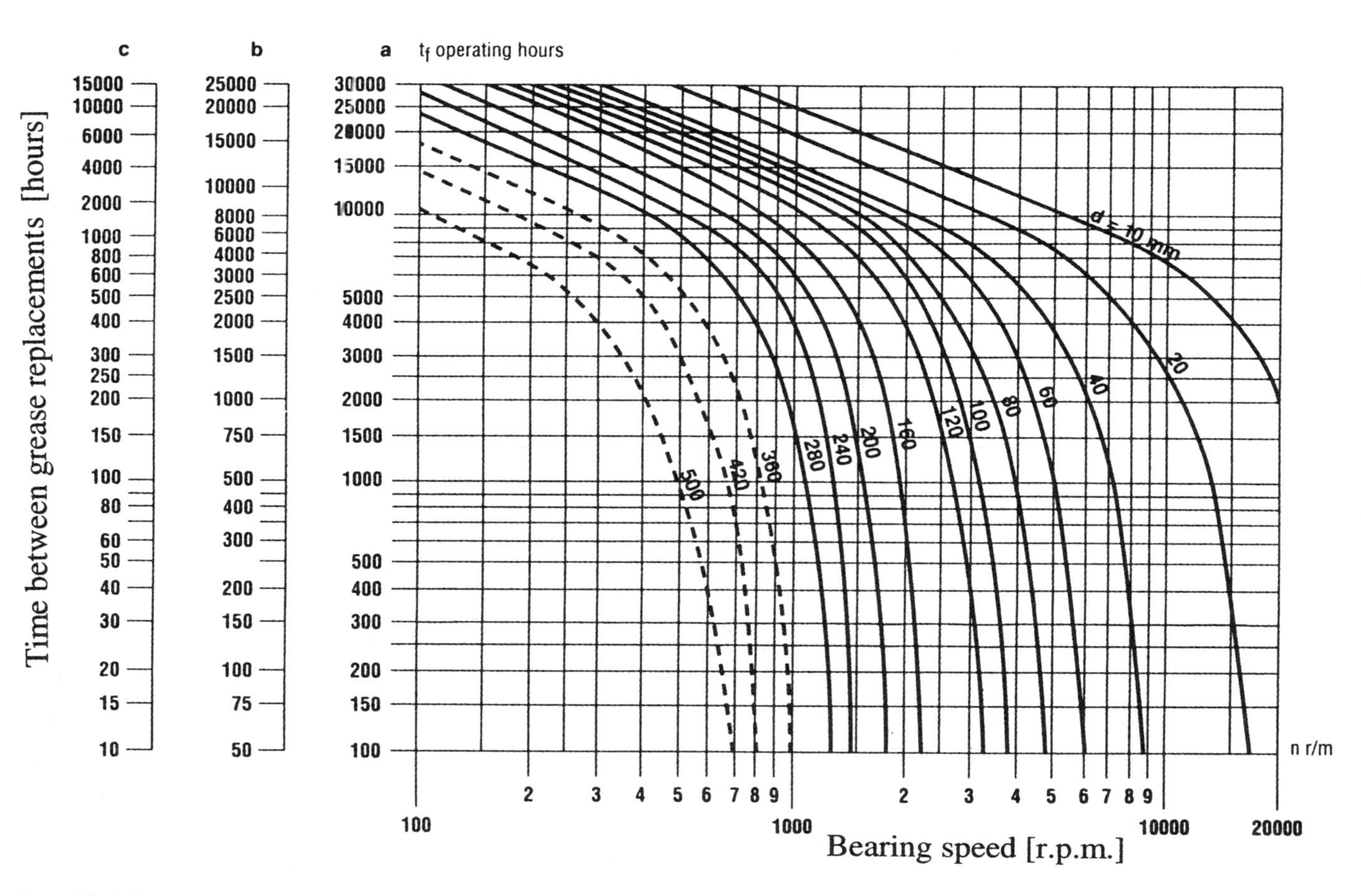

FIG. 13-16 Time between grease replacements, Δt (in hours) (from SKF, 1992, with permission from SKF, USA).

large bearing is determined by an empirical equation (SKF, 1992). The following empirical equation is for regular conditions, without any conduction of external heat into the bearing (the bearing temperature is only due to friction losses):

$$G = (0.3 - 0.5)DL \times 10^{-4} \quad (13\text{-}29)$$

Here,

G = continuous mass flow rate supply of grease *(g/h)*
D = bearing OD (mm)
L = bearing width (mm) [for thrust bearings use total height, H]

13.15.1 Topping-Up Intervals

In applications where the grease life is considerably shorter than the bearing life, either complete replacements (relubrication) or more frequent applications of topping-up grease (by grease guns) are required. Topping-up grease is much faster and it is preferred whenever possible. In most cases, during topping-up, the fresh grease replaces only part of the used grease, and more frequent applications are needed in comparison to complete grease replacements. The initial filling and subsequent topping-up and complete replacement of grease (after cleaning at main overhauls) is done as follows (SKF, 1992):

1. If the period between grease replacements, Δt (in hours) is less than 6 months of machine operation, the grease is topped-up at half the recommended Δt from Fig. 13.6. After three periods of topping-up, all grease is replaced by fresh grease.
2. If the period between grease replacements, Δt (in hours) is equivalent to more than 6 months of machine operation, topping-up should be avoided, and all the grease in the housing is replaced with fresh grease after each period.

13.15.2 Topping-Up Quantity

In the topping-up procedure, the grease in the bearing housing is only partially replaced by adding a small quantity of fresh grease after each period. The recommended grease quantity to be added can be obtained from the following empirical equation (SKF, 1992):

$$G_p(g) = 0.005D(\text{mm}) \times L(\text{mm}) \quad (13\text{-}30)$$

Here,

G_p = grease mass quantity to be added (grams)
D = bearing OD (mm)
L = total bearing width (mm) [for thrust bearings use total height, H]

13.16 LIQUID LUBRICATION SYSTEMS

Oil lubrication can be provided by several methods. For low and moderate speeds, an oil bath, also called an *oil sump*, is used. For low speeds, the oil level in an oil bath is the center of the lower rolling element. For heavy-duty large bearings cooling is necessary, and the oil is circulated in the oil bath. If the oil level is the center of the lower rolling element, it is referred to as a *wet sump*; if all the oil is drained, it is referred to as a *dry sump*. The level is determined by the height of the outlet. A pump feeds the oil through flow dividers to the bearing housing. The oil can be supplied also by gravitation. The major advantage of circulation lubrication is that it can cool the bearings. Circulation lubrication of many bearings is relatively inexpensive.

An additional method is mist lubrication. In this method, the oil is not recovered. The most important advantage is that the lubrication layer is very thin. It results in low viscous resistance to the motion of the rolling elements. For example, mist lubrication is used for machine tool spindles.

Several examples of the various methods of oil lubrications follow.

13.16.1 Bearing Housing with Oil Sump

Oil lubrication requires a special design of the bearing housing, often referred to as a *pillow block*. Various standard designs of pillow blocks are available from bearing manufacturers. It is possible to select a design based on the optimal oil level and rate of flow of lubricant that is appropriate for each application. For large bearings, a welded housing is less expensive than a cast housing.

An example is the housing of the propeller-ship shaft bearing shown in Fig. 13-17. In this example, the speed is 105 RPM and the shaft diameter is 560 mm. Contact seals protect the bearing from the corrosive seawater. The oil can be fed by circulation lubrication, and the pressure in the housing is kept above ambient pressure to prevent penetration of seawater.

In this arrangement, the fluid level is relatively high, and it can be applied only when the bearing speed is low. In order to minimize the viscous resistance at high speed, the oil level must be lower. For low speeds, the oil level should not be above the center of the lowest rolling element; but this level is too high for high-speed bearings. A drain is always provided for oil replacement.

The oil level is preferably checked when the machine is at rest, when all the oil is drained into the reservoir. There are always oil losses, and a sight-glass gauge is usually provided for checking oil level; oil is added as soon as the oil level is low. This method requires much individual attention to each bearing, and it can be expensive in manufacturing industries where a large number of bearings are maintained.

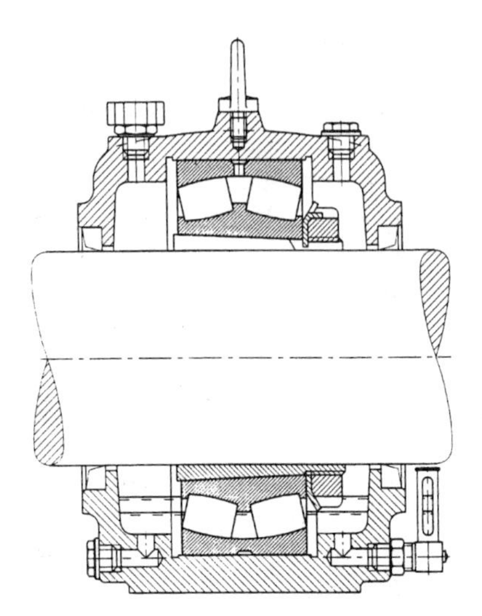

FIG. 13-17 Bearing housing of a ship shaft (from FAG, 1998, with permission of FAG and Handel AG).

13.16.2 Lubrication with Wick Arrangement

A better design for feeding a very low flow rate of oil into the bearing is the wick feed arrangement. A design for a vertical shaft is shown in Fig. 13-18. The wick siphons oil from a reservoir into the bearing. An important advantage is that the wick acts like a filter and supplies only clean oil to the bearing (solid particles are not siphoned). Viscous friction is minimized by this arrangement. The wick continues to deliver oil even when the machine is not operating.

An improved design where oil is fed only during bearing operation is shown in Fig. 13-19. A wick provides lubricant by capillary attraction to a rotating bearing. The bearing is above the fluid level, and the wick must be in contact with the collar for proper function of this arrangement. The oil is thrown off by centrifugal force, and the oil is continually siphoned. This system delivers oil only when required, i.e., when the bearing is rotating. The oil is drained back into the reservoir without losses.

Wick feed has an important advantage where the bearing operates at high speeds, because it can supply a continuous low flow rate of filtered oil to the bearing. With this wick feed system, there is no resistance to the motion of the rolling elements through the oil reservoir. For effective operation, the wicks should be properly maintained; they have to be replaced occasionally. During servicing, the wick should be dried and thoroughly saturated with oil before reinstallation. This prevents absorption of moisture, which would impair the oil-siphon action.

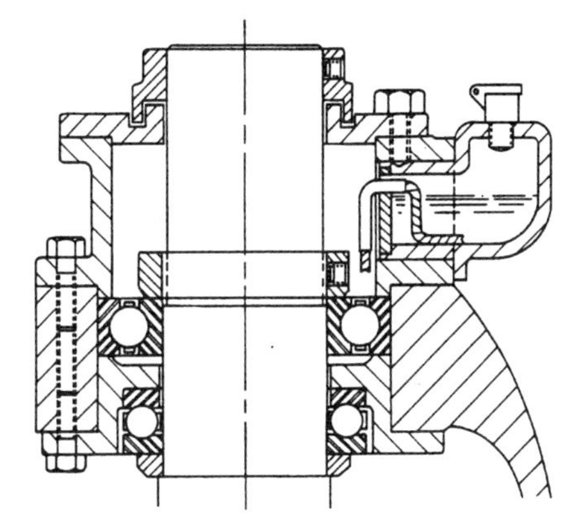

FIG. 13-18 Bearing housing with a wick for oil feeding (from SKF, 1992, with permission).

13.16.3 Oil Circulating Systems

There are several benefits in using oil circulation systems for rolling bearings, where a monitoring pump supplies a low flow rate of oil to each bearing. In certain applications, particularly in hot environments, the oil circulation plays an important role in assisting to transfer heat from the bearing. In addition, a circulating system simplifies maintenance, particularly for large industrial machines with many bearings. For oil circulation, a special design of the housing is used for controlling the oil level.

An example of a bearing housing for oil circulation is shown in Fig. 13-20. The level of the oil in the housing is controlled by the height of the outlet. For a

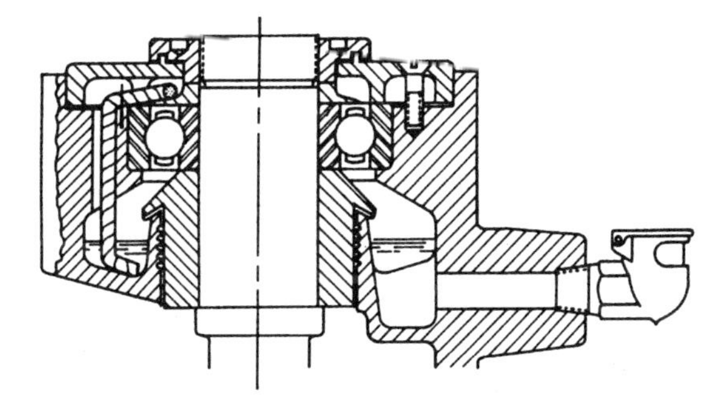

FIG. 13-19 Bearing housing with a wick and centrifugal oil feeding (from SKF, 1992, with permission).

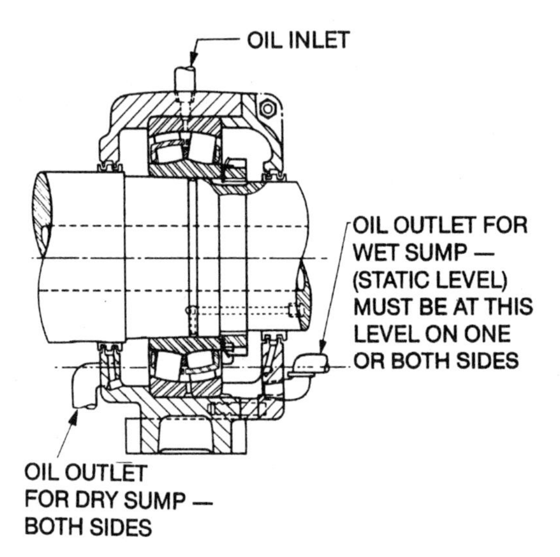

FIG. 13-20 Oil circulation for pulp and paper dryers (from SKF, 1992, with permission from SKF, USA).

wet sump, the oil level at a standstill should not be higher than the center of the lowest ball or roller. A sight-glass gauge is usually provided for easy monitoring.

As mentioned earlier, high-speed bearings require a dry sump, where the oil drains completely after passing through the bearing. In addition, a dry sump is used for bearings operating at high temperature because the lubricant must not be exposed for long to the high temperature (to minimize oxidation). For a dry sump, two outlets are located at the lowest points on both sides of the housing, as shown on the left side of Fig. 13-20.

For applications where bearing failure must be avoided at any cost, oil circulation systems require an automatic monitoring to indicate when oil flow is blocked through any bearing. Safety measures include electrical interlocking of the oil pump motor with the motor that is driving the machine.

13.16.4 Oil Mist Systems

This arrangement entails lubrication by a mixture of air and atomized oil. An atomizer device forms the oil mist. In order to have the required quantity of oil and appropriate viscosity at the bearings' rolling contacts, oil mist system manufacturers provide recommendations for system designs, capacities, and operating temperatures and pressures.

The bearing operating temperature is reduced by this method of lubrication, by means of air cooling. A thin oil layer is formed on the bearing surfaces due to

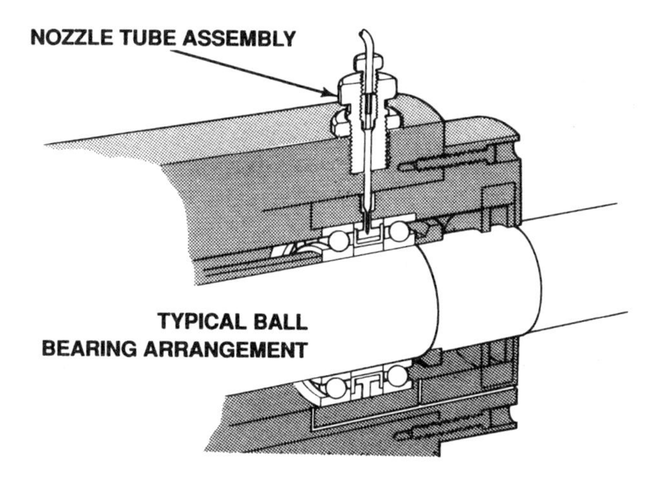

FIG. 13-21 Nozzle assembly of oil mist system. (Reprinted with permission from Lubriquip Inc.)

the air flow, which prevents accumulation of excess oil. The air is supplied under pressure, and it prevents moisture from the environment from penetrating into the bearing. An additional advantage is that oil mist lubrication supplies clean, fresh oil into the bearings (the oil is not recycled). These advantages increase the life expectancy of the bearing. Although the oil in the mist is lost after passing through the bearing, very little lubricant is used, so oil consumption is relatively low. The connection of the nozzle assembly in the bearing housing is shown in Fig. 13-21.

In Fig. 13-22, a mist lubrication system is shown that is widely used for grinding spindles. The air, charged with a mist of oil, is introduced in the housing

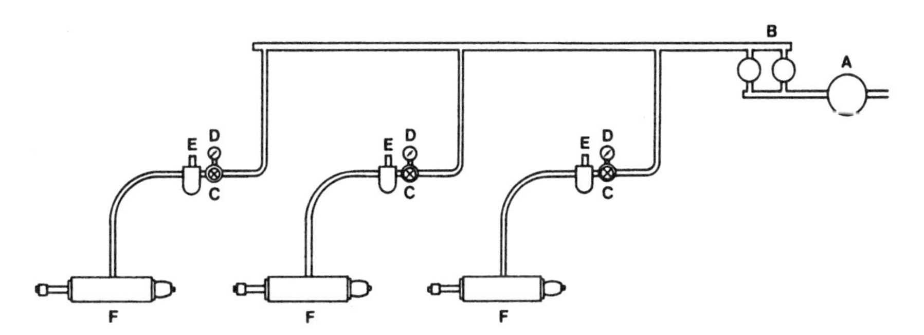

FIG. 13-22 Oil mist system for machine tool spindles (from SKF, 1992, with permission of SKF).

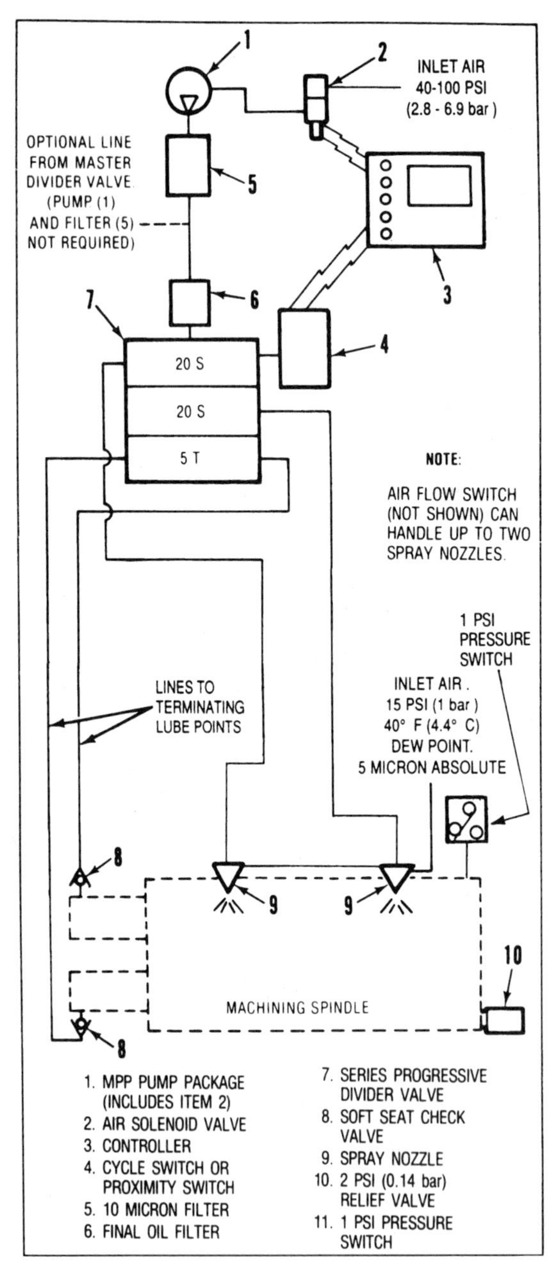

FIG. 13-23 Control of advanced oil mist system with flow dividers (reprinted with permission of Lubriquip).

between the bearings in order to ensure that the air passes through the bearings before escaping from the housing. Air from the supply line passes through a filter, B, then through a pressure reduction valve, D, and then through an atomizer, E, where the oil mist is generated. The air must be sufficiently dry before it is filtered, and a dehumidifier, A, is often used.

Advanced oil mist systems with precise control of the flow rate are often used in machining spindles. The systems include a series of flow dividers and an electronic controller. A schematic layout of a controlled system is shown in Fig. 13-23.

13.16.5 Lubrication of High-Speed Bearings

In bearings operating at very high speeds (high DN value) a considerable amount of heat is generated, and jet lubrication proved to be effective in transferring the heat away from the bearing, see a survey by Zaretzky (1997). Jet lubrication is used for high-speed bearings aircraft engines. Several nozzles are placed around the bearing, and the jet is directed to the rolling elements near the contact with the inner race. The centrifugal forces move the oil through the bearing for cooling and lubrication. Experiments have shown that in small bearings jet lubrication can be used successfully at very high speeds of 3 million DN, and speeds to 2.5 million DN for larger bearing of 120 mm bore diameter.

A more effective method of lubrication for very high-speed bearings is by means of under-race lubrication, see Zaretzky (1997). The lubricant is fed through several holes in the inner race. In addition, the lubricant is used for cooling in clearances (annular passages) between the inner and outer rings and their seats.

13.16.6 Oil Replacement in Circulation Systems

The time period between oil replacements depends on the operating conditions, particularly oil temperature, and the amount of contamination that is penetrating into the oil as well as the quantity of oil in circulation. In most cases, the reason for frequent oil replacements is the oxidation of the oil due to elevated temperatures or the penetration of dust particles into the oil.

If the bearing temperature is below 50°C (120°F) and the bearing is properly sealed from any significant contamination, the life of the oil is long and intervals of one year are adequate. At elevated temperatures, however the oil life is much shorter. For similar operating conditions, if the oil temperature is doubled and reaches 100°C (220°F), the oil life is reduced to only 3 months (a quarter of the time for 50°C (120°F).

In central lubrication systems, the oil is fed from an oil sump through a filter and than passes through the bearing and returns to the oil sump. In order to reduce the oil temperature, the system can include a cooler. There are many variable operating conditions that determine the oil temperature, including the rate of flow of the circulation and the presence of a cooling system, which reduces the oil temperature. Since there are many operating parameters, it is difficult to set rigid rules for the lubrication intervals. It is recommended to test the oil frequently for determining the optimum time period for oil replacement. The tests include measurement of the oxidation level of the oil, the amount of antioxidation additives left in the oil, and the level of contamination by dust particles.

13.17 HIGH-TEMPERATURE APPLICATIONS

In cases where heat is transferred into the bearings from outside sources, cooling of the oil in circulation is necessary to avoid excessive bearing temperatures and premature oxidation of the lubricant. Examples are combustion processes (such as car engines) and steam dryers. In addition, high temperatures reduce the viscosity and effectiveness of the oil. Various methods for controlling the oil temperature are used. In Fig. 13-24, a cooling disc is shown that is mounted on the shaft between the bearing and the heat source. The disc increases the convection area of heat transferred from the shaft (SKF, 1992).

An improved cooling system is shown in Fig. 13-25. It is a design of a pillow block with water-cooling coils. Water-cooled copper coils transfer the heat away from the oil reservoir in the pillow block. It is important to shut off the cooling water whenever the machine is stationary in order to prevent condensation, which generates rust.

Air is also used for cooling bearings. A direct stream of fresh air is usually created through the use of fans, blowers, or air ducts around the bearing that can

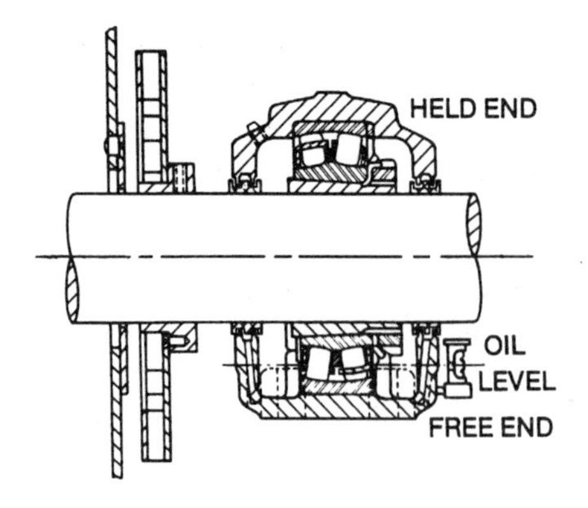

FIG. 13-24 Cooling disc mounted on the shaft (from SKF, 1992, with permission).

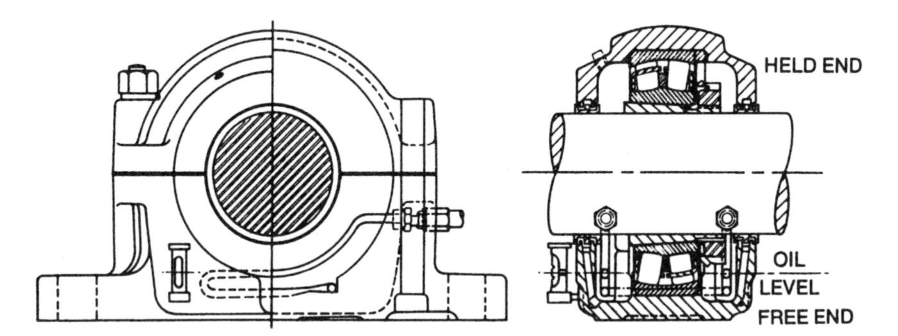

FIG. 13-25 Pillow block with water cooling (from SKF, 1992, with permission).

help in dissipating the heat. An additional method that is widely used is to cool the oil outside the bearing, via a heat exchanger. The circulating oil can pass through a radiator for cooling, such as in car engine oil circulation.

13.17.1 Moisture in Rolling Bearings

Lubricants do not completely protect the bearing against corrosion caused by moisture that penetrates into the bearing. In particular, the combined effect of acids (products of oxidation) and moisture are harmful to the bearing surfaces.

The design of bearing arrangement and lubrication systems must ensure that the bearing is sealed from moisture. Certain lubricants can reduce moisture effects, such as compound oils, which are more water repellent than regular mineral oils. Lithium-based greases are good water repellents and also provide an effective labyrinth seal. In all cases, the lubricant should completely cover the bearing surface to protect it. Nonoperating machines should be set in motion periodically in order to spread the lubricant over the complete bearing surfaces for corrosion protection.

13.18 SPEED LIMIT OF STANDARD BEARINGS

The standard bearing has a much lower speed limit than special steels. Bearing manufacturers recommend a speed limit to their standard bearings. The DN value is widely used for limiting the speed of various rolling bearings. This is defined as the product of bearing bore in mm and shaft speed in RPM.

The friction power loss in a rolling bearing is proportional to the rolling velocity, which is proportional to the bearing temperature rise above ambient temperature. The centrifugal force of the rolling elements is also a function of the DN value. Special steels have been developed for aircraft turbine engines that can operate at very high speeds of 2 million DN. There is continuous search for better

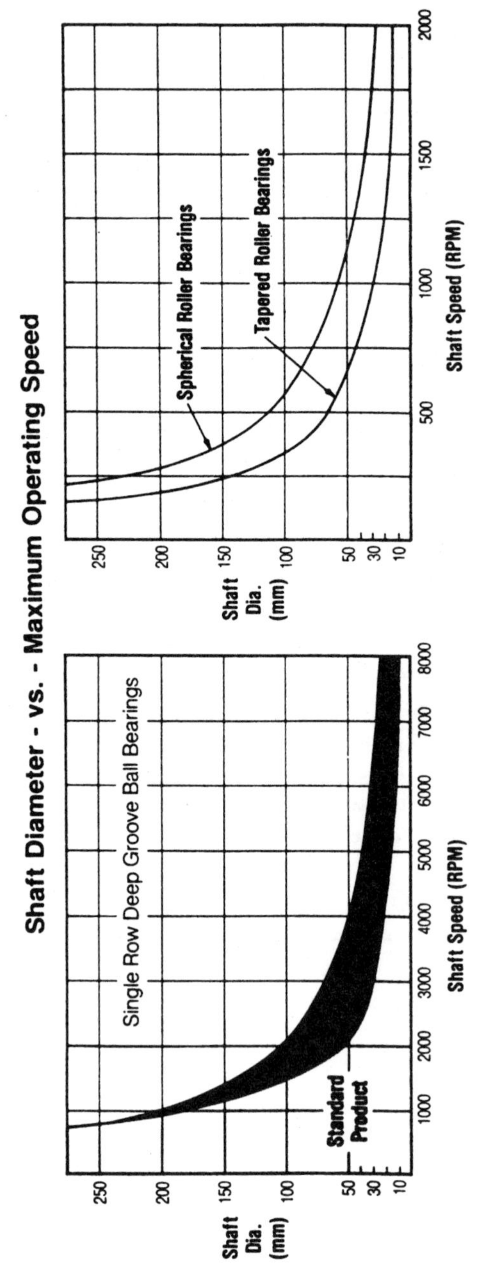

FIG. 13-26 Speed limit of standard bearings (from SKF, 1992, with permission).

materials, such as the introduction of silicon nitride rolling elements, and unique designs (see Chapter 18) to allow a breakthrough past the limit of 2 million DN.

However, for standard bearings, made of SAE 52100 steel, the maximum DN value is quite low, of the order of magnitude of 0.1 million. The reason for limiting the DN value of industrial bearings is in order to limit the temperature rise and, thus, to extend the fatigue life of the bearings.

Bearing manufacturers recommend low limits of the DN values. The speed limits for various bearing types can be obtained from Fig. 13-26. These limits are based on a temperature limit of 82°C (180°F) as measured on the outside bearing diameter. Standard steel at higher temperature starts to lose its hardness and fatigue resistance at that temperature. Standard bearing steel, SAE 52100, can operate at higher temperatures, up to 177°C (350°F). However, the bearing life (as well as lubricant life) is lower.

Figure 13-26 shows that the speed limit of standard bearings is quite low. In Sec. 13.19, special steels are discussed that are used for much higher speeds.

13.19 MATERIALS FOR ROLLING BEARINGS

In the United States, the standard steel for ball bearings is SAE 52100 (0.98% C, 1.3% Cr, 0.25% Mn, 0.15% Si). It is widely used for the rings and rolling elements of standard ball bearings as well as certain roller bearings. SAE 52100 is of the through-hardening type of steel. This steel can be hardened thoroughly to Rockwell C 65. In general, steels with carbon content above 0.8%, combined with less than 5% of other alloys, are of the *hypereutectoid* type, where the cross section of the rings can be hardened thoroughly.

However, large bearings with a large cross-sectional area of the rings are made of case-hardening (carburizing) steels. An example of a widely used case-hardening steel is SAE 4118 (0.18% C, 0.4% Cr, 0.4% Mn, 0.15% Si, 0.08% Mo). Case-hardening steels contain less than 0.8% carbon and are of the *hypoeutectoid* type. This means that they must be diffused with additional carbon in order to be hardened by heat treatment. The advantage of a case-hardening steel is that it is less brittle, because only the surface is hardened while the inside cross section remains relatively soft. In comparison, the through-hardening steels have high hardness over the complete cross section.

Rolling bearings made of these two types of steel can be used only at low temperature (below 350°F or 177°C). Above this temperature, these steels lose their hardness. For applications at higher temperatures, high-alloyed steels have been developed that maintain the required hardness at high temperature. Examples of special steels that provide better fatigue resistance at high temperatures appear in Sec. 13.19.2.

13.19.1 Stainless Steel AISI 440C

AISI 440C (1.1% C, 17% Cr, 0.75% Mn, 1% Si, 0.75% Mo) is a high-carbon stainless steel for rolling bearings. AISI 440C does not contain nickel and can be heat-treated and hardened to Rockwell C 60. In the United States, it became a standard stainless steel bearing material that is widely used in corrosive environments, particularly in instruments. A major disadvantage of this steel, in comparison to SAE 52100, is its shorter fatigue life. Therefore, for heavy loads, it is used only where there is no other way to protect the bearing from corrosion. AISI 440C is widely used in instrument ball bearings that must be rust free and where corrosion resistance is much more important than the load capacity. For certain applications, it is possible to combine the characteristics of corrosion resistance and high fatigue resistance by using chrome-coated bearings made of the standard SAE 52100 steel.

13.19.2 Special Steels for Aerospace Applications

For most applications, the preceding two types of steels provide adequate performance. For aerospace applications, however, there is a requirement for fatigue resistance for high-speed bearings operating at elevated temperatures in turbine engines. Special high-alloy-content steels were developed as well as higher purity by using better manufacturing processes such as vacuum induction melting (VIM) and vacuum arc remelting (VAR). The piston engine bearings of early aircraft used tool steels such as M_1 and M_2. During the 1950s, the turbine engine aircraft has been developed, and there was a requirement for better rolling-element bearings that can resist the high speed and high temperature in the aircraft turbine engine. For this purpose, the vacuum melting process was developed and used with high-alloy-content steel AISI M-50 and much later, the recently introduced casehardened steel M50NiL. These bearings are also used for other applications of high speed and elevated temperatures.

An interesting survey by Zaretszky (1997b) shows the major breakthroughs, which resulted in bearing fatigue-life improvement of approximately 200 times, between 1948 and 1988. The most important developments are high purity steel processing, composition of special steels, ultrasonic inspection techniques, improvement of bearing design, and better lubrication. After World War II, the requirement for reliable operation of jet engines and helicopter rotors was the major drive for research and development, which resulted in impressive improvements in the performance of rolling-element bearings for aerospace applications.

13.19.2.1 M50 Bearing Material for Aerospace Applications

AISI M-50 (0.8% C, 4% Cr, 0.1%Ni, 0.25% Mn, 0.25% Si, 4.25% Mo) was developed in the 1950, and it is used for rolling bearings in aerospace applications. In addition, it has industrial applications for rolling bearings operating at elevated temperatures up to 315°C (600°F). AISI M-50 is through-hardening steel, because it has relatively high carbon content. This material demonstrated significant improvement in fatigue life, in comparison to the earlier steels. However, the high demand in aircraft engines, with fatigue combined with high temperature and high centrifugal forces, can result in the initiation of cracks and even complete fracture of rings made of through-hardening steels such as M-50. For that reason, the speed of aircraft engines has been limited to 2.4 million DN.

In order to break through this limit, a lot of research has been conducted to improve bearing materials. The recent development (during the 1980s) of high-alloyed casehardened steel M50NiL significantly improved the fatigue resistance of jet engine bearings.

13.19.2.2 M-50NiL Bearing Steel for Aerospace Applications

During the 1980s, M50NiL has been developed and introduced into high-speed aerospace applications. M50NiL is casehardened steel, which has a softer core, and it is less brittle than the through-hardened steel AISI M-50. In turn, M50NiL has improved fracture toughness, better fatigue resistance, better impact resistance in high-speed bearings (and gears), and can operate at high temperatures similar to AISI M-50. Therefore, M50NiL gradually replaces AISI M-50 as the material of choice for jet engine bearings in aircraft.

M50NiL (0.15% C, 4% Cr, 3.5% Ni, 0.15% Mn, 1% V, 4.0% Mo) differs from AISI M-50 by its lower carbon content. M50NiL requires carburizing for getting hard surfaces. The low carbon content makes it casehardened steel with softer and less brittle material inside the cross section. M50NiL has less carbon and more nickel and vanadium in comparison to AISI M-50. These alloys increase hot hardness and form hard carbides that reduce wear. M50NiL has uniformly distributed carbides, which is less likely to initiate fatigue cracks.

The most important advantage of M50NiL is that it is casehardened steel with optimum fatigue properties under rolling contact. In rolling contact fatigue tests, M50NiL demonstrated approximately twice the fatigue life, L_{10}, of standard AISI M-50 (Bamberger, 1983). The two materials were processed by the same VIM-VAR process, and tested under identical conditions of load and speed.

An important characteristic in aircraft engines is that M50NiL allows sufficient time for the detection of spalling damage in the bearing before any catastrophic failure, because the tough core minimizes undesired crack propagation. In addition, M50NiL can operate at higher speeds, is more wear resistant, has higher tensile stress, higher fracture toughness, and lower boundary lubrication friction than AISI M-50.

13.19.2.3 DD400

For instrument ball bearings, corrosion resistance is very important. A stainless steel DD400 has been developed for precision miniature rolling bearings and small instrument rolling bearings. Corrosion resistance, combined with adequate hardness, has been achieved by increasing the quantity of dissolved chromium in the material. However, corrosion-resistant stainless steels have reduced fatigue resistance, and they are applied only for light-duty bearings. The composition of DD400 is 0.7% C, 13% Cr and it is martensitic stainless steel. DD400 replaced AISI 440C (1% C, 17% Cr), which was used for similar applications. DD400 demonstrated better performance in comparison to AISI 440C in small bearings. The most important advantages are: better surface finish of the races and rolling elements, better damping of vibrations, and improved fatigue life. These advantages are explained by the absence of large carbides in the heat-treated material.

13.20 PROCESSES FOR MANUFACTURING HIGH-PURITY STEEL

In addition to the chemical composition, the manufacturing process is very important for improving fatigue resistance, particularly at high temperatures. For critical applications, such as aircraft engines, there is a requirement for fatigue-resistant materials with a high degree of purity. It was realized that there are significant amount of impurities in the bearing rings and rolling elements, in the form of nonmetal particles as well as microscopic bubbles from gas released into the metal during solidification. In fact, these impurities have an adverse effect equivalent to small cracks in the material. These microscopic cracks propagate and cause early fatigue failure. Therefore, a lot of effort has been directed at developing ultrahigh-purity steels for rolling bearings.

An advanced method for high purity steel is the *vacuum induction method* (VIM). The melting furnace is inside a large vacuum chamber. The process uses steel of high purity, and the required alloys are added from hoppers into the vacuum chamber. A second method is the *vacuum arc remelting* (VAR) where a consumable electrode is melted by an electrical arc in a vacuum chamber. The two methods were combined and referred to as VIM-VAR. In the combined method, the steel from the vacuum induction method is melted again by the vacuum arc method. Successive vacuum arc remelting improves the bearing fatigue life.

Zaretzky (1997c) presented a detailed survey of the processing methods and testing of bearing materials for aerospace applications.

13.21 CERAMIC MATERIALS FOR ROLLING BEARINGS

There is an ever-increasing demand for better materials for rolling-element bearings in order to increase the speed and service life of machinery. In addition, machines are often exposed to corrosive environments and high temperatures that cause steel bearings to fail. In a corrosive environment, the life of regular rolling bearings made of steel is short. It would offer a huge economic benefit if an alternate material could be developed that would increase the life of rolling bearings.

For the last several decades, engineers have been searching for alternative materials for the roller bearing. Although there are significant improvements in the manufacturing processes and composition of steel bearings, scientists and engineers have been continually investigating ceramics as the most promising alternative materials.

In aviation, there is an ever-present need for the reduction of weight. It is possible to reduce the size and weight of engines by operating at higher speeds. In addition, weight reduction can be achieved if the engine efficiency can be improved by operating at higher temperatures. Let us recall that according to the basic principles of thermodynamics, the efficiency of the Carnot cycle is proportional to the process temperature. Therefore, there is a need for materials that can operate at high temperatures. It has been recognized that the bearings are one bottleneck that limits the speed and temperature of jet engines. A lot of research has been conducted in developing and testing ceramic materials that can endure higher temperatures in comparison to steel. In addition, ceramics have a low density, which is important in reducing the centrifugal force of the rolling elements, a limiting factor of speed.

Initially, tests were conducted with rolling elements made of aluminum oxide and silicon carbide. However, these tests indicated unacceptable early catastrophic failure, particularly at high speeds and under heavy loads. Better results were obtained later with silicon nitride, Si_3N_4.

The early manufacturing process for silicon nitride involved hot pressing. The parts did not have a uniform structure and had many surface defects. The parts required expensive finishing by diamond-coated tools. Moreover, the finished parts did not have the required characteristics for using them in rolling-element bearings.

Later, the development of a hot isostatically pressed (HIP) manufacturing process significantly improved the structure of silicon nitride. The most important properties of silicon nitride that make it suitable for rolling bearings is fatigue resistance under rolling contact and relatively high fracture toughness. Silicon nitride rolling elements showed a fatigue-failure mode by spalling, similar to steel. In addition, silicon nitride proved to be wear resistant under the high contact pressure of heavily loaded bearings. Most of the applications use silicon nitride ceramic rolling elements in steel rings, referred to as *hybrid ceramic bearings*.

13.21.1 Hot Isostatic Pressing (HIP) Process

The introduction of the HIP process offered many advantages over the previous hot-pressing process. The HIP process is done by applying a high pressure of inert gases—argon, nitrogen, helium—or air at elevated temperatures to all grain surfaces under a uniform temperature. Temperatures up to 2000°C (3630°F) and pressures up to 207 MPa (30,000 psi) are used. The temperature and pressure are accurately controlled. The term *isostatic* means that the static pressure of gas is equal in all directions throughout the part.

This process is already widely used for shaping parts of ceramic powders as well as other mixtures of metals and nonmetal powders. This process minimizes surface defects and internal voids in the parts. The most important feature of this process is that it results in strong bonds between the powder boundaries of similar or dissimilar materials, which improve the characteristics of the parts for many engineering applications.

In addition, the process reduces the cost of manufacturing because it forms net or near-net shapes (close to final shape) from various powders, such as metal, ceramic, and graphite. The cost is reduced because the parts are near the final shape and less expensive machining is needed.

There are also important downsides to ceramics in rolling bearings. The cost of manufacturing of ceramic parts is several times that of similar steel parts. In rolling bearings, a major problem is that the higher elastic modulus and lower Poisson ratios of silicon nitride result in higher contact stresses than for steel bearings (see Hertz equations in Chapter 12). It is obvious that silicon nitride's higher elastic modulus and hardness result in a small contact area between the balls and races. In turn, the maximum compression stress must be higher for ceramic on steel and even more in ceramic on ceramic. The high contact stresses can become critical and can cause failure of the ceramic rolling elements. This is particularly critical in all-ceramic bearings, because ceramic-on-ceramic contact results in higher stresses than ceramic balls on steel races.

13.21.2 Silicon Nitride Bearings

The most widely used type is the *hybrid bearing*. It combines silicon nitride balls with steel races. The second type is the *all-ceramic bearing*, often referred to as a *full-complement ceramic bearing*. The two types benefit from the properties of silicon nitride, which include low density, corrosion resistance, heat resistance, and electrical resistance.

13.21.3 Hybrid Bearings

The surfaces of steel races and ceramic rolling elements are compatible, in the sense that they have relatively low adhesive wear. Ceramics sliding or rolling on metals do not generate high adhesion force or microwelds at the asperity contacts. The ceramic rolling elements have high electrical resistance, which is important in electric motors and generators because they eliminate the problem of arcing in steel bearings. However, the most important advantage of silicon nitride rolling elements is their low density. The specific density of silicon nitride is 3.2, in comparison to 7.8 for steel (about 40% of steel). The centrifugal forces are proportional to the density of the rolling elements, and they become critical at high speeds. Since pressed silicon nitride rolling elements are lighter, the centrifugal forces are reduced.

Many experiments confirmed that hybrid bearings have a longer fatigue life than do M-50 steel rolling elements. At very high speeds, the relative improvement in the fatigue life of silicon nitride hybrid bearings is even higher, due to the lower density, which reduces the centrifugal forces.

The silicon nitride is very hard and has exceptionally high compressive strength, but the tensile strength is low. Low tensile strength is a major problem for mounting the rings on steel shafts; but hybrid bearings have steel rings, so this problem is eliminated.

Although research in hybrid bearings was conducted two decades earlier, it is only since 1990 that they have been in a wide use for precision applications, including machine tools. The high rigidity of silicon nitride balls was recognized for its potential for improvement in precision and reduction of vibrations. This property can be an advantage in high-speed rotors.

13.21.3.1 Fatigue Life of Hybrid Bearings

There is already evidence that hybrid bearings made of silicone nitride balls and steel rings have much longer fatigue life than do steel bearings of similar geometry. Examples of research work are by Hosang (1987) and Chiu (1995). The major disadvantage of hybrid bearings is their high cost.

However, the advantages of the hybrid bearing are expected to outweigh the high cost. Longer life at higher speeds and higher temperatures may end up saving money over the life cycle of the machine by reducing the need for maintenance and replacement parts. In addition, longer bearing life will result in reduced machine downtime, which results in the expensive loss of production. We have to keep in mind that the cost of bearing replacement is often much higher than the cost of the bearing itself.

13.21.3.2 Applications of Hybrid Bearings

Hybrid ceramic bearings have already been applied in high-speed machine tools, instrument bearings, and turbo machinery. Other useful applications of silicon nitride balls include small dental air turbines, food processing, semiconductors, aerospace, electric motors, and robotics.

In hybrid bearings, the ceramic balls prevent galling and adhesive wear even when no liquid lubricant is used. Nonlubricated hybrid bearings wear less than dry all-steel bearings. Operation of steel bearings without lubrication results in the formation of wear debris, which accelerates the wear process. Ceramic balls have a higher modulus of elasticity than steel, which makes the bearing stiffer, useful in reducing vibration and for precision applications.

Hybrid ceramic bearings demonstrated very good results in applications without any conventional grease or oil lubrication, but only a thin solid lubricant layer transferred from the cage material. Example of a successful application is in the propellant turbopump of the Space Shuttle, where grease or oil lubrication must be avoided due to the volatility of the propellants, see Gibson (2001).

For propulsion into orbit, the NASA Space shuttle has three engines. Each engine is fed propellants by four turbopumps, which were equipped with hybrid ceramic bearings with silicon nitride ceramic balls and a self-lubricating cage made of sintered PTFE and bronze powders. The PTFE is transferred as a third body of a thin film solid lubrication on the balls and races. The hybrid ceramic bearing in this severe application did not show any significant wear of the raceways. Tests indicated that various cage material combinations affected the life of the self-lubricated bearing in different ways. The best results were obtained by using silicon nitride ceramic balls and sintered PTFE and bronze cage. This combination was implemented successfully in all NASA Space shuttles.

The hybrid bearing is currently passing extensive tests for ultimate use in jet aircraft engines. However, at this time, it has not reached the stage of being actually used in aircraft engines. For safety reasons, the hybrid bearing must pass many strict tests before it can be approved for use in actual aircraft.

13.21.4 All-Ceramic Bearings

The most important advantage of all-ceramic bearings is that they resist corrosion, even in severe chemical and industrial environments where stainless steel bearings lack sufficient corrosion resistance.

Zaretzky (1989) published a survey of the research and development work in ceramic bearings during the previous three decades. He pointed out that since the elastic modulus of silicon nitride is higher than that of steel, the Hertz stresses are higher than for all-steel bearings. Zaretzky concluded that the dynamic capacity of the all-silicon-nitride bearing is only 5–12% of that of an all-steel bearing of similar geometry. In addition, there are problems mounting the ceramic ring on a steel shaft. The difference in thermal expansion results in high tensile stresses. Silicon nitride has exceptionally high compressive strength, but the tensile strength is low. Therefore, a ceramic ring requires a special design for mounting it on a steel shaft.

The most important advantage of all-silicon-nitride bearings is that they can operate at high temperature above the limits of steel bearings. However, at temperatures above 578 K (300°C), the available liquid lubricants cannot be used. Early tests indicated that all-ceramic bearings can operate with minimal or no lubrication. However, when tests were conducted at higher speeds, similar to those in gas turbine engines, catastrophic bearing failure resulted after a short time. In the future, solid lubricants may be developed to overcome this problem.

Another problem in the way of extending the operating temperature of all-ceramic bearings is that high-temperature cage materials were not available. Tests indicated that the best results could be achieved with graphite cages; see Mutoh et al. (1994).

An important advantage of the all-ceramic bearing remains that it can resist corrosion in very corrosive environments where steel bearings would be damaged. Moreover, regular bearings often fail because an industrial corrosive environment breaks down the lubricant. In such cases, the all-ceramic bearing can be a solution to these problems. It also can operate with minimal or no lubrication. In addition, it has high rigidity, important in precision machines.

The all-ceramic bearings are used in the etching process for silicon wafers, where sulfuric acid and other corrosives are used. Only ceramic bearings can resistant this corrosive environment. Another application is ultraclean vacuum systems. Liquid lubricants evaporate in a vacuum, and ceramic bearings are an alternative for this purpose. All-ceramic bearings can also be used in applications where nonmagnetic bearings are required. Hybrid bearings with stainless steel rings are also used for this purpose.

Sealed pumps driven by magnetic induction are widely used for pumping various corrosive chemicals. Most sealed pumps operate with hydrodynamic journal bearing with silicon carbide sleeve. The ceramic sleeves are used because of their corrosive resistance and for their nonmagnetic properties.

However, the viscosity of the process fluids is usually low, and the hydrodynamic fluid film is generated only at high speeds. For pumps that operate with frequent start-ups, there is high wear and the bearings do not last for a long time. All-ceramic rolling bearings made of silicon nitride proved to be a better selection for sealed pumps. The silicone nitride rolling bearings are not sensitive to frequent start-ups and have good chemical corrosion resistance as well as the desired nonmagnetic properties for this application.

13.21.5 Cage Materials for Hybrid Bearings

Different cage materials have been tested in ceramic hybrid bearings. Appropriate cage material is a critical problem in applications where solid lubrication or operation without lubricant is required. In such cases, the cage material provides the solid lubricant.

A graphite cage offered a low wear rate in high-temperature applications. Self-lubricating (soft) cage materials resulted in a longer bearing life with lower wear rate and lower friction in comparison to hard cage material. However, at high temperatures, self-lubricating cage materials resulted in excessive degradation of the cage material by high-temperature oxidation.

13.22 ROLLING BEARING CAGES

The rolling bearing cage, often referred to as a *separator* or a *retainer*, is mounted in the bearing in order to equally space the rolling elements (balls or rollers) and prevent contact friction between them. The cage rotates with the rolling elements, which are freely rotating in the confinement of the cage. In addition, the cage retains the grease to provide for effective lubrication. Cages made of porous materials, such as phenolic, absorb liquid lubricants and assist in providing a very thin layer of oil for a long time. Examples of rolling bearing cage designs are shown in Fig. 13-27 (FAG, 1998).

Cages are made of the following materials.

Cages made of brass are commonly used in medium and large roller bearings.
Cages made of nylon strengthened by two round strips of steel are commonly used in small ball bearings.
Cages made of steel are used in miniature ball bearings.
Cages made of phenolic are used in ultrahigh-precision bearings.

13.23 BEARING SEALS

Seals act as a barrier that prevents the loss of the lubricant from the bearing housing. In addition, seals restrict the entry of any foreign particles or undesired process liquids into the bearings. Reliable operation of the seals is very important. In the case of lubricant loss, it can result in bearing failure. Any penetration of foreign particles into the bearing will result in reduction of its service life. Thus seals are essential for the proper functioning of the bearing. Seals are generally classified into two types, contact seals and noncontact seals.

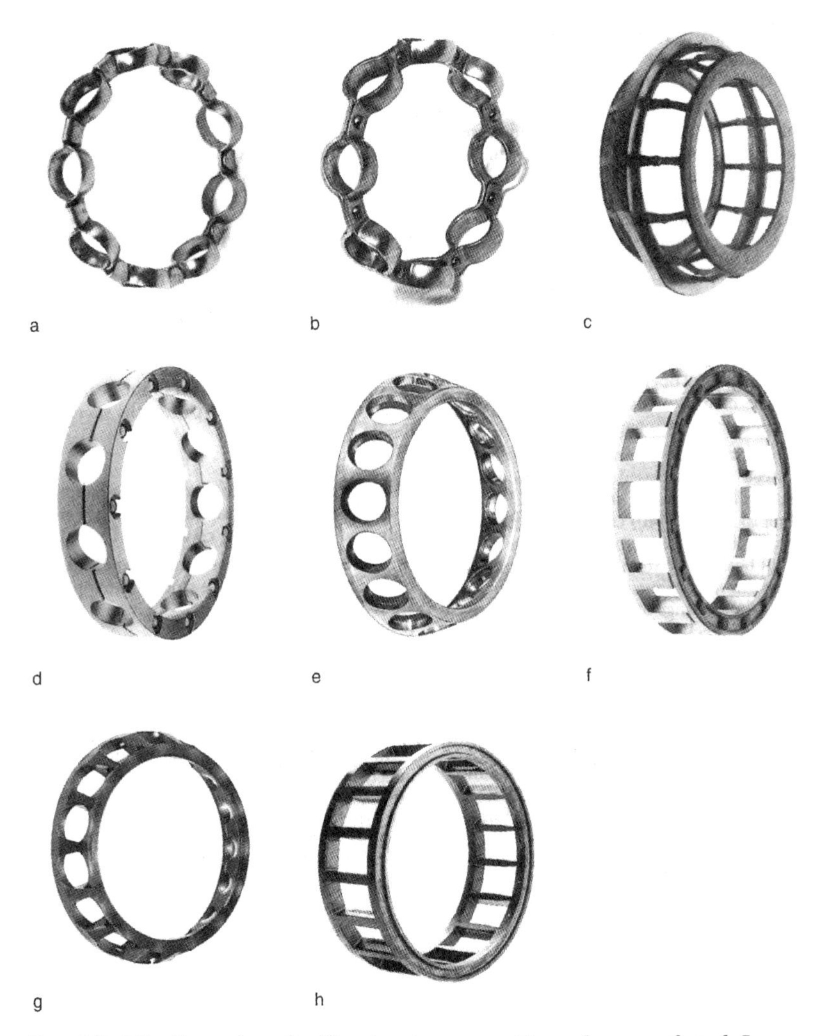

FIG. 13-27 Examples of rolling bearing cages. *Pressed cages of steel*: Lug cage (a) and rivet cage (b) for deep-groove ball bearings, window-type cage (c) for spherical roller bearings. *Machined brass cages*: Riveted machined cage (d) for deep-groove ball bearings, brass window-type cage (e) for angular contact ball bearings and machined brass cage with integral crosspiece rivets (f) for cylindrical roller bearings. *Molded cages* made of glass-fibre reinforced polyamide: window-type cage (g) for single-row angular contact ball bearings and window-type cage (h) for cylindrical roller bearings. (From FAG, 1998, with permission.)

13.23.1 Contact Seals

These seals remain in contact with the sliding surface, and thus they wear after a certain period of operation and need replacement. They are also referred to as rubbing seals. In order to make these seals effective; a certain amount of contact pressure should always be present between the seal and shaft. The wear of contact seals increases by the following factors:

Friction coefficient
Bearing temperature
Sliding velocity
Surface roughness
Contact pressure

Under favorable conditions, there is a very thin layer of lubricant that separates the seal and the shaft surfaces (similar to fluid film but much thinner). The film thickness can reach the magnitude of 500 nm, at shaft surface speed of 0.4 m/s (Lou Liming, 2001). A few examples of widely used contact seals are presented in Figs. 13-28a–f.

13.23.1.1 Felt Ring Seals

These seals (Fig. 13.28a) are widely used for grease lubrication. Felt ring seals are soaked in a bath of oil before installation, for reduction of friction. Felt seals provide excellent sealing without much contact pressure and are effective against penetration of dust. Therefore, they do not cause much friction power loss. The number of felt rings depends on the environment of the machine. The dimensions of felt seals are standardized.

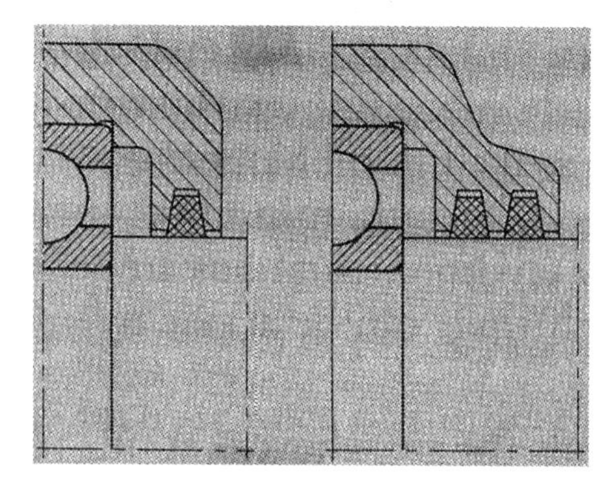

FIG. 13-28a Felt ring seal (from FAG, 1998, with permission).

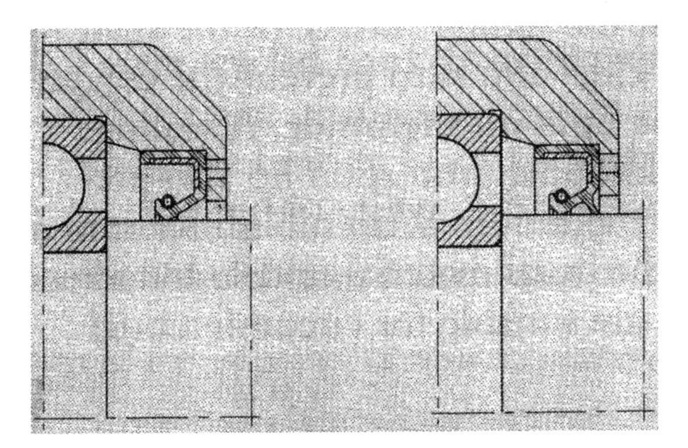

FIG. 13-28b Radial shaft seals (from FAG, 1998, with permission).

13.23.1.2 Radial Shaft Sealing Rings

These are the most widely used contact lip seals for liquid lubricant (Fig. 13-28b). The basic construction incorporates the lip of the seal pressed on the sliding surface of a shaft with the help of a spring.

13.23.1.3 Double-Lip Radial Seals

These seals (Fig. 13-28c) consist of two lips. The outer lip restricts any entry of foreign particle, and the inner lip retains the lubricant inside the bearing housing. When grease is applied between the two lips, the bearing life increases.

13.23.1.4 Axially Acting Lip Seals

The major advantage of this seal (Fig. 13-28d) is that it is not sensitive to radial misalignment. The seal is installed by pushing it on the surface of the shaft until its lip comes in contact with the housing wall. These seals are often used as extra

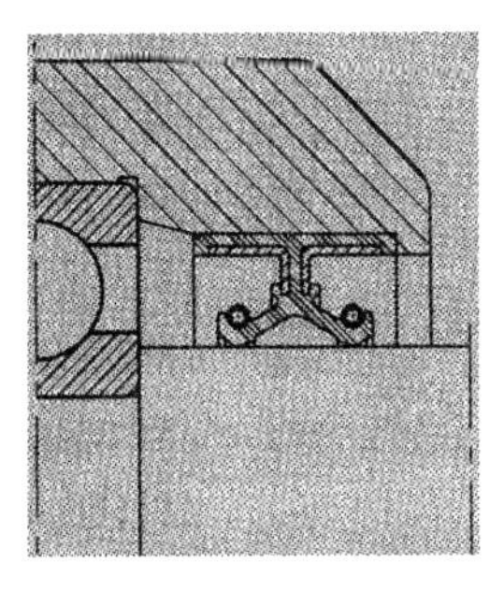

FIG. 13-28c Double-lip radial seal (from FAG, 1998, with permission).

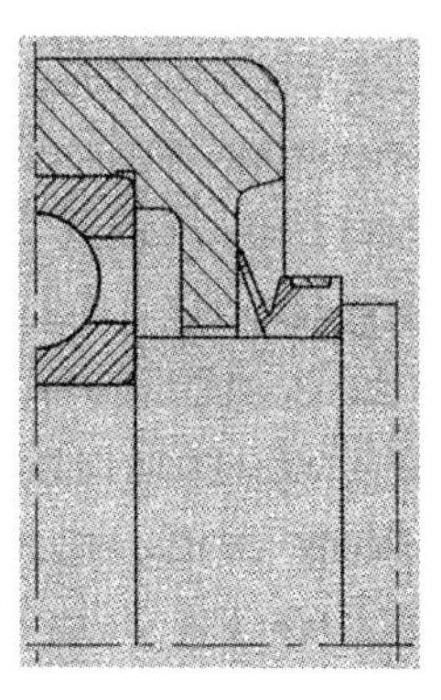

FIG. 13-28d Axially acting lip seal (from FAG, 1998, with permission).

seals in a contaminated environment. At very high speeds, these seals are not effective due to the centrifugal forces.

13.23.1.5 Spring Seals

These seals (Fig. 13.28e) are effective only for grease lubrication. A thin round sheet metal is clamped to the inner or outer ring, and provides a light contact pressure with the second ring.

13.23.1.6 Sealed Bearing

This seal (Fig. 13.28f) is manufactured with the bearing, and widely used for sealed for life bearings. The seal is made of oil resistant rubber, which is connected to the outer ring, and lightly pressed on the inner ring.

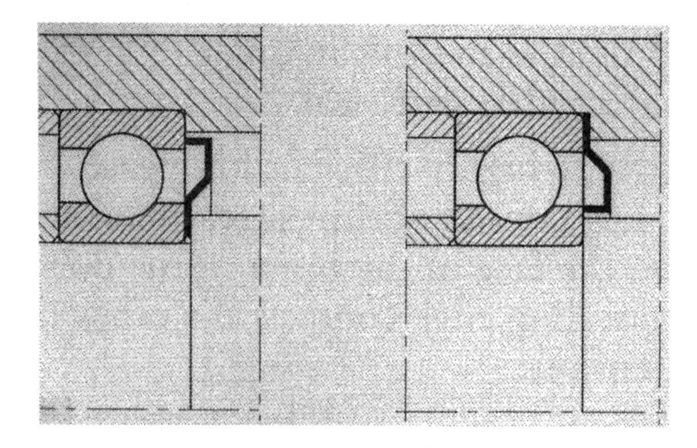

FIG. 13-28e Spring seals (from FAG, 1998, with permission).

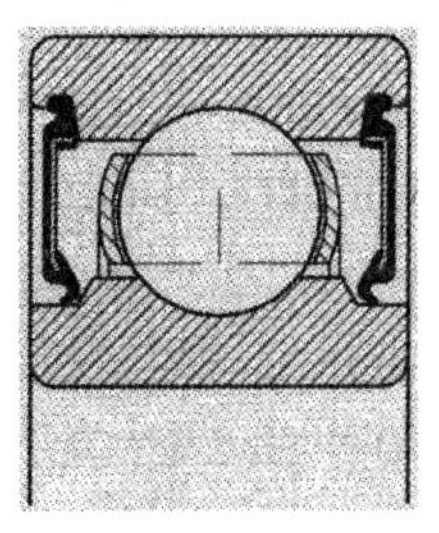

FIG. 13-28f Sealed bearing (from FAG, 1998, with permission).

13.23.2 Noncontact Seals

Noncontact seals are also known as nonrubbing seals. These seals are widely used for grease lubrication. In these seals there is only viscous friction, and thus they perform well for a longer time. In noncontact seals there is a small radial clearance between the housing and the shaft (0.1–0.3 mm). These seals are not so sensitive to radial misalignment of the shaft. Most important, since there is no contact, not much heat is generated by friction, which make it ideal for high-speed applications.

A number of grooves are designed into the housing, which contain grease. The grease filled grooves form effective sealing. If the environment is contaminated, the grease should be replaced frequently. If oil is used for lubrication, the grooves are bored spirally in the direction opposite to that of the rotation of the shaft. Such seals are also known as shaft-threaded seals.

Some examples of noncontact seals are shown in Fig. 13-29.

FIG. 13-29a Grooved labyrinth seal (from FAG, 1998, with permission).

FIG. 13-29b Axial webbed noncontact seal (from FAG, 1998, with permission).

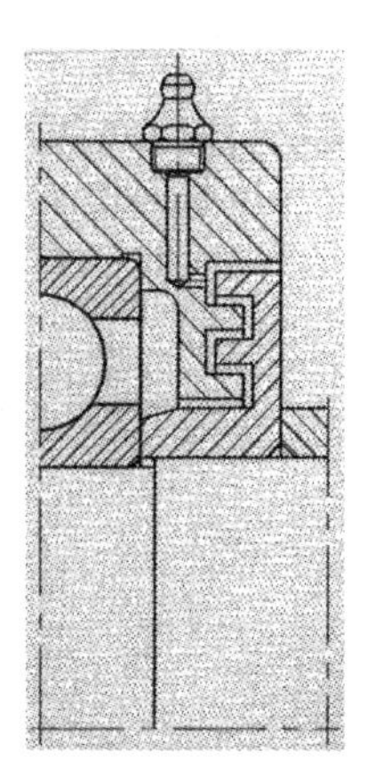

FIG. 13-29c Radial webbed noncontact seal (from FAG, 1998, with permission).

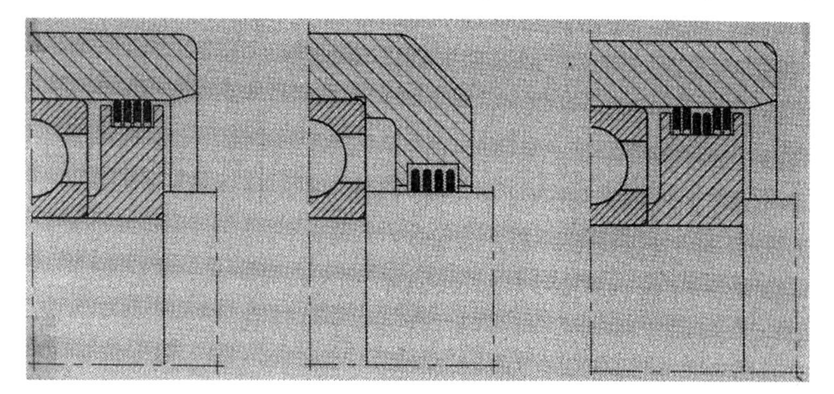

FIG. 13-29d Noncontact seal with lamellar rings (from FAG, 1998, with permission).

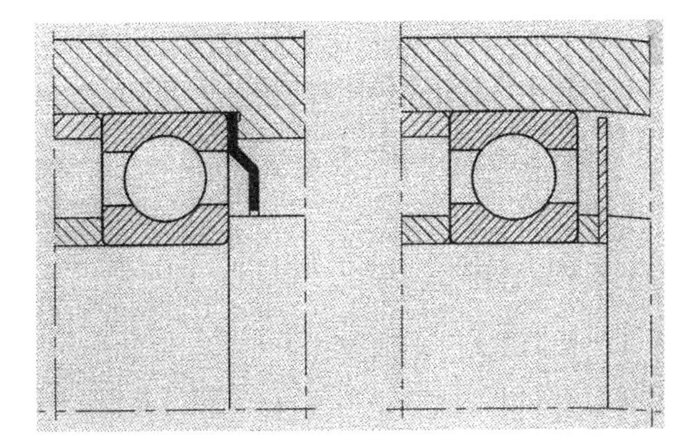

FIG. 13-29e Baffle plates seal (from FAG, 1998, with permission).

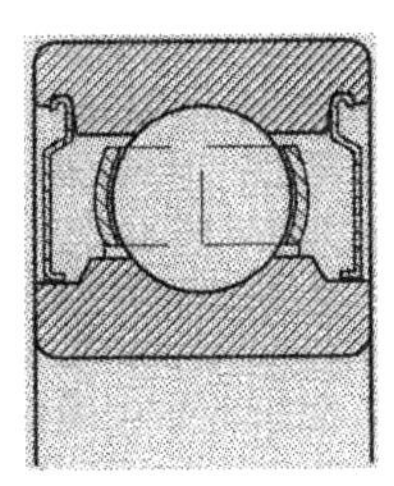

FIG. 13-29f Bearing with shields (from FAG, 1998, with permission).

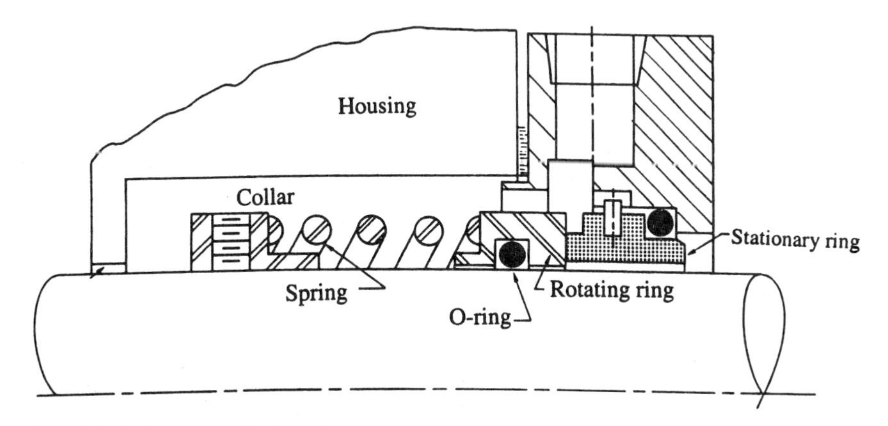

FIG. 13-30 Mechanical seal.

13.24 MECHANICAL SEALS

This seal is widely used in pumps. The sealing surfaces are normal to the shaft, as shown in Fig. 13-30. The concept is that of two rubbing surfaces, one stationary and one rotating with the shaft. The surfaces are lubricated and cooled by the process fluid. The normal force between the rubbing surfaces is from the spring force and the fluid pressure. The materials of the rubbing rings are a combination of very hard and very soft materials, such as silicon carbide and graphite. The lubrication film is very thin, and the leak is negligible.

Problems

13-1 A single-row, standard deep-groove ball bearing operates in a machine tool. It is supporting a shaft of 30-mm diameter. The bearing is designed for 90% reliability. The radial load on the bearing is 3000 N (no axial load). The shaft speed is 7200 RPM. The lubricant is SAE 20 oil, and the maximum expected surrounding (ambient) temperature is 30°C. Assume the oil operating temperature is 10°C above ambient temperature.

a. Find the life adjustment factor a_3.
b. Find the adjusted fatigue life L_{10} of a deep-groove ball bearing.
c. Find the maximum static radial equivalent load.

The deep groove bearing data, as specified in a bearing catalogue, is as follows:

Designation bearing:	No. 6006
Bore diameter:	$d = 30$ mm
Outside diameter:	$D = 55$ mm
Dynamic load rating:	$C = 2200$ lb
Static load rating:	$C_0 = 1460$ lb

13-2 In a gearbox, two identical standard deep-groove ball bearings support a shaft of 35-mm diameter. There is locating/floating arrangement where the floating bearing supports a radial load of 10,000N and the locating bearing supports a radial load of 4000N and a thrust load of 5000N. The shaft speed is 3600 RPM. The lubricant is SAE 30 oil, and the maximum expected surrounding (ambient) temperature is 30°C. Assume that the oil operating temperature is 5°C above ambient temperature. The two deep-groove bearings are identical. The data, as specified in a bearing catalogue, is as follows:

Designation bearing:	No. 6207
Bore diameter:	$d = 35$ mm
Outside diameter:	$D = 72$ mm
Dynamic load rating:	$C = 4400$ lb
Static load rating:	$C_0 = 3100$ lb

a. Find the life adjustment factor a_3 for the locating and floating bearings.
b. Find the adjusted fatigue life L_{10} of a deep-groove ball bearing for the locating and floating bearings.
c. Find the static radial equivalent load.
d. Find the radial static equivalent load for the locating and floating bearing.

13-3 Find the operating clearance (or interference) for a standard deep-groove ball bearing No. 6312 that is fitted on a shaft and inside housing as shown in Fig. 13-6. During operation, inner ring as well as shaft temperature is 8°C higher than the temperature of outer ring and housing. The bearing is of C3 class of radial clearance (radial clearance of 23–43 μm from Table 13-2).

The dimensions and tolerances of inner ring and shaft are

Bore diameter:	$d = 60$ mm $(-15/+0)$ μm
Shaft diameter:	$d_s = 60$ mm $(+21/+2)$ μm k6
OD of inner ring:	$d_1 = 81.3$ mm

The dimensions and tolerances of outer ring and housing seat are

OD of outer ring:	$D = 130$ mm $(+0/-18)$ μm
ID of outer ring:	$D_1 = 108.4$ mm
ID of housing seat:	$D_H = 130$ mm $(-21/+4)$ μm K6

Neglect the surface smoothing effect, and assume that the housing and shaft seats were measured, and the actual dimension is at 1/3 of the tolerance zone, measured from the tolerance boundary close to the surface where the machining started, e.g., the shaft diameter is 60 mm + [21–(21–3)/3] μm = 60.015 mm.

Consider elastic deformation and thermal expansion for the calculation of the two boundaries of the operating radial clearance tolerance zone.

Coefficient of thermal expansion of steel is $\alpha = 0.000011$ [1/C]

13-4 A standard deep-groove ball bearing No. 6312 that is mounted on a shaft and into a housing as shown in Fig. 13-6. The bearing width is $B = 31$ mm. The shaft and ring are made of steel $E = 2 \times 10^{11}$ Pa.

The dimensions and tolerances of inner ring and shaft are

Bore diameter:	$d = 60\,mm\ (-15/+0)\ \mu m$
Shaft diameter:	$d_s = 60\,mm\ (+24/+11)\ \mu m$, m5
OD of inner ring:	$d_1 = 81.3\,mm$

The dimensions and tolerances of outer ring and housing seat are

OD of outer ring:	$D = 130\,mm\ (+0/-18)\ \mu m$
ID of outer ring:	$D_1 = 108.4\,mm$
ID of housing seat:	$D_H = 130\,mm\ (-21/+4)\ \mu m$, K6

Neglect the surface smoothing effect, and assume a rectangular cross section of the bearing rings for all calculations.

1. Find the maximum and minimum pressure between the shaft and bore surfaces.
2. Find the minimum and maximum tensile stress in the inner ring after it is tightly fitted on the shaft.
3. If the friction coefficient is $f = 0.5$, find the maximum axial force (for the tightest tolerance), which is needed for sliding the inner ring on the shaft.
4. Find the minimum inertial torque (N–m), which can result in undesired rotation sliding of the shaft inside the inner ring during the start-up ($f = 0.5$).
5. The bearing is heated for mounting it on the shaft without any axial force. Find the temperature rise of the bearing (relative to the shaft), for all bearings and shaft within the specified tolerances. Coefficient of thermal expansion of steel is $\alpha = 0.000011$ [1/C].

13-5 Modify the design of the bearing arrangement of the NC–lathe main spindle in Fig. 13-10b. The modified design will be used for rougher machining at lower speeds. Adjustable bearing arrangement with two tapered roller bearings will replace the current bearing arrangement. For a rigid support, an adjustable bearing arrangement was selected with the apex points between the two bearings.

a. Design and sketch the cross-section view of the modified lathe main spindle.
b. Show the centerlines of the tapered rolling elements and the apex points, if the bearings preload must not be affected by temperature rise during operation.
c. Specify the tolerances for the seats of the two bearings.

13-6 Modify the design of the bearing arrangement of the NC–lathe main spindle in Fig. 13-10b to a locating/floating bearing arrangement. On the right hand (the locating side), the modified design entails three adjacent angular ball bearings, two in an adjustable arrangement, and the third in tandem arrangement to machining thrust force. On the left hand, two adjacent cylindrical roller bearings are the floating bearings that support only radial force.

a. Design and sketch the cross-section view of the modified design.
b. Specify tolerances for all the bearing seats.

14

Testing of Friction and Wear

14.1 INTRODUCTION

There is an increasing requirement for testing the performance of bearing materials, lubricants, lubricant additives, and solid lubricants. For bearings running on ideal full oil films, the viscosity is the only important lubricant property that affects the friction. However, in practice most machines are subjected to variable conditions, vibrations and disturbances and occasional oil starvation. For these reasons, even bearings designed to operate with a full fluid film will have occasional contact, resulting in a rubbing of surfaces under boundary lubrication conditions and, under certain circumstances, even under dry friction conditions. Many types of oil additives, greases, and solid lubricants have been developed to reduce friction and wear under boundary friction. Users require effective tests to compare the effectiveness of boundary lubricants as well as of bearing materials for their specific purpose.

It is already known that the best test is one conducted on the actual machine at normal operating conditions. However, a field test can take a very long time, particularly for testing and comparing bearing life for various lubricants or bearing materials. An additional problem in field testing is that the operation conditions of the machines vary over time, and there are always doubts as to whether a comparison is being made under identical operating conditions. For example, manufacturers of engine oils compare various lubricants by the average miles the car travels between engine overhauls (for expediting the field test, taxi service cars are used). It is obvious that the cars are driven by various drivers; and most probably, the cars are not driven under identical conditions. Field tests can be expensive if the bearings are periodically inspected for wear or any other damage. Concerning the measurement of friction-energy losses, in most cases it is impossible to test friction losses on an actual machine. Friction losses in a car engine can be estimated only by changes in the total fuel consumption. Obviously, this is a rough estimate because friction-energy loss is only a portion of the total energy consumption of the machine.

For all these reasons, various testing machines with accelerated tests have been developed and are used in laboratory simulations that are as close as possible to the actual conditions. The common commercial testing machines are intended for measuring friction and wear for various lubricants under boundary lubrication conditions or for comparing various solid lubricants under dry friction conditions. Most commercial testing machines operate under steady conditions of sliding speed and load.

14.2 TESTING MACHINES FOR DRY AND BOUNDARY LUBRICATION

Most commercial testing machines are for measuring friction and wear under high-pressure-contact conditions of point or line contact (nonconformal sliding contacts) (Fig. 14-1). These tests are primarily for rolling bearings and gear lubricants. In addition, there are many testing machines for journal bearings and thrust bearings (conformal contacts). For nonconformal contacts, a widely used test is the four-ball apparatus, where one ball rotates against three stationary balls at constant speed and under steady load. The operating parameters of wear, friction, and life to failure by seizure (when the balls weld together) are compared for various materials and lubricants. The friction torque is measured and the friction coefficient is calculated. In addition to friction, the time or number of revolutions to seizure can be measured as a function of load. Wear can also be compared by intermediate measurements of weight loss or changes in ball diameters, for various ball materials and lubricants.

The following are examples of friction and wear tests of various noncon formal contacts that have been introduced by various companies.

Four-ball machine (introduced by Shell Co.)
Pin on a disk (point contact because the edge of the pin is spherical)
Block on rotating ring (introduced by Timken Co.)
Reciprocating pad on a rotating ring
Shaft rotating between two V-shaped surfaces (introduced by Falex Co.)
SAE test of two rotating rings in line contact

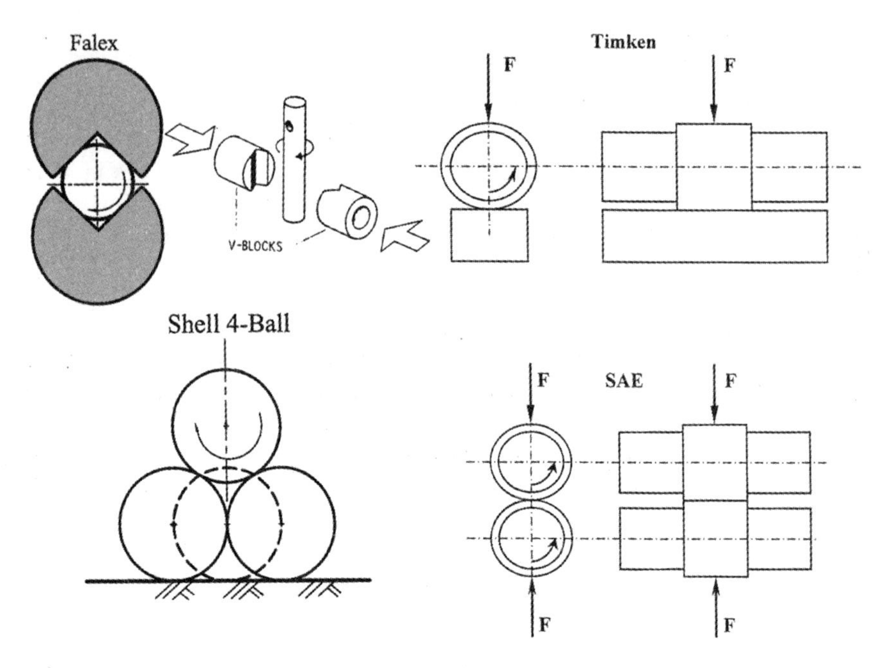

FIG. 14-1 Friction and wear tests of nonconformal contacts.

Although these testing machines are useful for evaluating the performance of solid lubricants and comparing bearing materials for dry friction, there are serious reservations concerning the testing accuracy of liquid lubricants for boundary friction or comparing various boundary friction lubricant additives. These reservations concern the basic assumption of boundary lubrication tests: that there is only one boundary lubrication friction coefficient, independent of sliding speed, that can be compared for different lubricants. However, measurements indicated that, in many cases, the friction coefficient is very sensitive to the viscosity or sliding speed. For example, certain additives can increase the viscosity, which will result in higher hydrodynamic load capacity and, in turn, reduction of the boundary friction.

The friction force has a hydrodynamic component in addition to the contact friction (adhesion friction). Therefore, it is impossible to completely separate the magnitude of the two friction components. Certain boundary additives to mineral oils may reduce the friction coefficient, only because they slightly increase the viscosity. Even for line or point contact, there is an EHD effect that increases with velocity and sliding speed. The hydrodynamic effect would reduce the boundary friction because it generates a thin film that separates the surfaces. This argument has practical consequences on the testing of boundary layer lubricants. These tests are intended to measure only the adhesion friction of boundary lubrication; however, there is an additional viscous component.

Currently, boundary lubricants are evaluated by measuring the friction at an arbitrary constant sliding speed (e.g., four-ball tester operating at constant speed). The current testing methods of liquid boundary lubricants should be reevaluated. Apparently, better tests would be obtained by measuring the complete friction versus velocity, *f-U*, curve. In Sections 14.6 and 14.7, dynamic testing machines are described that are better able to evaluate separately the contact friction at the start-up and the mixed and hydrodynamic friction. The friction is a function of speed, which can be measured by dynamic tests.

14.3 FRICTION TESTING UNDER HIGH-FREQUENCY OSCILLATIONS

It has already been mentioned that in real machines the contact stresses of mating parts in relative motion are not completely constant. There are always vibrations and time-variable conditions. Even when the load is constant, there are friction-induced vibrations, resulting in small high-frequency oscillations in the tangential direction (parallel to the surface). For these reasons, it was realized that testing machines with high-frequency oscillations would offer a better simulation of the actual conditions in machinery.

It is well known that rolling-element bearings operate under high-frequency oscillations, and there has been a need for testing machines that simulate these dynamic conditions. Tests under high-frequency oscillations have been adopted as standard tests, such as ASTM D 5706 EP and ASTM D 5707 EP for greases and oils for rolling bearings.

In the testing machine shown in Fig. 14-2, there is friction between the upper specimen and the lower disk. The upper specimen can be a ball or a cylinder, for point or line contact, or a ring, for area contact. The material and size can be adapted to the user's requirement (equivalent to the material used in the actual machine). During the friction test, the upper specimen has horizontal oscillations (parallel to the disk area). Force is applied mechanically to the upper specimen in a vertical direction (normal to the disk area). The friction force is measured by a piezoelectric sensor that is placed under the lower specimen holder. The friction coefficient is calculated and recorded on-line on a chart during the test. The environment in the test chamber (temperature and humidity) can be controlled.

This test has been adopted by the American Society of Testing Materials (ASTM) for testing greases or liquid lubricants operating under high contact pressure, such as point or line contact in rolling bearings and gears. The

FIG. 14-2 Testing apparatus for friction and wear under high-frequency oscillations (from SRV Catalogue, with permission from Optimal Instruments).

manufacturers of the dynamic testing apparatus claim that actual comparisons with the performance of the same greases in the field indicated that dynamic tests are more reliable than static tests for comparing the performance of various greases.

During the ASTM D 5706 EP standard friction test, the upper part has horizontal oscillations of 1-mm amplitude and a frequency of 50 Hz. The test is run with a very small amount of grease, 0.1–0.2 g grease. After 30 seconds break-in under a load of 50 N, the load is raised by increments of 100 N at 2-minute intervals until failure occurs. Failure is determined by seizure or by a significant sudden rise in the friction force. The tests are run at elevated temperature to simulate the actual operating conditions of rolling-element bearings.

A similar standard test is ASTM D 5707 EP. In this test, however, the friction coefficient, as well as wear, versus time is recorded on a chart. The test is run with a very small amount of grease, 0.1–0.2 g grease; the frequency of horizontal oscillations is 50 Hz and 1-mm amplitude. After 30 seconds break-in under a load of 50 N, the load is raised to 200 N for 2 hours. The lowest and highest values of friction on the chart are reported. In addition, after the test, the average wear scar diameter on the test ball is measured with the aid of a microscope and on the lower specimen with a profilometer. These readings are reported as wear test results. The test can be applied for comparing various liquid lubricants. The wear can be measured on-line during the test by measuring the depth of the wear scar, as shown in Fig. 14-3. Larger-amplitude vibrations can be applied to better simulate the conditions in an actual machine.

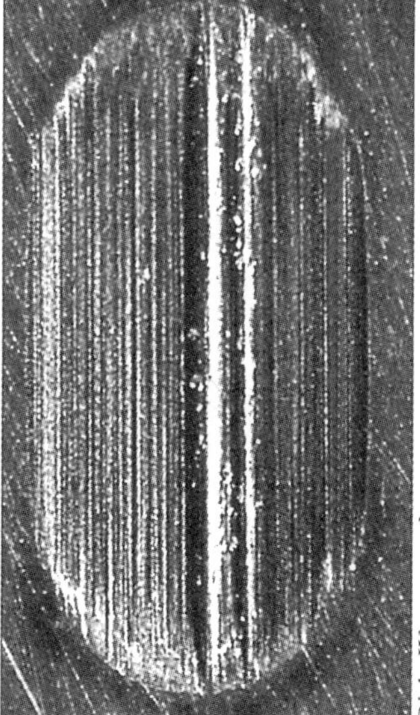

Disk: Wear scar after test

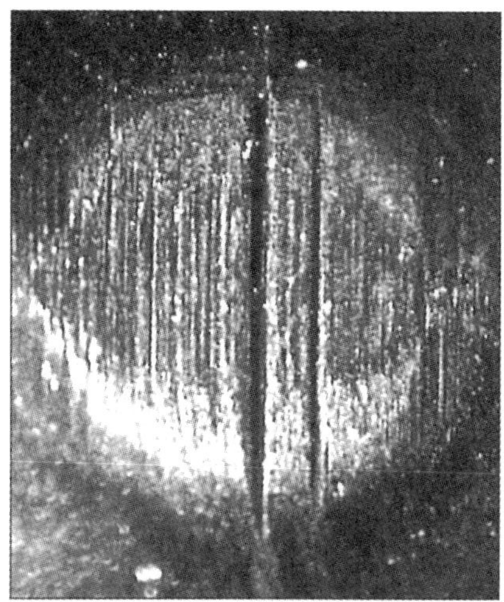

Ball: Wear scar after test

Test specimens:

Steel ball ø 10 mm

FIG. 14-3 Wear scars after a standard vibratory friction and wear test (from SRV Catalogue, with permission from Optimal Instruments).

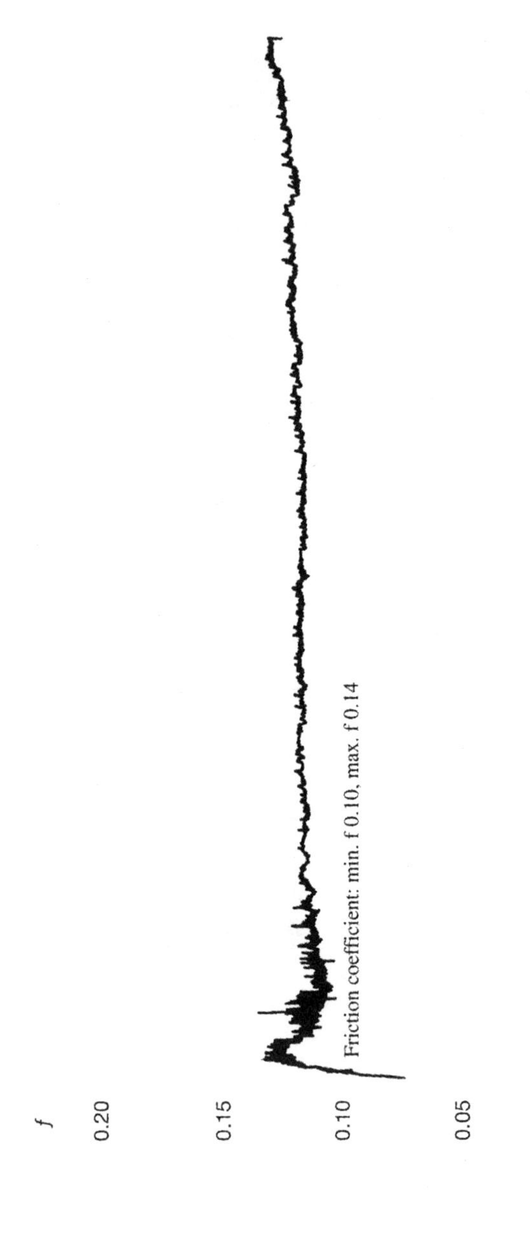

FIG. 14-4 Friction coefficient versus time (from SRV Catalogue, with permission from Optimal Instruments).

In Fig. 14-4, a test result is shown for a steel ball on a steel plate lubricated by synthetic oil. The result is a curve of friction coefficient versus time. The test time is 2 hours and the specimens are 10-mm steel balls on a lapped steel disk at a temperature of 200°C. The frequency of horizontal oscillations is 50 Hz and 1.5-mm amplitude. The reported friction coefficients are $f_{min} = 0.1$ and $f_{max} = 0.14$. The maximum wear measured during the test is 21 μm. The wear scars after the test on the disk and ball are shown in Fig. 14-3.

The reservations that have been raised for the steady tests are still valid for this vibratory test. Although these dynamic tests are effective in simulating the overall performance of real machines, a problem with the high-frequency test is that it does not test the pure effect of lubricant additives, such as antifriction and antiwear additives. The friction and wear are the combined effect of the viscosity of the lubricant as well as of the additives. In other words, there is no way to distinguish between the hydrodynamic and adhesion friction effects. Therefore, this would not be a good method to compare the effectiveness of various boundary lubrication additives.

In Sec. 14.4, a testing machine is described for testing the complete Stribeck curve. It offers a better distinction of the contact and viscous friction and the friction at each region. Therefore, the Stribeck curve can be a more useful test in developing and selecting lubricants. Nevertheless, the foregoing high-frequency test is very useful in testing solid lubricants and greases. For liquid lubricants, the test is useful for evaluating the combined effect under identical conditions of a specific application in the field.

14.4 MEASUREMENT OF JOURNAL BEARING FRICTION

The purpose of friction-testing machines is to measure the friction torque of a journal test-bearing friction or rolling-element test-bearing friction in isolation from any other source of friction in the system. There are several methods by which to measure the friction in bearings. The first method is based on the concept of the hydrostatic pad. It is designed for measuring the friction torque on the bearing housing by a load cell, while the bearing load is transferred to the bearing housing through a hydrostatic pad. Friction-testing machines with a hydrostatic pad can be designed for the measurement of static or dynamic friction. Dynamic friction is under time-variable conditions, such as oscillating velocity and time-variable load. Dynamic friction measurements require continuous recording or on-line data acquisition by a computer. All friction-testing machines for dynamic friction are universal, in the sense that they can be used under steady conditions as well as dynamic conditions. In most cases, however, machines for testing steady friction cannot be adapted for dynamic friction.

A relatively simple friction-testing machine is the pendulum tester. It can be applied for testing the friction coefficient of a journal bearing under steady conditions only.

The concept of this pendulum friction tester is to apply a load on the bearing by means of weight. The weights are placed on a rod connected to the bearing. During a steady operation under constant speed, the pendulum is tilted to an angle equal to the friction coefficient. An example of a pendulum tester is shown in Fig. 14-5. The angle is small, and the angle is measured by a dial gauge as shown.

This is a simple and low-cost tester. However, it has relatively low measurement precision in comparison to other machines. There are always small vibrations of the pendulum that make it difficult to get an average reading. This can be improved by damping the vibrations via a viscous damper. A second drawback that reduces the precision is that there is always some friction and it is

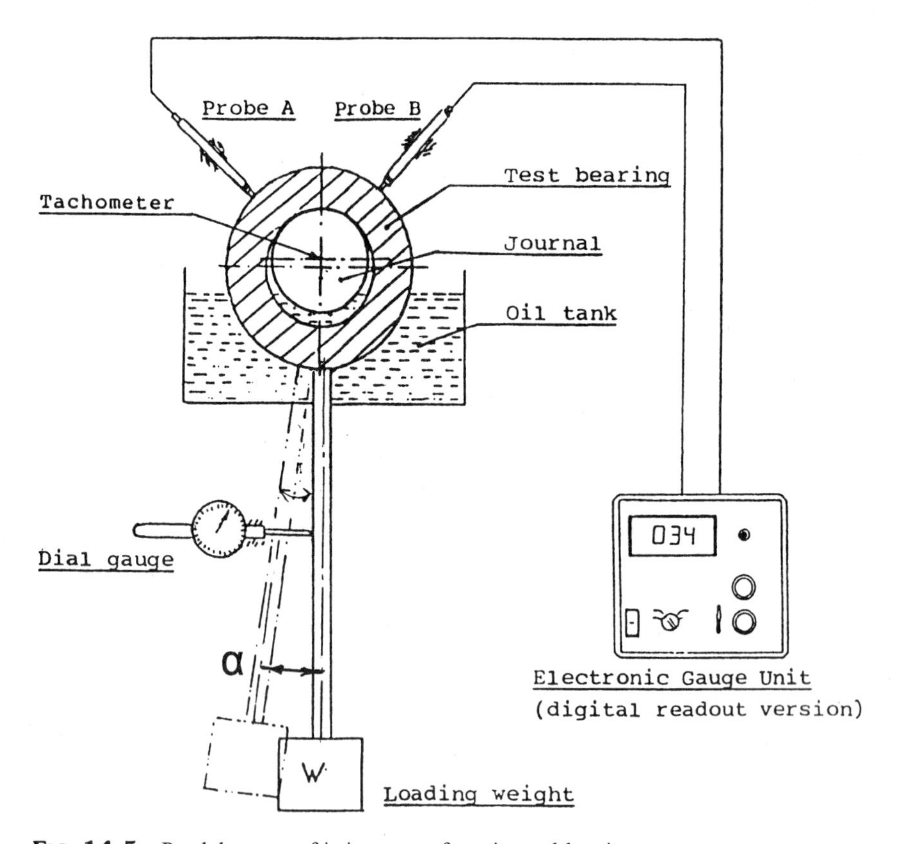

FIG. 14-5 Pendulum-type friction tester for a journal bearing.

impossible to adjust the zero position of the pendulum. A solution to this problem is to test in the two directions; namely, for each measurement the shaft is rotating in two directions. The pendulum-swing angle is measured for each direction and the average calculated. This is a relatively time-consuming test.

This tester is limited to friction measurements under steady speed. A variable-speed motor is used for measuring the f–U (friction versus velocity) curve. However, each point in this curve is measured under steady-state conditions. Since this tester is not for high-precision measurement, it is not suitable for comparing lubricants where the difference in friction is expected to be within a few percentage points.

14.5 TESTING OF DYNAMIC FRICTION

Most of the commercially available friction-testing machines have been designed for measurements under steady conditions. For the measurement of dynamic friction, under time-variable conditions, a unique design of the testing equipment, with strict requirements, is called for. In addition to a rigid design, on-line recording of the data and its processing is essential for time-variable conditions.

The most important principles in dynamic friction measurement are as follows.

1. The machine as well as the support for the test bearing must be very rigid. In addition, the load cell for measuring the friction force must be as rigid as possible.
2. Relative sliding is obtained by means of a stationary part and a moving part. The load cell for the friction measurement must always be connected to the stationary side. If the load cell were to be connected to the moving part, it would not read the correct friction force because it would read inertial forces as friction force, and the result would be a combined reading of friction and inertial forces.
3. The design must provide for a clear method of separating the measured friction in the test bearing from any other sources of friction in the system.
4. The testing system must provide the means for accurate measurement of the velocity and displacement of the sliding part relative to the stationary part.
5. The system must provide the means for on-line recording of the friction versus time and versus sliding velocity. This is currently done by a computer with a data-acquisition system. In addition, there is a requirement to measure friction versus a small displacement during the start-up.

6. The system must include the means to control the desired time-dependent sliding motion and load. This can be achieved by using a computer with direct current output and an amplifier that controls a servomotor for the required motions. The controller in the computer includes the algorithm for the control of motion and velocity. The motion and velocity are measured on-line to provide feedback to the computer controller for precision motion.

If the support of the steady part is not sufficiently rigid (including the load cell), there are several types of errors that are encountered in the measurements. Under dynamic operation, the stationary part will have a small variable displacement due to the elasticity in the system. This would result in reading errors in the load cell because small inertial forces would be added to the friction reading. This means that due to a variable elastic displacement there is a small acceleration, and the load cell will read inertial forces as friction force.

Moreover, if the system were not rigid, there would be friction-induced vibrations (stick-slip friction, see Sec. 16.1) at low velocity. In conclusion, the dynamics of the system can affect the friction measurement, and we are interested in a clean experiment where the bearing friction is measured in isolation from any other effect.

The examples in the following sections are of several universal testing machines for measuring rolling-element bearing friction or journal bearing friction under dynamic conditions. Although other designs of friction-testing machines are often used, all are based on similar concepts. The first two friction-testing machines can be applied for a journal bearing or a rolling-element bearing; the third machine is for friction in linear sliding motion.

14.6 FRICTION-TESTING MACHINE WITH A HYDROSTATIC PAD

A friction-testing machine with hydrostatic pad is shown in Fig. 14-6. It has a main shaft supported by two conical rolling bearings. The two bearings form an adjustable arrangement to eliminate undesirable clearance in these bearings. The shaft is driven by a variable-speed motor. In Fig. 14-6, the test bearing is a rolling-element bearing on the right side of the shaft, but it can be a journal bearing as well. The test bearing is housed in a cylindrical casing containing lubricant at a constant level.

The main shaft ends with a cone, on which a conical bore sleeve is mounted. The conical sleeve can by tightened by a nut, and in this way the outside diameter of the sleeve is slightly varied by elastic deformation. The test bearing, a journal bearing or rolling bearing, is mounted on this sleeve, and the clearance

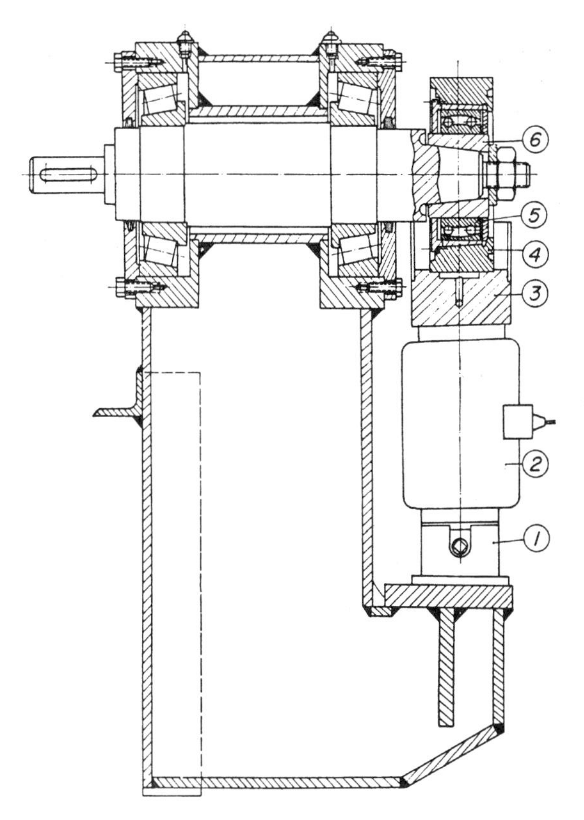

FIG. 14-6 Friction-testing machine with a hydrostatic pad. (From Harnoy, 1966).

(for a journal bearing) or the tight fit (for a rolling bearing) can be adjusted by tightening the nut on the conical sleeve.

The load is applied by means of a jack (1) and measured by a load cell (2). The bearing cylindrical casing is mounted on a hydrostatic pad (3). In this way, the load is transmitted through the hydrostatic fluid film. When the cylindrical casing is not turning, there are no shear stresses in the fluid film, and there is no additional viscous friction on the casing.

There are two symmetrical radial arms that are attached to the casing, on each side, and connected to load cells. The friction torque is measured by two calibrated load cells, which are connected to two symmetrical radial arms, thus preventing the casing from turning. Since there is no friction torque due to the hydrostatic pad, the torque on the casing that is read by the two load cells is equal to the friction torque of the test bearing.

This apparatus measures only the friction in the test bearing and not any other source of friction, such as the two conical bearings that support the shaft. This friction-testing machine is suitable for dynamic friction measurements as well as friction under steady conditions. For more details of this testing machine, see Lowey, Harnoy, and Bar-Nefi (1972).

The operation of this friction-testing machine under dynamic conditions requires a servomotor controlled by a computer and data-acquisition system, as described in Sections 14.7 and 14.8.

14.7 FOUR-BEARINGS MEASUREMENT APPARATUS

An apparatus for dynamic friction measurement has been designed, developed, and constructed in the bearing and lubrication laboratory of the Department of Mechanical Engineering at the New Jersey Institute of Technology. This apparatus can continuously measure the average dynamic friction of four equally loaded sleeve bearings in isolation from any other source of friction in the system, and the errors caused by inertial forces can be reduced to a negligible magnitude. In Fig. 14-7 a cross section of the apparatus is shown, and a photograph is shown in Fig. 14-8.

The design concept is to apply an internal load, action and reaction, between the inner housing (N) and the outer housing (K) by tightening the nut (P) on the bolt (R) to apply preload by deformation of the elastic steel ring (E). There are four equal test sleeve bearings (H), two bearings inside each housing. In this way, all four test bearings have approximately equal radial load, but in the opposite direction for each two of the four bearings, due to the preload in the elastic ring. The load on the bearings is measured by a calibrated, full strain gauge bridge bonded to the elastic ring. The total friction torque of all four bearings is measured by a calibrated rigid piezoelectric load cell, which prevents the rotation of the outer bearing housing (K). The load is transferred to the load cell by a radial arm attached to the external housing, as shown in the apparatus photograph in Fig. 14-8. Thus, the measured friction torque of the four bearings is isolated from any other sources of friction, such as friction in the ball bearings supporting the shaft. The time-variable friction measured by the load cell is stored in a computer with a data-acquisition system.

A lubricating oil reservoir is mounted above the mechanical apparatus in order to supply oil by gravity into the four bearings through four segments of flexible tubing. The oil is drained from the bearings through a hole in the external housing into a collecting vessel.

The shaft (C) is supported by two ball bearings (A) attached to the main support frame (B), and is driven by a computer-controlled DC servomotor. The

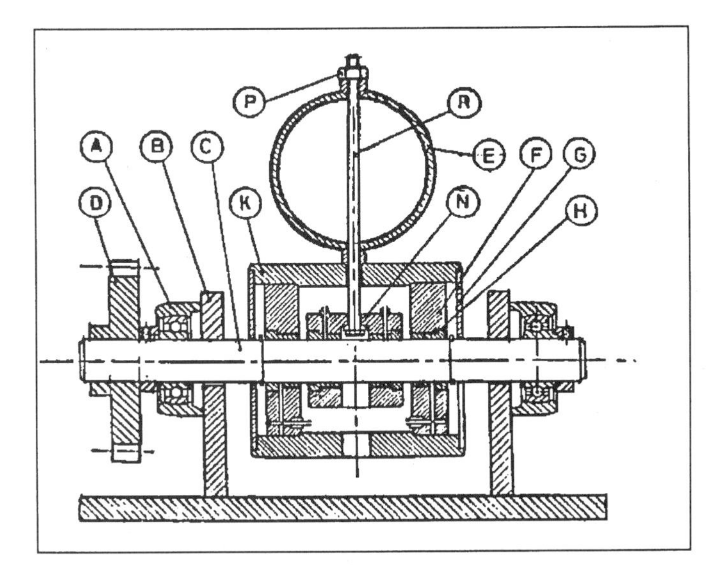

FIG. 14-7 Cross-sectional view of friction measurement apparatus. (From Bearing and Lubrication Laboratory, Department of Mechanical Engineering, New Jersey Institute of Technology.)

FIG. 14-8 Photograph of friction-testing apparatus. (From Bearing and Lubrication Laboratory, Department of Mechanical Engineering, New Jersey Institute of Technology.)

drive consists of a DC servomotor connected to the shaft through a timing belt and two pulleys (D). The rotational speed of the shaft is measured by an encoder, and this on-line measurement is fed into the computer, where the data is stored and analyzed. This arrangement forms a closed-loop control of the rotation of the shaft. In fact, the control algorithm includes a friction-compensation algorithm to generate the precise sinusoidal velocity or any other desired periodic velocity.

It is interesting to note that the measurement principle of four bearings was used by Mckee and Mckee as early as 1929. However, the early friction-testing apparatus used sliding weights for measuring the friction torque; and of course, this apparatus has been limited to bearings under steady conditions.

An improved version of the four-bearings friction-testing machine for bearings that require self-aligning is shown in Fig. 14-9. The self-aligning property is achieved by means of four self-aligning ball bearings. The self-aligning bearings are held from rotating by thin metal strips. At the same time, elastic bending of the strips allows a small angular rotation for self-aligning. In addition, this design can easily be adapted for measurement of steady and dynamic friction in rolling-element bearings.

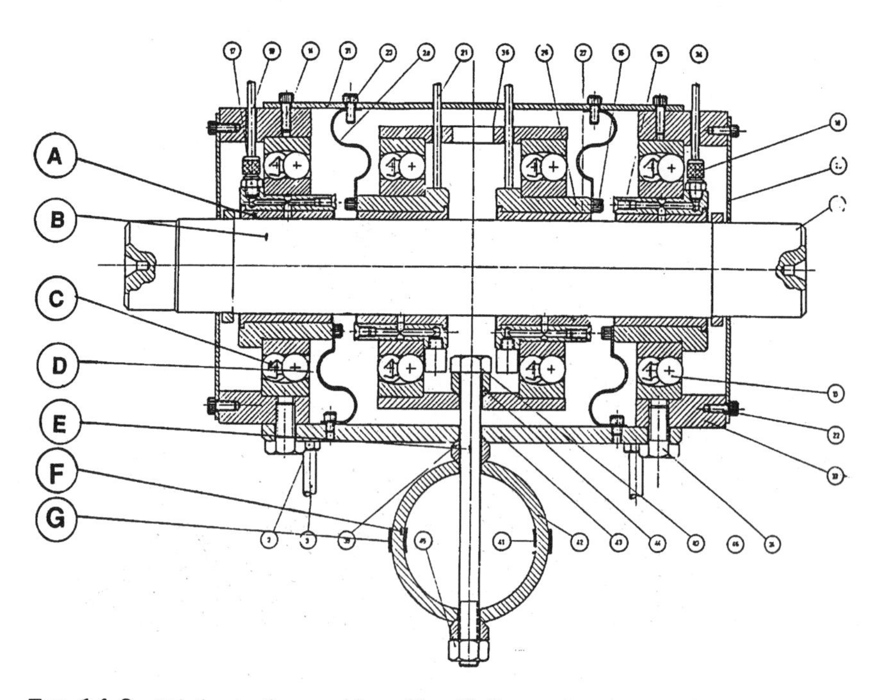

FIG. 14-9 Friction-testing machine with self-alignment arrangement.

14.7.1 Measurement Error Under Dynamic Conditions

It has already been mentioned that under dynamic conditions, there will always be a small, unsteady angular rotation of the bearing housing and sleeve due to the elasticity of its support (including elasticity of the load cell). This is the case in all the testing machines, because there is always a certain elasticity in the system. However, it is possible to minimize this elastic rotation of the bearing sleeve by a rigid design of the frame of the machine and by using a rigid load cell. As a result of this small angular elastic rotation of the housing, there will be a small angular acceleration, and the load cell, which keeps the bearing housing from rotating, has a small error because it reads inertial forces (or torque) as friction force.

For friction measurement under dynamic conditions, the magnitude of measurement error due to inertial forces has been evaluated by Harnoy et al. (1994). The condition for a negligible error is

$$\omega^2 \ll \frac{k_h}{I_h} \tag{14-1}$$

Here, k_h is the angular stiffness of the bearing housing support, ω is the freqency of oscillations under dynamic conditions, such as periodic load, and I_h is the moment of inertia of the bearing housing together with the bearing sleeve.

Piezoelectric load cells are much more rigid than strain gage, beam-type load-cells. However, for low-frequency tests, the piezoelectric cells have the disadvantage of an output drift. At the same time, at low frequency, our error analysis indicated that the elasticity of the load cell does not cause any significant measurement error. This discussion indicates that best results can be achieved by using a piezoelectric load cell for high-frequency tests and strain gage beam-type load cells for low-frequency oscillations.

14.8 APPARATUS FOR MEASURING FRICTION IN LINEAR MOTION

A cross-sectional view of the linear-motion friction measurement apparatus is shown in Fig. 14-10. The apparatus comprises of a lincar motion sliding table, driven by a servomotor and a ball screw drive. A closed-loop control system is provided via personal computer. Also, the computer system can store the experimental data for analysis. For regular precision, the sliding motion is measured by an encoder, which measures the rotation of the screw that drives the table. For high-precision measurement of the linear motion, an LVDT motion sensor can be used.

The design concept of the apparatus is a ball screw driven linear-positioning table (1), where the backlash can be eliminated, by preloading the screw drive.

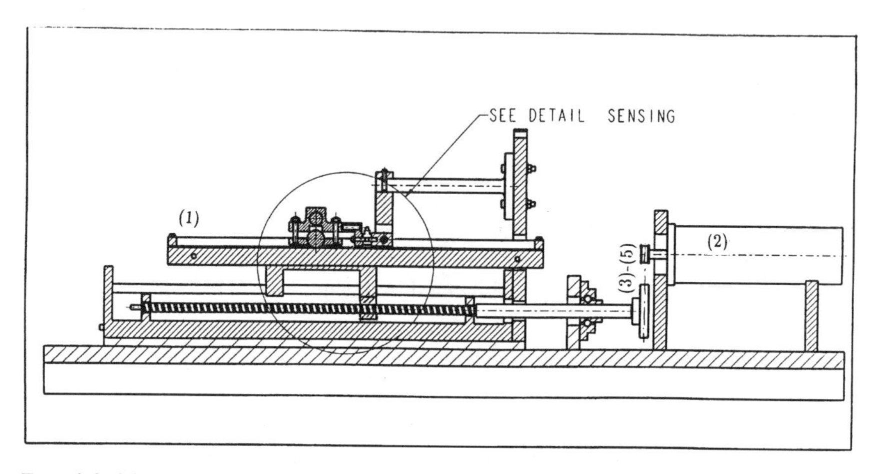

FIG. 14-10 Cross-sectional view of linear-motion friction measurement apparatus (from Bearing and Lubrication Laboratory Department of Mechanical Engineering, New Jersey Institute of Technology).

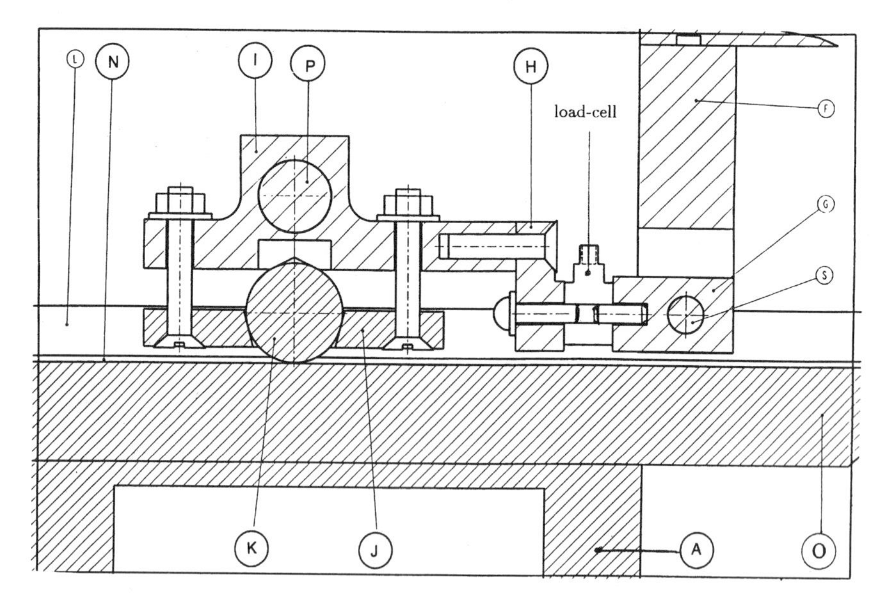

FIG. 14-11 Enlargement of cross-section contact area.

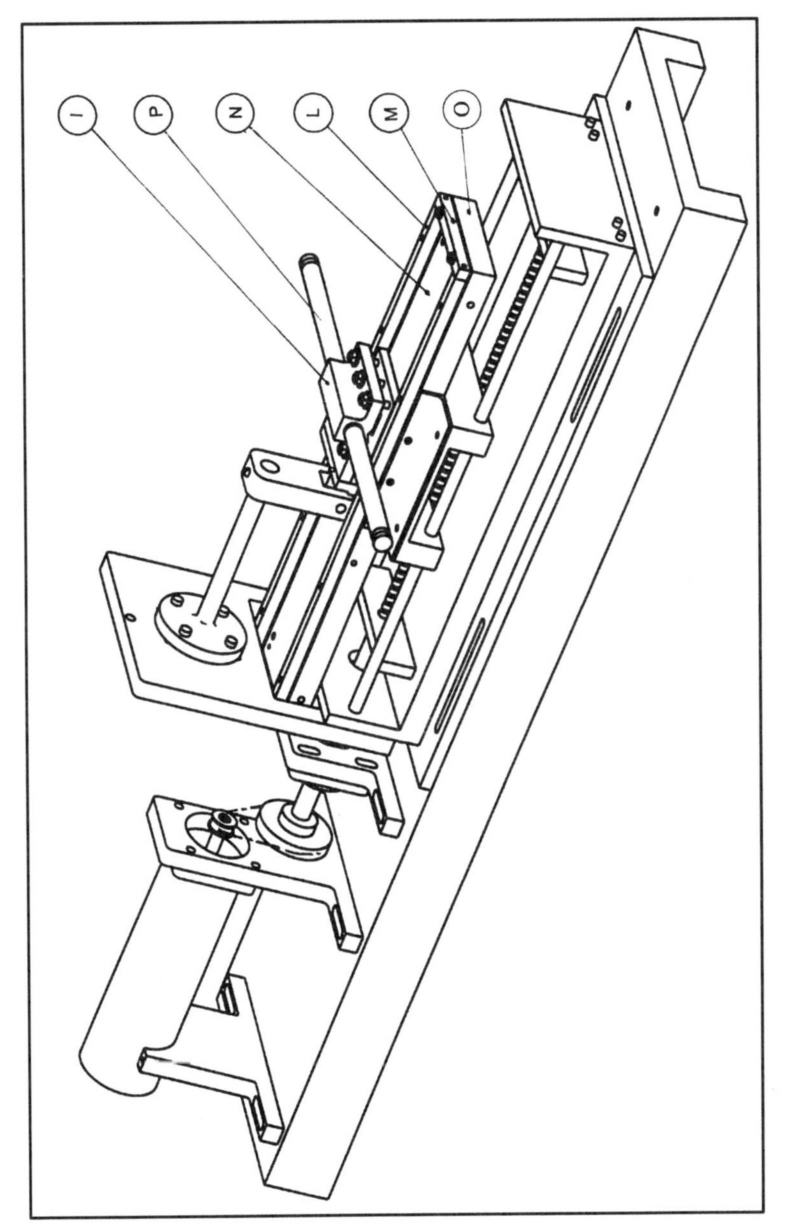

FIG. 14-12 Isometric of a linear-motion friction-measurement apparatus.

This apparatus is designed for measuring friction at very low sliding speeds. This is achieved by a speed reduction of considerable ratio by the screw drive. In addition, the speed of the motor is reduced by a set of pulleys and a timing belt (3-5). Closed-loop controlled motion is generated by a computer-controlled DC servomotor (2). Precise measurements of the motion is fed into a computer, which is equipped with a data-acquisition board.

In this linear apparatus, the contact geometry between the sliding surfaces can be replaced. Enlargement of the sensing area is shown in Fig. 14-11. It can test a sliding plane, a line or a point contact. The drawing shows a line contact between a nonrotating cylindrical shaft and a flat plate. The contact can be made of various material combinations. The line contact is created between a short, finely ground cylindrical shaft (K) and the flat friction surface (N). The shaft (K) is clamped in a housing assembly (I, J and H) designed to hold various shaft diameters. The normal load is centered above the line contact, and is supplied by a rod (P), which has weights attached (weights not shown). When the friction test surface moves, the friction force is transmitted through the housing assembly to a piezoelectric load cell. The load cell generates a voltage signal proportional to the friction force magnitude, which is fed to a data-acquisition system in a computer.

The design is shown in the isometric view in Fig. 14-12. The friction surface base (N) is attached to a moving platform (O) that is driven by the ball screw drive. The friction contact area can be dipped in lubricant, since the base has an attached railing (L and M), which serves to contain the lubricant.

For precise dynamic measurements, particularly with high-frequency oscillations, the load cell must be rigid as well as the support of the load cell. This is essential to prevent undesirable small linear displacement of the stationary cylinder (due to elastic deformation of the load-cell system and the support under the friction force). In the case of elastic displacement, the friction reading may include a small error of inertial force, acting on the load-cell (this is the major reason why most commercial friction testing devices are not suitable for dynamic tests).

Results of dynamic friction measurements performed by the last two testing machines are included in Chapter 17. The tests were conducted for oscillating motion at various frequencies, and the results are in the form of dynamic f–U curves.

15

Hydrodynamic Bearings Under Dynamic Conditions

15.1 INTRODUCTION

Hydrodynamic journal bearings under steady conditions were discussed in the previous chapters. The equations for pressure wave and load capacity were limited to a constant, steady load and constant speed. Under steady conditions of constant load and speed, the journal center is at a stationary point defined by a constant eccentricity and attitude angle. Under dynamic conditions, however, such as oscillating load or variable speed, the journal center moves relative to the bearing.

Under harmonic oscillations such as sinusoidal load, the journal center moves in a *trajectory* that repeats itself during each cycle. This type of trajectory is referred to as *journal center locus*. In practice, bearings in machines are always subjected to some dynamic conditions. In rotating machinery, there are always vibrations due to the shaft imbalance. The machine is a dynamic system that has a spectrum of vibration frequencies. Vibrations in a machine result in small oscillating forces (inertial forces) on the bearings at various frequencies, which are superimposed on the main, steady load. If the magnitude of the dynamic forces is very small in comparison to the main, steady force, the dynamic forces are disregarded.

However, there are many important cases where the dynamic bearing performance is important and must be analyzed. For example, the effects of bearing whirl near the critical speeds of the shaft can result in bearing failure. Dynamic analysis must always be performed in critical applications where bearing failure is expensive, such as the high cost of loss of production in generators or steam turbines or where there are safety considerations. In these cases, it is important for the design engineer to perform a dynamic analysis in order to predict undesired dynamic effects and prevent them by appropriate design.

In many machines the load is not steady. For example, the bearings in car engines are subjected to a cycle of a variable force that results from the combustion and inertial forces in the engine. There are many variable-speed machines that involve unsteady bearing performance, and even machines that operate at steady conditions are subjected to dynamic conditions during start-up and stopping.

In fact, most bearing failures result from an unexpected dynamic effect, such as a large vibration or severe disturbances. Engineers can improve the resistance of hydrodynamic bearings to unexpected dynamic effects by comparing the dynamic response of various bearing designs to scenarios of possible disturbances. An example of a unique design that improves the dynamic response is included in Chapter 18.

15.2 ANALYSIS OF SHORT BEARINGS UNDER DYNAMIC CONDITIONS

The following is a dynamic analysis of a short bearing. Short bearings are widely used in many applications under dynamic conditions, including car engines. The dynamic analysis of a short bearing is relatively simple because the bearing load can be expressed by a closed-form equation, as shown in Chapter 7. This analysis can be extended to a finite-length bearing, but the computations are more complex because the load capacity at each step must be determined by a numerical procedure.

The objective of a dynamic analysis is to solve for the trajectory of the journal center. The analysis involves the derivation of a set of differential equations and their solution by a finite-difference method with the aid of a computer program.

Dubois and Ocvirk (1953) solved the pressure distribution and load capacity of a short journal bearing under a steady load (see Chapter 7). This analysis is extended here to include unsteady conditions where the journal center, O_1, has an arbitrary velocity. It is shown in Fig. 15-1 that the journal center velocity is described by two components, $de/dt = C(d\varepsilon/dt)$ and $e\,(d\phi/dt)$, in the radial and tangential directions, respectively. The two assumptions of Dubois and Ocvirk for a steady short bearing are maintained here for dynamic conditions.

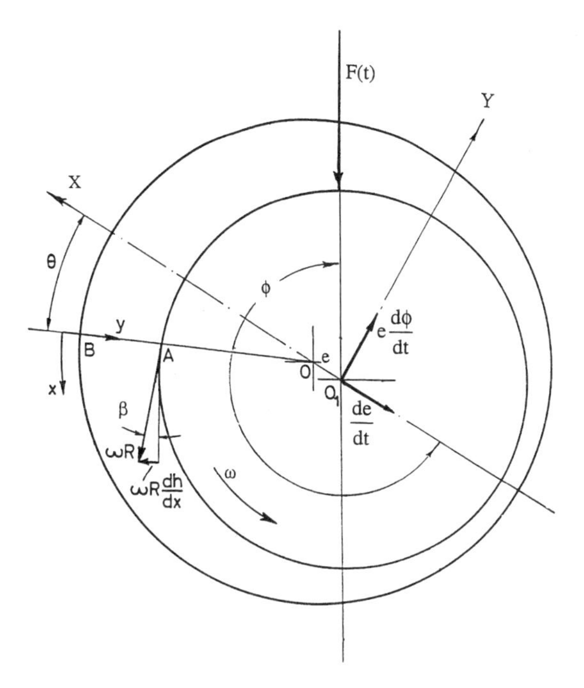

FIG. 15-1 Velocity components of the journal center under dynamic conditions.

First, pressure gradients in the x direction (around the bearing) are negligible in comparison to pressure gradients in the z (axial) direction: see Fig. 7-1. Second, only the pressure in the converging clearance region ($0 < \theta < \pi$) is considered for the force calculations, where the pressure is above atmospheric pressure. In addition, the assumptions of classical hydrodynamic theory are maintained: The viscosity is assumed to be constant (at an equivalent average temperature). Effects of the fluid mass (inertial forces) are neglected, as is fluid curvature. However, the journal mass is significant and must be considered for dynamic analysis.

The starting point of the analysis of a short journal bearing under dynamic conditions is the general Reynolds equation. Let us recall that the Reynolds equation for incompressible Newtonian fluid is

$$\frac{\partial}{\partial x}\left(\frac{h^3}{\mu}\frac{\partial p}{\partial x}\right) + \frac{\partial}{\partial z}\left(\frac{h^3}{\mu}\frac{\partial p}{\partial z}\right) = 6(U_1 - U_2)\frac{\partial h}{\partial x} + 12(V_2 - V_1) \tag{15-1}$$

The velocities on the right-hand side of this equation are in Fig. 5-2. If the bearing is stationary and the shaft rotates, the fluid film boundary conditions on the bearing surface are

$$U_1 = 0, \qquad V_1 = 0 \tag{15-2}$$

Under dynamic conditions, the velocity on the journal surface is a vector summation of the velocity of the journal center, O_1, and the journal surface velocity relative to that center. The journal center has radial and tangential velocity components, as shown in Fig. 15-1. The components are $d\varepsilon/dt = C(d\varepsilon/dt)$ in the radial direction and $e(d\phi/dt)$ in the tangential direction. Summation of the velocity components of O_1 with that of the journal surface relative to O_1 results in the following components, U_2 and V_2 (see Fig. 5-2 for the direction of the components):

$$U_2 = \omega R + \frac{de}{dt}\sin\theta - e\frac{d\phi}{dt}\cos\theta \tag{15-3}$$

$$V_2 = \omega R\frac{dh}{dx} + \frac{de}{dt}\cos\theta + e\frac{d\phi}{dt}\sin\theta \tag{15-4}$$

Here, h is the fluid film thickness around a journal bearing, given by the equation

$$h = C(1 + \varepsilon\cos\theta) \tag{15-5}$$

After substitution of U_2 and V_2 as well as U_1 and V_1 in the right-hand side of the Reynolds equation, Eq. (15-1), the pressure distribution can be derived in a similar way to that of a steady short bearing. The load components are obtained by integrating the pressure in the converging clearance only $(0 < \theta < \pi)$ as follows:

$$W_x = -2R\int_0^{\pi}\int_0^{L/2} p\cos\theta\, d\theta\, dz \tag{15-6}$$

$$W_y = 2R\int_0^{\pi}\int_0^{L/2} p\sin\theta\, d\theta\, dz \tag{15-7}$$

Converting into dimensionless terms, the equations for the two load capacity components (in the X and Y directions as shown in Fig. 15-1) become

$$\overline{W}_x = -0.5\varepsilon J_{12}|\overline{U}| + \varepsilon\dot{\phi}J_{12} + \dot{\varepsilon}J_{22} \tag{15-8}$$

$$\overline{W}_y = +0.5\varepsilon J_{11}\overline{U} - \varepsilon\dot{\phi}J_{11} - \dot{\varepsilon}J_{12} \tag{15-9}$$

Here, the dimensionless load capacity and velocity are

$$\overline{W} = \frac{C^2}{\mu U_0 L^3} W \qquad \text{and} \qquad \overline{U} = \frac{U}{U_0} \tag{15-10}$$

where $U = \omega R$ is the time-variable velocity of the journal surface and U_0 is a reference constant velocity used for normalizing the velocity. The integrals J_{ij} and their solutions are given in Eq. (7-13).

Under steady conditions, the external force, F, is equal to the bearing load capacity, W. However, under dynamic conditions, the resultant vector of the two forces accelerates the journal mass according to Newton's second law:

$$\vec{F} - \vec{W} = m\vec{a} \tag{15-11}$$

Here, m is the mass of the journal, $\vec{a}$ is the acceleration vector of the journal center, $\vec{F}$ is the external load, and $\vec{W}$ is the hydrodynamic load capacity. Under general dynamic conditions, $\vec{F}$ and $\vec{W}$ are not necessarily in the same direction, and both can be a function of time. In order to convert Eq. (15-11) to dimensionless terms, the following dimensionless variables are defined:

$$\overline{m} = \frac{C^3 U_0}{\mu L^3 R^2} m, \qquad \overline{F} = \frac{C^2}{\mu U_0 L^3} F \tag{15-12}$$

Dividing Eq. (15-11) into two components in the directions of $\overline{F}_x$ and $\overline{F}_y$ (along X and Y but in opposite directions) and substituting the acceleration components in the radial and tangential directions in polar coordinates, the following two equations are obtained:

$$\overline{F}_x - \overline{W}_x = \overline{m}\ddot{\varepsilon} - \overline{m}\varepsilon\dot{\phi}^2 \tag{15-13}$$

$$\overline{F}_y - \overline{W}_y = -\overline{m}\varepsilon\ddot{\phi} - 2\overline{m}\dot{\varepsilon}\dot{\phi} \tag{15-14}$$

The minus signs in Eq. 15-14 are minus because $\overline{F}_y$ is in the opposite direction to the acceleration. Here, the dimensionless time is defined as $\overline{t} = \omega t$, and the dimensionless time derivatives are

$$\dot{\varepsilon} = \frac{1}{\omega}\frac{d\varepsilon}{dt}, \qquad \dot{\phi} = \frac{1}{\omega}\frac{d\phi}{dt} \tag{15-15}$$

Substituting the values of the components of the load capacity, W_x and W_y, from Eqs. (15-8) and (15-9) into Eqs. (15-13) and (15-14), the following two differential equations for the journal center motion are obtained:

$$\overline{F}(t)\cos(\phi - \pi) = -0.5\varepsilon J_{12}|\overline{U}| + J_{12}\varepsilon\dot{\phi} + J_{22}\dot{\varepsilon} + \overline{m}\ddot{\varepsilon} - \overline{m}\varepsilon\dot{\phi}^2 \tag{15-16}$$

$$\overline{F}(t)\sin(\phi - \pi) = 0.5\varepsilon J_{11}\overline{U} - J_{11}\varepsilon\dot{\phi} - J_{12}\dot{\varepsilon} - \overline{m}\varepsilon\ddot{\phi} - 2\overline{m}\dot{\varepsilon}\dot{\phi} \tag{15-17}$$

Here, $\overline{F}(t)$ is a time-dependent dimensionless force acting on the bearing. The force (magnitude and direction) is a function of time. In the two equations, ε is the eccentricity ratio, ϕ is the attitude angle, and $\overline{m}$ is dimensionless mass, defined by Eq. (15-12). The definition of the integrals J_{ij} and their solution are in Chapter 7.

Equations (15-16) and (15-17) are two differential equations required for the solutions of the two time-dependent functions ε and ϕ. The variables ε and ϕ represent the motion of the shaft center, O_1, with time, *in polar coordinates*. The solution of the two equations as a function of time is finally presented as a plot of the trajectory of the journal center. If there are steady-state oscillations, such as sinusoidal force, after the initial transient, the trajectory becomes a closed locus that repeats itself each load cycle. A repeated trajectory is referred to as a *journal center locus*.

15.3 JOURNAL CENTER TRAJECTORY

The integration of Eqs. (15-16) and (15-17) is performed by finite differences with the aid of a computer program. Later, a computer graphics program is used to plot the journal center motion. The plot of the time variables ε and ϕ, in polar coordinates, represents the trajectory of the journal center motion relative to the bearing. The eccentricity ratio ε is a radial coordinate and ϕ is an angular coordinate.

Under harmonic conditions, such as sinusoidal load, the trajectory is a closed loop, referred to as a *locus*. Under harmonic oscillations of the load, there is initially a transient trajectory; and after a short time, a steady state is reached where the locus repeats itself during each cycle.

In heavily loaded bearings, the locus can approach the circle $\varepsilon = 1$, where there is a contact between the journal surface and the sleeve. The results allow comparison of various bearing designs. The design that results in a locus with a lower value of maximum eccentricity ratio ε is preferable, because it would resist more effectively any unexpected dynamic disturbances.

15.4 SOLUTION OF JOURNAL MOTION BY FINITE-DIFFERENCE METHOD

Equations (15-16) and (15-17) are the two differential equations that are solved for the function of ε versus ϕ. The two equations contain first- and second-order time derivatives and can be solved by a finite-difference procedure. The equations are not linear because the acceleration terms contain second-power time derivatives. Similar equations are widely used in dynamics and control, and commercial software is available for numerical solution. However, the reader will find it beneficial to solve the equations by himself or herself, using a computer and any programming language that he or she prefers. The following is a demonstration of a solution by a simple finite-difference method.

The principle of the finite-difference solution method is the replacement of the time derivatives by the following finite-difference equations (for simplifying the finite difference procedure, $\bar{F}$, $\bar{m}$ and $\bar{t}$ are renamed F, m and t):

$$\dot{\phi}_n = \frac{\phi_{n+1} - \phi_{n-1}}{2\,\Delta t}; \qquad \dot{\varepsilon}_n = \frac{\varepsilon_{n+1} - \varepsilon_{n-1}}{2\,\Delta t} \tag{15-18}$$

and the second time derivatives are

$$\ddot{\phi}_n = \frac{\phi_{n+1} - 2\phi_n + \phi_{n-1}}{\Delta t^2}; \qquad \ddot{\varepsilon}_n = \frac{\varepsilon_{n+1} - 2\varepsilon_n + \varepsilon_{n-1}}{\Delta t^2} \tag{15-19}$$

For the nonlinear terms (the last term in the two equations), the equation can be linearized by using the following backward difference equations:

$$\dot{\phi}_n = \frac{\phi_n - \phi_{n-1}}{\Delta t} \tag{15-20}$$

By substituting the foregoing finite-element terms for the time-derivative terms, the two unknowns ε_{n+1} and ϕ_{n+1} can be solved as two unknowns in two regular linear equations.

After substitution, the differential equations become

$$F_x + \frac{1}{2}\varepsilon_n J_{12} = \varepsilon_n J_{12}\left(\frac{\phi_{n+1} - \phi_{n-1}}{2\,\Delta t}\right) + J_{22}\left(\frac{\varepsilon_{n+1} - \varepsilon_{n-1}}{2\,\Delta t}\right) + m\left(\frac{\varepsilon_{n+1} - 2\varepsilon_n + \varepsilon_{n-1}}{\Delta t^2}\right) - m\varepsilon_n\left(\frac{\phi_n - \phi_{n-1}}{\Delta t}\right)^2 \tag{15-21}$$

$$F_y - \frac{1}{2}\varepsilon_n J_{11} = -\varepsilon_n J_{11}\left(\frac{\phi_{n+1} - \phi_{n-1}}{2\,\Delta t}\right) - J_{12}\left(\frac{\varepsilon_{n+1} - \varepsilon_{n-1}}{2\,\Delta t}\right) - m\varepsilon_n\left(\frac{\phi_{n+1} - 2\phi_n + \phi_{n-1}}{\Delta t^2}\right) - 2m\left(\frac{\varepsilon_{n+1} - \varepsilon_{n-1}}{2\Delta t}\right) \times \left(\frac{\phi_n - \phi_{n-1}}{\Delta t}\right) \tag{15-22}$$

Here, F_x and F_y are the external load components in the X and Y directions, respectively. Under dynamic conditions, the load components vary with time:

$$F_x = F(t)(\cos\phi - \pi) \tag{15-23}$$

$$F_y = F(t)(\sin\phi - \pi) \tag{15-24}$$

Equations (15-21) and (15-22) can be rearranged as two linear equations in terms of ε_{n+1} and ϕ_{n+1} as follows:

$$\text{Rearranging Eq. (15-21):}\quad A = B\varepsilon_{n+1} + C\phi_{n+1} \tag{15-25}$$

$$\text{Rearranging Eq. (15-22):}\quad P = R\varepsilon_{n+1} + Q\phi_{n+1} \tag{15-26}$$

In the following equations, F and m are dimensionless terms (the bar is omitted for simplification). The values of the coefficients of the unknown variables [in Eqs. (15-25) and (15-26)] are

$$A = F_X + \frac{\varepsilon_n J_{12}}{2} + \frac{\varepsilon_n J_{12}\phi_{n-1}}{2\,\Delta t} + \frac{J_{22}\varepsilon_{n-1}}{2\,\Delta t} + \frac{2m\varepsilon_n}{\Delta t^2} - \frac{m\varepsilon_{n-1}}{\Delta t^2} + m\varepsilon_n\left(\frac{\phi_n - \phi_{n-1}}{\Delta t}\right)^2 \tag{15-27}$$

$$B = \frac{J_{22}}{2\,\Delta t} + \frac{m}{\Delta t^2} \tag{15-28}$$

$$C = \frac{\varepsilon_n J_{12}}{2\,\Delta t} \tag{15-29}$$

$$P = F_y - \frac{\varepsilon_n J_{11}}{2} - \frac{\varepsilon_n J_{11}\phi_{n-1}}{2\,\Delta t} - \frac{J_{12}\varepsilon_{n-1}}{2\,\Delta t} - \frac{2m\varepsilon_n\phi_n}{\Delta t^2} + \frac{m\varepsilon_n\phi_{n-1}}{\Delta t^2} - \frac{m\varepsilon_{n-1}\phi_n}{\Delta t^2} + \frac{m\varepsilon_{n-1}\phi_{n-1}}{\Delta t^2} \tag{15-30}$$

$$Q = -\frac{\varepsilon_n J_{11}}{2\,\Delta t} - \frac{m\varepsilon_n}{\Delta t^2} \tag{15-31}$$

$$R = -\frac{J_{12}}{2\,\Delta t} - \frac{m}{\Delta t^2}\phi_n + \frac{m\phi_{n-1}}{\Delta t^2} \tag{15-32}$$

The numerical solution of the two equations for the two unknowns becomes

$$\varepsilon_{n+1} = \frac{AQ - PC}{BQ - RC} \tag{15-33}$$

$$\phi_{n+1} = \frac{AR - PB}{CR - QB} \tag{15-34}$$

The last two equations make it possible to march from the initial conditions and find ε_{n+1} and ϕ_{n+1} from any previous values, in dimensionless time intervals of $\Delta\bar{t} = \omega\Delta t$.

For a steady-state solution such as periodic load, the first two initial values of ε and ϕ can be selected arbitrarily. The integration of the equations must be conducted over sufficient cycles until the initial transient solution decays and a periodic steady-state solution is reached, i.e., when the periodic ε and ϕ will repeat at each cycle.

The following example is a solution for the locus of a short hydrodynamic bearing loaded by a sinusoidal force that is superimposed on a constant vertical load. The example compares the locus of a Newtonian and a viscoelastic fluid. The load is according to the equation

$$F(t) = 800 + 800 \sin 2\omega t \tag{15-35}$$

In this equation, ω is the journal angular speed. This means that the frequency of the oscillating load is twice that of the journal rotation. The direction of the load is constant, but its magnitude is a sinusoidal function. The dimensionless load is according to the definition in Eq. (15-12).

The dimensionless mass is $\bar{m} = 100$ and the journal velocity is constant. The resulting steady-state locus is shown in Fig. 15-2 by the full line for a Newtonian fluid. The dotted line is for a viscoelastic lubricant under identical

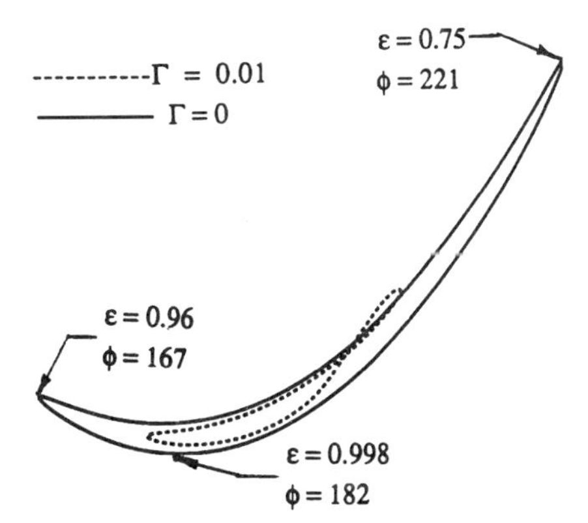

FIG. 15-2 Locus of the journal center for the load $F_t = 800 + 800 \sin 2\omega t$ and journal mass $\bar{m} = 100$.

conditions (see Chapt. 19). The viscoelastic lubricant is according to the Maxwell model in Chapter 2 [Eq. (2-9)]. The dimensionless viscoelastic parameter Γ is

$$\Gamma = \lambda\omega \tag{15-36}$$

where λ is the relaxation time of the fluid and ω is the constant angular speed of the shaft. In this case, the result is dependent on the ratio of the load oscillation frequency, ω_1, and the shaft angular speed, ω_1/ω.

16

Friction Characteristics

16.1 INTRODUCTION

The first friction model was the Coulomb model, which states that the friction coefficient is constant. Recall that the friction coefficient is the ratio

$$f = \frac{F_f}{F} \tag{16-1}$$

where F_f is the friction force in the direction tangential to the sliding contact plane and F is the load in the direction normal to the contact plane. Discussion of the friction coefficient for various material combinations is found in Chapter 11.

For many decades, engineers have realized that the simplified Coulomb model of constant friction coefficient is an oversimplification. For example, static friction is usually higher than kinetic friction. This means that for two surfaces under normal load F, the tangential force F_f required for the initial breakaway from the rest is higher than that for later maintaining the sliding motion. The static friction force increases after a rest period of contact between the surfaces under load; it is referred to as *stiction force* (see an example in Sec. 16.3). Subsequent attempts were made to model the friction as two coefficients of static and kinetic friction. Since better friction models have not been available, recent analytical studies still use the model of static and kinetic friction coefficients to analyze friction-induced vibrations and stick-slip friction effects in dynamic systems. However, recent experimental studies have indicated that this model of static and kinetic friction coefficients is not accurate. In fact, a better description of the friction characteristics is that of a continuous function of friction coefficient versus sliding velocity.

The friction coefficient of a particular material combination is a function of many factors, including velocity, load, surface finish, and temperature. Nevertheless, useful tables of constant static and kinetic friction coefficients for various material combinations are currently included in engineering handbooks. Although it is well known that these values are not completely constant, the tables are still useful to design engineers. Friction coefficient tables are often used to get an idea of the approximate average values of friction coefficients under normal conditions.

Stick-slip friction: This friction motion is combined of short consecutive periods of stick and slip motions. This phenomenon can take place whenever there is a low stiffness of the elastic system that supports the stationary or sliding body, combined with a negative slope of friction coefficient, f, versus sliding velocity, U, at low speed. For example, in the linear-motion friction apparatus (Fig. 14-10), the elastic belt of the drive reduces the stiffness of the support of the moving part.

In the stick period, the motion is due to elastic displacement of the support (without any relative sliding). This is followed by a short period of relative sliding (slip). These consecutive periods are continually repeated. At the stick period, the motion requires less tangential force for a small elastic displacement than for breakaway of the stiction force. The elastic force increases linearly with the displacement (like a spring), and there is a transition from stick to slip when the elastic force exceeds the stiction force, and vice versa. The system always selects the stick or slip mode of minimum resistance force.

In the past, the explanation was based on static friction greater than the kinetic friction. It has been realized, however, that the friction is a function of the velocity, and the current explanation is based on the negative $f - U$ slope, see a simulation by Harnoy (1994).

16.2 FRICTION IN HYDRODYNAMIC AND MIXED LUBRICATION

Hydrodynamic lubrication theory was discussed in Chapters 4–9. In journal and sliding bearings, the theory indicates that the lubrication film thickness increases with the sliding speed. Full hydrodynamic lubrication occurs when the sliding velocity is above a minimum critical velocity required to generate a full lubrication film having a thickness greater than the size of the surface asperities. In full hydrodynamic lubrication, there is no direct contact between the sliding surfaces, only viscous friction, which is much lower than direct contact friction. In full fluid film lubrication, the viscous friction increases with the sliding speed, because the shear rates and shear stresses of the fluid increase with that speed.

Below a certain critical sliding velocity, there is mixed lubrication, where the thickness of the lubrication film is less than the size of the surface asperities. Under load, there is a direct contact between the surfaces, resulting in elastic as well as plastic deformation of the asperities. In the mixed lubrication region, the external load is carried partly by the pressure of the hydrodynamic fluid film and partly by the mechanical elastic reaction of the deformed asperities. The film thickness increases with sliding velocity; therefore as the velocity increases, a larger portion of the load is carried by the fluid film. The result is that the friction decreases with velocity in the mixed region, because the fluid viscous friction is lower than the mechanical friction at the contact between the asperities.

The early measurements of friction characteristics have been described by f–U curves of friction coefficient versus sliding velocity by Stribeck (1902) and by McKee and McKee (1929). These f–U curves were measured under steady conditions and are referred to as *Stribeck curves*. Each point of these curves was measured under steady-state conditions of speed and load.

The early experimental f–U curves of lubricated sliding bearings show a nearly constant friction at very low sliding speed (boundary lubrication region). However, for metal bearing materials, our recent experiments in the Bearing and Bearing Lubrication Laboratory at the New Jersey Institute of Technology, as well as experiments by others, indicated a continuous steep downward slope of friction from zero sliding velocity without any distinct friction characteristic for the boundary lubrication region. The recent experiments include friction force measurement by load cell and on-line computer data acquisition. Therefore, better precision is expected than with the early experiments, where each point was measured by a balance scale.

An example of an f–U curve is shown in Fig. 16-1. This curve was produced in our laboratory for a short journal bearing with continuous lubrication. The experiment was performed under "quasi-static" conditions; namely, it was conducted for a sinusoidal sliding velocity at very low frequency, so it is equivalent to steady conditions. The curve demonstrates high friction at zero velocity (stiction, or static friction force), a steep negative friction slope at low velocity (boundary and mixed friction region), and a positive slope at higher velocity (hydrodynamic region). There are a few empirical equations to describe this curve at steady conditions. The negative slope of the f–U curve at low velocity is used in the explanation of several friction phenomena. Under certain conditions, the negative slope can cause instability, in the form of stick-slip friction and friction-induced vibrations (Harnoy 1995, 1996).

In the boundary and mixed lubrication regions, the viscosity and boundary friction additives in the oil significantly affect the friction characteristics. In

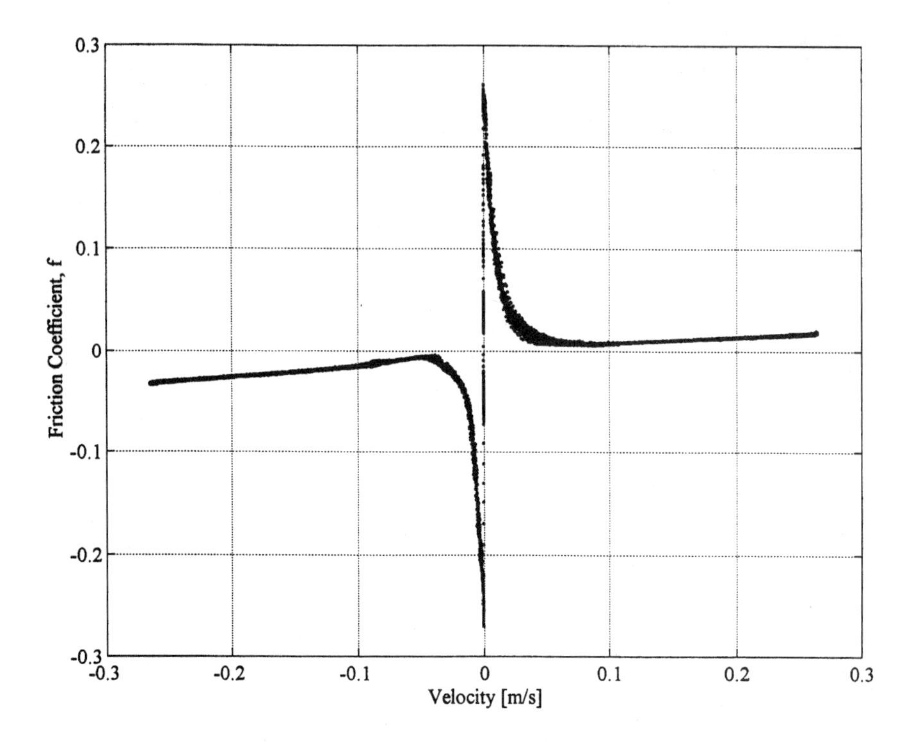

Fig. 16-1 f–U curve for sinusoidal velocity: oscillation frequency $= 0.0055$ rad/s, load $= 104$ N, 25-mm journal, $L/D = 0.75$, lubricant SAE 10W-40, steel on brass.

addition, the breakaway and boundary friction coefficients are higher with a reduced bearing load. For example, Fig. 16-2 is f–U curve for a low-viscosity lubricant without any additives for boundary friction reduction and lower bearing load. The curve indicates a higher breakaway friction coefficient than that in Fig. 16-1 lubricated with engine oil SAE 10W-40. The breakaway friction in Fig. 16-2 is about that of dry friction. However, for the two oils, the friction at the transition from mixed to full film lubrication is very low.

The steep negative slope in the mixed region has practical consequences on the accuracy of friction measurements that are widely used to determine the effectiveness of boundary layer lubricants. Currently such lubricants are evaluated by measuring the friction at an arbitrary constant sliding speed (e.g., a four-ball tester operating at constant speed). However, the f–U curve in Fig. 16-1 indicates that this measurement is very sensitive to the test speed. Apparently, a better evaluation should be obtained by testing the complete f–U curve. Similar to the journal bearing, the four-ball tester has a hydrodynamic fluid film; in turn, the

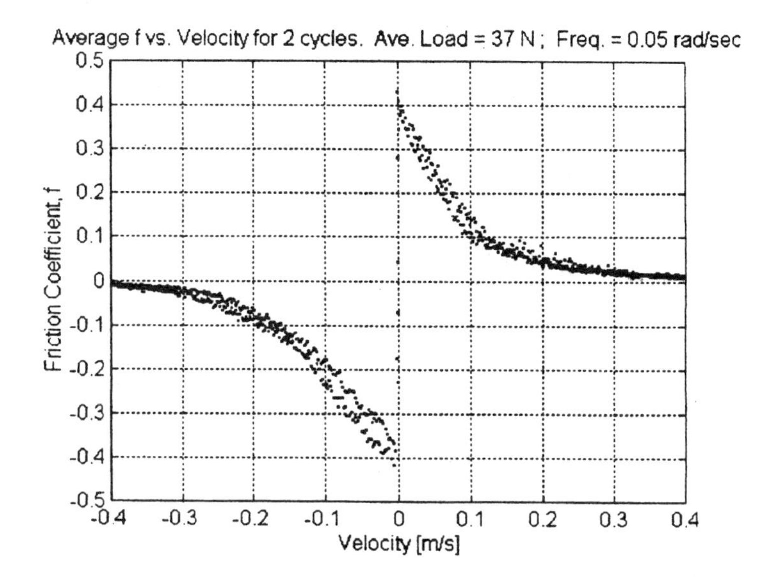

FIG. 16-2 f–U curve for sinusoidal velocity: oscillation frequency = 0.05 rad/s, load = 37 N, 25-mm journal, $L/D = 0.75$, low-viscosity oil, $\mu = 0.001\ \text{N-s/m}^2$, no additives, steel on brass.

friction torque is a function of the sliding speed (or viscosity). The current testing methods for boundary lubricants should be reevaluated, because they rely on the assumption that there is one boundary-lubrication friction coefficient, independent of sliding speed.

For journal bearings in the hydrodynamic friction region, the friction coefficient f is a function not only of the sliding speed but of the Sommerfeld number. Analytical curves of $(R/C)f$ versus Sommerfeld number are presented in the charts of Raimondi and Boyd; see Fig. 8-3. These charts are for partial journal bearings of various arc angles β. These charts are only for the full hydrodynamic region and do not include the boundary, or mixed, lubrication region. For a journal bearing of given geometry, the ratio C/R is constant. Therefore, empirical charts of friction coefficient f versus the dimensionless ratio $\mu n/P$, are widely used to describe the characteristic of a specific bearing. In the early literature, the notation for viscosity is z, and charts of f versus the variable zN/P were widely used (Hershey number—see Sec. 8.7.1).

16.2.1 Friction in Rolling-Element Bearings

Stribeck measured similar f–U curves (friction coefficient versus rolling speed) for lubricated ball bearings and published these curves for the first time as early as

1902. Rolling-element bearings operating with oil lubrication have a similar curve: an initial negative slope and a subsequent rise of the friction coefficient versus speed (due to increasing viscous friction). Although there is a similarity in the shapes of the curves, the breakaway friction coefficient of rolling bearings is much lower than that of sliding bearings, such as journal bearings. This is obvious because rolling friction is lower than sliding friction. The load and the bearing type affect the friction coefficient. For example, cylindrical and tapered rolling elements have a significantly higher friction coefficient than ball bearings.

16.2.2 Dry Friction Characteristics

Dry friction characteristics are not the same as for lubricated surfaces. The f–U curve for dry friction is not similar to that of lubricated friction, even for the same material combination. For dry surfaces after the breakaway, the friction coefficient can increase or decrease with sliding speed, depending on the material combination. For most metals, the friction coefficient has negative slope after the breakaway. An example is shown in Fig. 16-3 for dry friction of a journal bearing made of a steel shaft on a brass sleeve. This curve indicates a considerably higher friction coefficient at the breakaway from zero velocity (about 0.42 in comparison to 0.26 for a lubricated journal bearing—half of the breakaway friction). In addition, a dry bearing has a significantly greater gradual reduction of friction with velocity (steeper slope).

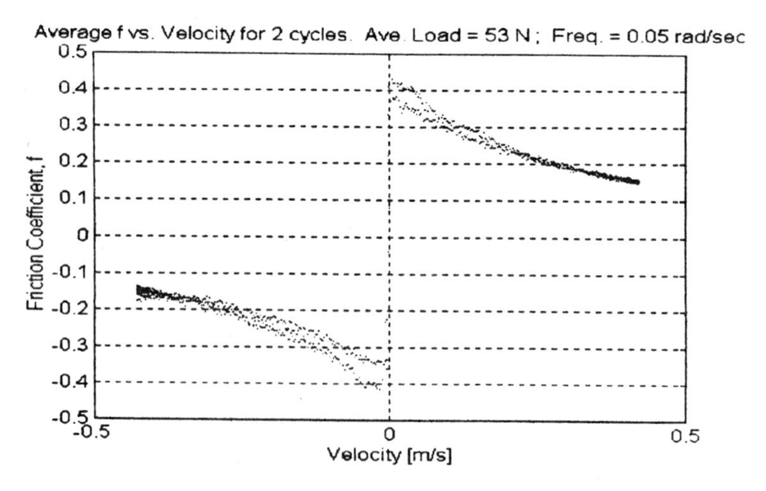

FIG. 16-3 f–U curve for sinusoidal velocity: oscillation frequency = 0.05 rad/s, load = 53 N, dry surfaces, steel on brass, 25-mm journal, $L/D = 0.75$.

16.2.3 Effects of Surface Roughness on Dry Friction

As already discussed, smooth surfaces are desirable for hydrodynamic and mixed lubrication. However, for dry friction of metals with very smooth surfaces there is adhesion on a larger contact area, in comparison to rougher surfaces. In turn, ultrasmooth surfaces adhere to each other, resulting in a higher dry friction coefficient. For very smooth surfaces, surface roughness below 0.5 μm, the friction coefficient f reduces with increasing roughness. At higher roughness, in the range of about 0.5–10 μm (20–40 microinches), the friction coefficient is nearly constant. At a higher range of roughness, above 10 μm, the friction coefficient f increases with the roughness because there is increasing interaction between the surface asperities (Rabinovitz, 1965).

16.3 FRICTION OF PLASTIC AGAINST METAL

There is a fundamental difference between dry friction of metals (Fig. 16-3) where the friction goes down with velocity, and dry friction of a metal on soft plastics (Fig. 16-4a) where the friction coefficient increases with the sliding velocity. Figure 16-4a is for sinusoidal velocity of a steel shaft on a bearing made of ultrahigh-molecular-weight polyethylene (UHMWPE). This curve indicates that there is a considerable viscous friction that involves in the rubbing of soft plastics. In fact, soft plastics are viscoelastic materials.

In contrast, for lubricated surfaces, the friction reduces with velocity (Fig. 16-4b) due to the formation of a fluid film. In Fig. 16-4b, the dots of higher friction coefficient are for the first cycle where there is an example of relatively higher *stiction* force, after a rest period of contact between the surfaces under load.

16.4 DYNAMIC FRICTION

Most of the early research in tribology was limited to steady friction. The early f–U curves were tested under steady conditions of speed and load. For example, the f–U curves measured by Stribeck (1902) and by McKee and McKee (1929) do not describe "dynamic characteristics" but "steady characteristics", because each point was measured under steady-state conditions of speed and load.

There are many applications involving friction under unsteady conditions, such as in the hip joint during walking. Variable friction under unsteady conditions is referred to as *dynamic friction*. Recently, there has been an increasing interest in dynamic friction measurements.

Dynamic tests, such as oscillating sliding motion, require on-line recording of friction. Experiments with an oscillating sliding plane by Bell and Burdekin

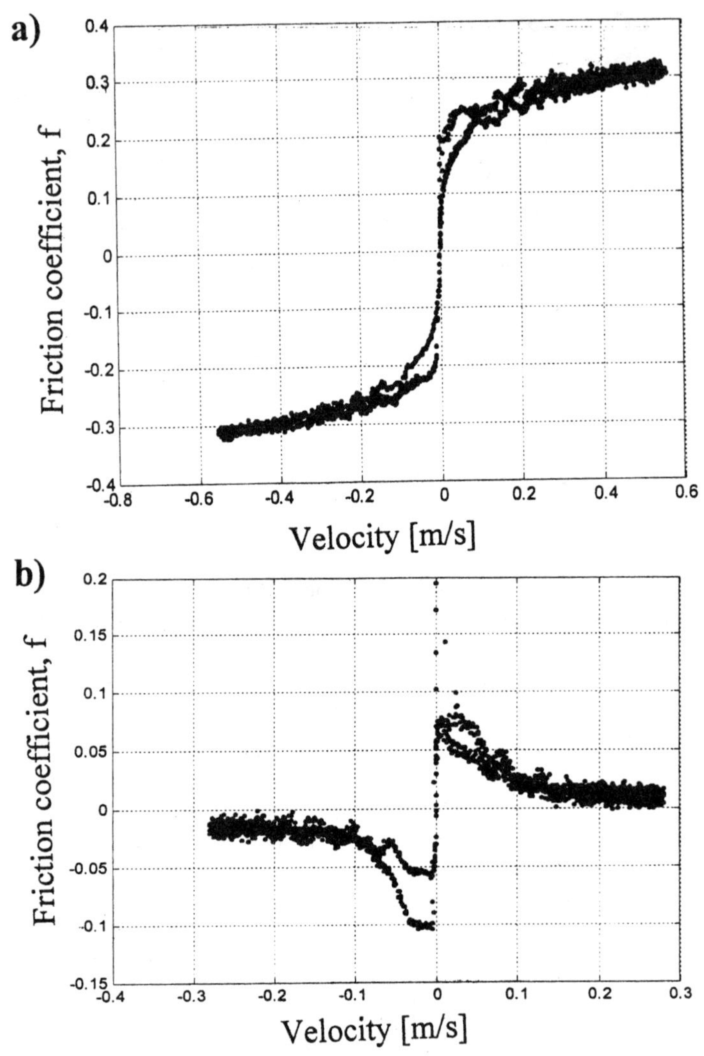

FIG. 16-4 f–U curve for sinusoidal velocity: frequency = 0.25 rad/s, load = 215 N, ultrahigh-molecular-weight polyethylene (UHMWPE) on steel, journal diameter 25 mm, $L/D = 0.75$ for rigid bearing at low frequency. (a) Dry surfaces. (b) Lubrication with SAE 5W oil.

(1969) and more recent investigations of line contact by Hess and Soom (1990), as well as recent measurements in journal bearings (Harnoy et al. 1994) revealed that the phenomenon of dynamic friction is quite complex. The f–U curves have a considerable amount of hysteresis that cannot be accounted for by any steady-friction model. The amount of hysteresis increases with the frequency of oscillations. At very low frequency, the curves are practically identical to curves produced by measurements under steady conditions.

For sinusoidal velocity, the friction is higher during acceleration than during deceleration, particularly in the mixed friction region. In the recent literature, the hysteresis effect is often referred to as *multivalued friction*, because the friction is higher during acceleration than during deceleration. For example, the friction coefficient is higher during the start-up of a machine than the friction during stopping. This means that the friction is not only a function of the instantaneous sliding velocity, but also a function of velocity history. Examples of $f-U$ curves under dynamic conditions are included in Chap. 17.

17

Modeling Dynamic Friction

17.1 INTRODUCTION

Early research was focused on bearings that operate under steady conditions, such as constant load and velocity. Since the traditional objectives of tribology were prevention of wear and minimizing friction-energy losses in steady-speed machinery, it is understandable that only a limited amount of research effort was directed to time-variable velocity. However, steady friction is only one aspect in a wider discipline of friction under time-variable conditions. Variable friction under unsteady conditions is referred to as *dynamic friction*. There are many applications involving dynamic friction, such as friction between the piston and sleeve in engines where the sliding speed and load periodically vary with time.

In the last decade, there was an increasing interest in dynamic friction as well as its modeling. This interest is motivated by the requirement to simulate dynamic effects such as friction-induced vibrations and stick-slip friction. In addition, there is a relatively new application for dynamic friction models—improving the precision of motion in control systems.

It is commonly recognized that friction limits the precision of motion. For example, if one tries to drag a heavy table on a rough floor, it would be impossible to obtain a high-precision displacement of a few micrometers. In fact, the minimum motion of the table will be a few millimeters. The reason for a low-precision motion is the negative slope of friction versus velocity. In comparison, one can move an object on well-lubricated, slippery surfaces and obtain much better precision of motion.

In a similar way, friction limits the precision of motion in open-loop and closed-loop control systems. This is because the friction has nonlinear characteristics of negative slope of friction versus velocity and discontinuity at velocity reversals. Friction causes errors of displacement from the desired target (hang-off) and instability, such as a stick-slip friction at low velocity.

There is an increasing requirement for ultrahigh-precision motion in applications such as manufacturing, precise measurement, and even surgery. Hydrostatic or magnetic bearings can minimize friction; also, vibrations are used to reduce friction (dither). These methods are expensive and may not be always feasible in machines or control systems.

An alternate approach that is still in development is model-based friction compensation. The concept is to include a friction model in the control algorithm. The control is designed to generate continuous on-line timely torque by the servomotor, in the opposite direction to the actual friction in the mechanical system. In this way, it is possible to approximately cancel the adverse effects of friction. Increasing computer capabilities make this method more and more attractive. This method requires a dynamic friction model for predicting the friction under dynamic conditions.

There is already experimental verification that displacement and velocity errors caused by friction can be substantially reduced by friction compensation. This effect has been demonstrated in laboratory experiments; see Amin et al. (1997). Friction compensation has been already applied successfully in actual machines. For example, Tafazoli (1995) describes a simple friction compensation method that improves the precision of motion in an industrial machine.

Another application of dynamic friction models is the simulation of friction-induced vibration (stick-slip friction). The simulation is required for design purposes to prevent these vibrations.

Stick-slip friction is considered a major limitation for high-precision manufacturing. In addition to machine tools, stick-slip friction is a major problem in measurement devices and other precision machines. A lot of research has been done to eliminate the stick-slip friction, particularly in machine tools. Some solutions involve hardware modifications that have already been discussed. These are expensive solutions that are not feasible in all cases. Attempts were made to reduce the stick-slip friction by using high-viscosity lubricant that improves the damping, but this would increase the viscous friction losses. Moreover, high-viscosity oil results in a thicker hydrodynamic film that reduces the precision of the machine tool. This undesirable effect is referred to as *excessive float*.

Various methods have been tried by several investigators to improve the stability of motion in the presence of friction. However, a model-based approach has the potential to offer a relatively low-cost solution to this important problem.

Armstrong-Hélouvry (1991) summarized the early work in friction modeling of the Stribeck curve by empirical equations. Also, Dahl (1968) introduced a model to describe the presliding displacement during stiction. However, these models are "static," in the sense that the friction is represented by an instantaneous function of sliding velocity and load. In recent years, several empirical equations were suggested to describe the phase lag and hysteresis in dynamic friction. Hess and Soom (1990) and Dupont and Dunlap (1993) developed such models.

17.2 DYNAMIC FRICTION MODEL FOR JOURNAL BEARINGS*

Harnoy and Friedland (1994) suggested a different modeling approach for lubricated surfaces, based on the physical principles of hydrodynamics. In the following section, this model is compared to friction measurements. This approach is based on the following two assumptions:

1. The load capacity, in the boundary and mixed lubrication regions, is the sum of a contact force (elastic reaction between the surface asperities) and hydrodynamic load capacity.
2. The friction has two components: a solid component due to adhesion in the asperity contacts and a viscous shear component.

This modeling approach was extended to line-contact friction by Rachoor and Harnoy (1996). Polycarpou and Soom (1995, 1996) and Zhai et al. (1997) extended this approach and derived a more accurate analysis for the complex elastohydrodynamic lubrication of line contact.

Under steady-state conditions of constant sliding velocity, the friction coefficient of lubricated surfaces is a function of the velocity. However, under dynamic conditions, when the relative velocity varies with time, such as oscillatory motion or motion of constant acceleration, the instantaneous friction depends not only on the velocity at that instant but is also a function of the velocity history.

The existence of dynamic effects in friction was recognized by several investigators. Hess and Soom (1995, 1996) observed a hysteresis effect in oscillatory friction of lubricated surfaces. They offered a model based on the steady f–U curve with a correction accounting for the phase lag between friction and velocity oscillations. The magnitude of the phase lag was determined empirically. A time lag between oscillating friction and velocity in lubricated

*This and subsequent sections in this chapter are for advanced studies.

surfaces was observed and measured earlier. It is interesting to note that Rabinowicz (1951) observed a friction lag even in dry contacts.

The following analysis offers a theoretical model, based on the physical phenomena of lubricated surfaces, that can capture the primary effect and simulate the dynamic friction. The result of the analysis is a dynamic model, expressed by a set of differential equations, that relates the force of friction to the time-variable velocity of the sliding surfaces. A model that can predict dynamic friction is very useful as an enhancement of the technology of precise motion control in machinery. For control purposes, we want to find the friction at oscillating low velocities near zero velocity.

Under classical hydrodynamic lubrication theory, (see Chapters 4–7) the fluid film thickness increases with velocity. The region of a full hydrodynamic lubrication in the f–U curve (Fig. 16-1) occurs when the sliding velocity is above the transition velocity, U_{tr} required to generate a lubrication film thicker than the size of the surface asperities. In Fig. 16-1, U_{tr} is the velocity corresponding to the minimum friction. In the full hydrodynamic region, there is only viscous friction that increases with velocity, because the shear rates and shear stresses are proportional to the sliding velocity.

Below the transition velocity, U_{tr}, the Stribeck curve shows the mixed lubrication region where the thickness of the lubrication film is less than the maximum size of the surface asperities. Under load, there is a contact between the surfaces, resulting in elastic as well as plastic deformation of the asperities. In the mixed region, the external load is carried partly by the pressure of the hydrodynamic fluid film and partly by the mechanical elastic reaction of the deformed asperities. The film thickness increases with velocity; therefore, as the velocity increases, a larger part of the external load is carried by the fluid film. The result is that the friction decreases with velocity in the mixed region, because the fluid viscous friction is lower than the mechanical friction at the contact between the asperities.

This discussion shows that the friction force is dependent primarily on the lubrication film thickness, which in turn is an increasing function of the steady velocity. However, for time-variable velocity, the relation between film thickness and velocity is much more complex. The following analysis of unsteady velocity attempts to capture the physical phenomena when the lubrication film undergoes changes owing to a variable sliding velocity. As a result of the damping in the system and the mass of the sliding body, there is a time delay to reach the film thickness that would otherwise be generated under steady velocity.

17.3 DEVELOPMENT OF THE MODEL

Consider a hydrodynamic journal bearing under steady conditions, when all the variables, such as external load and speed, are constant with time. Under these steady conditions, the journal center O_1 does not move relative to the bearing sleeve, and the friction force remains constant. In practice, these steady conditions will come about after a transient interval for damping of any initial motion of the journal center. When there is a motion of the journal center O_1, however, the oil film thickness and the friction force are not constant, which explains the dynamic effects of unsteady friction.

Before proceeding with the development of the dynamic model, the model for steady friction in the mixed lubrication region is presented. Dubois and Ocvirk (1953) derived the equations for full hydrodynamic lubrication of a short bearing. The following is an extension of this analysis to the mixed lubrication region. In the mixed region there is direct contact between the surface asperities combined with hydrodynamic load capacity. The theory is for a short journal bearing, because it is widely used in machinery, and because the steady performance of a short bearing in the full hydrodynamic region is already well understood and can be described by closed-form equations.

The mixed lubrication region is where the hydrodynamic minimum film thickness, h_n, is below a certain small transition magnitude, h_{tr}. Under load, the asperities are subject to elastic as well as plastic deformation due to the high-pressure contact at the tip of the asperities. Although the load is distributed unevenly between the asperities, the average elastic part of the deformation is described by the elastic recoverable displacement, δ, of the surfaces toward each other, in the direction normal to the contact area. The reaction force between the asperities of the two surfaces is an increasing function of the elastic, recoverable part of the deformation, δ. The normal reaction force of the asperities as a function of δ is similar to that of a spring; however, this springlike behavior is not linear.

In a journal bearing in the mixed lubrication region, the average normal elastic deformation, δ of the asperities is proportional to the difference between the transition minimum film thickness, h_{tr}, and the actual lower minimum film thickness, h_n:

$$\delta = h_{tr} - h_n \tag{17-1}$$

The elastic reaction force, W_e, of the asperities is similar to that of a nonlinear spring:

$$W_e = k_n(\delta)\,\delta \tag{17-2}$$

where $k_n(\delta)$ is the stiffness function of the asperities to elastic deformation in the direction normal to the surface. The contact areas between the asperities increase with the load and deformation δ. Therefore, $k_n(\delta)$ is an increasing function of δ.

If the elastic reaction force, W_e, between the asperities is approximated by a contact between two spheres, Hertz theory indicates that the reaction force, W_e, is proportional to $\delta^{3/2}$:

$$W_e \propto \delta^{3/2} \tag{17-3}$$

and Eq. (17-2) becomes

$$W_e = k_n\delta \Rightarrow k_n = k_0\delta^{1/2} \tag{17-4}$$

Here, k_0 is a constant which depends on the geometry and the elastic modulus of the two materials in contact. In fact, an average asperity contact is not identical to that between two spheres, and a better modeling precision can be obtained by determining empirically the two constants, k_0 and n, in the following expression for the normal stiffness:

$$k_n = k_0\delta^n \tag{17-5}$$

The two constants are selected for each material combination to give the best fit to the steady Stribeck curve in the mixed lubrication region.

For a journal bearing, the average elastic reaction of surface asperities in the mixed region, W_e in Eq. (17-2), can be expressed in terms of the eccentricity ratio, $\varepsilon = e/C$. In addition, a transition eccentricity ratio, ε_{tr}, is defined as the eccentricity ratio at the point of steady transition from mixed to hydrodynamic lubrication (in tests under steady conditions). This transition point is where the friction is minimal in the steady Stribeck curve.

The elastic deformation, δ (average normal asperity deformation), in Eq. (17-1) at the mixed lubrication region can be expressed in terms of the difference between ε and ε_{tr}:

$$\delta = C(\varepsilon - \varepsilon_{tr}) \tag{17-6}$$

and the expression for the average elastic reaction of the asperities in terms of the eccentricity ratio is

$$W_e = \kappa_n(\varepsilon)(\varepsilon - \varepsilon_{tr})\Delta \tag{17-7}$$

The elastic reaction force, W_e, is only in the mixed region, where the difference between ε and ε_{tr} is positive. For this purpose, the notation Δ is defined as

$$\Delta = \begin{cases} 1 & \text{if } (\varepsilon - \varepsilon_{tr}) > 0 \\ 0 & \text{if } (\varepsilon - \varepsilon_{tr}) \leq 0 \end{cases} \tag{17-8}$$

In a similar way to Eq. (17-5), $\kappa_n(\varepsilon)$ is a normal stiffness function, but it is a function of the difference of ε and ε_{tr},

$$\kappa_n(\varepsilon) = \kappa_0(\varepsilon - \varepsilon_{tr})^n \tag{17-9}$$

In the case of a spherical asperity, $n = 0.5$ and κ_0 is a constant. In actual contacts, the magnitude of the two constants n and κ_0 is determined for the best fit to the steady (Stribeck) f–U curve. In Eq. (17-7), $\Delta = 0$ in the full hydrodynamic region, and the elastic reaction force W_e, is also zero. But in the mixed region, $\Delta = 1$, and the elastic reaction force W_e is an increasing function of the eccentricity ratio ε.

In the mixed region, the total load capacity vector $\vec{W}$ of the bearing is a vector summation of the elastic reaction of the asperities, $\vec{W}_e$ and the hydrodynamic fluid film force, $\vec{W}_h$:

$$\vec{W} = \vec{W}_e + \vec{W}_h \tag{17-10}$$

The bearing friction force, F_f, in the tangential direction is the sum of contact and viscous friction forces. The contact friction force is assumed to follow Coulomb's law; hence, it is proportional to the normal contact load, W_e, while the hydrodynamic, viscous friction force follows the short bearing equation; see Eq. (7-27). Also, it is assumed that the asperities, in the mixed region, do not have an appreciable effect on the hydrodynamic performance.

Under these assumptions, the equation for the total friction force between the journal and sleeve of a short journal bearing over the complete range of boundary, mixed, and hydrodynamic regions is

$$F_f = f_m \, \kappa_n(\varepsilon) \, C(\varepsilon - \varepsilon_{tr}) \Delta \, sgn(U) + \frac{LR\mu}{C^2} \frac{2\pi}{(1 - \varepsilon^2)^{0.5}} U \tag{17-11}$$

Here, f_m is the static friction coefficient, L and R are the length and radius of the bearing, respectively, C is the radial clearance, and μ is the lubricant viscosity. The friction coefficient of the bearing, f, is a ratio of the friction force and the external load, $f = F_f/F$. The symbol sgn(U) means that the contact friction is in the direction of the velocity U.

17.4 MODELING FRICTION AT STEADY VELOCITY

The load capacity is the sum of the hydrodynamic force and the elastic reaction force. The equations for the hydrodynamic load capacity components of a short journal bearing [Eq. (7–16)] were derived by Dubois and Ocvirk, 1953. The following equations extend this solution to include the hydrodynamic components and the elastic reaction force:

$$F\cos(\phi - \pi) = \kappa_n(\varepsilon)\, C\,(\varepsilon - \varepsilon_{tr})\Delta + \frac{\varepsilon^2}{(1-\varepsilon^2)^2}\,\frac{\mu L^3}{C^2}\,|U| \tag{17-12}$$

$$F\sin(\phi - \pi) = \frac{\pi\varepsilon^2}{(1-\varepsilon^2)^2}\,\frac{\mu L^3}{C^2}\,U \tag{17-13}$$

The coordinates ϕ and e (Fig. 15-1) describe the location of the journal center in polar coordinates. The direction of the elastic reaction W_e is in the direction of X. In Eq. (17-12), the external load component F_x is equal to the sum of the hydrodynamic force component due to the fluid film pressure and the elastic reaction W_e, at the point of minimum film thickness. In Eq. (17-13), the load component F_y is equal only to the hydrodynamic reaction, because there is no contact force in the direction of Y.

For any steady velocity U in the mixed region, $(\varepsilon > \varepsilon_{tr})$ and for specified C, L, F, μ and $\kappa_n(\varepsilon)$, Eqs. (17-12 and 17-13) can be solved for the two unknowns, ϕ and ε. Once the relative eccentricity, ε, is known, the friction force F_f can be calculated from Eq. (17-11), and the bearing friction coefficient, f, can be obtained for specified R and f_m. By this procedure, the Stribeck curve can be plotted for the mixed and hydrodynamic regions.

For numerical solution, there is an advantage in having Eqs. (17-12) and (17-13) in a dimensionless form. These equations can be converted to dimensionless form by introducing the following dimensionless variables:

$$\overline{U} = \frac{U}{U_{tr}}; \qquad \overline{F} = \frac{C^2}{\mu U_{tr} L^3}\,F; \qquad \overline{\kappa} = \frac{C^3}{\mu U_{tr} L^3}\,\kappa_n \tag{17-14}$$

Here $\overline{\kappa}$ is a dimensionless normal stiffness to deformation at the asperity contact. The deformation is in the direction normal to the contact area. The velocity U_{tr} is at the transition from mixed to hydrodynamic lubrication (at the point of minimum friction in the f–U chart). The dimensionless form of Eqs. (17-12) and (17-13) is

$$\overline{F}\cos(\phi - \pi) = \overline{\kappa}(\varepsilon)(\varepsilon - \varepsilon_{tr})\Delta - 0.5 J_{12}\varepsilon\,|\,\overline{U}\,| \tag{17-15}$$

$$\overline{F}\sin(\phi - \pi) = 0.5 J_{11}\varepsilon\,\overline{U} \tag{17-16}$$

The integrals J_{11} and J_{12} are defined in Eqs. (7-13). Equations (17-15) and (17-16) apply to the mixed as well as the hydrodynamic lubrication regions in the Stribeck curve.

17.5 MODELING DYNAMIC FRICTION

For the purpose of developing the dynamic friction model, the existing hydrodynamic short bearing theory of Dubois and Ocvirk is extended to include the mixed region and dynamic conditions.

The assumptions of hydrodynamic theory of steady short bearings are extended to dynamic conditions. The pressure gradients in the x direction (around the bearing) are neglected, because they are very small in comparison with the gradients in the z (axial) direction (for directions, see Fig. 7-1). Similar to the analysis of a steady short bearing (see Chapter 7), only the pressure wave in the region $0 < \theta < \pi$ is considered for the fluid film force calculations. In this region, the fluid film pressure is higher than atmospheric pressure. In addition, the conventional assumptions of Reynolds' classical hydrodynamic theory are maintained. The viscosity, μ, is assumed to be constant (at an equivalent average temperature). The effects of fluid inertia are neglected, but the journal mass is considered, for it is of higher order of magnitude than the fluid mass.

Recall that under dynamic conditions the equations of motion are (see Chapter 15)

$$\vec{F} - \vec{W} = m\vec{a} \tag{17-17}$$

Writing Eq. (17-17) in components in the direction of W_x and W_y (i.e., the radial and tangential directions in Fig. 15-1), the following two equations are obtained in dimensionless terms:

$$\overline{F}_x - \overline{W}_x = \overline{m}\ddot{\varepsilon} - \overline{m}\varepsilon\dot{\phi}^2 \tag{17-18}$$

$$\overline{F}_y - \overline{W}_y = -\overline{m}\varepsilon\ddot{\phi} - 2\overline{m}\dot{\varepsilon}\dot{\phi}, \tag{17-19}$$

where the dimensionless mass and force are defined, respectively, as

$$\overline{m} = \frac{C^3}{\mu L^3 R^2}\, m; \qquad \overline{F} = \frac{C^2}{\mu U_{tr} L^3}\, F \tag{17-20}$$

Under dynamic conditions, the equations for the hydrodynamic load capacity components of a short journal bearing are as derived in Chapter 15. These equations are used here; in this case, however, the velocity is normalized by the transition velocity, U_{tr}. In a similar way to steady velocity, the load capacity components are due to the hydrodynamic pressure and elastic reaction force

$$\overline{W}_x = \overline{\kappa}(\varepsilon)(\varepsilon - \varepsilon_{tr})\Delta - 0.5 J_{12}\varepsilon \mid \overline{U} \mid + J_{12}\varepsilon\dot{\phi} + J_{22}\dot{\varepsilon} \tag{17-21}$$

$$\overline{W}_y = 0.5\varepsilon J_{11}\overline{U} - J_{11}\varepsilon\dot{\phi} - J_{12}\dot{\varepsilon} \tag{17-22}$$

Substituting these hydrodynamic and reaction force in Eqs. (17-18) and (17-19) yields

$$\overline{F}(t)\cos(\phi - \pi) = \overline{\kappa}(\varepsilon - \varepsilon_{tr})\Delta - 0.5\varepsilon J_{12}|\overline{U}| + J_{12}\varepsilon\dot{\phi} + J_{22}\dot{\varepsilon} + \overline{m}\ddot{\varepsilon} - \overline{m}\dot{\phi}^2 \tag{17-23}$$

$$\overline{F}(t)\sin(\phi - \pi) = 0.5\varepsilon J_{11}\overline{U} - J_{11}\varepsilon\dot{\phi} - J_{12}\dot{\varepsilon} - \overline{m}\varepsilon\ddot{\phi} - 2\overline{m}\dot{\varepsilon}\dot{\phi} \tag{17-24}$$

Here $\overline{F}(t)$ is a time-dependent dimensionless force acting on the bearing. The magnitude of this external force, as well as its direction is a function of time. In the two equations, ε is the eccentricity ratio, ϕ is defined in Fig. 15-1, and $\overline{m}$ is dimensionless mass, defined by Eq. (17-20). The definition of the integrals J_{ij} and their solutions are in Eqs. (7-13).

Equations (17-23) and (17-24) are two differential equations, which are required for the solution of the two time-dependent functions ε and ϕ. The solution of the two equations for ε and ϕ as a function of time allows the plotting of the trajectory of the journal center O_1 *in polar coordinates.*

These two differential equations yield the time-variable $\varepsilon(t)$, which in turn can be substituted into Eq. (17-11) for the computation of the friction force. For numerical computations, it is convenient to use the following dimensionless equation for the friction force obtained from Eq. (17-11):

$$\overline{F}_f = f_m\,\overline{\kappa}(\varepsilon)\,C\,(\varepsilon - \varepsilon_{tr})\,\Delta\,sgn(\overline{U}) + \frac{RC}{L^2}\,\frac{2\pi}{(1-\varepsilon^2)^{0.5}}\,\overline{U} \tag{17-25}$$

The dimensionless friction force and velocity are defined in Eq. (17-14). The friction coefficient of the bearing is the ratio of the dimensionless friction force and external load:

$$f = \frac{\overline{F}_f}{\overline{F}} \tag{17-26}$$

The set of three equations (17-23), (17-24) and (17-25) represents the dynamic friction model. For any time-variable shaft velocity $U(t)$ and time-variable load, the friction coefficient can be solved as a function of time or velocity.

This model can be extended to different sliding surface contacts, including EHD line and point contacts as well as rolling-element bearings. This can be done by replacing the equations for the hydrodynamic force of a short journal bearing with that of a point contact or rolling contact. These equations are already known from elastohydrodynamic lubrication theory; see Chapter 12.

17.6 COMPARISON OF MODEL SIMULATIONS AND EXPERIMENTS

Dynamic friction measurements were performed with the four-bearing measurement apparatus, which was described in Sec. 14.7. A computer with on-line data-acquisition system was used for plotting the results and analysis.

The model coefficients are required for comparing model simulations and experimental f–U curves under dynamic conditions. The modeling approach is to determine the model coefficients from the steady Stribeck curve. Later, the model coefficients are used to determine the characteristics under dynamic conditions.

In order to simplify the comparison, Eq. (17-25) has been modified and the coefficient γ introduced to replace a combination of several constants:

$$\overline{F}_f = f_m \, \overline{\kappa}(\varepsilon) \, C(\varepsilon - \varepsilon_{tr}) \, \Delta \, sgn(\overline{U}) + \gamma \, \frac{2\pi}{(1-\varepsilon^2)^{0.5}} \, \overline{U} \tag{17-27}$$

Here, f_m is the stiction friction coefficient and γ is a bearing geometrical coefficient. The friction force has two components: The first term is the contact component due to asperity interaction, and the second term is the viscous shear component. The normal stiffness constant, k_0, is selected by iterations to result in the best fit with the Stribeck curve in the mixed region.

A few examples are presented of measured curves of a test bearing (Table 17-1) as compared to theoretical simulations. The experiments were conducted under constant load and oscillating sliding velocity.

Friction measurements for bidirectional sinusoidal velocity were conducted under loads of 104 N and 84 N for each of the four test sleeve bearings. The analytical model was simulated for the following periodic velocity oscillations:

$$U = 0.127 \sin(\omega t) \tag{17-28}$$

Here, ω is the frequency (rad/s) of sliding velocity oscillations and U is the sliding velocity of the journal surface. The four-bearing apparatus was used to measure the dynamic friction between the shaft and the four sleeve bearings. Multigrade oil was applied, because the viscosity is less sensitive to variations of temperature, but it still varied initially by dissipation of friction energy during the

TABLE 17-1 Data from Friction Measurement Apparatus

Diameter of bearing ($D = 2R$)	$D = 0.0254$ m
Length of bearing	$L = 0.019$ m
Radial clearance in bearing	$C = 0.05$ mm
Mass of journal	$m = 2.27$ kg
Bearing material	Brass
Oil	SAE 10W-40

TABLE 17-2 Model Parameters for a Load of 84 N

$f_m = 0.26$	$k_0 = 7.5 \times 10^5$	$\mu = 0.02$ N-s/m^2
$U_{tr} = 0.06$ m/s	$F = 104$ N	$C = 5.08e^{-5}$ m
$\varepsilon_{tr} = 0.9727$	$m = 2.27$ kg	$\gamma = 0.0011$

test. After several cycles, however, a steady state was reached in which repeatability of the experiments was sustained.

For each bearing load, the Stribeck curve was initially produced by our four-bearing testing apparatus and used to determine the optimal coefficients required for the dynamic model in Eqs. (17-23), (17-24), and (17-27). The stiction friction coefficient, f_m and velocity at the transition, U_{tr}, were taken directly from the experimental steady Stribeck curve. The geometrical coefficient, γ, was determined from the slope in the hydrodynamic region, while the coefficient k_0 was determined to obtain an optimal fit to the experimental Stribeck curve in the mixed region. All other coefficients in Table 17-2, such as viscosity and bearing dimensions, are known. These constant coefficients, determined from the steady f–U curve, were used later for the simulation of the following f–U curves under dynamic conditions.

17.6.1 Bearing Load of 104 N (Table 17-2, Figs. 17-1, 17-2, 17-3)

17.6.2 Bearing Under Load of 84 N (Table 17-3,

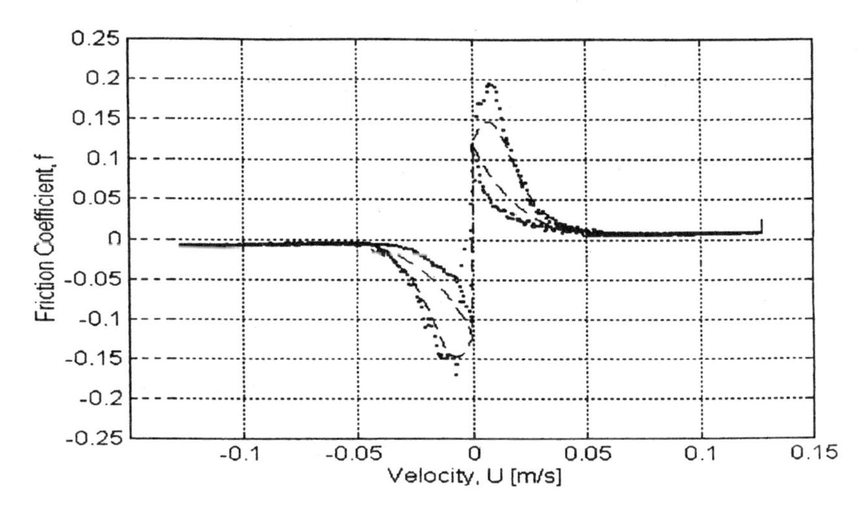

FIG. 17-1 Comparison of measured and theoretical f–U curves for sinusoidal sliding velocity: load = 104 N, $U = 0.127 \sin(0.045t)$ m/s, oscillation frequency = 0.045 rad/s (measurement..., simulation —).

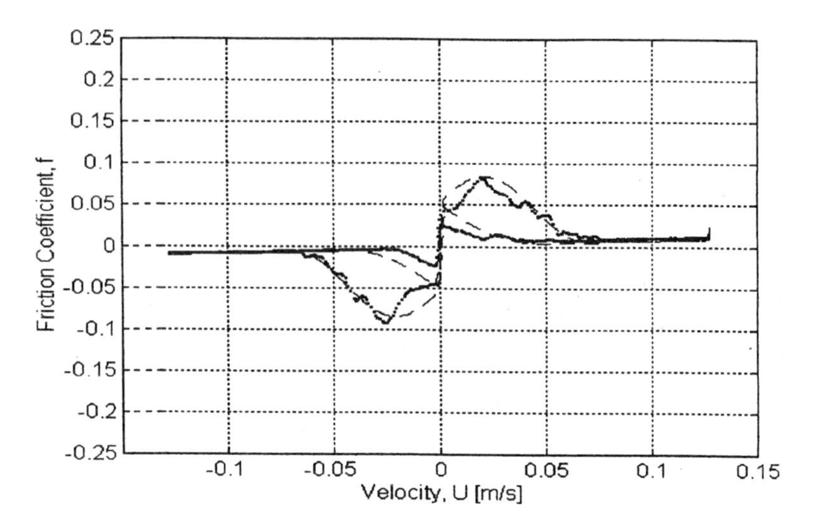

FIG. 17-2 Comparison of measured and theoretical f–U curves for sinusoidal sliding velocity: load = 104 N, $U = 0.127\sin(0.25t)$ m/s, oscillation frequency = 0.25 rad/s (measurement. . ., simulation —).

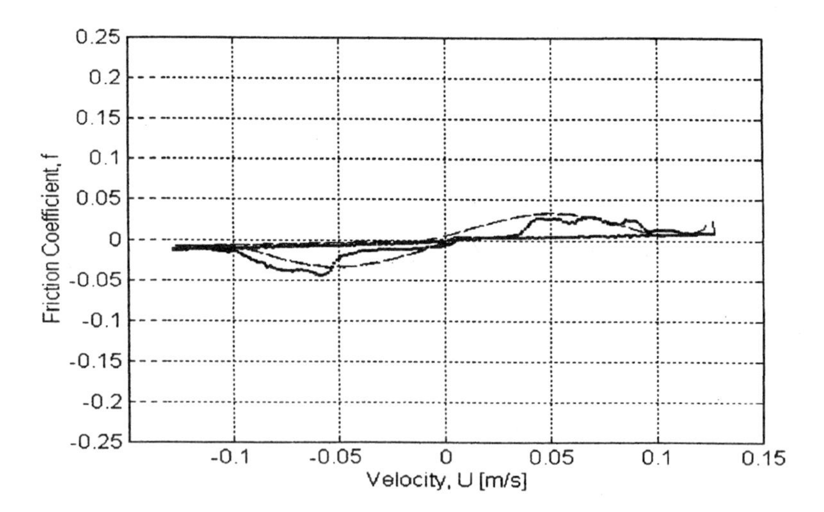

FIG. 17-3 Comparison of measured and theoretical f–U curves for sinusoidal sliding velocity: load = 104 N, $U = 0.127\sin(t)$ m/s, oscillation frequency = 1 rad/s (measurement. . ., simulation —).

TABLE 17-3 Model Parameters for a Load of 84 N

$f_m = 0.26$	$k_0 = 6.25 \times 10^5$	$\mu = 0.02$ N-s/m^2
$U_{tr} = 0.05$ m/s	$F = 84$ N	$C = 5.08e^{-5}$ m
$\varepsilon_{tr} = 0.9718$	$m = 2.27$ kg	$\gamma = 0.0011$

Figs. 17-4, 17-5, 17-6)

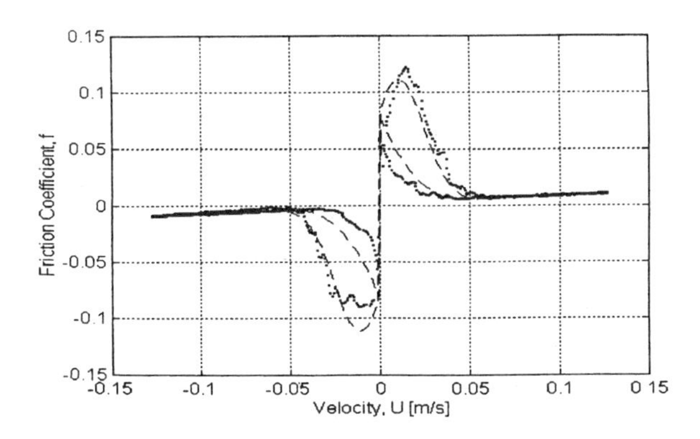

FIG. 17-4 Comparison of measured and theoretical f–U curves for sinusoidal sliding velocity: load = 84 N, $U = 0.127 \sin(0.1t)$ m/s, oscillation frequency = 0.1 rad/s (measurement . . . , simulation —).

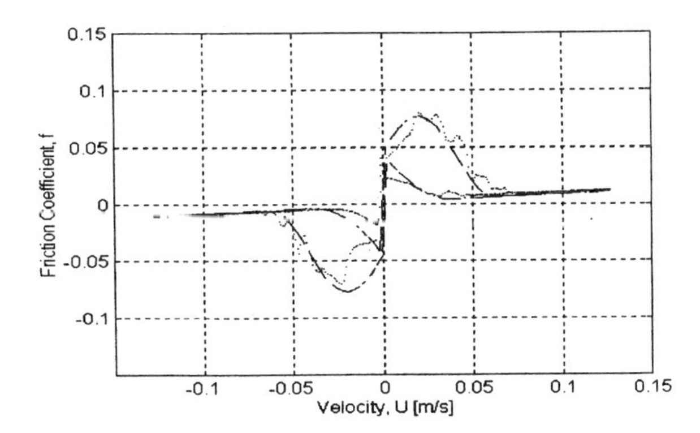

FIG. 17-5 Comparison of measured and theoretical f–U curves for sinusoidal sliding velocity: load = 84 N, $U = 0.127 \sin(0.25t)$ m/s, oscillation frequency = 0.25 rad/s (measurement . . . , simulation —).

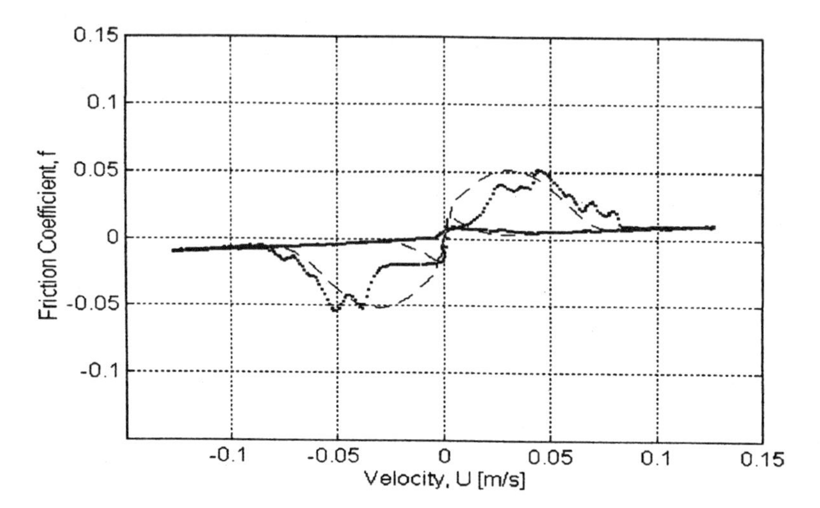

FIG. 17-6 Comparison of measured and theoretical f–U curves for sinusoidal sliding velocity: load = 84 N, $U = 0.127\sin(0.5t)$ m/s, oscillation frequency = 0.5 rad/s (measurement . . . , simulation —).

17.6.3 Conclusions

In conclusion, the f–U curves indicate reasonable agreement between experiments and simulation. At low frequency of velocity oscillations, the curves reduce to the steady Stribeck curve and do not demonstrate any significant hysteresis. At higher frequency, both analytical and experimental curves display similar hysteresis characteristics, which increase with the frequency. This phenomenon was detected earlier in experiments of unidirectional velocity oscillations.

In addition to the hysteresis, the experiments, as well as the simulation, detected several new dynamic friction characteristics that are unique to bidirectional oscillations with velocity reversals.

1. The magnitude of the friction discontinuity (and stiction friction) at zero velocity reduces when the oscillation frequency increases.
2. The stiction friction reduces to zero above a certain frequency of velocity oscillations.
3. The discontinuity at velocity reversals in the experimental curves is in the form of a vertical line. This means that the Dahl effect (presliding displacement) in journal bearings is relatively small, because the discontinuity is an inclined line wherever presliding displacement is of higher value.

The explanation for the reduction in the magnitude of the stiction force at higher frequencies is as follows: At high frequency there is insufficient time for the fluid film to be squeezed out. As the frequency increases, the fluid film is thicker, resulting in lower stiction force at velocity reversals.

18

Case Study

Composite Bearing—Rolling Elements and Fluid Film in Series

18.1 INTRODUCTION

A composite bearing of rolling and hydrodynamic components in series is a unique design that was proposed initially to overcome two major disadvantages of hydrodynamic journal bearings: Severe wear during start-up and stopping, and risk of catastrophic failure during any interruption of lubricant supply.

18.1.1 Start-Up and Stopping

Hydrodynamic bearings are subjected to severe wear during the starting and stopping of journal rotation. In addition, in variable-speed machines, when a bearing operates at low-speed, there is no full fluid film, resulting in wear. In these cases, there is also a risk of bearing failure due to overheating, which is a major drawback of hydrodynamic journal bearings.

In theory, there is a very thin fluid film even at low journal speeds. But in practice, due to surface roughness, vibrations, and disturbances, a critical minimum speed is required to generate adequate fluid film thickness for complete separation of the sliding surfaces. During start-up, wear is more severe than during stopping, because the bearing accelerates from zero velocity, where there is relatively high static friction. In certain cases, there is stick-slip friction during bearing start-up (see Harnoy 1966). During start-up, as speed increases, the fluid film builds up and friction reduces gradually.

18.1.2 Interruption of Oil Supply

A hydrodynamic bearing has a high risk of catastrophic failure whenever the lubricant supply is interrupted, even for a short time. The operation of a hydrodynamic journal bearing is completely dependent on a continuous supply of lubricant, particularly at high speed. If the oil supply is interrupted, this can cause overheating and catastrophic (sudden) bearing failure. At high speed, heat is generated at a fast rate by friction. Without lubricant, the bearing can undergo failure in the form of melting of the bearing lining. The lining is often made of a white metal of low melting temperature. Under certain conditions, interruption of the oil supply can result in bearing seizure (the journal and bearing weld together).

Interruption of the oil supply can occur for several reasons, such as a failure of the oil pump or its motor. In addition, the lubricant can be lost due to a leak in the oil system. This risk of failure prevents the use of hydrodynamic bearings in critical applications where safety is a major concern, such as in aircraft engines.

Replacing the hydrodynamic journal bearing with an externally pressurized hydrostatic journal bearing can eliminate the severe wear during starting and stopping. But a hydrostatic journal bearing is uneconomical for many applications because it needs a hydraulic system that includes a pump and an electric motor. For many machines, the use of hydrostatic bearings is not feasible. In addition, an externally pressurized hydrostatic bearing does not eliminate the risk of catastrophic failure in the case of oil supply interruption.

18.1.3 Limitations of Rolling Bearings

Rolling bearings are less sensitive than hydrodynamic bearings to starting and stopping. However, rolling bearing fatigue life is limited, due to alternating rolling contact stresses, particularly at very high speed. This problem is expected to become more important in the future because there is a continuous trend to increase the speed of machines. Manufacturers continually attempt to increase machinery speed in order to reduce the size of machines without reducing power.

It was shown in Chapter 12 that at very high speeds, the centrifugal forces of the rolling elements increase the contact stresses. At high speeds, the temperature of a rolling bearing rises and the fatigue resistance of the material deteriorates. The centrifugal forces and temperature exacerbate the problem and limit the speed of reliable operation. Thus the objective of long rolling bearing life and that of high operating speeds are in conflict. In conclusion, the optimum operation of the rolling bearing occurs at relatively low and medium speeds, while the best performance of the hydrodynamic bearing happens at relatively high speeds.

Over the years, there has been considerable improvement in rolling bearing materials. By using bearings made of high-purity specialty steels, fatigue life has been extended. High-quality rolling bearings made of specialty steels involve higher cost. These bearings are used in aircraft engines and other unique applications where the high cost is justified. However, since there is a continual requirement for faster speeds, the fatigue life of rolling bearings will continue to be a bottleneck in the future for the development of faster machines.

It would offer considerable advantage if the bearing could operate in a rolling mode at low speed and at higher speed would convert to hydrodynamic fluid film operation. In fact, this is the purpose of the composite bearing that utilizes the desirable features of both the hydrodynamic and the rolling bearing by combining them in series. In addition, if the oil supply is interrupted, the bearing will work in the rolling mode only and thus eliminate the high risk of failure of the common fluid film bearing.

In the following discussion, it is shown that it is possible to mitigate the drawbacks of the hydrodynamic journal bearing by using a *composite bearing*, which is a unique design of hydrodynamic and rolling bearings in series. In previous publications, this design was also referred to as the *series hybrid bearing*, the *angular-compliant bearing* and *hydro-roll*.

18.2 COMPOSITE-BEARING DESIGNS

The combination was tested initially (Harnoy 1966; Lowey, Harnoy, and Bar-Nefi 1972) by inserting the journal directly in the rolling-element inner ring bore; see Fig. 18-1. They used a radial clearance commonly accepted in hydrodynamic journal bearings of the order of magnitude $C \approx 10^{-3} \times R$. Later, this combination was improved (see Harnoy 1966), by inserting a sleeve at a tight fit into the bore of the rolling bearing; see Fig. 18-2. The journal runs on a fluid film in a free-fit clearance inside the bore of this sleeve. In this way, the desired sleeve material and surface finish can be selected as well as the ratio of the length and diameter, L/D, of the sleeve. In many applications, a self-aligning rolling element is desirable to ensure parallelism of the fluid film surfaces. The lubrication is an oil bath arrangement. The oil is fed in the axial direction of the clearance to form a fluid film between the journal and the sleeve; see Fig. 18-2.

Anderson* (1973) suggested a practical combination for use in gas turbines; see Fig. 18-3. This is a combination of a conical hydrodynamic bearing

* It is interesting that the work by the NASA group headed by Anderson and that of Lowey, Harnoy and Bar-Nefi in the Technion–Israel Institute of Technology were performed independently, without any knowledge of each other's work.

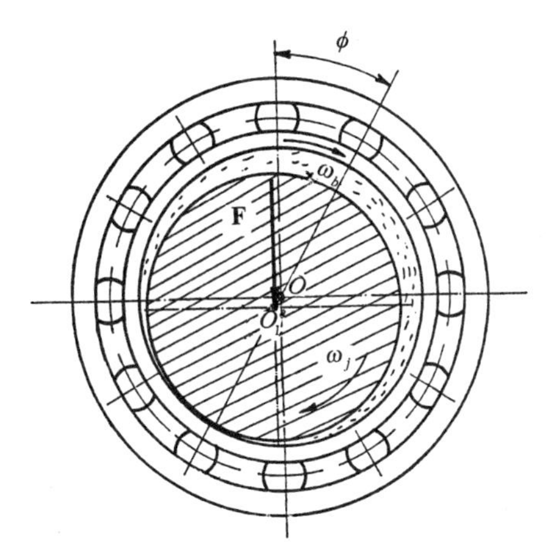

FIG. 18-1 Composite bearing arrangement of hydrodynamic and rolling bearings in series. (From Harnoy, 1966.)

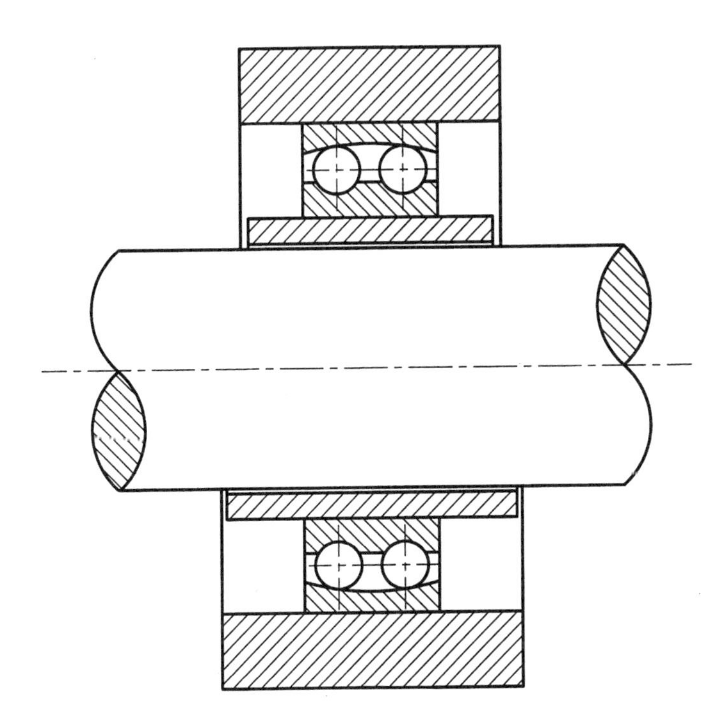

FIG. 18-2 Composite-bearing design with inner sleeve. (From Harnoy, 1966.)

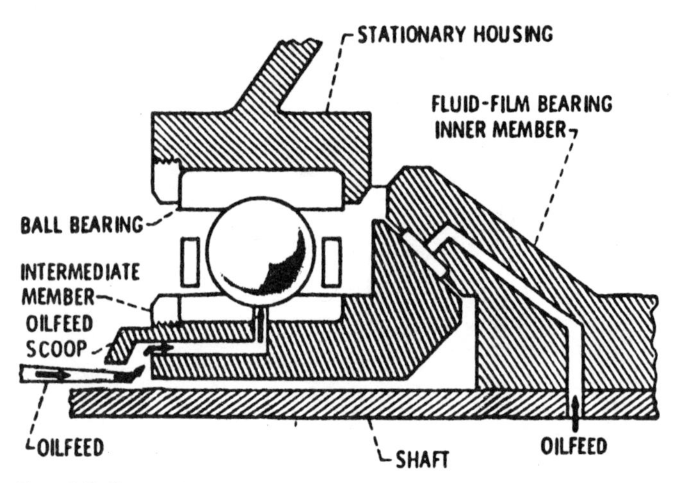

FIG. 18-3 Anderson composite bearing for radial and thrust loads. (From Anderson, 1973.)

and a rolling bearing in series, to provide for thrust and radial loads in gas turbine engines.

In the foregoing combinations of hydrodynamic and rolling bearings in series, the rolling-element bearing operates in rolling mode at low speed, including starting and stopping, while a sliding mode of the hydrodynamic fluid film is initiated at higher speed. The benefits of this combination are reduction of friction and wear and longer bearing life due to reduction of rolling speed.

It was mentioned earlier that the risk of catastrophic failure is the reason that hydrodynamic bearings are not applied in critical applications where safety is involved, such as aircraft engines. In fact, the composite bearing can overcome this problem, because, in the case of oil supply interruption, the composite bearing would continue to operate in the rolling mode, which requires only a very small amount of lubricant.

It is interesting to note that there are also considerable advantages in a hybrid bearing in which the rolling and hydrodynamic bearings are combined in parallel. Wilcock and Winn (1973) suggested the parallel combination.

18.2.1 Friction Characteristics of the Composite Bearing

In Fig. 18-4, f–U curves (friction coefficient versus velocity) are shown of a rolling bearing and of a fluid film bearing. These are the well-known Stribeck curves. Discussion of the various regions of the fluid film friction curve is included

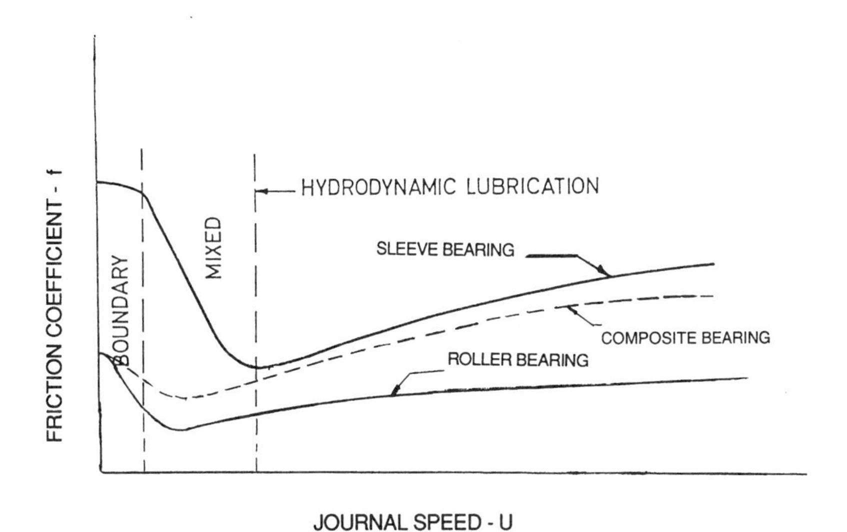

FIG. 18-4 Friction coefficient as a function of speed. (From Harnoy and Khonsari, 1996.)

in Chapter 8; measurement methods are covered in Chapter 14. The following discussion shows that the composite bearing, in fact, improves the friction characteristics by eliminating the high start-up friction of a fluid-film bearing.

The sleeve bearing friction curve in Fig. 18-4 has high friction in the boundary and mixed lubrication regions because the sliding surface asperities are in direct contact at low speed. In the hydrodynamic lubrication region, the sliding surfaces are separated by a fluid film and viscous friction is increasing almost linearly with speed. The curve for rolling bearing friction is similar, but start-up friction and high-speed friction are much lower than that of the common sleeve bearing.

The purpose of the composite bearing is to avoid the high friction in the boundary lubrication region and most of the mixed region of a sleeve bearing. In Fig. 18-4, the dotted line shows the expected friction characteristic of a properly designed composite bearing. During start-up, the composite bearing operates as a rolling bearing and the starting friction is as low as in a rolling bearing. The friction coefficient at the high rated speed is expected to be somewhat lower than for a regular journal bearing. This is because the viscous friction is proportional to the sliding speed only and the total speed of a composite bearing is divided into rolling and sliding parts.

18.2.2 Composite-Bearing Start-Up

During start-up, the sliding friction of a hydrodynamic bearing is higher than that of a rolling bearing. Therefore, sliding between the journal and the sleeve is replaced by a rolling action (similar to that in an internal gear mechanism). Thus the surface velocity of the shaft, $R\omega_j$, is equal to the velocity of the sleeve bore surface, $R_1\omega_b$. The velocities are shown in Fig. 18-1. The difference between the journal and bore surface radii is small and negligible, so we can assume $R_1 = R$.

An important aspect in the operation of a composite bearing is that the friction during the transition from rolling to sliding is significantly lower than for a regular start-up of a regular hydrodynamic journal bearing. The friction is lower because the initial rolling generates a fluid film between the rolling surfaces of the journal and the sleeve bore. This effect is explained next according to hydrodynamic theory.

18.2.3 Analysis of Start-up

For bearings under steady conditions, if the bearing sleeve and the journal are rotating at different speeds, the Reynolds equation for incompressible and isothermal conditions reduces to the following form [see Eq. (6-21b)]:

$$\frac{\partial}{\partial x}\left(\frac{h^3}{\mu}\frac{\partial p}{\partial x}\right)+\frac{\partial}{\partial z}\left(\frac{h^3}{\mu}\frac{\partial p}{\partial z}\right)=6R(\omega_j+\omega_b)\frac{\partial h}{\partial x} \tag{18-1}$$

The surface velocities of bearing and journal, $R\omega_j$ and $R_1\omega_b$, respectively, are shown in Fig. 18-1. In Eq. (18-1), p is the pressure and h is the fluid film thickness. For a regular journal bearing, there is only journal rotation, i.e., one surface has velocity $R\omega_j$ while the sleeve is stationary. After integration of Eq. (18-1), the pressure distribution in the fluid film and the load capacity are directly proportional to the sum $R(\omega_j + \omega_b)$.

During start-up, there is only the rolling mode, and the boundary conditions of the fluid film are

$$R\omega_j = R_1\omega_b \tag{18-2}$$

In comparison, in a regular journal bearing of a stationary sleeve, $\omega_b = 0$. Therefore, in the case of pure rolling, the sum of the velocities is double that of pure sliding in a common journal bearing. This means that during start-up, the fluid film pressure of a composite bearing is double that in a common hydrodynamic journal bearing, where $\omega_b = 0$. In the rolling mode, only half of the journal speed is required to generate the film thickness of a regular bearing with a stationary sleeve. This film of the rolling mode prevents wear and high friction at the transition from rolling to sliding.

The physical explanation is that the fluid is squeezed faster by the rolling action than by sliding. Doubling the pressure via rolling action is well known for those involved in the analysis of EHD lubrication of rolling elements.

18.3 PREVIOUS RESEARCH IN COMPOSITE BEARINGS

Experiments by Harnoy (1966) demonstrated that the composite bearing operates as a rolling element during starting and stopping, while hydrodynamic sliding is initiated at higher speeds. At the high rated speed, the rolling element rotates at a reduced speed because the speed is divided between rolling and sliding modes according to a certain ratio. The reduction of the rolling-element speed offers the important advantage of extending rolling bearing life. The composite bearing has a longer life than either a rolling bearing or fluid film bearing on its own. In addition, if the oil supply is interrupted, the composite bearing converts to rolling bearing mode, and the risk of a catastrophic failure is eliminated.

Developments in aircraft turbines generated a continual need for bearings that can operate at very high speeds. As discussed earlier, only rolling bearings are used in aircraft engines, because of the risk of oil supply interruptions in fluid film bearings. The centrifugal forces of the rolling elements is a major bottleneck limiting the speed of aircraft gas turbines.

The centrifugal forces dramatically increase with the DN value (the product of rolling bearing bore in millimeters and shaft speed in revolutions per minute). The centrifugal force of the rolling elements is a reason for limiting aircraft turbine engines to 2 million DN. This was NASA's motivation for initiating a research program to find a better bearing design for high-speed applications. Several ideas were tested to break through the limit of 2 million DN. Ball bearings with hollow balls were tested to reduce the mass of the rolling elements. Later, the introduction of silicone nitride rolling elements proved to be more effective in this direction (see Chapter 13).

In the early 1970s, a research team at the NASA Lewis Research Center did a lot of research and development work on the performance of the composite bearing (for example, Anderson, Fleming, and Parker 1972, and Scribbe, Winn, and Eusepi 1976). The NASA team refers to the composite bearing as a *series hybrid bearing*. The objective was to reach a speed of 3 million DN. The idea was to reduce the rolling-element speed by introducing a fluid film bearing in series that would participate in a portion of the total speed of the shaft. In fact, this work was successful, and operation at 3 million DN was demonstrated. This work proved that the composite bearing is a feasible alternative to conventional rolling bearings in aircraft turbines. Ratios of rolling-element speed to shaft speed (ω_b/ω_j) of a series hybrid bearing were tested by Anderson, Fleming, and Parker (1972). The results, a function of the shaft speed, are shown for two thrust loads in Fig. 18-5.

However, the composite bearing never reached the stage of actual application in aircraft engines, because better rolling-element bearings were developed that satisfied the maximum-speed requirement. In addition, the actual speed of aircraft engines did not reach the high DN values that had been expected earlier.

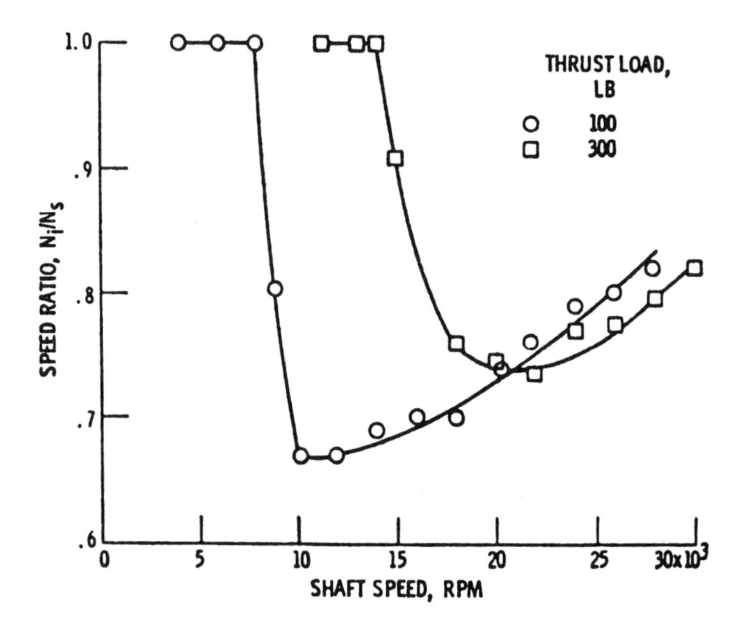

FIG. 18-5 Ratio of inner race speed to shaft speed vs. shaft speed for the composite bearing. (From Anderson, Fleming, and Parker, 1972.)

However, the requirement for higher speeds is increasing all the time. In the future, should the speed requirement increase above the limits of conventional rolling bearings, the composite bearing can offer a ready solution. Moreover, the composite bearing can significantly reduce the high cost of aircraft maintenance that involves frequent-replacement of rolling bearings.* Although the composite bearing has not yet been used in actual aircraft, it can be expected that this low-cost design will find many other applications in the future. The advantages of the composite bearing justify its use in a variety of applications as a viable low-cost alternative to the hydrostatic bearing.

18.4 COMPOSITE BEARING WITH CENTRIFUGAL MECHANISM

The composite arrangement always reduced the rolling element's speed. However, the results are not always completely satisfactory, because the rolling speed is not low enough. Experiments have indicated that in many cases the rolling speed in the composite bearing in Fig. 18-2 is too high for a significant improvement in

---

* The U.S. Air Force spends over $20 million annually on replacing rolling-element bearings (Valenti, 1995).

fatigue life. Whenever the friction of the rolling-element bearing is much lower than that of the hydrodynamic journal bearing, the rolling element rotates at relatively high speed. To improve this combination, a few ideas were suggested to control the composite bearing and to restrict the rotation of the rolling elements to a desired speed.

In Fig. 18-6a, a design is shown where the sleeve is connected to a mechanism similar to a centrifugal clutch; see Harnoy and Rachoor (1993). A design based on a similar principle was suggested by Silver (1972). A disc with radial holes is tightly fitted on the sleeve and pins slide along radial holes. Due to the action of centrifugal force, a friction torque is generated between the pins and the housing that increases with sleeve speed. This friction torque restricts the rolling speed and determines the speed of transition from rolling to sliding. The centrifugal design allows the sleeve to rotate continuously at low speed. This offers additional advantages, such as enhanced heat transfer from the lubrication film, (Harnoy and Khonsari, 1996) and improved performance under dynamic conditions, (Harnoy and Rachoor, 1993). Long life of the rolling element is maintained because the rolling speed is low. This design has considerable advantages, in particular for high-speed machinery that involves frequent start-ups. Figure 18-6b is a design of a composite bearing for radial and thrust loads with adjustable arrangement.

It is possible to increase the speeds $(\omega_b + \omega_j)$ during the transition from rolling to sliding, resulting in a thicker fluid film at that instant. This can be achieved by means of a unique design of a delayed centrifugal mechanism where the motion of the pins is damped as shown in Fig. 18-7. The purpose of this mechanism is to delay the transition from rolling to sliding during start-up, resulting in higher speeds $(\omega_b + \omega_j)$ at the instant of transition. The delayed action is advantageous only during the start-up, when the wear is more severe than that during the stopping period, since a certain time is required to form a lubricant film or to squeeze it out.

18.4.1 Design for the Desired Rolling Speed

The following derivation is required for the design of a centrifugal mechanism with the desired rolling speed, ω_b. The derivation is for a short journal bearing and a typical ball bearing.

The steady rolling speed ω_b can be solved from the friction torque balance, acting on the sleeve system—a combination of the sleeve and the centrifugal mechanism. The hydrodynamic torque, M_h, of a short bearing is:

$$M_h = \frac{L\mu R^3}{C}(\omega_j - \omega_b)\frac{2\pi}{(1-\varepsilon^2)^{0.5}} \tag{18-3}$$

Here, $R(\omega_j - \omega_b)$ replaces U in Eq. (7-29).

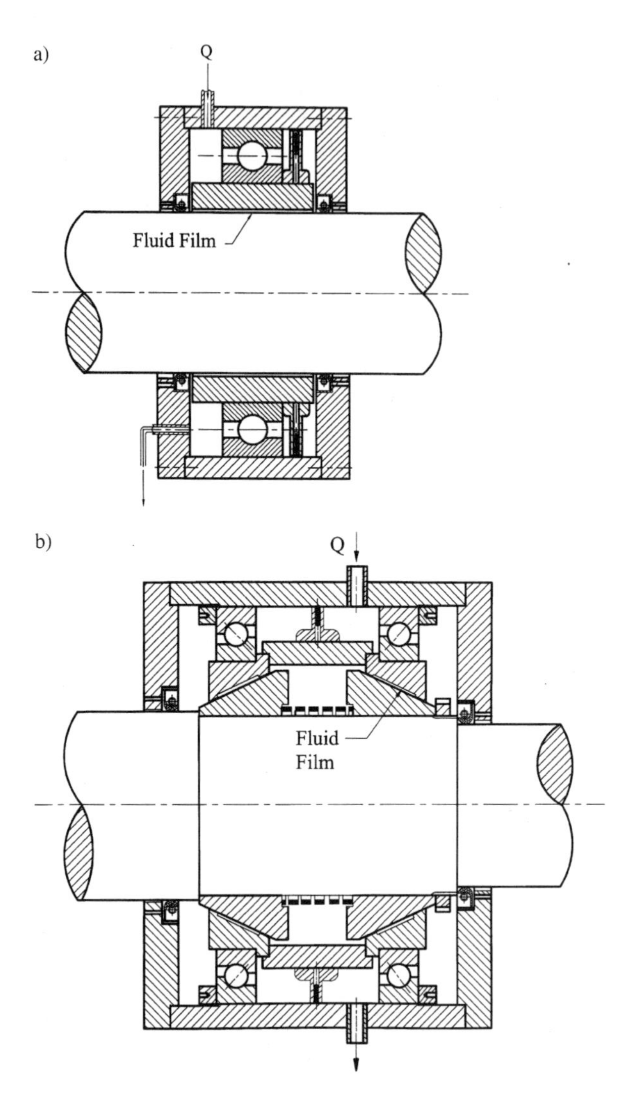

FIG. 18-6 (a) Centrifugal mechanism to control rolling speeds. (b) Composite bearing with centrifugal restraint for radial and thrust loads.

The mechanical friction torque on the sleeve is due to the centrifugal force of the pins, F_c, and the friction coefficient, f_c, between the pins and the housing at radius R_h; see Fig. 18-7:

$$M_f = F_c R_h f_c \tag{18-4}$$

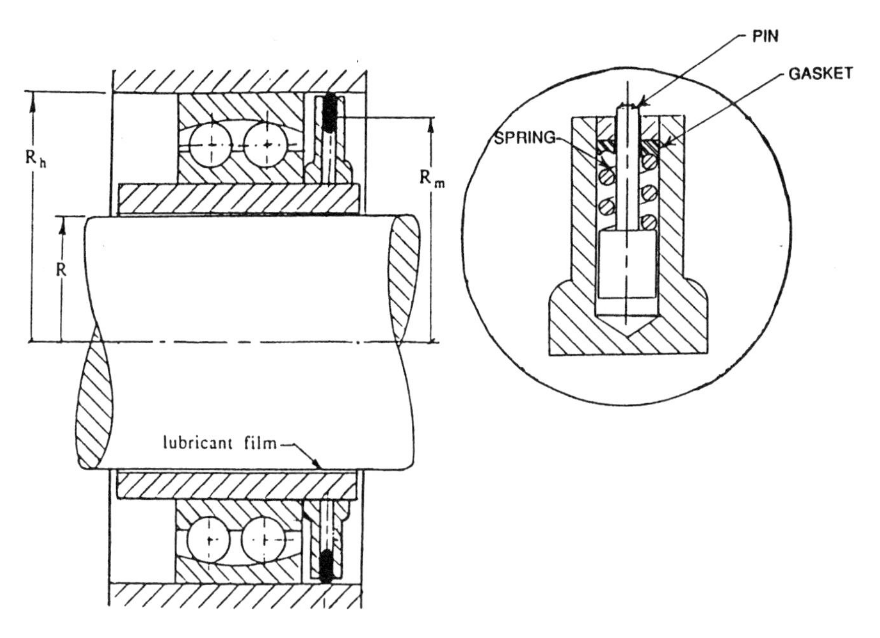

FIG. 18-7 Composite bearing with delayed centrifugal constraint.

The centrifugal contact force, F_c, between the small pins of total mass m_c and the housing is

$$F_c = m_c R_m \omega_b^2 \tag{18-5}$$

Here, m_c is the total mass of the centrifugal pins and R_m is the radius of the circle of the center of the pins when they are in contact with the housing. After substitution of this F_c in Eq. (18-4), the equation of the friction torque becomes

$$M_f = m_c R_m R_h \omega_b^2 f_c \tag{18-6}$$

The contact area between the pins and the housing is small, so boundary lubrication can be expected at all speeds. Thus, the friction coefficient f_c is effectively constant.

The friction torque due to centrifugal action of the pins, M_f, acts in the direction opposite to the hydrodynamic torque. If the composite bearing operates under steady conditions, there is no inertial torque and the equilibrium equation is

$$\frac{2\pi L \mu R^3}{C} \frac{\omega_j - \omega_b}{(1 - \varepsilon^2)^{0.5}} = f_c m_c R_m R_h \omega_b^2 + M_r \tag{18-7}$$

The friction torque M_r of a ball bearing at low speeds is generally much lower than the hydrodynamic friction torque at high speeds, so M_r can be neglected in

Eq. (18-7); see Fig. 18-4. However, in certain cases, such as in a tightly fitted conical bearing, the rolling friction is significant and should be considered. Equation (18-7) yields the following solution for the rolling speed ω_b:

$$\omega_b = \frac{(n^2 + mq)^{0.5} - n}{m} \tag{18-8}$$

where

$$m = f_c m_c R_m R_h \tag{18-9}$$

$$n = \pi L \mu R^3 C (1 - \varepsilon^2)^{0.5} \tag{18-10}$$

$$q = 2n\omega_j \tag{18-11}$$

The speed ω_b can be determined by selecting the mass of the pins m_c.

18.5 PERFORMANCE UNDER DYNAMIC CONDITIONS

The advantages of the composite bearing are quite obvious under steady constant load. However, the composite bearing did not gain wide acceptance, because there were concerns about possible adverse effects under unsteady or oscillating loads (dynamic loads). In rotating machinery, there are always vibrations and the average load is superimposed by oscillating forces at various frequencies. Harnoy and Rachoor (1993) analyzed the response of a composite bearing with a centrifugal mechanism, as shown in Fig. 18-6a and b, under dynamic conditions of a steady load superimposed with an oscillating load. The analysis involves angular oscillations of the sleeve, time-variable eccentricity, and unsteady fluid film pressure.

This analysis is essential for predicting any possible adverse effects of the composite arrangement on the bearing stability. Most probably, the unstable region is not identical to that of the common fluid film bearing. Nevertheless, there are reasons to expect improved performance within the stable region.

The following is an explanation of the criteria for improved bearing performance under dynamic loads and why composite bearings are expected to contribute to such an improvement. Unlike operation under steady conditions, where the journal center is stationary, under dynamic conditions, such as sinusoidal force, the journal center, O_1 is in continuous motion (trajectory) relative to the sleeve center O, and the eccentricity e varies with time. For a periodic load, such as in engines, the journal center O_1 reaches a steady-state trajectory referred to as *journal locus* that repeats in each time period. If the maximum eccentricity e_m of this locus (the maximum distance O–O_1 in Fig. 18-1) were to be reduced by the composite arrangement in comparison to the common journal bearing, it would mean that there is an important improvement in bearing performance. When the eccentricity ratio $\varepsilon = e/C$ approaches 1, there is contact and wear of the journal and sleeve surfaces. As discussed in previous chapters, due to surface roughness, dust, and disturbances, ε_m must be kept low (relative to 1) to prevent bearing wear.

Of course, one can reduce the maximum eccentricity of the locus by simply increasing the oil viscosity, μ; however, this is undesirable because it will increase the viscous friction. If it can be shown that a composite bearing can reduce the maximum eccentricity e_m, for the same viscosity and dynamic loads, then there is a potential for energy savings. In that case, it would be possible to reduce the viscosity and viscous losses without increasing the wear.

There is a simple physical explanation for expecting a significant improvement in the performance of a composite bearing under dynamic conditions, namely, the relative reduction of e_m under oscillating loads. Let us consider a bearing under sinusoidal load. During the cycle period, the critical time is when the load approaches its peak value. At that instant, the journal center, O_1 is moving in the radial direction (away from the bearing center O) and the eccentricity e approaches its maximum value e_m. At that instant, the fluid film is squeezed to its minimum thickness.

Under dynamic load, a significant part of the load capacity of the fluid film is proportional to the sum of the journal and sleeve rotations $(\omega_b + \omega_j)$ [see Eq. (18-1)]. As the external force increases, the fluid film is squeezed and the hydrodynamic friction torque, M_h, increases as well, causing the sleeve to rotate faster (ω_b increases). At that critical instant, the fluid film load capacity increases, due to a rise in $(\omega_b + \omega_j)$, in the direction directly opposing the journal motion toward the sleeve surface, resulting in reduced e_m. The sleeve oscillates periodically as a pendulum due to the external harmonic load.

However, it will be shown that the complete dynamic behavior is more complex. The inertia and damping of the sleeve motion cause a phase lag between the sleeve and the force oscillations. In certain cases, depending on the design parameters, one can expect adverse effects. If the phase lag becomes excessive, it would result in unsynchronized sleeve rotation, opposite to the desired direction. This discussion emphasizes the significance of a full analysis, not only to predict behavior but also to provide the tools for proper design.

18.5.1 Equations of Motion

The following analysis is for a composite bearing operating at the rated constant journal speed, with the centrifugal restraint (Fig. 18-6). The length L of the internal bore of most rolling bearings is short relative to the diameter D. For this reason, the following is for a short journal bearing, which assumes $L \ll D$. The analysis can be extended to a finite-length journal bearing; however, it is adequate for our purpose—to compare the dynamic behavior of a composite bearing to that of a regular one.

The first step is a derivation of the dynamic equation that describes the rotation of the composite bearing *sleeve unit*, consisting of the sleeve, the inner ring of the rolling bearing, and the centrifugal disc system. The three parts are tightly fitted and are rotating together at an angular speed ω_b, as shown in Fig. 18-8. This sleeve unit has an equivalent moment of inertia I_{eq}. The degree of freedom of sleeve rotation, which is involved with I_{eq}, includes the rolling elements that rotate at a reduced speed. It is similar to an equivalent moment of inertia of meshed gears.

A periodic load results in a variable hydrodynamic friction torque, and in turn there are angular oscillations of the sleeve unit (the angular velocity ω_b varies periodically). The sleeve unit oscillations are superimposed on a constant

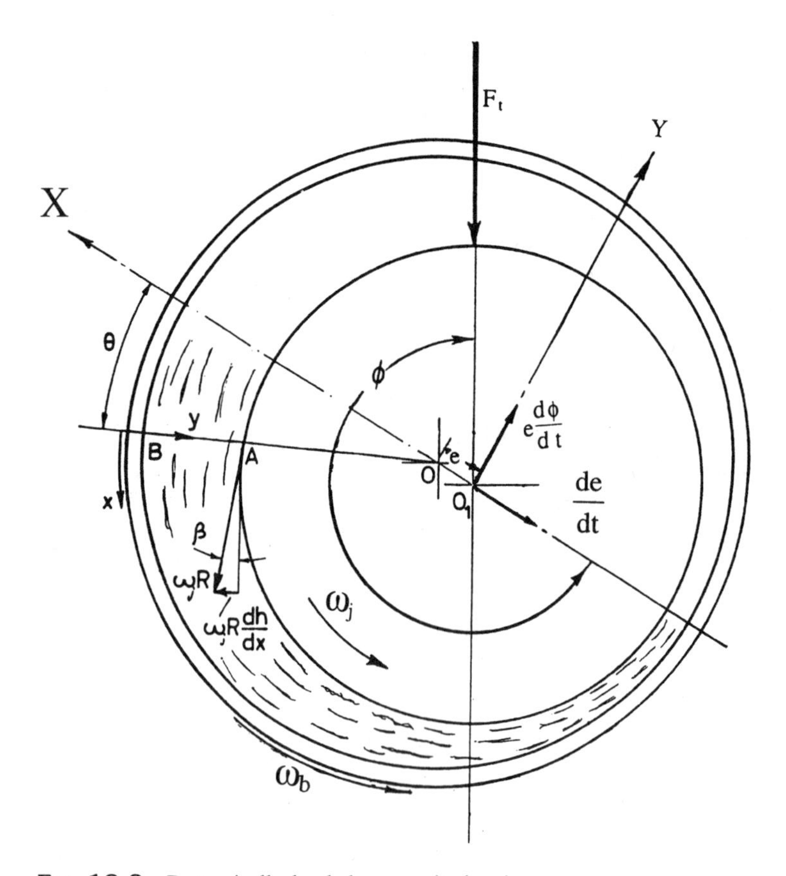

FIG. 18-8 Dynamically loaded composite bearing.

speed of rotation. At the same time, the mechanical friction between the pins and the housing damps these oscillations.

The difference between the hydrodynamic (viscous) friction torque M_h and the mechanical friction torque of the pins M_f is the resultant torque that accelerates the sleeve unit. The rolling friction torque M_r is small and negligible. The equation of the sleeve unit motion becomes

$$M_h - M_f = I_{\text{eq}} \frac{d\omega_b}{dt} \tag{18-12}$$

Substituting the values of the hydrodynamic torque and the mechanical friction torque from Eqs. (18-3) and (18-6) into Eq. (18-12) results in the following equation for the sleeve motion:

$$\frac{L\mu R^3}{C}(\omega_j - \omega_b)\frac{2\pi}{(1-\varepsilon^2)^{0.5}} - m_c R_m R_h \omega_b^2 f_c = I_{\text{eq}} \frac{d\omega_b}{dt} \tag{18-13}$$

This equation is converted to dimensionless form by dividing all the terms by $I_{\text{eq}}\omega_j^2$. The final dimensionless dynamic equation of the sleeve unit motion is

$$(1-\xi)H_1 \frac{2\pi}{(1-\varepsilon^2)^{0.5}} - \xi^2 H_2 = \dot{\xi} \tag{18-14}$$

Here, ξ is the ratio of the sleeve unit angular velocity to the journal angular velocity:

$$\xi = \frac{\omega_b}{\omega_j} \tag{18-15}$$

The time derivative $\dot{\xi} = d\xi/d\bar{t}$ is with respect to the dimensionless time, $\bar{t} = \omega_j t$, and the dimensionless parameters H_1 and H_2 are design parameters of the composite bearing defined by

$$H_1 = \frac{L\mu R^3}{C I_{\text{eq}} \omega_j}; \qquad H_2 = m_c R_m R_h f_c \tag{18-16}$$

18.5.2 Equation of Journal Motion

Chapter 7 presented the solution of Dubois and Ocvirk (1953) for the pressure distribution of a short journal bearing under steady conditions. This derivation was extended in Chapter 15 to a short bearing under dynamic conditions. In this chapter, this derivation is further extended to a composite bearing where the sleeve unit rotates at unsteady speed.

It was shown in Chapter 15 that in a journal bearing under dynamic conditions, the journal center O_1 has an arbitrary velocity described by its two components, de/dt and $e\,d\phi/dt$, in the radial and tangential directions, respectively. The purpose of the following analysis is to solve for the journal center trajectory of a composite bearing.

Let us recall that the Reynolds equation for the pressure distribution p in a thin incompressible fluid film is

$$\frac{\partial}{\partial x}\left(\frac{h^3}{\mu}\frac{\partial p}{\partial x}\right)+\frac{\partial}{\partial z}\left(\frac{h^3}{\mu}\frac{\partial p}{\partial z}\right)=6(U_1-U_2)\frac{\partial h}{\partial x}+12(V_2-V_1) \tag{18-17}$$

Similar to the derivation in Sec. 15.2, the journal surface velocity components, U_2 and V_2 are obtained by summing the velocity vector of the surface velocity, relative to the journal center O_1 (velocity due to journal rotation), and the velocity vector of O_1 relative to O (velocity due to the motion of the journal center O_1). At the same time, the sleeve surface has only tangential velocity, $R\omega_b$, in the x direction. In a composite bearing, the fluid film boundary conditions on the right-hand side of Eq. (18-17) become

$$V_1=\omega_j R\frac{dh}{dt}+\frac{de}{dt}\cos\theta+e\frac{d\phi}{dt}\sin\theta \tag{18-18}$$

$$V_2=0 \tag{18-19}$$

$$U_1=\omega_b R \tag{18-20}$$

$$U_2=\omega_j R+\frac{de}{dt}\sin\theta-e\frac{d\phi}{dt}\cos\theta \tag{18-21}$$

According to our assumptions, $\partial p/\partial x$ on the left-hand side of Eq. (18-17) is negligible. Considering only the axial pressure gradient and substituting Eqs. (18-18)–(18-21) into Eq. (18-17) yields

$$\frac{\partial}{\partial z}\left(h^3\frac{\partial p}{\partial z}\right)=6\mu\frac{\partial}{\partial x}R(\omega_j+\omega_b)+6\mu\left(\frac{de}{dt}\cos\theta+e\frac{d\phi}{dt}\sin\theta\right) \tag{18-22}$$

Integrating Eq. (18-22) twice with the following boundary conditions solves the pressure wave:

$$p=0 \quad \text{at} \quad z=\pm\frac{L}{2} \tag{18-23}$$

In the case of a short bearing, the pressure is a function of z and θ. The following are the two equations for the integration of the load capacity components in the directions of W_x and W_y:

$$W_x = -2R \int_0^{\pi} \int_0^{L/2} p \cos\theta \, d\theta \, dz \tag{18-24}$$

$$W_y = 2R \int_0^{\pi} \int_0^{L/2} p \sin\theta \, d\theta \, dz \tag{18-25}$$

The dimensionless load capacity W and the external dynamic load $F(t)$ are defined as follows:

$$\overline{W} = \frac{C^2}{\mu R \omega_j L^3} W; \qquad \overline{F}(t) = \frac{C^2}{\mu R \omega_j L^3} F(t) \tag{18-26}$$

where the journal speed ω_j is constant. After integration and conversion to dimensionless form, the following fluid film load capacity components are obtained:

$$\overline{W}_x = -\frac{1}{2} J_{12} \varepsilon (1 + \xi) + \varepsilon \dot{\phi} J_{12} + \dot{\varepsilon} J_{22} \tag{18-27}$$

$$\overline{W}_y = \frac{1}{2} J_{11} \varepsilon (1 + \xi) - \varepsilon \dot{\phi} J_{12} - \dot{\varepsilon} J_{22} \tag{18-28}$$

The integrals J_{ij} and their solutions are defined according to Eq. (7-13). The resultant of the load and fluid film force vectors accelerates the journal according to Newton's second law:

$$\vec{F}(t) + \vec{W} = m\vec{a} \tag{18-29}$$

Here, $\vec{a}$ is the acceleration vector of the journal center O_1 and m is the journal mass. Dimensionless mass is defined as

$$\overline{m} = \frac{C^3 \omega_j R}{L^3 R^2} m \tag{18-30}$$

After substitution of the acceleration terms in the radial and tangential directions (directions X and Y in Fig. 18-8) the equations become

$$\overline{F}_x(t) - \overline{W}_x = \overline{m}\ddot{\varepsilon} - \overline{m}\varepsilon\dot{\phi}^2 \tag{18-31}$$

$$\overline{F}_y(t) - \overline{W}_y = -\overline{m}\varepsilon\ddot{\phi} - 2\overline{m}\dot{\varepsilon}\dot{\phi} \tag{18-32}$$

Substituting the load capacity components of Eqs. (18-27) and (18-28) into Eqs. (18-31) and (18-32) yields the final two differential equations of the journal motion:

$$\overline{F}(t)\cos(\phi-\pi) = -0.5J_{12}\varepsilon(1+\xi) + \varepsilon\dot{\phi}J_{12} + \dot{\varepsilon}J_{22} + \overline{m}\ddot{\varepsilon} - \overline{m}\varepsilon\dot{\phi}^2 \tag{18-33}$$

$$\overline{F}(t)\sin(\phi-\pi) = 0.5J_{11}\varepsilon(1+\xi) - \varepsilon\dot{\phi}J_{11} - \dot{\varepsilon}J_{12} - \overline{m}\varepsilon\ddot{\phi} - 2\overline{m}\dot{\varepsilon}\dot{\phi} \tag{18-34}$$

Equations (18-33), (18-34), and (18-14) are the three differential equations required to solve for the three time-dependent functions ε, ϕ, and ξ. These three variables represent the motion of the shaft center O_1 with time, *in polar coordinates*, as well as the rotation of the sleeve unit.

18.5.3 Comparison of Journal Locus under Dynamic Load

In machinery there are always vibrations and bearing under steady loads are usually subjected to dynamic oscillating loads. The following is a solution for a composite bearing under a vertical load consisting of a sinusoidal load superimposed on a steady load according to the equation (in this section, $\bar{F}$ and $\bar{m}$ are renamed F and m)

$$F(t) = F_s + F_o \sin \alpha\omega_j t \tag{18-35}$$

Here, F_s is a steady load, F_o is the amplitude of a sinusoidal force, ω is the load frequency, and α is the ratio of the load frequency to the journal speed:

$$\alpha = \frac{\omega}{\omega_j} \tag{18-36}$$

Equations (18-33), (18-34), and (18-14) were solved by finite differences. By selecting backward differences, the nonlinear terms were linearized. In this way, the three differential equations were converted to three regular equations. The finite difference procedure is presented in Sec. 15.4.

Examples of the loci of a composite bearing and a regular journal bearing are shown in Fig. 18-9 for $\alpha = 2$ and in Fig. 18-10 for $\alpha = 2$ and $\alpha = 4$. Any reduction in the maximum eccentricity ratio, ε_m, represents a significant improvement in lubrication performance. The curves indicate that the composite bearing (dotted-line locus) has a lower ε_m than a regular journal bearing (solid-line locus). An important aspect is that the relative improvement increases whenever ε_m increases (the journal approaches the sleeve surface); thus, the composite bearing plays an important role in wear reduction. For example, in the heavily loaded

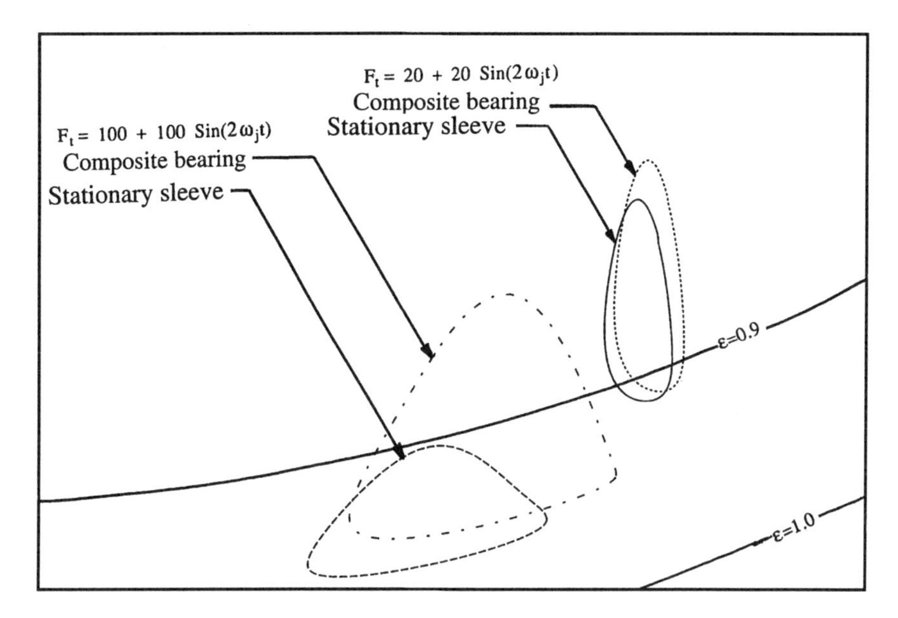

FIG. 18-9 Journal loci of a regular bearing and a composite bearing; $F(t) = 100 + 100\sin(2\omega_j t)$ and $F(t) = 20 + 20\sin(2\omega_j t)$. The journal mass is $m = 50$. The design parameters are $H_1 = 0.1$ and $H_2 = 1.0$.

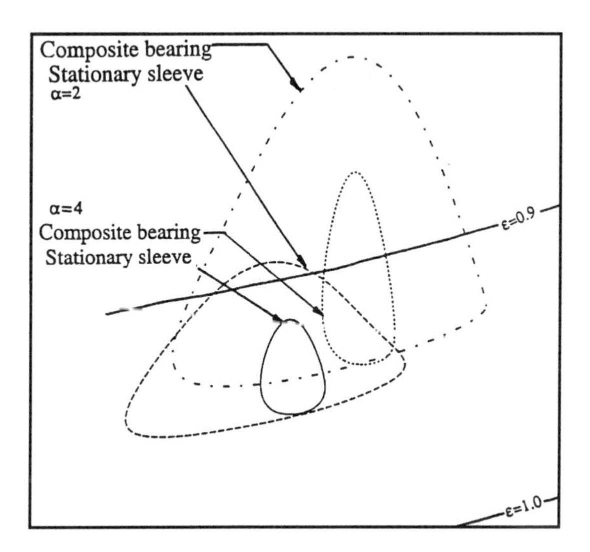

FIG. 18-10 Journal loci of rigid and compliant sleeve bearings. The load $F(t) = 100 + 100\sin(\alpha\omega_j t)$, for $\alpha = 2$ and $\alpha = 4$. The journal mass $m = 50$. $H_1 = 0.1$ and $H_2 = 1.0$.

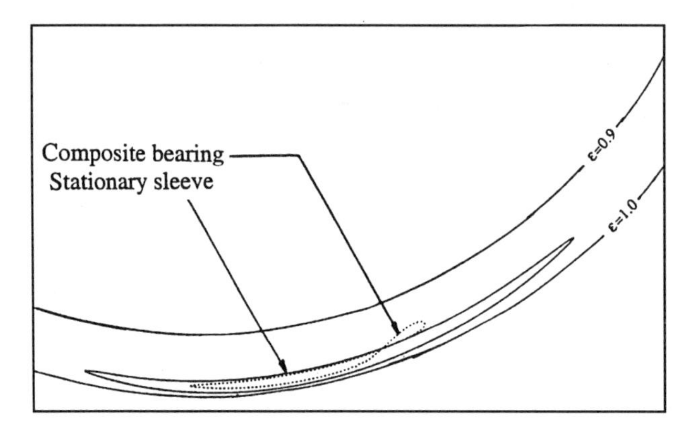

FIG. 18-11 Journal loci of rigid and compliant sleeve bearings under heavy load. $F(t) = 800 + 800\sin(2\omega_j t)$. The journal mass is $m = 100$, $H_1 = 0.1$ and $H_2 = 100$.

bearing in Fig 18-11, the composite bearing nearly doubles the minimum film thickness ε_m of a regular journal bearing. This can be observed by the distance between the two loci and the circle $\varepsilon = 1$.

If there is a relatively large phase lag between the load and sleeve unit oscillations, the lubrication performance of the composite bearing can deteriorate. In order to benefit from the advantages of a composite bearing, in view of the many design parameters, the designer must in each case conduct a similar computer simulation to determine the dynamic performance.

18.6 THERMAL EFFECTS

The peak temperature, in the fluid film and on the inner surface of the sleeve (near the minimum film thickness) was discussed in Sec. 8.6. Excessive peak temperature T_{max} can result in bearing failure, particularly in large bearings with white metal lining. Therefore, in these cases, it is necessary to limit T_{max} during the design stage.

With a properly designed composite bearing, a much more uniform temperature distribution is expected; since the sleeve unit rotates, the severity of the peak temperature is reduced.

The heat transfer from the region of the minimum film thickness to the atmosphere is affected by the rotation of the sleeve as well as many other parameters, such as bearing materials, lubrication, heat conduction at the contact between the rolling elements and races, the design of the bearing housing, and its connection to the body of the machine.

In order to elucidate the effect of the rotation of the sleeve on heat transfer, Harnoy and Khonsari (1996) studied the effect of sleeve rotation in isolation from any other factor that can affect the rate of heat removal from the hydrodynamic oil film. For this purpose, the heat transfer problem of a hydrodynamic bearing at steady-state conditions is studied and a comparison made between the temperature distributions in stationary and rotary sleeves while all other parameters, such as geometry and materials, are identical for the two cases. For comparison purposes, a model is presented where the sleeve loses heat to the surroundings at ambient temperature T_{amb}. It has been shown that such a model can yield practical conclusions concerning the thermal effect of the rotating sleeve in the composite bearing.

An example of a typical hydrodynamic bearing is selected. The purpose of the analysis is to determine the temperature distributions inside the rotating and stationary sleeves. The geometrical parameters and operating conditions of the two hydrodynamic bearings are summarized in Table 18.1.

18.6.1 Thermal Solution for Stationary and Rotating Sleeves

The temperature distribution in the fluid film is solved by the Reynolds equation, together with the equation of viscosity variation versus temperature. The viscous friction losses are dissipated in the fluid as heat, which is transferred by convection (fluid flow) and conduction through the sleeve. The shaft temperature

TABLE 18-1 Bearing and Lubrication Specifications

Outer sleeve radius, R_o	0.095 m
Shaft radius, R_j	0.05 m
Shaft speed, ω_j	3500 RPM
Sleeve wall thickness, b	0.01 m
Sleeve length, L	0.1 m
Sleeve thermal diffusivity, α_b	$1.5 \times 10^{-5}\,\mathrm{m^2/s}$
Sleeve speed, ω_b	200 RPM
Clearance, C	0.00006 m
Eccentricity ratio, ε	0.5
Length-to-diameter ratio, L/D	1
Thermal conductivity of sleeve material, K_b	45 W/m-K
Density of bush material, ρ_b	$8666\,\mathrm{kg/m^3}$
Specific heat of sleeve material, C_{pb}	0.343 kJ/kg-K
Thermal conductivity of oil, K_o	0.13 W/m-K
Density of oil, ρ_o	$860\,\mathrm{kg/m^3}$
Ambient temperature, T_{amb}	24.4°C
Viscosity of the oil at the inlet temperature, μ	0.03 kg/m-s
Viscosity–temperature coefficient, β	0.0411/°K
Oil thermal diffusivity, α_o	$7.6 \times 10^{8}\,\mathrm{m^2/s}$

From Harnoy and Khonsari, 1996.

is assumed to be constant. The following equation, in a cylindrical coordinate system (r, θ), was used for solving the temperature distribution in the sleeve (the coordinate system is fixed to the solid sleeve and rotating with it):

$$\frac{\partial^2 T}{\partial r^2} + \frac{1}{r}\frac{\partial T}{\partial r} + \frac{1}{r^2}\frac{\partial^2 T}{\partial \theta^2} = \frac{1}{\alpha}\frac{dT}{dt} \tag{18-37}$$

where α is the thermal diffusivity of the solid. For a rotating sleeve in stationary (Eulerian) coordinates (the sleeve rotates relative to the stationary coordinates) this equation can be expressed as

$$\frac{\partial^2 T}{\partial r^2} + \frac{1}{r}\frac{\partial T}{\partial r} + \frac{1}{r^2}\frac{\partial^2 T}{\partial \theta^2} = \frac{\omega_b}{\alpha}\frac{\partial T}{\partial \theta} + \frac{1}{\alpha}\frac{dT}{dt} \tag{18-38}$$

where ω_b is the angular speed of the sleeve.

The following order of magnitude analysis intends to show that when the sleeve rotates above a certain speed, its maximum temperature difference in the circumferential direction, ΔT_c, becomes negligible compared with the maximum temperature difference, ΔT_r, in the radial direction. The order of magnitude of all terms in Eq. (18-38) are:

$$\begin{aligned} \frac{\partial^2 T}{\partial r^2} &= O\left(\frac{\Delta T_r}{b^2}\right) \\ \frac{1}{r}\frac{\partial T}{\partial r} &= O\left(\frac{\Delta T_r}{R_b b}\right) \\ \frac{1}{r^2}\frac{\partial^2 T}{\partial \theta^2} &= O\left(\frac{\Delta T_c}{\pi R_b^2}\right) \end{aligned} \tag{18-39a}$$

$$\frac{\omega_b}{\alpha}\frac{\partial T}{\partial \theta} = O\left(\frac{\omega_b R_b}{\alpha}\right)\left(\frac{\Delta T_c}{R_b}\right) \tag{18-39b}$$

Here, b represents the sleeve wall thickness, $b = R_o - R_i$. The radius R is taken as the average value of the outer and inner radii of the bushing, $R_b = (R_o + R_i)/2$. Substituting these orders in Eq. (18-38) and assuming $b \ll R$, the order of the ratio of the temperature gradients is

$$\frac{\dfrac{\partial T}{\partial r}}{\dfrac{1}{R}\dfrac{\partial T}{\partial \theta}} = O\left(\frac{\omega_b R_b b}{\alpha_b}\right) \tag{18-40}$$

The dimensionless parameter on the right-hand side of Eq. (18-40) is a modified Peclet number (Pe). Equation (18-40) indicates that when $\text{Pe} \gg 1$, the radial temperature gradient is much higher than the temperature gradient in the circumferential direction, and the temperature distribution can be assumed to be uniform around the sleeve. In fact, in the circumferential direction, most of the heat is effectively transferred by the moving mass of the rotating sleeve and only a negligible amount of heat is transferred by conduction. In the example (Table 18-1), if the sleeve speed is 200 RPM, the Peclet number is

$$\text{Pe} = \frac{\omega_b R_b b}{\alpha_b} = 692 \tag{18-41}$$

This number indicates that the circumferential temperature gradient is relatively low, and only heat conduction in the radial direction needs to be considered in solving for the temperature distribution. It is interesting to note that there would be no significant change in the composite bearing thermal characteristics even at much lower sleeve speeds. For example, for $\omega_b = 30$ RPM, the resulting Pe is above 100, and the assumption of negligible circumferential temperature gradients should still hold. It should be noted that a composite bearing design operating at a low sleeve speed might not be desirable. Elastohydrodynamic lubrication in the rolling bearing requires a certain minimum speed below which

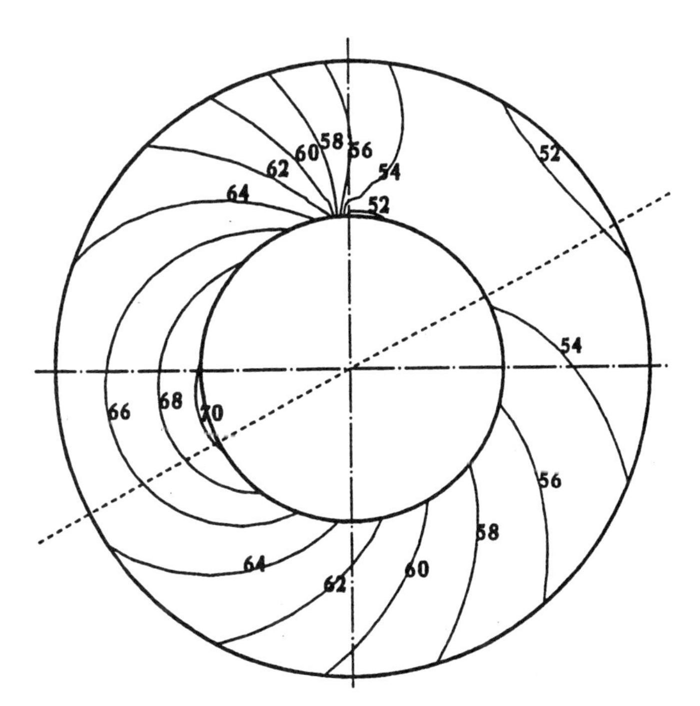

FIG. 18-12 Thermohydrodynamic solution showing the isotherm contours plot in a stationary sleeve of a journal bearing. $L/D = 1$, $\varepsilon = 0.5$, $N_{\text{shaft}} = 3500$ RPM. (From Harnoy and Khonsari, 1996.)

the friction is somewhat higher, as the rolling bearing friction–velocity curve presented in Fig. 18-4 demonstrates.

A full thermohydrodynamic analysis was performed with the bearing specifications listed in Table 18-1 assuming a stationary sleeve. The solution for the temperature profile in the stationary sleeve is presented by isotherms in Fig. 18-12.

Hydrodynamic lubrication theory indicates that the amount of heat dissipated in the oil film is proportional to the average shear rate and, in turn, proportional to the difference between the journal and sleeve speeds $(\omega_j - \omega_b)$. Therefore, it is reasonable to assume that the heat flux from the oil film to the surroundings is also proportional to $(\omega_j - \omega_b)$. Therefore, the ratio of the radial heat fluxes of rotating and stationary is

$$Q_{\text{rotating sleeve}} = Q_{\text{rigid sleeve}} \frac{(\omega_j - \omega_b)}{\omega_j} \tag{18-42}$$

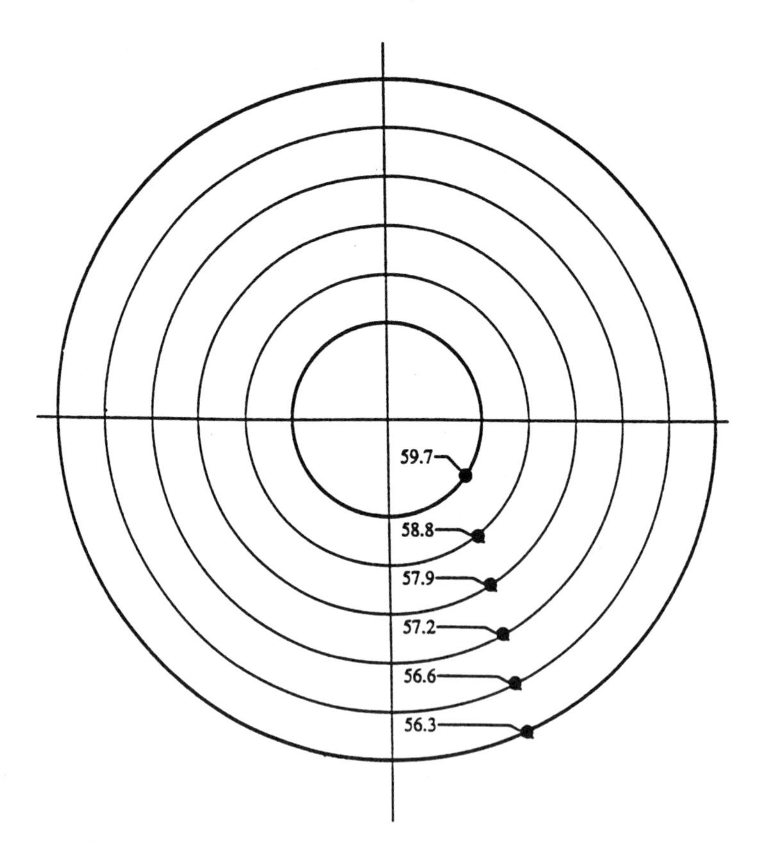

FIG. 18-13 Isotherm contours plot of a rotating sleeve unit. (From Harnoy and Khonsari, 1996.)

The surface temperatures T_i (inner wall) and T_o (outside wall) of the sleeve are solved by the following equations:

$$T_i = T_{\text{amb}} + Q_{\text{rotating sleeve}} \left\{ \frac{\ln(R_o/R_i)}{2\pi k_b L} + \frac{1}{2\pi R_o L h} \right\} \tag{18-43}$$

$$T_o = T_i - Q_{\text{rotating sleeve}} \left\{ \frac{\ln(R_o/R_i)}{2\pi k_b L} \right\} \tag{18-44}$$

Here, h is the correction coefficient. The temperature distribution in the sleeve is obtained from

$$\frac{T - T_i}{T_o - T_i} = \frac{\ln(r/R_i)}{\ln(R_o/R_i)} \tag{18.45}$$

The results are circular isotherms, as shown in Fig. 18-13. The uniformity in the temperature profile, together with a reduction in the maximum temperature (59.7°C for composite bearing versus 71°C for a conventional hydrodynamic bearing), is indicative of the superior thermal performance.

19

Non-Newtonian Viscoelastic Effects

19.1 INTRODUCTION

The previous chapters focused on Newtonian lubricants such as regular mineral oils. However, non-Newtonian multigrade lubricants, also referred to as VI (viscosity index) improved oils are in common use today, particularly in motor vehicle engines. The multigrade lubricants include additives of long-chain polymer molecules that modify the flow characteristics of the base oils. In this chapter, the hydrodynamic analysis is extended for multigrade oils.

The initial motivation behind the development of the multigrade lubricants was to reduce the dependence of lubricant viscosity on temperature (to improve the viscosity index). This property is important in motor vehicle engines, e.g., starting the engine on cold mornings. Later, experiments indicated that multigrade lubricants have complex non-Newtonian characteristics. The polymer-containing lubricants were found to have other rheological properties in addition to the viscosity. These lubricants are *viscoelastic* fluids, in the sense that they have viscous as well as elastic properties.

Polymer additives modify several flow characteristics of the base oil.

1. The polymer additives increase the viscosity of the base oil.
2. The polymer additives moderate the reduction of viscosity with temperature (improve the viscosity index).

3. The viscosity becomes a decreasing function of shear rate (shear-thinning property).
4. Normal stresses are introduced. In simple shear flow, $u = u(y)$, there are normal stress differences $\sigma_x - \sigma_y$ (first difference) and $\sigma_y - \sigma_z$ (second difference). The first difference is much higher than the second difference.
5. The polymer additives introduce stress-relaxation characteristics into the fluid, exemplified by a phase lag between the shear stress and a periodic shear rate. This property is what is meant by the term *viscoelasticity*; namely, the fluid becomes elastic as well as viscous.

Although multi grade oils were developed to improve the viscosity index, later experiments revealed a significant improvement in the lubrication performance of journal bearings that cannot be explained by changes of viscosity. Dubois et al. (1960) compared the performance of mineral oils and VI improved oils in journal bearings under static load. They used high journal speeds and measured load capacity, friction and eccentricity. The results indicated a superior performance of the multigrade oils with polymer additives. Additional conclusion of this investigation (important for comparison with analytical investigations) is that the relative improvement in load capacity of the VI improved oils becomes greater as the eccentricity increases. Okrent (1961) and Savage and Bowman (1961) found less friction and wear in the connecting-rod bearing in a car engine (dynamically loaded journal bearing).

Analytical investigations showed that the improvements in the lubrication performance of VI improved oils are not due to changes in the viscosity. Horowitz and Steidler (1960) performed analytical investigation and showed that the improvement in the lubrication performance could not be accounted for by the different function of viscosity versus shear rate and temperature. In fact, they found that the non-Newtonian viscosity increases the friction coefficient (opposite trend to the experiments of Dubois et al., 1960).

A survey of the previous analytical investigation by Harnoy (1978) shows that the measured order of magnitude of the first and second normal stresses is too low to explain any significant improvement in the lubrication performance. This discussion indicates that the elasticity of the fluid (stress-relaxation effect) is the most probable explanation of the improvement in performance of *viscoelastic* lubricants.

The criterion for improvement of the lubrication performance is very important. For example, polymer additives increase the viscosity of mineral oils; in turn, the load capacity increases. However, our basis of comparison is the load capacity at equivalent viscosity and bearing geometry. Higher viscosity on its own is not considered as an improvement in the lubrication performance, because the friction losses as well as load capacity are both proportional to the viscosity. Moreover, it is possible to use higher viscosity oils without resorting to oil additives of long chain polymer molecules. An appropriate criterion for an improvement of the lubrication performance is the ratio between the friction force and the load capacity (bearing friction coefficient).

19.2 VISCOELASTIC FLUID MODELS[1]

For the analysis of viscoelastic fluids, various models have been developed. The models are in the form of *rheological equations*, also referred to as *constitutive equations*. An example is the Maxwell fluid equation (Sec. 2.9).

Multi-grade lubricants are predominantly viscous fluids with a small elastic effect. Therefore, in hydrodynamic lubrication, the viscosity has a dominant role in generating the pressure wave, while the fluid elasticity has only a small (*second order*) effect. In such cases, the flow of non-Newtonian viscoelastic fluids can be analyzed by using *differential type constitutive equations*. The main advantage of these equations is that the stress components are explicit functions of the strain-rate components. In a similar way to Newtonian Navier-Stokes equations, viscoelastic differential-type equations can be directly applied for solving the flow. Differential type equations were widely used in the theory of lubrication for bearings under steady and particularly unsteady conditions.

Differential type constitutive equations are restricted to a class of flow problems where the Deborah number is low, $De \ll 1$. The ratio De is of the relaxation time of the fluid, λ, to a characteristic time of the flow, Δt; $De = \lambda/\Delta t$. Here, Δt is the time for a significant change in the flow; e.g., in a sinusoidal flow, Δt is the oscillation period.

The early analytical work in hydrodynamics lubrication of viscoelastic fluids is based on the second-order fluid equation of Rivlin and Ericksen (1955) or on the equation of Oldroyd (1959). These early equations are referred to as *conventional*, differential-type rheological equations. Coleman and Noll (1960) showed that the Rivlin and Ericksen equation represents the first perturbation from Newtonian fluid for slow flows, but its use has been extended later to high shear rates of lubrication.

An analysis based on the conventional second order equation (Harnoy and Hanin, 1975) indicated significant improvements of the viscoelastic lubrication performance in journal bearings under steady and dynamic loads. Moreover, the improvements increase with the eccentricity (in agreement with the trends observed in the experiments of Dubois et al., 1960).

---

[1] This section and the following viscoelastic analysis are for advanced studies.

An important feature of these conventional equations for viscoelastic fluids is that they describe the unsteady stress-relaxation effect and the first normal-stress difference $(\sigma_x - \sigma_y)$ in a steady shear flow by the same parameter. In many cases, the relaxation time that describes dynamic (unsteady) flow effects was determined by normal-stress measurements in steady shear flow between rotating plate and cone (Weissenberg rheometer).

In conventional rheological equations, the normal stresses are proportional to the second power of the shear-rate. Hydrodynamic lubrication involves very high shear-rates, and the conventional equations predict unrealistically high first-normal-stress differences. Moreover, when the actual measured magnitude of the normal stresses was considered in lubrication, its effect is negligibly small in comparison to the stress-relaxation effect. It was realized that for high shear-rate flows, the two effects of the first normal stress difference and stress relaxation must be described by means of two parameters capable of separate experimental determination.

For high shear rate flows of lubrication, the forgoing arguments indicated that there is a requirement for a different viscoelastic model that can separate the unsteady relaxation effects from the normal stresses.

19.2.1 Viscoelastic Model for High Shear-Rate Flows

A rheological equation that separates the normal stresses from the relaxation effect was developed and used for hydrodynamic lubrication by Harnoy (1976). For this purpose, a unique convective time derivative, $\delta/\delta t$, is defined in a coordinate system that is attached to the three principal directions of the derived tensor. This rheological equation can be derived from the Maxwell model (analogy of a spring and dashpot in series). The Maxwell model in terms of the deviatoric stresses, τ', is

$$\tau'_{ij} + \lambda \frac{\delta}{\delta t} \tau'_{ij} = \mu e_{ij} \tag{19-1}$$

Here, λ is the relaxation time and the strain-rate components, e_{ij}, are

$$e_{ij} = \frac{1}{2}\left(\frac{\partial v_i}{\partial x_j} + \frac{\partial v_j}{\partial x_i}\right) \tag{19-2}$$

where v_i are the velocity components in orthogonal coordinates x_i. The deviatoric stress tensor can be derived explicitly as

$$\tau'_{ij} = \mu\left(1 + \lambda \frac{\delta}{\delta t}\right)^{-1} e_{ij} \tag{19-3}$$

Expanding the operator in terms of an infinite series of increasing powers of λ results in

$$\tau'_{ij} = \mu\left(e_{ij} - \lambda\frac{\delta}{\delta t}e_{ij} + \lambda^2\frac{\delta^2}{\delta t^2}e_{ij} + \cdots + (-\lambda)^{n-1}\frac{\delta^{n-1}}{\delta t^{n-1}}e_{ij}\right) \tag{19-4}$$

For low-Deborah number, $\mathrm{De} = \lambda/\Delta t$, where Δt is a characteristic time of the flow, second-order and higher powers of λ are negligible. Therefore, only terms with the first power of λ are considered, and the equation gets the following simplified form:

$$\tau'_{ij} = \mu\left(e_{ij} - \lambda\frac{\delta}{\delta t}e_{ij}\right) \tag{19-5}$$

The tensor time derivative is defined as follows (see Harnoy 1976):

$$\frac{\delta e_{ij}}{\delta t} = \frac{\partial e_{ij}}{\partial t} + \frac{\partial e_{ij}}{\partial x_\alpha}v_\alpha - \Omega_{i\alpha}e_{\alpha j} + e_{i\alpha}\Omega_{\alpha j} \tag{19-6}$$

The definition is similar to that of the Jaumann time derivative (see Prager 1961). Here, however, the rotation vector Ω_{ij} is the rotation components of a rigid, rectangular coordinate system (1, 2, 3) having its origin fixed to a fluid particle and moving with it. At the same time, its directions always coincide with the three principle directions of the derived tensor. The last two terms, having the rotation, Ω_{ij}, can be neglected for high-shear-rate flow because the rotation of the principal directions is very slow. Equations (19-5) and (19-6) form the viscoelastic fluid model for the following analysis.

19.3 ANALYSIS OF VISCOELASTIC FLUID FLOW

Similar to the analysis in Chapter 4, the following derivation starts from the balance of forces acting on an infinitesimal fluid element having the shape of a rectangular parallelogram with dimensions dx and dy, as shown in Fig. 4-1. The following derivation of Harnoy (1978) is for two-dimensional flow in the x and y directions. In an infinitely long bearing, there is no flow or pressure gradient in the z direction. In a similar way to that described in Chapter 4, the balance of forces results in

$$d\tau\, dx = dp\, dy \tag{19-7}$$

Remark: If the fluid inertia is not neglected, the equilibrium equation in the x direction for two-dimensional flow is [see Eq. (5-4b)]

$$\rho \frac{Du}{Dt} = -\frac{\partial p}{\partial x} + \frac{\partial \sigma_x'}{\partial x} + \frac{\partial \tau_{xy}}{\partial y} \tag{19-8}$$

After disregarding the fluid inertia term on the left-hand side, the equation is equivalent to Eq. (19-7). In two-dimensional flow, the continuity equation is

$$\frac{\partial u}{\partial x} + \frac{\partial v}{\partial y} = 0 \tag{19-9}$$

For viscoelastic fluid, the constitutive equations (19-5) and (19-6) establishes the relation between the stress and velocity components. Substituting Eq. (19-5) in the equilibrium equation (19-8) yields the following differential equation of steady-state flow in a two-dimensional lubrication film:

$$\frac{dp}{dx} = \mu \frac{\partial^2 u}{\partial y^2} - \lambda\mu \frac{\partial}{\partial y}\left(\frac{\partial^2 u}{\partial y\, \partial x} + \frac{\partial^2 u}{\partial y^2} v\right) \tag{19-10}$$

Converting to dimensionless variables:

$$\bar{u} = \frac{u}{U}; \qquad \bar{v} = \frac{R}{C}\frac{v}{U}; \qquad \bar{x} = \frac{x}{R}; \qquad \bar{y} = \frac{y}{C} \tag{19-11}$$

The ratio Γ, often referred to as the Deborah number, De, is defined as

$$\mathrm{De} = \Gamma = \frac{\lambda U}{R} \tag{19-12}$$

The flow equation (19-10) becomes

$$\frac{\partial^2 \bar{u}}{\partial \bar{y}^2} - \mu \frac{\partial}{\partial y}\left(\frac{\partial^2 \bar{u}}{\partial \bar{y}\, \partial \bar{x}} \bar{u} + \frac{\partial^2 \bar{u}}{\partial \bar{y}^2} \bar{v}\right) = 2F(\bar{x}) \tag{19-13}$$

where

$$2F(\bar{x}) = \frac{C^2}{\eta U R} \frac{dp}{d\bar{x}} \tag{19-14}$$

In these equations, λ is small in comparison to the characteristic time of the flow, Δt. The characteristic time Δt is the time for a significant periodic flow to take place, such as a flow around the bearing or the period time in oscillating flow. It results that De is small in lubrication flow, or $\Gamma \ll 1$.

19.3.1 Velocity

The flow $\bar{u} = \bar{u}(\bar{y})$ can be divided into a Newtonian flow, $\bar{u}_0$, and a secondary flow, $\bar{u}_1$, owing to the elasticity of the fluid:

$$\bar{u} = \bar{u}_0 + \Gamma \bar{u}_1 \tag{19-15}$$

In the flow equations, the secondary flow terms include the coefficient Γ.

19.3.2 Solution of the Differential Equation of Flow

In order to solve the nonlinear differential equation of flow for small Γ, a perturbation method is used, expanding in powers of Γ and retaining the first power only, as follows:

$$\bar{u} = \bar{u}_0 + \Gamma \bar{u}_1 + 0(\Gamma^2) \tag{19-16}$$

$$\bar{v} = \bar{v}_0 + \Gamma \bar{v}_1 + 0(\Gamma^2) \tag{19-17}$$

$$F(\bar{x}) = F_0(\bar{x}) + \Gamma F_1(\bar{x}) + 0(\Gamma^2) \tag{19-18}$$

Introducing Eqs. (19-16)–(19-18) into Eq. (19-13) and equating terms with corresponding powers of Γ yields two linear equations:

$$\frac{\partial^2 \bar{u}_0}{\partial^2 \bar{y}^2} = 2F_0(\bar{x}) \tag{19-19}$$

$$\frac{\partial^2 \bar{u}_1}{\partial \bar{y}^2} - \frac{\partial}{\partial \bar{y}}\left(\frac{\partial^2 \bar{u}_0}{\partial \bar{x}\, \partial \bar{y}}\, \bar{u}_0 + \frac{\partial^2 \bar{u}_0}{\partial \bar{y}^2}\, \bar{v}_0\right) = 2F_1(\bar{x}) \tag{19-20}$$

The boundary conditions of the flow are:

$$\text{at } \bar{y} = 0, \qquad \bar{u} = 0 \tag{19-21}$$

$$\text{at } \bar{y} = \frac{h}{c}, \qquad \bar{u} = 1 \tag{19-22}$$

Because there is no side flow, the flux q is constant:

$$\int_0^h u\, dy = q = \frac{h_e U}{2} \tag{19-23}$$

For the first velocity term, $\bar{u}_0$, the boundary conditions are:

$$\text{at } \bar{y} = 0 \qquad \bar{u}_0 = 0 \tag{19-24}$$

$$\text{at } \bar{y} = \frac{h}{c} \qquad \bar{u}_0 = 1 \tag{19-25}$$

Expanding the flux into powers of Γ:

$$q = q_0 + \Gamma q_1 + 0(\Gamma^2 q) = \frac{h_e U}{2} \tag{19-26}$$

and we denote

$$h_i = \frac{2q_i}{U} \qquad \text{for } i = 0 \text{ and } 1 \tag{19-27}$$

The flow rate of the zero-order (Newtonian) velocity is

$$\int_0^{h/c} \bar{u}_0 \, d\bar{y} = \frac{q_0}{CU} = \frac{h_0}{2C} \tag{19-28}$$

After integrating Eq. (19-19) twice and using the boundary conditions (19-24), (19-25), and (19-28), the zero-order equations result in the well-known Newtonian solutions:

$$\bar{u}_0 = M\bar{y}^2 + N\bar{y} \tag{19-29}$$

where

$$M = 3C^2\left(\frac{1}{h^2} - \frac{h_0}{h^3}\right) \tag{19-30}$$

$$N = C\left(\frac{3h_0}{h^2} - \frac{2}{h}\right) \tag{19-31}$$

The velocity component in the y direction, v_0, is determined from the continuity equation. Substituting $\bar{u}_0$ and $\bar{v}_0$ in Eq. (19-20) enables solution of the second velocity, $\bar{u}_1$. The boundary conditions for $\bar{v}_1$ are:

$$\text{at } \bar{y} = 0, \qquad \bar{u}_1 = 0 \tag{19-32}$$

$$\text{at } \bar{y} = \frac{h}{c}, \qquad \bar{u}_1 = 0 \tag{19-33}$$

$$\int_0^{h/c} \bar{u}_1 \, d\bar{y} = \frac{q_1}{CU} = \frac{h_1}{2C} \tag{19-34}$$

The resulting solution for the velocity in the x direction is

$$\bar{u} = \bar{u}_0 + \Gamma u_1 = \alpha\bar{y}^4 + \beta\bar{y}^3 + \gamma\bar{y}^2 + \delta\bar{y} \tag{19-35}$$

where

$$\alpha = 3\Gamma C^4\left(-\frac{2}{h^5}+\frac{5h_e}{h^6}-\frac{3h_e^2}{h^7}\right)\frac{dh}{d\bar{x}} \tag{19-36}$$

$$\beta = \Gamma C^3\left(18\frac{h_e^2}{h^6}-\frac{24h_e}{h^5}+\frac{8}{h^4}\right)\frac{dh}{d\bar{x}} \tag{19-37}$$

$$\gamma = 3C^2\left(\frac{1}{h^2}-\frac{h_e}{h^3}\right)+\Gamma C^2\left(-\frac{6}{5h^3}+\frac{9h_e}{h^4}-\frac{54h_e^2}{5h^5}\right)\frac{dh}{d\bar{x}} \tag{19-38}$$

$$\delta = C\left(\frac{3h_e}{h^2}-\frac{2}{h}\right)+\Gamma C\left(\frac{9h_e^2}{5h^4}-\frac{4}{5}\frac{1}{h^2}\right)\frac{dh}{d\bar{x}} \tag{19-39}$$

where $h_e = h_o + \Gamma h_1$ is an unknown constant.

19.4 PRESSURE WAVE IN A JOURNAL BEARING

In a similar way to the solution in Chapter 4, the following pressure wave equation is obtained from Eq. (19-10) and the fluid velocity:

$$p = 6R\mu U\int_0^x\left(\frac{1}{h^2}-\frac{h_e}{h^3}\right)dx + \Gamma R\mu U\left(-\frac{4}{5}\frac{1}{h^2}+\frac{9}{10}\frac{h_e^2}{h^4}\right)+k \tag{19-40}$$

The last constant, k, is determined by the external oil feed pressure. The constant h_e is determined from the boundary conditions of the pressure p around the bearing.

The analysis is limited to a relaxation time λ that is much smaller than the characteristic time, Δt of the flow. In this case, the characteristic time is $\Delta t = O(U/R)$, which is the order of magnitude of the time for a fluid particle to flow around the bearing. The condition becomes $\lambda \ll U/R$.

For a journal bearing, the pressure wave for a viscoelastic lubricant was solved and compared to that of a Newtonian fluid; see Harnoy (1978). The pressure wave was solved by numerical integration. Realistic boundary conditions were applied for the pressure wave [see Eq. (6-67)]. The pressure wave starts at $\theta = 0$ and terminates at θ_2, where the pressure gradient also vanishes. The solution in Fig. 19-1 indicates that the elasticity of the fluid increases the pressure wave and load capacity.

19.4.1 Improvements in Lubrication Performance of Journal Bearings

The velocity in Eq. (19-35) allows the calculation of the shear stresses, and friction torque on the journal. The results indicated (Harnoy, 1978) that the elasticity of the fluid has a very small effect on the viscous friction losses of a

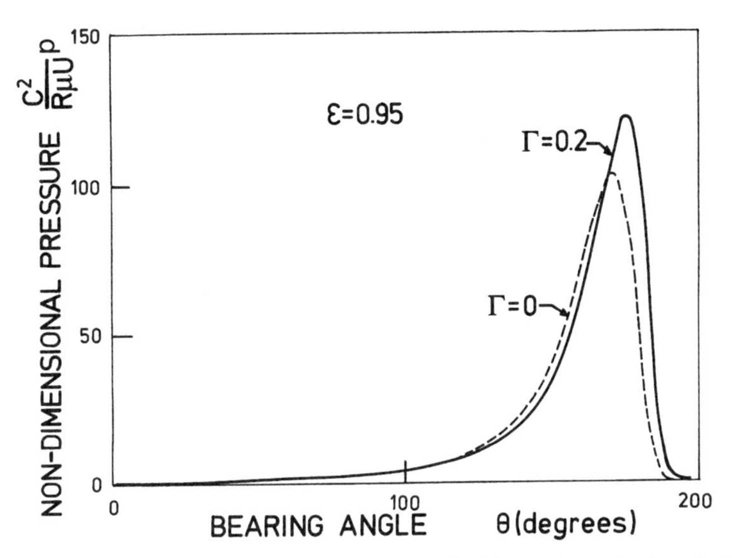

FIG. 19-1 Journal bearing pressure wave for Newtonian and viscoelastic lubricants. (From Harnoy, 1978.)

journal bearing, and the reduction in the friction coefficient is mostly due the load capacity improvement. As mentioned in Sec. 19.1, the friction coefficient is a criterion for the improvement in the lubrication performance under static load. In short hydrodynamic journal bearings, e.g., in car engines, the elasticity of the fluid reduces the friction coefficient by a similar order of magnitude (Harnoy, 1977).

Harnoy and Zhu (1991) conducted dynamic analysis of short hydrodynamic journal bearings based on the same viscoelastic fluid model. The results show that viscoelastic lubricants play a significant role in improving the lubrication performance under heavy dynamic loads, where the eccentricity ratio is high; see Fig. 15-2. For a viscoelastic lubricant, the maximum eccentricity ratio ε_{min} of the locus of the journal center is significantly reduced in comparison to that of a Newtonian lubricant. In conclusion, analytical results based on the viscoelastic fluid model of Eqs. (19-5) and (19-6) indicated significant improvements of lubrication performance under steady and dynamic loads. Moreover, the improvement increases with the eccentricity. These results are in agreement with the trends obtained in the experiments of Dubois et al. 1960.

However, similar improvements in performance were obtained by using the conventional second-order equation. Therefore, the results for journal bearings cannot indicate the appropriate rheological equation, which is in better agreement with experimentation. It is shown in Sec. 19.5 that squeeze-film flow can be used for the purpose of validation of the appropriate rheological equation, because the solutions of two theoretical models are in opposite trends.

Viscoelastic lubricants play a significant role in improving lubrication performance under heavy dynamic loads, where the eccentricity ratio is high; see Fig. 15-2. For a viscoelastic lubricant, the maximum locus eccentricity ratio ε_{min} is significantly reduced in comparison to that of a Newtonian lubricant.

19.4.2 Viscoelastic Lubrication of Gears and Rollers

Harnoy (1976) investigated the role of viscoelastic lubricants in gears and rollers. In this application, there is a pure rolling or, more often, a rolling combined with sliding. For rolling and sliding between a cylinder and plane (see Fig. 4-4) the solution of the pressure wave for Newtonian and viscoelastic lubricants is shown in Fig. 19-2. The viscoelastic fluid model is according to Eqs. (19-5) and (19-6). The results of the numerical integration are presented for different rolling-to-sliding ratios ξ. The relative improvement of the pressure wave and load capacity due to the elasticity of the fluid are more pronounced for rolling than for sliding (the relative rise of the pressure wave increases with ξ).

19.5 SQUEEZE-FILM FLOW

Squeeze-film flow between two parallel circular and concentric disks is shown in Fig. 5-5 and 5-6. Unlike experiments in journal bearings, squeeze-film experiments can be used for verification of viscoelastic models. In fact, the viscoelastic fluid model described by Eqs. (19-5) and (19-6) resulted in agreement with squeeze-film experiments, while the conventional second-order equation resulted in conflict with experiments.

Two types of experiments are usually conducted:

1. The upper disk has a constant velocity V toward the lower disk, and a load cell measures the upper disk load capacity versus the film thickness, h.
2. There is a constant load W on the upper disk, and the film thickness h is measured versus time. Experiments were conducted to measure the *descent time*, namely, the time for the film thickness to be reduced to half of its initial height.

For Newtonian fluids, the solution of the load capacity in the first experiment is presented in Sec. 5.7. If the upper disk has a constant velocity V toward the lower disk (first experiment), the load capacity of the squeeze-film of

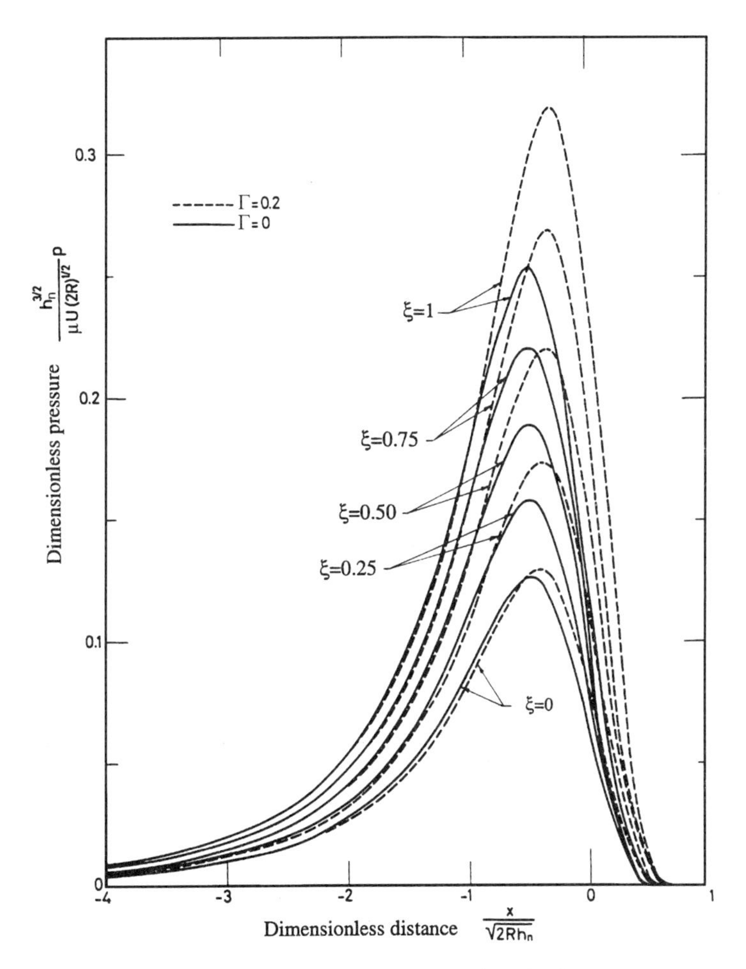

FIG. 19-2 Comparison of Newtonian and viscoelastic pressure waves in rollers for various rolling-to-sliding ratios ξ.

viscoelastic fluid is less than its Newtonian counterpart. In Sec. 5.7, it was shown that the squeeze-film load capacity of a Newtonian fluid is

$$W_o = \frac{3\pi\mu V R^4}{2h^3} \tag{19-41}$$

Here, W_o is the Newtonian load capacity, h is the clearance, and V is the disk velocity when the disks are approaching each other. If the fluid is viscoelastic, under constant velocity V, the equation for the load capacity W becomes (Harnoy, 1987)

$$\frac{W}{W_o} = 1 - 2.1\ \mathrm{De} \tag{19-42}$$

Here, De is the ratio

$$\mathrm{De} = \frac{\lambda V}{h} \tag{19-43}$$

and h is the clearance. This result is in agreement with the physical interpretation of the viscoelasticity of the fluid. In a squeeze action, the stresses increase with time, because the film becomes thinner. For viscoelastic fluid, the stresses are at an earlier, lower value. This effect is referred to as a *memory effect*, in the sense that the instantaneous stress is affected by the history of previous stress. In this case, it is affected only by the recent history of a very short time period.

For the first experiment of load under constant velocity, all the viscoelastic models are in agreement with the experiments of small reduction in load capacity. However, for the second experiment under constant load, the early conventional models (the second order fluid and other models) are in conflict with the experiments. Leider and Bird (1974) conducted squeeze-film experiments under a constant load. For viscoelastic fluids, the experiments demonstrated a longer squeezing time (*descent time*) than for a comparable viscous fluid. Grimm (1978) reviewed many previous experiments that lead to the same conclusion.

Tichy and Modest (1980) were the first to analyze the squeeze-film flow based on Harnoy rheological equations (19.5) and (19.6). Later, Avila and Binding (1982), Sus (1984), and Harnoy (1987) analyzed additional aspects of the squeeze-film flow of viscoelastic fluid according to this model. The results of all these analytical investigations show that Harnoy equation correctly predicts the trend of increasing descent time under constant load, in agreement with experimentation. In that case, the theory and experiments are in agreement that the fluid elasticity improves the lubrication performance in unsteady squeeze-film under constant load.

Brindley et al. (1976) solved the second experiment problem of squeeze-film under constant load using the second order fluid model. The result predicts an opposite trend of decreasing descent time, which is in conflict with the experiments. In this case, the second dynamic experiment can be used for validation of rheological equations.

An additional example where the rheological equations (19.5) and (19.6) are in agreement with experiments, while the conventional equations are in conflict with experiments is the boundary-layer flow around a cylinder. These experiments can also be used for similar validation of the appropriate viscoelastic

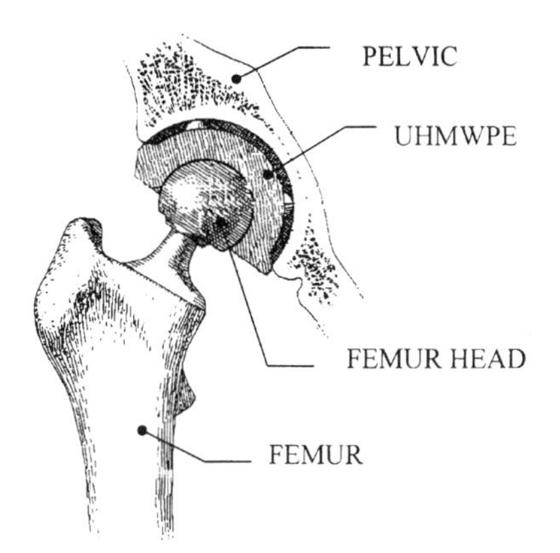

FIG. 20-1 Hip replacement joint.

The combinations with UHMWPE have relatively low friction and wear in comparison to earlier designs with metal sockets. Later, the stainless steel femur was replaced with titanium or cobalt alloys for better compatibility with the body. It proved to be a good design and material combination, with a life expectancy of 10–15 years. This basic hip joint design is still commonly used today.

For comparison with the implant joint, an example of a natural joint (synovial joint) is shown in Fig. 20-2. The cartilage is a soft, compliant material, and together with the synovial fluid as a lubricant, it is considered to be superior in performance to any manmade bearing (Dowson and Jin, 1986, Cooke et al., 1978, and Higginson, 1978).

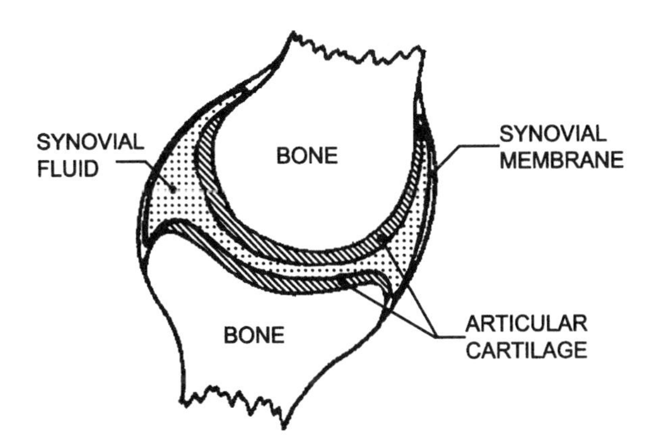

FIG. 20-2 Example of a natural joint.

Although significant progress has been made, there are still two major problems in the current design that justify further research in this area. The most important problems are

1. A major problem is that particulate wear debris is undesirable in the body.
2. A life of 10–15 years is not completely satisfactory, particularly for young people. It would offer a significant benefit to patients if the average life could be extended.

Previous studies, such as those by Willert et al. 1976, 1977, Mirra et al. 1976, Nolan and Phillips, 1996, and Pappas et al., 1996, indicate that small-size wear debris of UHMWPE is rejected by the body. Furthermore, there are indications that the wear debris contributes to undesirable separation of the metal from the bone. There is no doubt that any improvement in the life of the implant would be of great benefit.

20.2 ARTIFICIAL HIP JOINT AS A BEARING

The artificial hip joint is a heavily loaded bearing operating at low speed and with an oscillating motion. The maximum dynamic load on a hip joint can reach five times the weight of an active person. During walking or running, the hip joint bearing is subjected to a dynamic friction in which the velocity as well as load periodically oscillate with time. The oscillations involve start-ups from zero velocity. The joint is considered a lubricated bearing in the presence of body fluids, although the lubricant is of low viscosity and inferior to the natural synovial fluid.

For a lubricated sleeve or socket bearings, a certain minimum product of viscosity and speed, μU, is required to generate a full or partial fluid film that can reduce friction and wear. At very low speed, there is only boundary lubrication with direct contact between the asperities of the sliding surfaces.

Dry friction of polymers (such as UHMWPE) against hard metals is unique, because the friction coefficient increases with sliding velocity (Fig. 16-4). Friction of metals against metals has an opposite trend of a negative slope of friction versus sliding velocity. For polymers against metals, the start-up dry friction is the minimum friction, whereas it is the maximum friction in metal against metal. However, for lubricated surfaces, there is always a negative slope of friction versus sliding velocity, and the start-up friction is the maximum friction for polymers against metals as well as metals against metals.

From a tribological perspective, the performance of artificial joints is inferior to that of synovial joints. The reciprocating swinging motion of the hip joint means that the velocity will be passing through zero, where friction is highest, with each cycle. In its present design, the maximum velocity reached in an artificial joint is not sufficient (or sustained long enough) to generate full hydrodynamic lubrication. Under normal activity, much of the motion associated with joints is of low velocity and frequency. In artificial joints this means that lubrication is characterized by boundary lubrication, or at best mixed lubrication. In contrast, natural, synovial joints are characterized by a mixture of a full fluid film and mixed lubrication. Experiments by Unsworth et al. (1974, 1988) and O'Kelly et al. (1977) suggest that hip and knee synovial joints operate with an average friction coefficient of 0.02. In comparison, the friction coefficient measured in artificial joints ranges from 0.02–0.25. High friction causes the loosening of the implant. In addition, wear rate of artificial joints is much higher than in synovial joints.

The synovial fluid provides lubrication in the natural joint. It is highly non-Newtonian, exhibiting very high viscosity at low shear rates; however, it is only slightly more viscous than water at high shear rates. Dowson and Jin (1986, 1992) have attempted to analyze the lubrication of natural joints. In their work, they couple overall elastohydrodynamic analysis with a study of the local, micro-elastohydrodynamic action associated with surface asperities. Their analysis indicates that microelastohydrodynamic action smooths out the initial roughness of cartilage surfaces in the loaded junctions in articulating synovial joints.

In natural joints a cartilage is attached to the bone surfaces. This cartilage is elastic and porous. The elastic properties of the cartilage allow for some compliance that extends the fluid film region. This is in contrast to artificial joints, which are relatively rigid and consequently exhibit poor lubrication in which ideal separation of the surfaces does not occur. Contact between the plastic and metal surfaces increases the friction and leads to wear. The problem is compounded due to the fact that synovial fluid in implants is much less viscous than that in natural joints (Cooke et al. 1978). Therefore, any future improvement in design which extends the fluid film regime would be very beneficial in reducing friction and minimizing wear in artificial joints.

20.3 HISTORY OF THE HIP REPLACEMENT JOINT

Dowson (1992, 1998) reviewed the history of artificial joint implants. The following is a summary of major developments of interest to design engineers.

Unsuccessful attempts at joint replacement were performed* as early as 1891. These attempts failed due to incompatible materials, and infections. In 1938, Phillip Wiles designed and introduced the first stainless steel artificial hip

*The German surgeon Gluck (1891) replaced a diseased hip joint with an ivory ball and socket held in place with cement and screws. Two years later, a French surgeon, Emile Pean replaced a shoulder joint with an artificial joint made of platinum rods joined by a hard rubber ball.

joint (see Wiles, 1957). The prosthesis consisted of an acetebular cup and femoral head (both made of stainless steel held in place by screws). The matching surfaces of the cup and femoral head were ground and fitted together accurately. The basic design of Phillip Wiles was successful and did not change much over time; however, the steel-on-steel combination lacks tribological compatibility (see Chapter 11), resulting in high friction and wear. The high friction caused the implants to fail by loosening of the cup that had been connected to the pelvis by screws. Failure occurred mostly within one year; therefore, only six joints were implanted.

In the 1950s, there were several interesting attempts to improve the femoral head material. For example, the Judet brothers in Paris used acrylic for femoral head replacement in 1946 (Judet and Judet, 1950); however, there were many failures due to fractures and abrasion of the acrylic head. In 1950, Austin Moore in the United States used *Vitallium*, a cobalt–chromium–molybdenum alloy, for femoral head replacement (see Moore, 1959).

Between 1956 and 1960, the surgeon G.K. McKee replaced the stainless steel with Vitallium; in addition, McKee and Watson-Farrar introduced the use of methyl-methacrylate as a cement to replace the screws. The design consisted of relatively large-diameter femoral head, and the outer surface of the cup had studs to improve the bonding of the cup to the bone by cement (see McKee and Watson-Farrar, 1966, and McKee, 1967). They used a metal-on-metal, closely fitted femoral head and acetabular cup. These improvements significantly improved the success rate to about 50%. However, the metal-on-metal combination loosened due to fast wear, and it was recognized that there is a need for more compatible materials.

Dr. Sir John Charnley developed the successful modern replacement joint (see Charnley, 1979). Charnley conducted research on the lubrication of natural and artificial joints, and realized that the synovial fluid in natural joints is a remarkable lubricant, but the body fluid is not as effective in the metal-on-metal artificial joint. He concluded that a self-lubricating material would be beneficial in this case. In 1969, Dr. Charnley replaced the metal cup with a polytetraflouro-ethelyne (PTFE) cup against a stainless steel femoral head. The design consisted of a stainless steel, small-diameter femoral head and a PTFE acetebular cup. The PTFE has self-lubricating characteristics, and very low friction against steel. However, the PTFE proved to have poor wear resistance and lacked the desired compatibility with the body (implant's life was only 2–3 years).

In 1961, Dr. Charnley replaced the PTFE cup with UHMWPE, which was introduced at that time. The wear rate of this combination was 500 to 1800 times lower than for PTFE cup. In addition, he replaced the screws and bolts with methyl-methacrylate cement (similar to the technique of McKee and Watson-Farrar). A study that followed 106 cases for ten years, and ended in 1973, showed a success rate of 92%. This design remains (with only a few improvements) the most commonly used artificial joint today.

The use of cement in place of screws, UHMWPE, ceramics, and metal alloys with super fine surface finish has led to the remarkable success of orthopedic joint implants; this is one of the important medical achievements.

However, there are still a few problems. Wear debris generated by the rubbing motion is released into the area surrounding the implant. Although UHMWPE is compatible with the body, a severe foreign-body response against the small wear debris has been observed in some patients. Awakened by the presence of the debris, the body begins to attack the cement, resulting in loosening of the joint. Recently, complications resulting from UHMWPE wear debris have renewed some interest in metal-on-metal articulating designs (Nolan and Phillips, 1996).

Wear is still a major problem limiting the life of joint implants. With the current design and materials, young recipients outlive the implant. With the average life span increasing, recipients will outlive the life of the joint. Unlike natural joints, the implants are rigid, the lubrication is inferior, and there is no soft layer to cushion impact. Further improvements are expected in the future; new implants will likely be more similar to natural joints.

20.4 MATERIALS FOR JOINT IMPLANTS

The materials in the prostheses must be compatible with the body. They must not induce tumors or inflammation, and must not activate the immune system. The materials must have excellent corrosion resistance and, ideally, high wear resistance and low coefficient of friction against the mating material. Publications by Sharma and Szycher (1991) and Williams (1982) deal with materials compatible with the body.

For the femoral head, low density is desirable, and high yield strength is very important. Common materials used are cobalt-chromium-molybdenum alloys and titanium alloy (6Al-4V). Cobalt alloys have excellent corrosion resistance (much better than stainless steel 316). The titanium alloy has high strength and low density but it is relatively expensive. Titanium alloys have a low toxicity and a strong resistance to pitting corrosion, but its wear resistance is somewhat inferior to cobalt alloys. Titanium alloy is considered a good choice for patients with sensitivity to cobalt debris. Aluminum oxide ceramic is also used in the manufacture of femoral heads. It has excellent corrosion resistance and compatibility with the body.

20.4.1 Ceramics

Aluminum oxide ceramic femoral heads in combination with UHMWPE cups have increasing use in prosthetic implants. Fine grain, high density aluminum oxide has the required strength for use in the heavily loaded femoral heads, high corrosion resistance, and wear resistance, and it has the advantage of self-anchoring to the human body through bone ingrowth. Most important, femoral ball heads with fine surface-finish ceramics reduce the wear rate of UHMWPE cups. Dowson and Linnett (1980) reported a reduction of 50% in the wear rate of UHMWPE against aluminum-oxide ceramic, in comparison to UHMWPE against steel (observed in laboratory and in vivo tests).

The apparent success of the ceramic femoral head design led to experiments with ceramic-on-ceramic joint (the UHMWPE cup is replaced with a ceramic cup). However, the results showed early failure due to fatigue and surface fracture. Ceramic-on-ceramic designs require an exceptional surface finish and precise manufacturing to secure close fit. Surgical implantation of the all-ceramic joint is made more difficult by the necessity to maintain precise alignment. In addition, the strength requirements must be carefully considered during the design (Mahoney and Dimon, 1990, Walter and Plitz, 1984, and McKellop et al., 1981).

20.5 DYNAMIC FRICTION

Most of the previous research on friction and wear of UHMWPE against metals was conducted under steady conditions. It was realized, however, that friction characteristics under dynamic conditions (oscillating sliding speed) are different from those under static conditions (steady speed).

Under dynamic condition, the friction is a function of the instantaneous sliding speed as well as a memory function of the history of the speed. It would benefit the design engineers to have an insight into the dynamic friction characteristics of UHMWPE used in implant joints. During walking, the hip joint is subjected to oscillating sliding velocity. Dynamic friction experiments were conducted at New Jersey Institute of Technology, Bearing and Bearing Lubrication Laboratory. The testing apparatus is similar to that shown in Fig. 14-7, and the test bearing is UHMWPE journal bearing against stainless steel shaft. The oscillation sliding in the hip joint is approximated by sinusoidal motion, obtained by a computer controlled DC servomotor. The friction and sliding velocity are measured versus time, and the readings are fed on-line into a computer with a data acquisition system, where the data is stored, analyzed and plotted.

Figures 20-3 and 20-4 are examples of measured f–U curves for simulation of the walking velocity and frequency. The frequency of normal walking is approximated by sinusoidal sliding velocity $\omega = 4\,\mathrm{rad/s}$, and a maximum sliding velocity of $\pm 0.07\,\mathrm{m/s}$. The shaft diameter is 25 mm, with $L/D = 0.75$ and a constant load of 215 N.

For dry friction (Fig. 20-3), the friction increases with sliding velocity. At the start-up (acceleration) the friction is higher than for stopping (deceleration).

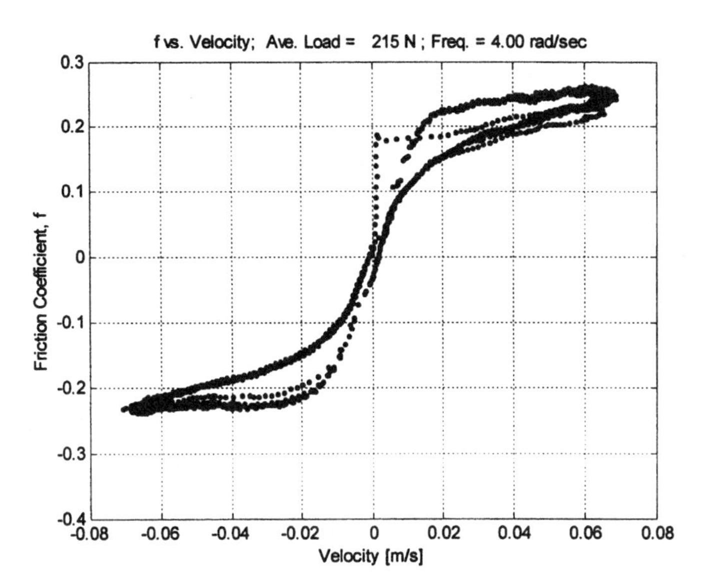

FIG. 20-3 Friction–velocity curve for dry friction, UHMWPE against stainless steel, frequency = 4 rad/s, maximum velocity = ± 0.07 m/s, load = 215 N (Bearing and Bearing Lubrication Laboratory, NJIT).

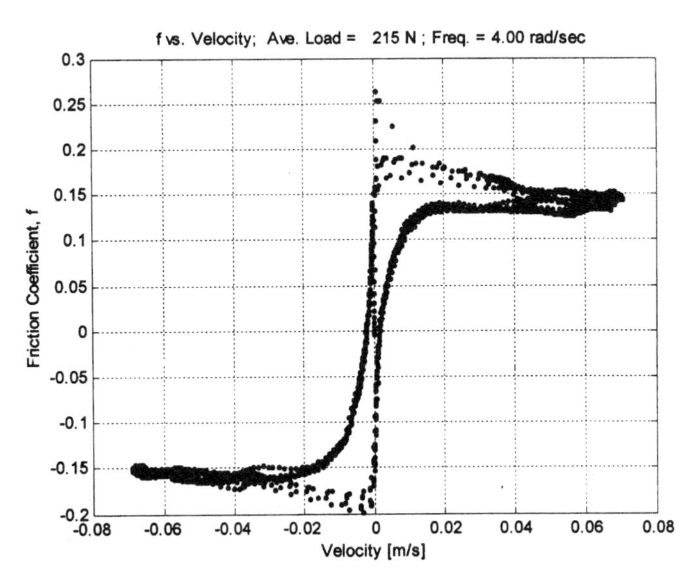

FIG. 20-4 Friction–velocity curve, lubrication with low viscosity oil, $\mu = 0.001$ N-s/m^2, UHMWPE against stainless steel, frequency = 4 rad/s, maximum velocity = ± 0.07 m/s, load = 215 N (Bearing and Bearing Lubrication Laboratory, NJIT).

Unlike the metal-on-metal curve, the dynamic f–U curve with UHMWPE has considerable hysteresis for dry friction. This effect suggests that the friction of polymers involves viscous friction.

Several cycles are measured, and the curve shows a good repeatability of the cycles—except for the first cycle (dotted line), which has a higher stiction force. Unlike what we see with metal-on-metal testing, the friction coefficient increases with the velocity, reaching a maximum of $f = 0.26$. In this case, the breakaway friction at each cycle approaches zero. However, at the first cycle of the experiment (dotted line), there is a higher stiction force, and the breakaway friction coefficient is near 0.2.

Figure 20-4 is for identical conditions, but lubrication is provided with a very light (low viscosity) oil, $\mu = 0.001$ N-s/m^2. This curve simulates the friction in an actual joint implant. The curve indicates that even for a low viscosity and speed, the bearing operates in the boundary and mixed lubrication regime, and the friction decreases versus sliding velocity. This curve also shows a considerable hysteresis. For lubricated surfaces, the first cycle (dotted line in Fig. 20-4) also demonstrates a higher stiction force of $f = 0.25$ while the following cycles have a reduced maximum breakaway coefficient of $f = 0.2$.

Appendix A

Units and Definitions of Material Properties

A.1 UNIT SYSTEMS

The traditional unit system in the United States has been the Imperial system, often referred to as the British system, although in United Kingdom the Imperial system was replaced by the SI International System (Système Internationale, French). In the United States, the engineering societies are in favor of adopting SI, and most engineering publications and textbooks currently use SI units. Many engineering companies are in transition from Imperial to SI units, so engineers must be familiar with the two systems. For this reason, this text uses both systems, although most of the example problems are presented in SI units.

The SI is based on three units: mass, length, and time. The unit of mass is the kilogram (kg), that of length is the meter (m), and that for time is the second (s). The unit of force is the Newton (N), which is defined by Newton's second law as the force required to accelerate 1 kg of mass at the rate of 1 m^2/s.

Gravitational acceleration is $g = 9.81\ m^2/s$, so the weight (force exerted by gravity at the earth's surface) of 1 kg mass is

$$F = mg = 1 \times 9.81 = 9.81\ N \tag{A-1}$$

The unit of energy (or work) is the Joule (J), which is equivalent to N-m. The unit of power, which is energy per unit of time, is the watt (W). The watt is equivalent to J/s, or, in basic SI units, N-m/s.

Pressure or stress is force per unit area. The SI unit is the pascal (Pa), which is equivalent to N/m^2. This is a small unit, and prefixes such as kPa (10^3 Pa) and MPa (10^6 Pa) are often used.

In SI units, very large or very small numbers are often needed in practical problems, and the following prefixes serve to indicate multiplication of units by various powers of 10:

μ (micro-) $= 10^{-6}$ k (kilo-) $= 10^3$
m (milli-) $= 10^{-3}$ M (mega-) $= 10^6$
c (centi-) $= 10^{-2}$ G (giga-) $= 10^9$

For example, the well-known Imperial unit of pressure is psi ($lb_f/in.^2$).

1 psi is $= 6895\ N/m^2$ (Pa) $= 6.895$ kPa.

A second example is the modulus of elasticity of the steel:

$$E = 2.05 \times 10^{11}\ \text{Pa}\ (N/m^2) = 2.05 \times 10^5\ \text{MPa}, = 2.05 \times 10^2\ \text{GPa}.$$

A.2 DEFINITIONS OF MATERIAL PROPERTIES

A.2.1 Density, ρ

Material density ρ is mass per unit volume. The SI unit of density is kg/m^3. In Imperial units, the density is lb_m/ft^3, or lb_m/in^3. For example, the density of water at 4°C is 1000 kg/m^3, and in imperial units it is 62.43 $lb_m/in.^3$.

The conversion is

$$1\ kg/m^3 = 0.06243\ lb_m/ft^3.$$

A.2.2 Specific Weight, γ

Specific weight, γ, is the gravity force (weight) per unit volume of the material

$$\gamma = \rho g \tag{A-2}$$

The SI unit of density is N/m^3. For example, the specific weight γ of water at 4°C is 9810 N/m^3, obtained by the equation

$$\gamma_{water} = \rho g = 1000 \times 9.81 = 9810\ N/m^3.$$

The Imperial unit of specific density is lb_f/ft^3, or lb_f/in^3. For example, the specific weight γ of water at 4°C is 62.4 lb_f/ft^3. The conversion is

$$1\ lb_f/ft^3 = 157.1\ N/m^3$$
$$1\ N/m^3 = 0.00636\ lb_f/ft^3$$

A.2.3 Specific Gravity, S

Specific gravity, S, of a material is the ratio of its specific weight to the specific weight of water at 4°C. It is also the ratio of its density to the density of water at 4°C. For example, if the density of a steel is 7800 kg/m^3, its specific density is $7800/1000 = 7.8$ (specific gravity is a dimensionless ratio).

A.2.4 Specific Heat, c

Specific heat, c, is the amount of heat that must be transferred to a unit of mass of a material to raise its temperature by one degree. For gas, the specific heat depends if the unit of mass has a constant pressure, c_p, or if the unit of mass has a constant volume c_v. The specific heat of a material is a function of its temperature. The SI unit of specific heat is J/Kg-°C (a widely used unit is KJ/Kg-°C), and the Imperial unit is BTU/lb_m°F.

The conversion ratio is

$$1 \text{ BTU}/\text{lb}_m{}^\circ\text{F} = 2326 \text{ J}/\text{Kg-}^\circ\text{C}$$

$$1 \text{ BTU}/\text{lb}_m{}^\circ\text{F} = 2.326 \text{ KJ}/\text{Kg-}^\circ\text{C}$$

A.2.5 Thermal Conductivity, k

The thermal conductivity is a measure of the rate of heat transfer through a material. It is the coefficient k in the Fourier Law of heat conduction

$$q = -kA\frac{\partial T}{\partial x} \tag{A-3}$$

where q is the rate of heat transfer, A is the area normal to the temperature gradient $\partial T/\partial x$.

The SI unit of thermal conductivity is Watt per meter per Celsius degree, W/m-C. The Imperial unit of thermal conductivity is BTU/h-ft-°F. The conversion ratio is

$$1 \text{ W/m-C} = 0.57782 \text{ BTU/h ft }^\circ\text{F}.$$

A.2.6 Absolute Viscosity, μ

The absolute viscosity, μ, is a measure of the fluid resistance to flow. The viscosity and its units are presented in Chap. 2. The SI unit of absolute viscosity is N-s/m^2 (or Pa-s). An additional widely used unit is the poise, (P) (after Poiseuille), which is dyne-s/cm^2, and a smaller traditional unit is centipoise (cP).

$$1 \text{ centipoise, (cP)} = 10^{-2} \times \text{poise}$$

An Imperial unit for the viscosity is the reyn (after Osborne Reynolds), which is lbf-s/in.^{2}.

Conversions

1. 1 centipoise is equal to 1.45×10^{-7} reyn
2. 1 centipoise is equal to 0.001 N-s/m^2
3. 1 centipoise is equal to 0.01 poise
4. 1 reyn is equal to 6.895×10^3 N-s/m^2
5. 1 reyn is equal to 6.895×10^6 centipoise
6. 1 N-s/m^2 is equal to 10^3 centipoise
7. 1 N-s/m^2 is equal to 1.45×10^{-4} reyn

A.2.7 Kinematic Viscosity, ν

The kinematic viscosity, ν, is the ratio of the absolute viscosity and density

$$\nu = \frac{\mu}{\rho} \tag{A-4}$$

The SI unit of kinematic viscosity is m^2/s. Additional widely used traditional unit is the stokes (St) (after Stokes), which is cm^2/s, and a smaller unit is the centistokes (cSt), which is mm^2/s.

The common Imperial unit is in.^{2}/s.

Conversions

1 centistokes, cSt $= 10^6$ m^2/s
1 stokes, St $= 10^4$ m^2/s
1 m^2/s $= 6.452 \times 10^{-4}$ in.^{2}/s
1 m^2/s $= 10^{-4}$ stokes
1 in.^{2}/s $= 0.00155$ cSt

Appendix B

Numerical Integration

The pressure wave along the bearing is solved by integration of Eq. 4-13. Although some integrals can be solved analytically, complex functions can be solved by numerical integration. This appendix is a survey of the various methods for numerical integration, and examples are presented. A simple numerical integration is demonstrated by means of a spreadsheet computer program, which is favored by engineers and students for its simplicity, and because the spreadsheet program can be used for graphic presentation of the pressure wave.

The methods of approximate numerical integration are based on a summation of small areas of width Δx below the curve, which are approximated by various methods that include the midpoint rule, rectangle rule, trapezoidal rule, and Simpson rule.

B.1 MIDPOINT RULE

Integration by midpoint rule is an approximation. The area below the curve is approximated by the sum of the rectangular areas, as shown in Fig. B-1.

The integral is approximated by the following equation:

$$\int_a^b f(x)dx \approx \frac{b-a}{n}\sum_{i=1}^{n} f(x_j)$$

$$x_j = \frac{x_{i-1} + x_i}{2}$$

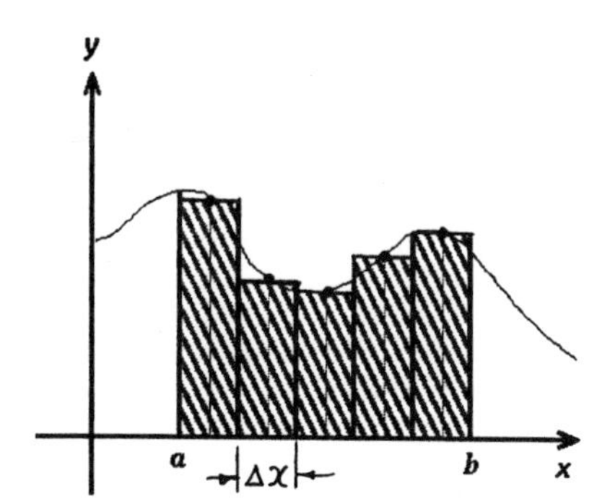

FIG. B-1 Integration by midpoint rule.

where a is the lower limit and b is upper limit of the integration, and n is the number of columns.

B.2 RECTANGLE RULE (FIG. B-2)

The integral is approximated by the following equation:

$$\int_a^b f(x)dx \approx \sum_{i=1}^{n} f(x_i)\Delta x$$

$$\Delta x = \frac{b-a}{n}$$

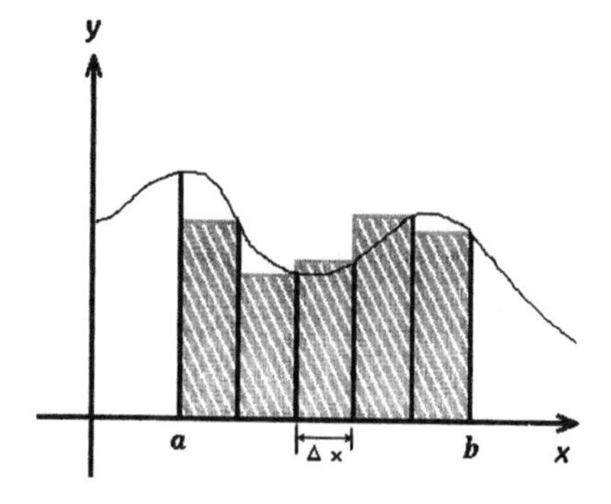

FIG. B-2 Rectangle rule.

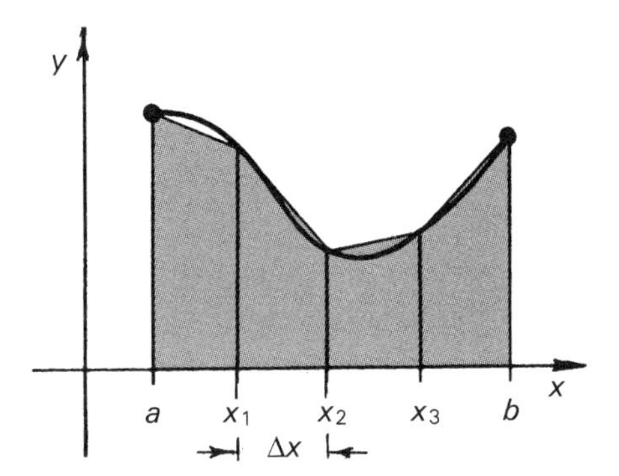

FIG. B-3 Integration by the trapezoidal rule.

B-3 TRAPEZOIDAL RULE (FIG. B-3)

The integral is approximated by the following equation:

$$\int_a^b f(x)dx \approx T$$

$$T = \frac{\Delta x}{2}[f(x_0) + 2f(x_1) + 2f(x_2) + \ldots 2f(x_{i-1}) + f(x_i)]$$

$$\Delta x = \frac{b-a}{n}$$

The endpoints, at points a and b, are counted only once, while all the other points have the coefficient 2.

B-4 SIMPSON RULE (FIG. B-4)

The Simpson rule is based on approximating the graph by parabolas rather than straight lines. The parabola is determined each time by the three consecutive points through which it passes.

$$f(x_{i-1}), f(x_i) \text{ and } f(x_{i+1})$$

The area from (x_{i-1}) to (x_{i+1}) is

$$A_i = \frac{\Delta x}{3}(f(x_{i-1}) + 4f(x_i) + f(x_{i+1})$$

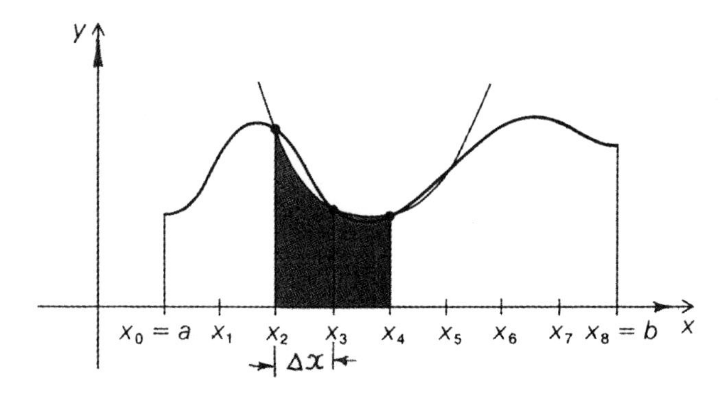

FIG. B-4 Integration by the Simpson rule.

If this procedure is repeated for every three adjacent points, the following Simpson rule for approximate integration is obtained:

$$\int_a^b f(x)dx \approx S$$

$$S = \frac{\Delta x}{3}[f(x_0) + 4f(x_1) + 2f(x_2) + 4f(x_3) + 2f(x_4) \ldots 2f(x_{i-2})$$
$$+ 4f(x_{i-1}) + f(x_i)]$$

$$\Delta x = \frac{b - a}{n}$$

$$x_{i-1} = a + (n - 1)\Delta x$$

$$x_i = b$$

The endpoints, at points a and b, are counted only once, while all the others are counted according to the coefficients 4 and 2 in the Simpson rule. For the Simpson rule, the n must be an even number. The Simpson rule yields the best approximation.

Example Problem B-1

Numerical Integration Using a Spreadsheet Program

Integrate the function $f(x) = 3x^2$ in the boundaries $0 \leq x \leq 2$. Use the approximate rectangle rule, and solve the summation with the aid of a spreadsheet.

Compare with the trapezoidal and Simpson rules.

Solution

The concept of numerical integration is to section the area under the curve into a large number of rectangles, calculating the area of each individual rectangle, and

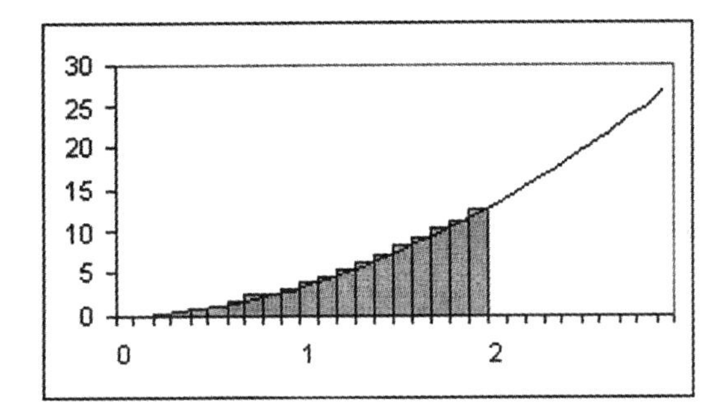

FIG. B-5 Approximate integration by summation of rectangles.

finally summing the rectangular areas to obtain the total area under the curve (see Fig. B-5).

The numerical integration is according to the equation

$$\int_a^b f(x)dx \approx \sum_{i=1}^{n} f(x_i)\Delta x_i$$

$$\Delta x_i = \frac{b-a}{n}$$

In this problem, the function is

$$f(x) = 3x^2$$

$$0 \leq x \leq 2$$

$$\int_0^2 3x^2 dx = \sum_{i=1}^{n} f(x_i)\Delta x_i$$

$$x_i = x_{i-1} + \Delta x$$

The summation is performed with the aid of a spreadsheet program (Table B-1). The first and second rows are added for explanation. The number of rectangles is selected ($n = 200$), resulting in uniform $\Delta x_i = 0.01$. The third column shows the values x_i, and the fourth column shows the respective values of the function $f(x_i)$. The fifth column lists the areas of the rectangles obtained by the product $f(x_i)\Delta x_i$. The sixth (last) column lists the sum of the rectangles to the last x_i. The solution of this numerical integration is at the bottom of this column. The exact solution of this integration is 8, and the errors of the various methods are compared in Table B-2. The best precision is obtained using the Simpson method.

TABLE B-1 Numerical integration with a spreadsheet program

Column ↓ Row →

Rectangular Method

Number	Rectangle Width	Distance from coordinate origin	Height of rectangle	Area of rectangle	Summation of rectangular Areas
Number	**Δx**	**x_i**	**$f(x) = 3x_i^2$**	**$f(x_i)(\Delta x)$**	**$\sum f(x_i)\,\Delta x$**
=A2 + 1	=B2	=C2 + B2	=3 * (C2^2)	=D2 * B2	=SUM( )
1	0.01	0.01	0.0003	0.000003	0.000003
2	0.01	0.02	0.0012	0.000012	0.000015
3	0.01	0.03	0.0027	0.000027	0.000042
4	0.01	0.04	0.0048	0.000048	0.00009
--------------	--------------	--------------	--------------	--------------	--------------
--------------	--------------	--------------	--------------	--------------	--------------
199	0.01	1.99	11.8803	0.118803	7.9401
200	**0.01**	**2**	**12**	**0.12**	**8.0601**

Trapezoidal Method

Number	Rectangle Width	Distance from coordinate origin	Height of rectangle	Trapezoidal Coefficient C		Summation of rectangular areas
i	**Δx**	**X_i**	**$f(x) = 3X_i^2$**	**C**	**$C\,f(X_i)$**	**$(\Delta x / 2) \sum f(X_i)\,C$**
1	0.01	0.01	0.0003	1	0.0003	0.0000015
2	0.01	0.02	0.0012	2	0.0024	0.0000135
3	0.01	0.03	0.0027	2	0.0054	0.0000405
4	0.01	0.04	0.0048	2	0.0096	0.0000885
------	------	------	------	------	------	------
------	------	------	------	------	------	------
199	0.01	1.99	11.8803	2	23.7606	7.9400985
200	**0.01**	**2**	**12**	**1**	**12**	**8.0000985**

Simpson Method

Number	Rectangle Width	Distance from coordinate origin	Height of rectangle	Simpson Coefficient C		Summation of rectangular areas
i	**Δx**	**X_i**	**$f(x) = 3X_i^2$**	**C**	**$C\,f(X_i)$**	**$(\Delta x / 3) \sum f(X_i)\,C$**
1	0.01	0.01	0.0003	1	0.0003	0.000001
2	0.01	0.02	0.0012	2	0.0024	0.000009
3	0.01	0.03	0.0027	4	0.0108	0.000045
4	0.01	0.04	0.0048	2	0.0096	0.000077
------	------	------	------	------	------	------
------	------	------	------	------	------	------
199	0.01	1.99	11.8803	4	47.5212	7.959997
200	**0.01**	**2**	**12**	**1**	**12**	**7.999997**

TABLE B-2 Errors of Various Numerical Integration Methods

Method	Solution	Percent Error
Actual solution	8.000	0%
Rectangular method	**8.06010**	0.751%
Trapezoidal method	**8.0000985**	0.00123%
Simpson method	**7.999997**	0.0000375%

Bibliography

Air Force Aero Propulsion Laboratory (1977): "Gas Lubricated Foil Bearing Development for Advanced Turbomachines", Report AF APL-TR-76-114, Vol. I and II.

ASTM, B23 (1990): ASTM Standards, ASTM Philadelphia, Vol. 02.04, pp. 9–11.

Allaire, P. E., and Flack, R. D. (1980): Journal Bearing Design for High Speed Turbomachinary. Bearing Design – Historical Aspects, Present Technology and Future Problems. W. J. Anderson (editor). ASME publication, New York, pp. 111–160.

Amin J., Friedland B., and Harnoy, A. (1997): "Implementation of a Friction Estimation and Compensation Technique" IEEE Control Systems, Vol. 17, No. 4, pp. 71–76.

Anderson, W. J., Fleming, D. P., and Parker, R. J. (1972): "The Series-Hybrid Bearing – A New High-Speed Bearing Concept", ASME Trans. J. of Lubrication Technology, Series F, Vol. 94, No. 2, pp. 117–124.

Anderson, W. J. (1973): "High Speed Hybrid Bearing, A Fluid Bearing and Rolling Bearing Connected in Series", U. S. Patent No. 3,759,588.

Armstrong-Hélouvry, B. (1991): "Control of Machines with Friction", Kluwer Academic Publishers, Boston, Massachusetts.

Arsenius, H. C., and Goran R. (1973): "The Design and Operational Experience of a Self Adjusting Hydrostatic Shoe Bearing for Large Size Runners", Instn. Mech. Engrs. C303, pp. 361–367.

Avila, F., and Binding, D. M. (1982): "Normal and Reverse Squeezing Flows", Journal of Non-Newtonian Fluid Mechanics, 11, pp. 111–126.

Bamberger, E. N., (1983): "Status of Understanding of Bearing Materials" Tribology in the '80s. Vol. 2, NASA CP-2300-VOL-2, Editor: W. F. Loomis, pp. 773–794.

Barus, C., (1893): "Isothermals, Isopiestics, and Isometrics Relative to Viscosity". Amer. Journal of Science, Vol. 45, pp. 87–96.

Bassani, R. and Piccigallo, B. (1992): Hydrostatic Lubrication, Elsevier.

Barwell, F. T. (1979): Bearing Systems, Principles and Practice, Oxford University Press.

Bell, R., and Burdekin, M., (1969): "A Study of the Stick-Slip Motion of Machine Tool Feed Drives", Inst. of Mech. Engr. 184, 29, pp. 543–560.

Blau, J. P. (1992): ASM Handbook, Friction, Lubrication, and Wear Technology, Vol. 18, ASM International.

Blau, J. P. (1995): "Friction Science and Technology", Marcel Dekker, Inc. NY.

Booker, J. F. (1965): "A Table of the Journal–Bearing Integrals", J. Basic Eng., Vol. 87, no. 2, pp. 533–535

Booser, E. R., Ryan, F., and Linkinhoker, C. (1970): "Maximum temperature for hydrodynamic bearings under steady load", Lubrication Engineering, Vol. 26, pp. 226–235.

Booser, E. R. (1992): "Bearing materials", Encyclopedia of Chemical Technology, 4th Ed. Vol. 4, John Wiley, pp. 1–21.

Bowden, F. P., and Tabor, D. (1956): Friction and Lubrication, John Wiley.

Bowden, F. P., and Tabor, D. (1986): The Friction and Lubrication of Solids, Oxford Science Press, England.

Brewe, D. E., and Hamrock, B. J. (1977): Simplified Solution for Elliptical Contact Deformation Between Two Elastic Solids. J. Lubr. Technol., Vol. 99, no. 4, pp 485–487.

Brough, D., Bell, W. F., and Rowe, W. B. (1980): "Achieving and monitoring high rate centerless grinding," Proc. of 21 Int. Con. on Machine Tool Design and Research, University of Swansea.

Charnley, J. (1979): "Low Friction Arthroplasty of the Hip", Springer-Verlag, pp. 1–376.

Chiu Y. P., Pearson, P. K., and Daverio, H. (1995): "Fatigue Life and Performance Testing of Hybrid Ceramic Ball Bearings" Lubrication Engineering, Vol, 52, No. 3, pp. 198–204.

Cole, J. A., and Hughes C. J. (1956): Oil Flow and Film Extent in Complete Journal Bearings, Proc. Inst. Mech. Engrs. (London), Vol. 170, no. 17.

Coleman, B. O., and Noll, W. (1960): "An Approximation Theorem for Functionals with Applications in Continuum Mechanics", Arch. Tat. Mech. Anal. Vol. 6, p. 355.

Cooke, A. F., Dowson, D., and Wright, V. (1978): "The Rheology of Synovial Fluid and Some Potential Synthetic Lubricants." Engineering and Medicine, Vol. 7, pp. 66–72.

Crankshaw, E., and Menrath, J. (1949): Mechanical Features of Steel-Backed Bearings, Sleeve Bearing Materials, American Society of Materials, pp. 150–164.

Dahl, P. R. (1968): "A Solid Friction Model", AFO 4695-67-C-D158, The Aerospace Corporation, El Segundo, California.

Dahl, P. R. (1977): "Measurement of Solid Friction Parameters of Ball Bearings", Proc. of 6th Annual Symp. On Incremental Motion, Control Systems and Devices. Champaign, IL, May 24–27.

Decker O. and Shapiro, W. (1968): "Computer-Aided Design of Hydrostatic Bearings for Machine Tool Applications", Proc. of 9th Int. Con. on Machine Tool Design and Research, University of Birmingham, pp. 797–839.

Deutschman, A. D., Michels W. J., and Wilson C. E. (1975): Machine Design Theory and Practice, Macmillan Publishing Co. NY.

Donaldson, R. R. and Patterson, S. R. (1983): "Design and Construction of a Large Vertical Axis Diamond Turning Machine," Lawrence Livermore, preprint UCRL 89738.

Dowson, D., and Higginson, G. R. (1966): Elastohydrodynamic Lubrication, The Fundamentals of Roller and Gear Lubrication. Pergamon, Oxford.

Dowson, D. and Zhong-Min Jin (1986): "Micro-Elastohydrodynamic Lubrication of Synovial Joints." Engineering in Medicine, Vol. 5 (No. 2), pp. 63–65.

Dowson, D. (1992): ASM Handbook, Friction, Lubrication, and Wear Technology, Vol. 18, ASM International, pp. 656–662.

Dowson, D. (1998): "Advances in Medical Tribology", Mechanical Engineering Publications Limited, London.

Dubois, G. B., and Ocvirk, E. W. (1953): "Analytical Derivation and Experimental Evaluation of Short Bearing Approximation for Full Journal Bearings," NASA Report 1157.

Dubois, G. B., Ocvirk, E. W., and Wehe, R. L. (1960): "Study of Effect of a Non-Newtonian Oil on Friction and Eccentricity Ratio of a Plain Journal Bearing," NASA TN D 427.

Dupont, P. E., and Dunlap, E. (1993): Friction Modeling and Control in Boundary Lubrication, Proc. American Control Conference, AACC, San Francisco, CA, pp. 1915–1919.

Earles, L. L., Palazzolo, A. B., and Armentrout, R. W. (1989): "A Finite Element Approach to Pad Flexibility Effects in Tilt Pad Journal Bearings," ASME/STLE Tribology Conference, Oct. 17–18, 1989, Fort Lauderdale, Florida.

Ernst, H., and Merchant, M. E. (1940): "Chip Formation, Friction and Finish," Cincinnati Milling Machine Co. Report, Aug. 24, 1940.

FAG (1986): "Rolling Bearing and their Contribution to the Progress of Technology", p. 191.

FAG (1998): "The Design of Rolling Bearing Mountings", Publication No. WL 00 200/5 EA.

FAG (1999): "FAG Rolling Bearing Catalogue", WL 41520/3 EA.

Fisher T. E., and Tomizawa, H. (1985): "Interaction of Tribochemistry and Microfracture in the Friction and Wear of Silicon Nitride", Wear, Vol. 105, p. 29.

Gibson, H. (2001): "Lubrication of Space Shuttle Main Engine Turbopump Bearings", Lubrication Engineering, Vol. 57, no. 8, pp. 10–12.

Girard, L. D., (April 1851): Comptes Rendus, Vol. 28.

Gluck, T. (1891): Arch Klin Chir. Vol. 41, p. 186.

Grimm, R. J. (1978): AIChE J. Vol. 24 No. 3, p. 427.

Hamrock, B. J., and Anderson, W. J. (1973): Analysis of an Arched Outer-Race Ball Bearing Considering Centrifugal Forces. J. Lubr. Technol., Vol 95, no. 3, pp. 265–276.

Hamrock, B. J., and Brewe, D. E. (1983): "Simplified Solution for Stresses and Deformations", J. Lubr. Technol., Vol. 105, no. 2, pp. 171–177.

Hamrock, B. J., and Dowson, D. (1977): Isothermal Elastohydrodynamic Lubrication of Point Contact, part III—Fully Flooded Results, J. of Lubrication Technology, Vol. 98, no. 3, pp. 375–383.

Hamrock, B. J., and Dowson, D. (1981): Ball Bearing Lubrication – The Elastohydrodynamics of Elliptical Contacts. Wiley-Interscience, New York.

Hamrock, B. J. (1994): Fundamentals of Fluid Film Lubrication, McGraw-Hill Inc. NY.

Harnoy, A. (1966): "Composite Bearing-Rolling Element with Hydrodynamic Journal," M.S thesis, Technion, Israel Inst. of Technology.

Harnoy, A. (1974): "The Effects of Stress Relaxation and Cross Stresses in Lubricants with Polymer Additives", CNRS publication, No. 233, pp. 37–44.

Harnoy, A., and Hanin, M. (1974): "Second Order, Elastico-Viscous Lubricants in Dynamically Loaded Bearings", ASLE Transactions, Vol. 17, pp. 166–171.

Harnoy, A., and Phillippoff, W. (1976): "Investigation of Elastico-Viscous Lubrication of Sleeve Bearing", ASLE Transactions, Vol. 19, pp. 301–308.

Harnoy, A. (1976): "Stress Relaxation Effect in Elastico-Viscous Lubricants in Gears and Rollers", Journal of Fluid Mechanics, Cambridge U.K., Vol. 19, pp. 501–517.

Harnoy, A. (1977): "Three Dimensional Analysis of the Elastico-Viscous Lubrication of Journal Bearings", Rheologica Acta, Vol. 16, pp. 51–60.

Harnoy, A. (1978): "An Investigation into the flow of Elastico-Viscous Fluids Past a Circular Cylinder", Rheologica Acta, Vol. 26, pp. 493–498.

Harnoy, A. (1978): "An Analysis of Stress Relaxation in Elastico-Viscous Fluid Lubrication of Journal Bearings", ASME Transactions, Journal of Lubrication Technology, Vol. 100, pp. 287–295.

Harnoy, A. (1979): "The role of the Fluid Relaxation Time in Laminar Elastico-Viscous Boundary Layers", Rheologica Acta, Vol. 18, pp. 220–216.

Harnoy, A. (1987): "Squeeze Film Flow of Elastic Fluids at Steady Motion and Dynamic Loads", ASME Transactions, Journal of Tribology, Vol. 109, pp. 691–695.

Harnoy, A., and Sood, S. S. (1989): "Viscoelastic Effects in the Performance of Dynamically Loaded, Short Journal Bearings," ASME Transactions, Journal of Tribology, Vol. 111, pp. 555–556.

Harnoy, A. (1989): "Normal Stresses and Relaxation Effects in Viscoelastic Boundary Layer Flow Past Submerged Bodies." Journal of Rheology, Vol. 33, pp. 93–117.

Harnoy, A., and Zhu, H. (1991): "The Role of the Elasticity of the Lubricant in Rotor Dynamics," Tribology Transactions, Vol. III, No. 4, pp. 353–360.

Harnoy, A., and Rachoor, H. (1993): "Angular-Compliant Hydrodynamic Bearing Performance Under Dynamic Loads," ASME, Transaction, Journal of Tribology, Vol. 115, pp. 342–347.

Harnoy, A., Friedland, B., Semenock, R., Rachoor, H., and Aly, A. (1994): "Apparatus for Empirical Determination of Dynamic Friction," Proceedings of ACC Conference, Baltimore, Maryland, June 29–July 1, 1994, Vol. 1, pp. 545–550.

Harnoy, A., and Friedland, B. (1994): "Dynamic Friction Model of Lubricated Surfaces for Precise Motion Control," Tribology Transactions, Vol. 37, No. 3, pp. 608–614.

Harnoy, A., and Friedland, B., and Rachoor, H. (1994): "Modeling and Simulation of Elastic and Friction Forces in Lubricated Bearings for Precise Motion Control," Wear, J. of science and Technology, Oxford UK, vol. 12, p.p. 155–165.

Harnoy, A. (1995): "Model Based Investigation of Friction During the Start-up of Hydrodynamic Journal Bearings," ASME Transactions, Journal of Tribology, Vol. 117, pp. 667–672.

Harnoy, A. (1996): "Simulation of stick-slip in control systems," Tribology Transactions, Vol. 40, pp. 360–366.

Harnoy, A., and Khonsari, M. M. (1996): "Hydro-Roll: A Novel Bearing Design With Superior Thermal Characteristics," Tribology Transactions, Vol. 39, No. 2, pp. 455–461.

Harnoy, A. (1996): "Simulation of Stick–Slip in Control Systems", Tribology Transactions, Vol. 40, pp. 360–366.

Harris, T. A. (1984): Rolling Bearing Analysis. John Wiley, New York.

Hertz, H. (1881): The Contact of Elastic Solids. J. Reine Angew Math., Vol. 92, pp. 156–171.

Heshmat, H., Shapiro, W., and Gray, S. (1982): Development of Foil Journal Bearings for High Load Capacity and High Speed Whirl Stability, Trans ASME, J. Lub. Tech., pp. 149–156.

Hess, D. P, and Soom, A., (1990): "Friction at a Lubricated Line Contact Operating at Oscillating Sliding Velocities", ASME Transactions, Journal of Tribology, Vol. 112, pp. 147–152.

Higginson, G. R. (1978): "Elastohydrodynamic Lubrication in Human Joints." Engineering in Medicine, Vol. 7 (No. 1), pp. 35–41.

Hodgekinson, F. (1923): "Improvements relating to lubrication of bearings," British Pat. 18595.

Horowitz, H. H., and Steidler, F. E. (1960): "The Calculated Journal Bearing Performance of Polymer Thickened Lubricants," ASLE Trans., Vol. 3, p. 124.

Hosang G. W., Heshmat, H. W. (1987): "Result and Design Techniques from the Application of Ceramic Ball Bearings to the MREADCOM 10 kW Turbine", AIAA, SAE, ASME, ASEE, 23rd Joint Propulsion Conference, June 29–July 2, 1987, San Diego.

Jones, A. B. (1946): "New Departure Engineering Data; Analysis of Stresses and Deflections". Vols. I and II, General Motors, Inc., Detroit, Michigan.

Jones, W. R. et al. (1975): "Pressure-viscosity Measurements for Several Lubricants", ASLE Trans., Vol. 18, no. 4, pp. 249–262.

Judet, J., and Judet, R. (1950): "The Use of an Artificial Femoral Head for Arthroplasty of the Hip Joint", J. Bone Joint Surg., Vol. 32B, p. 166.

Kaufman, H. N. (1980): "Tribology; Friction, Lubrication and Wear" Szeri, A. Z., Hemisphere, pp. 477–505.

Kennedy, F E., Booser, E.R., and Wilcock, D. F. (1998): "Tribology, Lubrication, and Bearing Design," Handbook of Mechanical Engineering CRC Press, Boca Raton, FL.

Kher, A. K., and Cowley, A. (1967): "The Design and Performance Characteristic of a Capillary Compensated Hydrostatic Journal Bearing", Proc. of 8th Int. Con. on Machine Tool Design and Research, University of Birmingham, pp. 397–418.

Khonsari, M. (1987): "A Review of Thermal Effects in Hydrodynamic Bearings, Part I and II", ASLE Trans., pp. 19–26.

Kingsbury, G. R. (1997): "Oil Film Bearing Materials", Tribology, Data Handbook, CRC Press, Boca Raton, FL, pp. 503–525.

Leider, P. J., and Bird, R. B. (1974): "Squeezing Flow Between Parallel Disks. I. Theoretical Analysis. II. Experimental Results," Ind. Eng. Fundam., Vol. 13, pp. 336–346.

Licht, L. (1972): The Dynamic Characteristics of a Turborotor Simulator on Gas-Lubricated Foil Bearings, ASME Trans., J. Lub. Tech., Vol. 94, pp. 211–222.

Loewy, K., Harnoy, A., and Bar-Nefi, S. (1972): "Composite Bearing, Rolling Element with Hydrodynamic Journal", Israel Journal of Technology, Vol. 10, pp. 271–278.

Ludema, K. C. (1996): Friction, Wear, Lubrication, CRC Press, Boca Raton FL.

Lundberg, G., and Palmgren, A. (1947): Dynamic Capacity of Rolling Bearings. Acta Polytech., Mech. Eng. Scr., Vol. 1, no. 3 (Stockholm) pp. 6–9.

Mahoney, O. M., and Dimon, J. H. (1990): "Unsatisfactory Results With a Ceramic Total Hip Prosthesis." Journal of Bone and Joint Surgery, Vol 72A, p 663.

McKee, S. A. (1928): "The Effect of Running-In on Journal Bearing Performance", Mech. Eng., Vol. 50, pp. 528–533.

McKee, S. A., and McKee, T. R. M. (1929): "Friction of Journal Bearings as Influenced by Clearance and Length", Trans. Am. Soc. Mech. Eng, Vol. 51, pp. 161–171.

McKee, G. K., and Watson-Farrar, J. (1966): "Replacement of the Arthritis Hips by the McKee-Farrar Prostethesis" J. Bone Joint Surg., Vol. 48B (No. 2), p. 245.

McKee, G. K. (1967): "Developments of Total Hip Replacement", Proc. Inst. Mech. Engr. Vol. 181, pp. 85–89.

McKellop, H., Clarke, I., Markolf, K., and Amstut, H. (1981): "Friction and Wear Properties of Polymer, Metal, and Ceramic Prosthetic Joint Materials Evaluated on a Multichannel Screening Device." Journal of Biomedical Materials Research, Vol. 15, p 619–653.

Mirra, J. M., Amstutz, H. C., Matos, M., and Gold, R. (1976): "The Pathology of Joint Tissues and its Clinical Relevance in Prosthesis Failure", Clin. Orthop., Vol 117, pp. 221–240.

Moore, A. T. (1959): "The Moore Self-Locking Vitallium Prosthesis in Fresh Femoral Neck Frames. A New Low Posterior Approach", Am Acad. Orthoped. Surg., Vol. 16, p. 166.

Mukherjee, A., Bhattacharyya, R., Rao Dasary, A. M. (1985): "A Theoretical Study of Stability of a Rigid Rotor Under the Influence of Dilute Viscoelastic Lubricants", ASME Transactions, J. of Lubrication Technology, Vol. 107, pp. 75–81.

Mukherjee, A., Bhattacharyya, R., Rao Dasary, A. M. (1984): "A Theoretical Inquiry into the Molecular Origin of Harnoy's Idea for Constitutive Relations for Viscoelastic Lubricants at High Shear rates", DSC Thesis I.I.T., ME Dept., Kharagpur 72130, India.

Mutoh, Y., Tanaka, K., and Uenohara, M. (1994): Retainer-Dependent Wear of Silicone Nitride Bearings at High Temperatures, ASME Transactions, Journal of Tribology Vol. 116, pp. 463–469.

Norton, R. L. (1996): Machine Design. An Integrated Approach, Prentice-Hall.

Nolan, J. F., and Phillips, H. (1996): "Joint Replacement and Particulate Wear Debris." Lancet, 348 (9031), pp. 839–840.

Ogawa, H., and Aoyama, T. (1991): "Application of the Advanced Ceramics to Sliding Guideways: Friction Wear Characteristics Under Water Based Lubrication" PED Vol. 54, TRIB Vol. 2, Tribological Aspects in Manufacturing, ASME, pp. 51–60.

Okrent, E. H. (1961): "The Effect of Lubricant Viscosity and Composition on Engine Friction and Bearing Wear," ASLE Trans., Vol. 4, p. 97, and Vol. 7, 1964, p. 147.

O'Kelly, J., Unsworth, A., Dowson, D., Jobbins, B., and Wright, V. (1977): "Pendulum and Simulator for Studies of Friction in Hip Joints." Evaluation of Artificial Joints, Ed. D. Dowson and V. Wright, Biological Engineering Society, pp. 19–29.

Oldroyd, J. G. (1959): Proc. Roy. Soc. A, 245, pp. 278–297.

Opitz, H. (1967): Pressure pad bearings, Proc. Inst. of Mech. Engrs., Lubrication and Wear, London.

Orndorff, R. L., and Tiedeman, C. R. (1977): "Water-Lubricate Rubber Bearings, History and New Developments," Society of Naval Architects and Marine Engineers, San Diego, pp. 49.

Pan, P., and Hamrock, B. J. (1989): Simple Formulae for Performance Parameters Used in Elastohydrodynamically Lubricated Line Contacts, J. of Tribology, Vol. 111, no. 2, pp. 246–251.

Pappas, J. M., Schmidt, A. C., Shanbhag, A. S., Whiteside, T. A., Rubash, H. E., Herndon, J. H. (1996): "Biological Response to Particulate Debris from Nonmetallic Orthopedic Implants", Human Biomaterials Applications, Humana Press, Totowa, NJ, pp. 115–135.

Peterson, M. B., and Winer, W. O., Eds. (1980): Wear Control Handbook, ASME, NY.

Pinkus, O. and Sternlicht, B. (1961): Theory of Hydrodynamic Lubrication, McGraw-Hill.

Polycarpou, A., and Soom, A. (1995): "Two dimensional Models of Boundary and Mixed Friction of a Line Contact", ASME J. of Tribology, Vol. 117, pp. 178–184.
Polycarpou, A., and Soom, A. (1996): "A Two Component Mixed Friction Model for a Lubricated Line Contact", ASME J. of Tribology, Vol. 117, pp. 178–184.
Prager, W. (1961): "An Elementary Discussion of Definitions of Stress Rate", Quart. Appl. Math. Vol. 18, p. 403.
Rabinovitz, E. (1951): "The Nature of the Static and Kinematic co efficient of Friction", J. of Applied Physics, Vol. 22, no. 1, pp. 668–675.
Rabinowicz, E. (1965): Friction and Wear of Materials, John Wiley, NY.
Rachoor, H. and Harnoy, A. (1996): "Modeling of Friction in Lubricated Line Contacts for Precise Motion Control," Tribology Transactions, Vol. 39, pp. 476–482.
Raimondi, A. A., and Boid, J. (1954): An Analysis of Capillary Compensated Hydrostatic Journal Bearings. Westinghouse Research Laboratories, paper No. 60-94451-8-PI.
Raimondi, A. A., and Boyd, J. (1955): Applying Bearing Theory to the Analysis and Design of Pad-Type Bearing. ASME Trans., Vol. 77, No. 3, pp. 287–309.
Raimondi, A. A., and Boid, J. A. (1958): Solution for the Finite Journal Bearing and its Application to Analysis and Design, Part III, ASLE Transactions, Vol. 1, no. 1 pp. 159–209.
Rumberger, J. H. and Wertwijn, G. (April 1968): "Hydrostatic Lead Screws", Machine Design, pp. 218–224.
Reiner, M., Hanin, M., and Harnoy, A. (1969): "An Analysis of Lubrication with Elastico-Viscous Liquid", Israel Journal of Technology, Vol. 7, pp. 273–279.
Reynolds, O. (1886): "On the Theory of Lubrication and its Application to Mr. Beauchamp Tower's Experiments, Including an Experimental Determination of the Viscosity of Olive Oil," Phil. Trans. Roy. Soc., London, Vol. 177, Part I, pp. 157–234.
Rightmire, G. K., (1969): "On Rubber to Metal Bonding Techniques with Reference to Compliant Surface Bearings," Technical report No. 10, Columbia University, Nov. 15, ONR Contract Nonr-4256 (06).
Rivlin, R. S., and Ericksen, J. L. (1955): "Stress Deformation Relations for Isotropic Materials", J. Rat. Mech. Anal. Vol. 4, p. 323.
Roach, A. E., Goodzeit, C. L. and Hunnicutt, R. P. (1956): "Scoring Characteristics of 38 Different Elemental Metals in High-Speed Sliding Contact with Steel," Trans. Am. Soc. Mech. Engrs., Vol. 78, pp. 1659–1667.
Roelands, C. J. A. (1966): "Correlational Aspects of the Viscosity-Temperature-Pressure Relationship of Lubricating Oils", Druk, V. R. B., Groingen, Netherlands.
Robertson, W. S. (1984): Lubrication in Practice, Marcel Dekker, NY.
Rowe, W. B. (1989): "Advances in Hydrostatic and Hybrid Bearing Technology." Proc. Inst, of Mech. Engrs., pp. 225–241.
Rowe, W. B. (1967): "Experience with Four Types of Grinding Machine Spindle," Proc. 8th Int. Conf. on Machine Tool Design and Research, Manchester University, pp. 453–476.
Savage, M. W., and Bowman, L. D. (1961): "Radioactive Tracer Measurements of Engine Bearing Wear", ASLE Trans., Vol. 4, p. 322.
Scribbe, W. H., Winn, L. W., and Eusepi, M. (1976): "Design and Evaluation of a 3 Million DN Series-Hybrid Thrust Bearing", ASME Trans., J. of Lubrication Technology, pp. 586–595.

Sharma, C., and Szycher, M. (1991): "Blood Compatible Materials and Devices", Techmionic Publishing Company, Inc., Lancaster.

Shigley, J. E., and Mitchell, L. D. (1983): Mechanical Engineering Design, 4th ed. McGraw-Hill, New York.

Silver, A. (1972): "Hybrid Bearing", U.S patent No. 3,642,331, Feb 15.

SKF, (February 1992): "Bearing Installation and Maintenance Guide, Includes Shaft and Housing Fits", Publication No. 140–710.

Sommerfield, A. (1904): Zur Hydrodynamischen Theories der Schmiermittelreibung. Z. Angew. Math. Phys., Vol. 50, pp. 97–155.

Stribeck, R. (1902): "Die wesentlichen Eigenschaften der Gleit-und Rollenlager (The Important Qualities of Sliding and Roller Bearings)", Zeitschrift des Vereines Deutscher Ingenieure, 46, (39), pp. 1432–1437.

Suriano, F. J., Dayton R. D., and Woessner, F. G., (1983): Test Experience with Turbine-End Foil-Bearing-Equipped Gas Turbine Engines, ASME, Paper 83-GT-73.

Sus, D. (1984): "Relaxations und Normalspannungseffekte in der Quetschstromung", Rhelogica Acta, Vol. 23, pp. 489–496.

Szeri, A. Z. (1980): "Tribology; Friction, Lubrication and Wear", Hemisphere. pp. 47–55.

Szeri, A. Z. (1998): "Fluid Film Lubrication", Cambridge University Press.

Tafazoli, S., De Silva, C. W., and Lawrence, P. D. (1994): "Friction Estimation in a Planar Electrohydraulic Manipulator", Proc. American Control Conference. Baltimore, MD. June 29–July 1.

Taylor, G. I. (1923): "Stability of a Viscous Liquid Contained Between two Rotating Cylinders", Philos. Trans. Royal Society, London, Series A, Vol. 223, pp. 289–343.

Tichy, J. A., and Winer, W. O. (1978): "An Investigation into the Influence of Fluid Viscoelastisity in a Squeeze Film Bearing," ASME Trans. Journal of Lubrication Tech., Vol. 100, pp. 56–64.

Tichy, J. A. and Modest, M. F. (1980): "A Simple Low Deborah Number Model for Unsteady Hydrodynamic Lubrication, Including Fluid Inertia", Journal of Rheology, Vol. 24, pp. 829–845.

Trygg, B., and McIntyre, B. (1982): "Hydrostatic Shoe Bearing Eliminates Problems for Large Mills", CIM Bulletin, pp. 105–109.

Tomizawa, H., and Fisher T. E. (1987): "Friction and Wear of Silicon Nitride and Silicon Carbide in Water: Hydrodynamic Lubrication at Low Sliding Speed Obtained by Tribochemical Wear'" STLE Trans., Vol. 30, p. 41.

Tower, B. (1885): "Second Report on Friction Experiments," Proc. Inst. Mech. Engrs., Vol. 36, pp. 58–70.

Ueki, M. (1993): "R&D on Functional Structural Ceramics (Application to Sliding Parts)", Nippon Steel Technical Report, No. 59, October 1993.

Underwood, A., F. (1949): The Selection of Bearing Materials, Sleeve Bearing Materials, American Society for Metals, pp. 210–222.

Unsworth, A., Dowson, D., Wright, V., and Koshal, D. (1974): "The Frictional Behavior of Human Synovial Joints – Part II: Artificial Joints." Trans. ASME, Journal of Lubrication Technology, Vol. 97 (No. 3), pp. 377–382.

Unsworth, A., Pearcy, M. J., White, E. F. T., and White, G. (1988): "Frictional Properties of Artificial Hip Joints." Engineering in Medicine, Vol. 17, (No. 3), pp. 101–104.

Valenti, M. (Dec. 1995): “Upgrading Jet Turbine Technology”, Mechanical Engineering, Vol. 56.

Wang, Q., Cheng, H. S., Fine, M. E., (1994): “Frictional Behavior of Some Nitrogen Ceramics in Conformal Contact with Tin Coated Al-Si Alloy, Steel and MMC”, Tribology Transactions, Vol. 37, pp. 587–593.

Walter, A., and Plitz, W. (1984): “Wear Mechanism of Alumina-Ceramic Bearing Surfaces of Hip-Joint Prosthesis.” Biomaterials and Biomechanics, Elsevier, Amsterdam, pp. 55–60.

White, F. M. (1999): “Fluid Mechanics”, WCB/McGraw-Hill, p. 215

Wilcock, D. F., and Booser, E. R. (1957): Bearing Design and Application, McGraw-Hill.

Wilcock, D. F., and Winn, L.W. (1973): “Hybrid Boost Bearing Assembly”, U.S patent No. 3,708,215.

Wiles, P. (1957): “The Surgery of the Osteo-Arthritic Hip”, British J. Surgery, Vol. 32B, p. 488.

Willert, H. G., and Semelitch, M. (1977): “Reactions of the Articular Capsule to Wear Products of Artificial Joint Prostheses”, J. Biomed. Mater. Res., Vol 11, pp 157–164.

Willert, H. G., and Semelitch, M. (1976): “Tissue Reaction to Plastic and Metallic Wear Products of Joint Endoprostheses in Total Hip Prosthesis”, Williams and Wilkins, Baltimore, pp. 205–239.

Williams, D. (1982): Biocompatibility of Orthopedic Implants. Vol. 1 and 2. CRC Press, Boca Raton.

Zaretzky, E. V. (1989): “Ceramic Bearings for Use in Gas Turbines”, ASME Trans., Journal of Gas Turbines and Power, Vol. 111, pp. 146–157.

Zaretzky, E. V. (1997a) “Tribology for Aerospace Applications” STLE Publication SP-37, pp. 231–232.

Ibid (1997b): pp. 325–237.

Ibid (1997b): pp. 345–391.

Zhai, X., Needham, G., and Chang, L. (1997): “On the Mechanism of Multi-Valued Friction in Unsteady Sliding Line Contacts Operating in the Regime of Mixed-Film Lubrication.” ASME Trans., J. of Tribology, Vol. 119, pp. 149–155.

Index